Principles of Traveling Wave Tubes

For a complete listing of the *Artech House Radar Library*,
turn to the back of this book

Principles of Traveling Wave Tubes

A. S. Gilmour, Jr.

Artech House
Boston • London

Library of Congress Cataloging-in-Publication Data
Principles of traveling wave tubes/A. S. Gilmour
Includes bibliographical references and index.
ISBN 0-89006-720-1
1. Traveling-wave tubes. I. Title
TK7871.77.G55 1994 94-7670
621.381'335–dc20 CIP

A catalogue record for this book is available from the British Library

685 Canton Street
Norwood, MA 02062

International Standard Book Number: 0-89006-720-1
Library of Congress Catalog Card Number: 94-7670

10 9 8 7 6 5 4 3 2 1

Contents

Preface

This book is based on material assembled for courses, short courses and seminars on traveling wave tubes and microwave tubes that I have given nearly 100 times at various Navy, Air Force, Army and NASA facilities; Varian Associates, Hughes Electron Dynamics Division, Teledyne Electronic Technologies, Northrop Defense Systems Division, Texas Instruments, the French Ministry of Defense, the University of California at Los Angeles, The University at Buffalo and other locations. There have been nearly 2000 students with backgrounds ranging from 3rd-class Navy electronic technicians through traveling wave tube experts to individuals at the chief scientist level. I am indebted to the many participants who made substantial contributions and recommendations to the material of the courses and to this book.

The book contains a balance of practical and theoretical material and so those new to TWTs as well as experienced microwave and TWT technicians, engineers and managers can benefit from its use. To a large extent, each chapter builds on material from earlier chapters and so, for the newcomers, it is advisable to work through the book starting at the beginning. For those with TWT experience, most of the chapters can be used without referring to others. For those who are interested in further study and research, the references to the works of over 200 authors dating from the invention of the TWT to the present time should be of value.

Several individuals have provided especially valuable contributions to this book. Included are Eugene (Joe) Dutkowski of the Crane Division of the Naval Surface Warfare Center; Jon Christensen, James Hansen and Dr. Ivo Tammaru of the Hughes Electron Dynamics Division; David Zavidil, Phil Lally and Robert Dipple of Teledyne Electronic Technologies; Edward Jones and Bruce Dudley of Rome Laboratory; Dr. Richard True of Litton Electron Devices Division; George Miram of Varian Associates and Dr. Stanley Kaisel. Finally, I wish to thank my wife for her patience and understanding during the long nights and weekends required for writing and preparing the camera-ready manuscript for this book.

Chapter 1

Introduction

The traveling wave tube (TWT) is one of two major categories of microwave devices known as linear-beam or O-type tubes. The other is the klystron. While many different RF circuits have been devised for use in TWTs, two predominate. These are

1. Helix, for broadband applications,
2. Coupled cavity, for high power applications.

Other circuits are mentioned in this book, but the majority of the discussions center on helix and coupled cavity TWTs.

Traveling wave tubes are used from frequencies below 1 GHz to over 100 GHz. Power generation capabilities range from watts to megawatts. For helix TWTs, bandwidths may be as high as two octaves or more. For coupled cavity TWTs, bandwidths in the 10–20% range are common.

Over 50% of the sales volume of all microwave tubes is for TWTs. Applications for TWTs are varied. They serve as the final amplifier in nearly all communications satellites. In many radar systems, one or more TWTs is used as the high power amplifier that generates the transmitted RF pulse. In others, a TWT may be used as the driver for some other high power RF amplifier such as a crossed field amplifier.

The largest application for TWTs is in electronic countermeasures systems. After a received enemy signal has been suitably processed, it is fed to one or more TWTs for transmission as a deceptive signal back to the enemy.

1.1 EARLY HISTORY OF THE TWT [1]

The possibility of a traveling wave interaction between an electron beam and a nearby RF circuit was recognized in 1933 by Haeff [2, 3]. In patents filed in that year, Haeff described electron beam deflection tubes to be used as detectors or oscillographs that contained some of the features of helix traveling wave tubes. In the Haeff devices, an RF signal propagating on a helical structure was used to deflect a nearby hollow electron beam. The velocity of the electron beam was equal to the velocity of propagation of the traveling wave on the helical structure. There was no indication by Haeff that amplification of the traveling RF wave would occur.

In 1935, Posthumus [4] first described and built a cavity-type magnetron oscillator. He explained its operation as being due to an interaction between the tangential component of a traveling RF wave rotating with a velocity equal to that of the average velocity of the electrons. As a result of the interaction, electron energy is converted to amplification of the RF wave.

In May, 1940, Lindenblad [5] was the first to describe helix traveling wave amplifiers that were similar to a helix traveling wave tube. He was the first to explain that a synchronous interaction between an electron beam and the RF wave on a helix could produce amplification of a signal on the helix. In a form of the tube described in the patent and shown schematically in Figure 1-1, Lindenblad produced useful amplification over a 30 MHz band at a carrier frequency of 390 MHz. In this tube the signal was applied to the beam with a grid in the gun. In other experiments, the signal was applied to the helix.

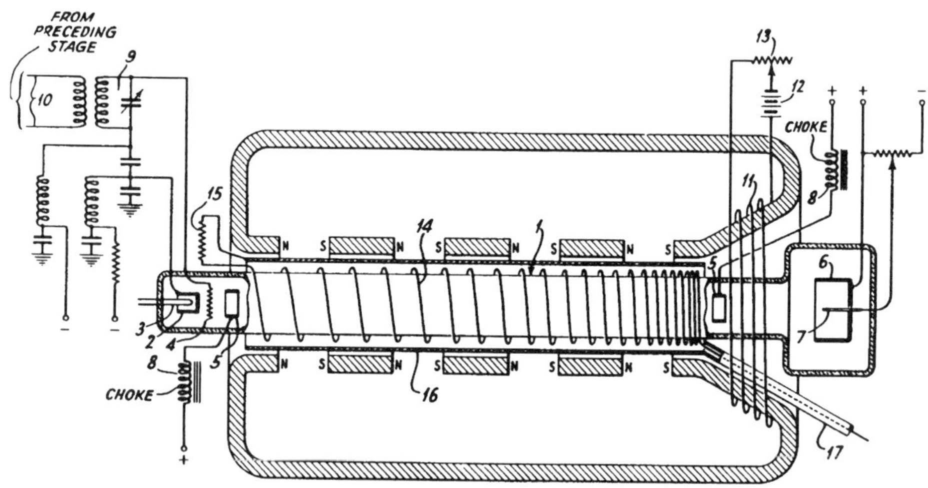

Figure 1-1 N.E. Lindenblad's traveling wave amplifier; patent filed May 1940.

The first Lindenblad tube was, apparently, a modified version of Haeff's inductive output tube (which in 1982 became the klystrode). Lindenblad extended the glass vacuum envelope of the Haeff tube and replaced the cavity resonator with a helix. The helix was wound around the outside of the glass envelope and had a pitch such that the axial velocity of a wave on the helix was equal to the velocity of an electron beam inside the envelope. It is interesting that Lindenblad recognized the value of tapering the pitch of the helix to maintain synchronism with the electron beam, the velocity of which is reduced as energy is extracted to produce amplification.

In his patent, in addition to the helix, Lindenblad describes the use of a helical waveguide as the slow wave circuit. He also states that the helical conductors can, if desired, be placed within the envelope of the tube and surrounding the electron beam.

Kompfner [6–10], working in England, was apparently unaware of Lindenblad's earlier work. He reported that he realized in late 1942 (shortly after Lindenblad's patent was issued) that the basic growing wave principle of the magnetron could be used for amplification of RF signals. His objective was to develop an amplifier capable of a sensitivity and noise figure comparable to the best of the crystal-mixer receivers available at that time. Shown in Figure 1-2 is Kompfner's first sketch of a traveling wave tube.

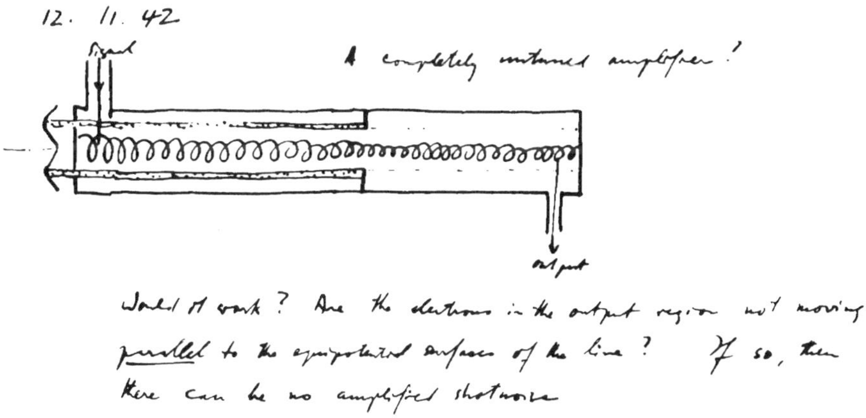

Figure 1-2 Sketch of traveling wave tube from Kompfner's notebook. From *The Invention of the Traveling Wave Tube* by R. Kompfner, copyright 1963 by San Francisco Press, reproduced with permission.

Kompfner built has first traveling wave tube in early 1943. He summarized the operation of the tube as follows:

> "When the radio frequency power emerging from the helix with the beam switched on was compared with the radio frequency power without the beam, it was found that, at a beam voltage of 2440 volts, there was an increase of 49%, while at a beam voltage of 2200 volts, there was a decrease of 40%."

After various measurements and experiments, Kompfner built and tested the tube shown in Figure 1-3. With an electron beam current of 110 μA at 1830 V a power amplification of 6 was achieved at a frequency of 3.3 GHz. The noise factor was 14 dB and was observed to depend on the fraction of the total current that arrived at the collector electrode. A further improvement in design increased power amplification to 14 and reduced noise factor to 11 dB. In theoretical studies completed before 1944, Kompfner developed first order theories for power amplification and noise factor. The important result was that the wave tends to grow exponentially with distance along the transmission line.

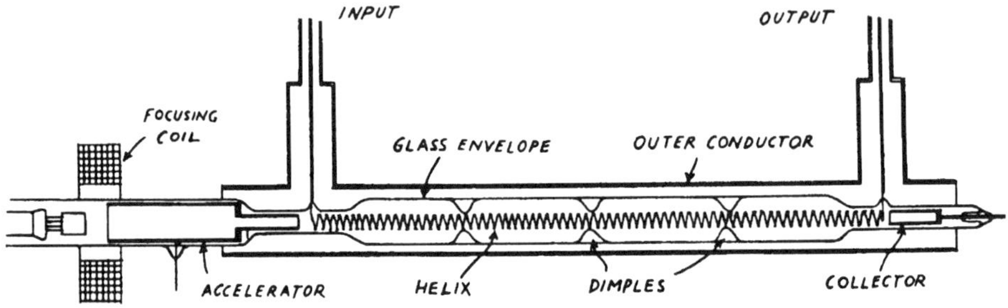

Figure 1-3 Kompfner's experimental helix amplifier. From Rudolf Kompfner, *Proc. IRE*, February 1947. © 1947 IRE (now IEEE).

The first public announcement of the British wartime work on helix traveling wave tubes was made at the Fourth Institute of Radio Engineer's Electron Tube Conference at Yale on June 27 and 28, 1946. At the same conference, work in the United states by J.R. Pierce and L.M. Field at Bell Telephone Laboratories was described. Subsequently [11], Pierce and Field described helix traveling wave tubes like that shown in Figure 1-4 (figure taken from Patent 2,602,148 by J.R. Pierce [12]). One unique feature of this device, described by Pierce, was the use of longitudinal insulating rods for supporting and accurately positioning the helix. Another was the use of a system of solenoids to produce a uniform magnetic field for focusing the electron beam. Pierce also described techniques for providing loss to suppress backward traveling waves and oscillations. In the patent by Field [13] the suppression of oscillations through the use of thin layers of colloidal graphite on the ceramic support rods for the helix is described. Near the midpoint of the helix, the conductivity of the coating was increased to provide the proper dissipation of reflected energy with a minimum of reduction in the gain of the forward growing wave.

During the postwar years from 1946 to 1950, there was intense activity in the development of adequate theories to explain the operation of the TWT. In addition to Pierce's theoretical work [14], noteworthy contributions [15, 16] were made by A. Blanc-Lapierre and P. Lapostolle in France in December 1946 that contained an analysis of traveling wave interaction using a structure consisting of resonant cavities coupled by axially aligned apertures.

It seems appropriate to conclude this brief early history by citing the publication by J.R. Pierce [17] of *Traveling Wave Tubes* in 1950. This book provided a unified coordination and exposition of the theory of the traveling wave tube and has been used as the standard reference for the theory of traveling wave tubes ever since.

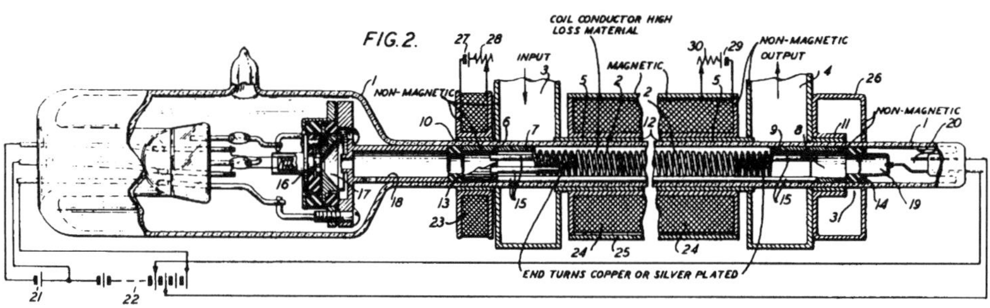

Figure 1-4 Helix amplifier shown in U.S. Patent 2,602,148 by J.R. Pierce.

1.2 BASIC OPERATION OF THE TWT

There are two basic types of TWTs. The helix TWT, the main elements of which are shown in Figure 1-5, is a relatively low-power (usually tens to hundreds of watts) broad-band (over two octaves is possible) device. The coupled-cavity TWT is capable of powers of megawatts but bandwidth is limited to about 10-20%. Conceptually, the electron gun, electron beam and collector are the same for all TWTs. However, there are substantial differences in the RF circuits. In spite of the differences, the operating principles are the same. We will discuss the operation of the helix TWT here.

For the helix TWT, it is helpful in explaining the RF behavior to first consider a single-wire transmission line above a ground plane as shown in Figure 1-6. Charges

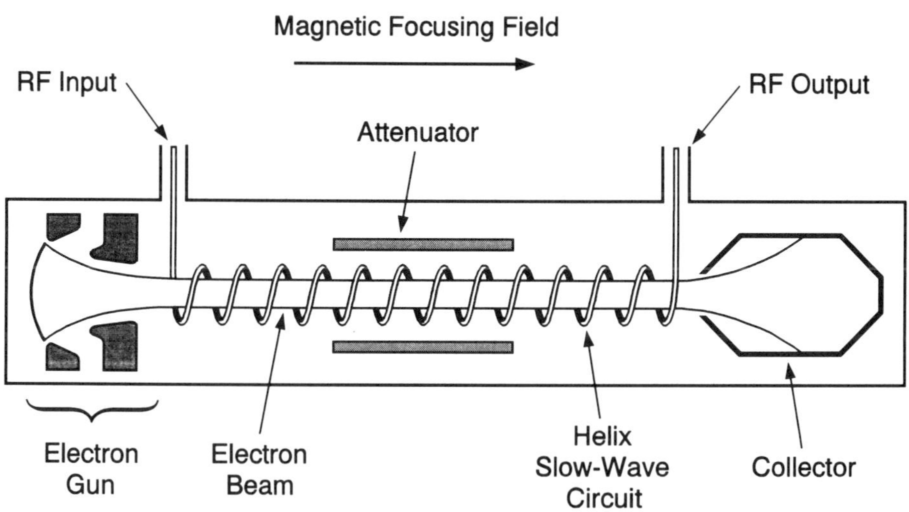

Figure 1-5 Basic helix TWT.

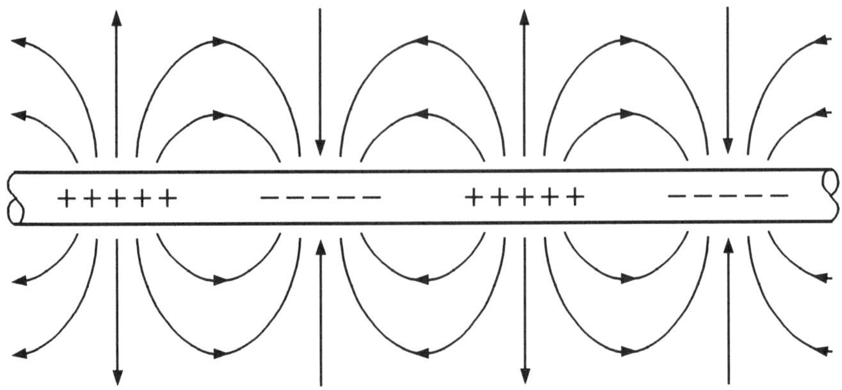

Figure 1-6 RF charge and electric field patterns for a single-wire transmission line above a ground plane.

on the transmission line and the RF electric field pattern are shown at an instant in time. The RF magnetic field pattern is not shown and will not be considered because the RF magnetic forces on the electron beam in a microwave tube are insignificant compared with the forces produced by the electric field. If the RF generator is on the left end of the transmission line in Figure 1-6, and if a matched load is on the right end, then the charge and field patterns move to the right with constant amplitude. The velocity of propagation is that of light and is independent of frequency, so the transmission line is completely nondispersive.

Next, assume that the single-wire transmission line is formed into a helix as shown in Figure 1-7. An RF signal applied to the left end of the helix will travel at a velocity near that of light in a helical path along the conductor. The average velocity in the axial direction will be the velocity in the helical direction reduced by the pitch of the helix. The polarity of the signal will alternate every half-wavelength along the helical conductor. In Figure 1-7, two complete turns are required for each half-wavelength.

Electric field lines extend from regions of positive charge to regions of negative charge with a configuration similar to that shown for the transmission line in Figure 1-6. Other than the reduced velocity, the primary difference between the field pattern for the helix and that for a single-wire transmission line is that there is an electric field inside the helix with large axial components.

When an electron beam is injected along the axis of the helix, the axial electric field components accelerate some electrons and decelerate others. In Figure 1-7, the forces on the electrons would be toward the region denoted by (A) and away from the regions denoted by (B). The field pattern would vary sinusoidally in the axial direction, as indicated in Figure 1-8. If the axial velocity of the electric field and the beam are the same, then electrons will experience a continual force toward region (A) as the beam travels through the helix. As a result, an electron bunch will start to form in region (A) (also indicated in Figure 1-8(a)).

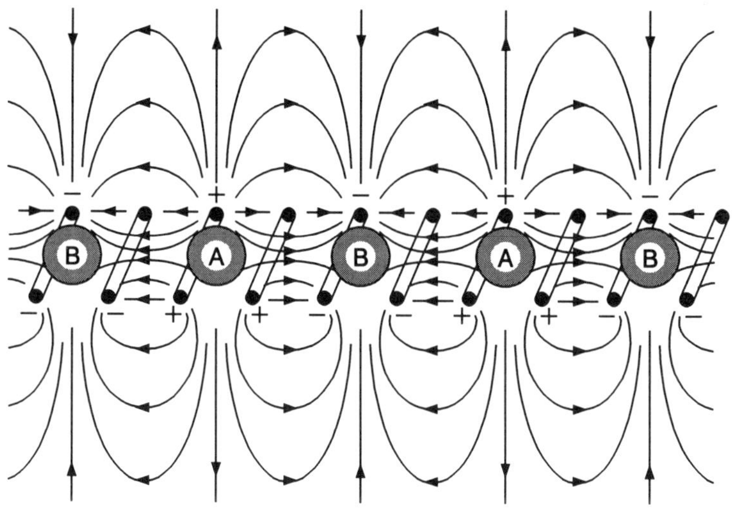

Figure 1-7 RF charge and electric field patterns for a helix.

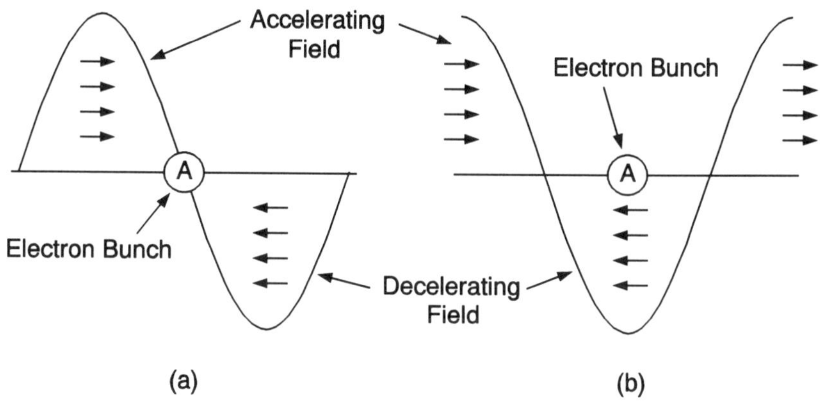

Figure 1-8 Axial field that bunches and extracts energy from beam (a) as beam enters circuit and (b) after interaction has occurred.

The fields produced by the bunching electrons in the beam will cause electrons on the helix to move away from the region adjacent to (A) and toward the (B) regions. This will produce two changes in the field on the helix:

1. The electron current flowing to the left from the helix from (A) corresponds, of course, to conventional current flowing to the right. This current, in turn, produces a positive voltage region on the helix to the left of (A). Similarly, the

electron current flowing to the right on the helix form (A) produces a negative voltage to the right of (A). The net result is that the induced voltage waveform on the helix is phase shifted 90° to the left of the initial voltage waveform.

2. As the beam-wave interaction proceeds, the induced waveform rapidly becomes much larger that the initial waveform.

As the voltage waveform on the helix shifts to the left, decelerating field regions move into the regions where the bunches are located (shown in Figure 1-8(b)). The energy extracted from the decelerating bunches is transferred to the circuit field, thereby producing amplification of that field. The mutual interaction of the beam and circuit result in an exponential growth of the circuit voltage as shown by the left half of the waveforms in Figure 1-9.

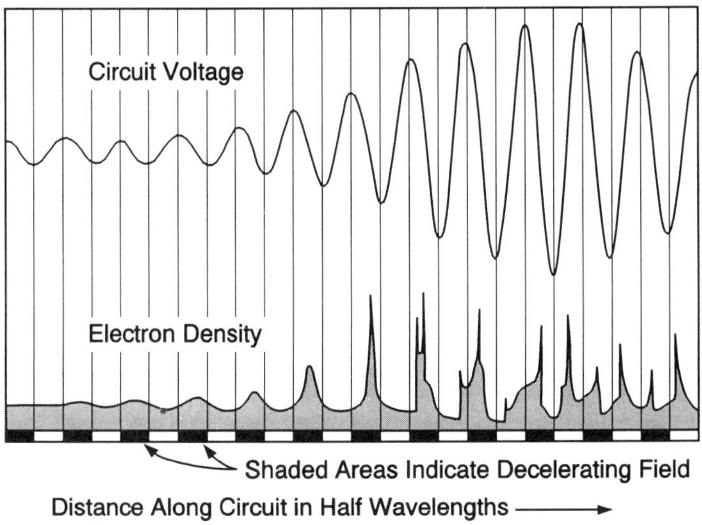

Figure 1-9 Voltage and charge density build-up in a TWT. Adapted from P. Hess, Ph.D. Dissertation, University of California at Berkeley, 1960.

As the interaction process continues, electron velocities continue to decrease. Also, space charge forces within the bunches continue to increase. As a result, a portion of each bunch is retarded in phase enough so that it leaves the decelerating field region and enters an accelerating field (see electron density plots in the right third of Figure 1-9). Electrons in the acceleration portion of each bunch extract energy from the circuit field. Eventually, as the bunches continue to fall back in phase, energy extracted from the circuit wave becomes equal to the energy supplied and the wave on the circuit stops growing. At this point, "saturation" is said to occur. For maximum amplification, the signal should be removed from the circuit at this point.

1.3 OVERVIEW OF BOOK

In selecting the subject matter for this book (and the short courses on which the book is based), a serious attempt has been made to include at least basic discussions of all the technologies that are essential to an understanding of the operation of traveling wave tubes. The order of presentation has been selected so that the reader, in working through the book from the beginning, can develop, in a logical sequence, a working knowledge of TWTs.

Thus, after a review of electron dynamics, the source of electrons (the cathode) and the components (electron gun and focusing structures) required to form the electrons into a beam are discussed. Next, the behavior of the beam after being subjected to a localized RF electric field is analyzed. This is followed by the analysis of the interaction of the beam with a traveling RF electric field. Perhaps the most significant result of this analysis is the importance of the relative velocities of the traveling RF field and the electron beam to the amplification of the RF field. This leads directly to a general discussion and analysis of the variation of velocity with RF frequency (dispersion) of TWT circuits. With an understanding of the significance of dispersion characteristics, it is possible to analyze in some detail helix TWTs and then coupled cavity TWTs.

The remainder of the book is devoted to subjects not related to the amplification processes in TWTs but, nevertheless, important to the understanding of the total operation of a TWT. For example, many modern TWTs use sophisticated electron beam collectors. These enhance the efficiency but do not directly affect the interaction of the beam with the circuit. Noise, nonlinearities and distortion are important in most applications of TWTs. When TWTs fail it sometimes results from degradation of the vacuum environment in the tube and so a discussion of vacuum requirements is included. A degradation in vacuum environment along with many other factors can lead to electrical breakdown in a TWT and so breakdown phenomena along with requirements for protecting TWTs against the effects of breakdown are discussed. Finally, since the invention of the TWT, reliabilty has been a major issue in many applications. In the chapter on reliability, causes for failures are examined and reliability trends in various applications are discussed. As a result of improvements in manufacturing technology and in user understanding and protection, TWT reliability has improved significantly and continues to improve.

REFERENCES

[1] Some of the material for this early history was obtained from R.L. Wathen, "A history of the traveling wave tube," Report No. 5202-5020, Sperry Gyroscope Company, May 18, 1953.

[2] A.V. Haeff, U.S. Patent 2,064,469, filed October 23, 1933, issued December 15, 1936.

[3] A.V. Haeff, U.S. Patent 2,233,126, filed October 23, 1933, issued February 25, 1941.

[4] K. Posthumus, "Oscillations in split-anode magnetron," *Wireless Engineer and Experimental Wireless,* 12, March 1935, pp. 126–132.

[5] N.E. Lindenblad, U.S. Patent 2,300,052, filed May 4, 1940, issued October 27, 1942.
[6] R. Kompfner, "The traveling wave valve," *Wireless World*, 53, November 1946, pp. 369–372.
[7] R. Kompfner, "The traveling wave tube as an amplifier at microwaves," *Proc. IRE*, 35, February 1947, pp. 124–127.
[8] R. Kompfner, "The traveling wave tube," *Wireless Engineer*, 52, September 1947, pp. 255–266.
[9] R. Kompfner, "On the operation of the traveling wave tube at low level," *British Institution of Radio Engineers*, August–September 1950, pp. 283–289.
[10] R. Kompfner, "The invention of the traveling wave tube," from a lecture series at the University of California at Berkeley, published by San Francisco Press, 1963.
[11] J.R. Pierce and L.M. Field, "Traveling wave tubes," *Proc. IRE*, 52, February 1947, pp. 108–111.
[12] J.R. Pierce, U.S. Patent 2,602,148, filed October 22, 1946, issued July 1, 1952.
[13] L.M. Field, U.S. Patent 2,575,383, filed October 22, 1946, issued November 20, 1951.
[14] J.R. Pierce, "Theory of the beam type traveling wave tube," *Proc. IRE*, 35, February 1947, pp. 111–123.
[15] A. Blanc-Lapierre and P. Lapostolle, "Contribution a l'etude des amplificateurs a ondes progressives," *Annales des Telecommunications*, 1, December 1946, pp. 283–302.
[16] A. Blanc-Lapierre and P. Lapostolle, "Sur l'interaction entre une onde progressive et un faisceau d'electrons de vitesse voisine de celle de l'onde," *Comptes Rendue Rebdomadaires des Seances de l'Academie des Sciences*, 224, January 13, 1947, pp. 104–105.
[17] J.R. Pierce, *Traveling Wave Tubes*, Princeton, N.J.: Van Nostrand, 1950.

Chapter 2

Static Fields Produced by Electrons

2.1 ELECTRIC FIELD

At many points in the analysis of electron motion in a TWT it is necessary to include the effects of the electric field produced by the electrons. Several techniques are available for calculating the electric field produced by a charge distribution. The techniques that we will make use of employ either Poisson's equation (known as Laplace's equation when no charge is present) or Gauss's law.

2.1.1 Laplace's and Poisson's Equations

Laplace's equation is

$$\nabla^2 V = 0 \tag{2-1}$$

and Poisson's equation is

$$\nabla^2 V = -\frac{\rho}{\epsilon_o} \tag{2-2}$$

where V is the potential, ρ is the charge density (assumed to be positive) and ϵ_o is the permittivity of vacuum. In rectangular (cartesian) coordinates, the Laplacian operator ∇^2 is as follows:

$$\nabla^2 V = \frac{\partial^2}{\partial x^2} V + \frac{\partial^2}{\partial y^2} V + \frac{\partial^2}{\partial z^2} V \tag{2-3}$$

for cylindrical coordinates it is

$$\nabla^2 V = \frac{1}{r}\frac{\partial}{\partial r} V + \frac{\partial^2}{\partial r^2} V + \frac{1}{r^2}\frac{\partial^2}{\partial \theta^2} V + \frac{\partial^2}{\partial z^2} V \tag{2-4}$$

and for spherical coordinates

$$\nabla^2 V = \frac{1}{r^2}\frac{\partial}{\partial r}\left[r^2 \frac{\partial}{\partial r} V\right] + \frac{1}{r^2 \sin\theta}\frac{\partial}{\partial \theta}\left[\sin\theta \frac{\partial}{\partial \theta} V\right] + \frac{1}{r^2 \sin^2\theta}\frac{\partial^2}{\partial \phi^2} V \qquad (2\text{-}5)$$

The operator ∇^2 appears to be extremely complex (and is in most cases). For the geometries that we will be considering, the symmetries and approximations used will result in simplifications that make this operator quite manageable.

2.1.2 Gauss's Law

Gauss's law relates the electric field produced by a charge distribution to that charge distribution. The differential form of Gauss's law is

$$\nabla \cdot \boldsymbol{E} = \frac{\rho}{\epsilon_o} \qquad (2\text{-}6)$$

This form of Gauss's law will be used in Chapter 9 in the wave equation, which is needed to describe electron bunching when the electron charge is taken into consideration.

The integral form of Gauss's law is

$$\oint_S \boldsymbol{E} \cdot \boldsymbol{ds} = \frac{Q}{\epsilon_o} \qquad (2\text{-}7)$$

That is, the integral over a closed surface area, S, of the electric field, E, is the total positive charge, Q (divided by ϵ_0), contained within that area. When advantage can be taken of geometrical symmetries, the use of Gauss's law provides an easy and powerful way to find the electric field.

As an example of the application of the integral form of Gauss's law, consider a sphere of radius $b = 1$ mm containing electrons with a uniform density of $3 \times 10^{11}/\text{cm}^3$ (approximately the density of an electron bunch in an RF bunched beam in a TWT or klystron). The problem is to find the potential distribution inside and outside the sphere. In this case, only a radial component, E_r, of the electric field exists.

Inside the sphere, on a surface with radius r,

$$\oint_S \boldsymbol{E}_r \cdot \boldsymbol{ds} = \frac{Q}{\epsilon_o} = -\frac{\rho \frac{4}{3}\pi r^3}{\epsilon_o} \qquad (2\text{-}8)$$

The electric field is constant on the surface with radius r. Also, the direction of the electric field is the same as the surface vector, ds, and so the dot product becomes a simple multiplication. As a result, E_r may be taken outside the integral, and the integral is simply the surface area of the sphere. Now,

$$4\pi r^2 E_r = -\frac{\rho \frac{4}{3}\pi r^3}{\epsilon_o} \tag{2-9}$$

or

$$E_r = -\frac{\rho r}{3\epsilon_o} \tag{2-10}$$

Of course, the minus sign shows that the electric field is in the -*r* direction, which means that it is pointed radially inward, toward the center of the sphere of electrons.

The potential, *V*, at position, *r*, is the integral of the electric field; that is,

$$V = -\int_0^r E_r dr = \int_0^r \frac{\rho r}{3\epsilon_o} dr = \frac{\rho r^2}{6\epsilon_o} \tag{2-11}$$

At $r = b = 1$ mm,

$$V = \frac{\rho b^2}{6\epsilon_o} = \frac{1.6\times 10^{-19}\mathrm{C}\times 3\times 10^{11}\mathrm{e/cm^3}\times 10^6\mathrm{cm^3/m^3}\times (10^{-3})^2\mathrm{m^2}}{6\times 8.85\times 10^{-12}\mathrm{F/m}} \tag{2-12}$$

or

$$V = 904 \text{ V} \tag{2-13}$$

Thus, the potential at the surface of the sphere with respect to that at the center of the sphere is 904 V! The importance of this simple problem is that it illustrates the magnitude of the potential depression that can exist in the center of an electron bunch in a beam.

Outside the sphere of electrons, the electric field is found from

$$E_r \times 4\pi r^2 = -\frac{\rho \frac{4}{3}\pi b^3}{\epsilon_o} \tag{2-14}$$

and is

$$E_r = -\frac{\rho b^3}{3\epsilon_o r^2} \tag{2-15}$$

The potential is

$$V = -\int_b^r E_r dr + C = \int_b^r \frac{\rho b^3}{3\epsilon_o r^2} dr + C \tag{2-16}$$

where the constant, *C*, is the potential at $r = b$. When the integration is carried out,

$$V = -\frac{\rho b^3}{3\epsilon_o r} + \frac{\rho b^2}{3\epsilon_o} + \frac{\rho b^2}{6\epsilon_o} = \frac{\rho b^2}{2\epsilon_o} - \frac{\rho b^3}{3\epsilon_o r} \tag{2-17}$$

When r is $2b$, that is 2 mm, the potential is 1,808 V with respect to that at the center of the sphere. This is of significance in a microwave tube because it shows (as would be expected) that, as the drift tube walls move away from an electron beam, a significant increase in the potential depression in the center of the beam occurs.

Figure 2-1 shows the potential distribution inside and outside the sphere of electrons of density $3\times10^{11}/\text{cm}^3$.

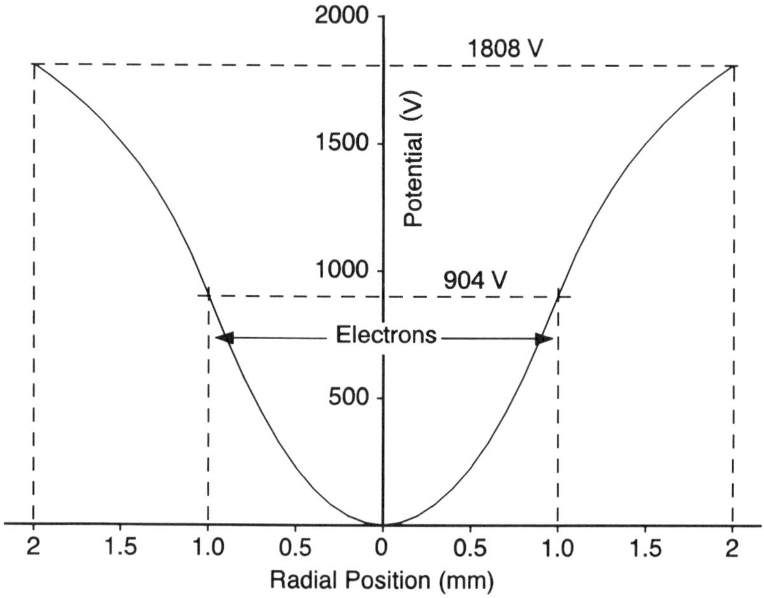

Figure 2-1 Potential distribution for 2 mm diameter, 3×10^{11} e/cm^3 sphere of electrons.

2.2 MAGNETIC FIELD

The electron beam in a TWT produces a magnetic field, but this field is small enough so that it is usually neglected. The magnitude of the field is easily determined by using Ampere's law. This law states that the integral of the magnetic field around a closed path is the current contained within that path. That is,

$$\oint H \cdot d\ell = I \tag{2-18}$$

As an example, assume that a TWT contains a 1 A electron beam with a radius,

b, of 1 mm as shown in Figure 2-2. In any circular path concentric with the beam, the magnetic field is constant and so H may be taken outside the integral. Thus, at the outer edge of the beam,

$$\oint H \cdot d\ell = H \oint d\ell = H 2\pi b \tag{2-19}$$

so

$$H = \frac{I}{2\pi b} = \frac{1}{2\pi} 10^3 \text{ A/m} \tag{2-20}$$

The magnetic flux density is

$$B = \mu H = 4\pi \times 10^{-7} H \tag{2-21}$$

or

$$B = 2 \times 10^{-4} \text{ T} = 2 \text{ G} \tag{2-22}$$

This flux density is about 3 orders of magnitude smaller than the flux density used to focus the electron beam and so its effect is usually neglected.

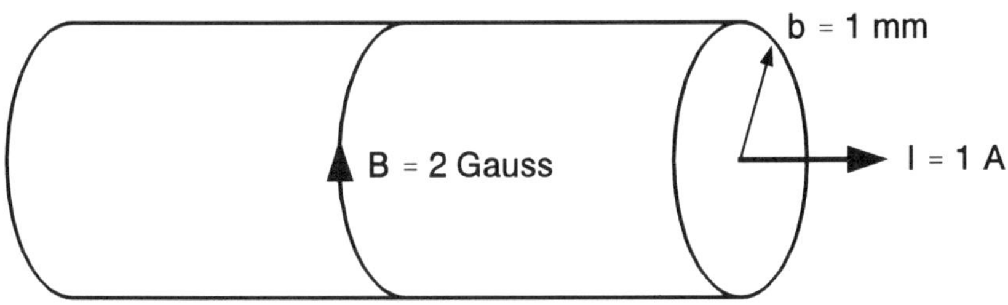

Figure 2-2 Magnetic flux generated by electron beam.

Chapter 3

Electron Motion in Static Electric Fields

In the electron gun and the collector of a TWT, dc voltages are applied to produce fields that accelerate or decelerate electrons. The ways in which those fields influence electrons are described in this chapter. Included is a discussion of the relativistic velocity correction, which is important in very high power devices.

The very simple case of electron deflection by transverse fields in a cathode ray tube is followed by an examination of electron motion in electric lenses. The focusing relation derived here will be of use in the analysis of electron guns.

Finally, the influence of the electric fields produced by the electrons on the electron trajectories in an axisymmetric configuration is examined. This will also be of use in the analysis of electron guns as well as in explanations of electron beam behavior.

3.1 MOTION PARALLEL TO FIELD

Assume that an electron is moving in the $+z$ direction and that the electric field, E_z, is in the $-z$ direction. Then, the force on the electron is

$$F = ma = m\frac{d^2z}{dt^2} = -eE \tag{3-1}$$

If E_z is constant, then

$$u_z = u_o - \frac{e}{m}E_z t \tag{3-2}$$

and

$$z = z_o + u_o t - \frac{e}{2m}E_z t^2 \tag{3-3}$$

where z_o is the initial position of the electron and u_o is the initial velocity.

If the electron starts at $z = 0$ at rest, then the position at any later time is

$$z = -\frac{e}{2m}E_z t^2 \tag{3-4}$$

and its velocity is

$$u_z = -\frac{e}{m}E_z t \tag{3-5}$$

The kinetic energy of the electron is

$$\frac{1}{2}mu^2 = \frac{1}{2}m\left[\frac{e}{m}E_z\right]^2 t^2 = -eE_z z \tag{3-6}$$

but, since E_z is constant, $-E_z z$ is the potential V through which the electron has been accelerated. Thus,

$$\frac{1}{2}mu^2 = eV \text{ J} \tag{3-7}$$

or

$$u = \left[\frac{2eV}{m}\right]^{1/2} = 5.93 \times 10^5 \sqrt{V} \text{ m/s} \tag{3-8}$$

The energy of the electron, eV, is often expressed in electron volts (eV) rather than in joules. The relation between electron volts and joules (J) is

$$1 \text{ eV} = 1.6 \times 10^{-19} \text{ J} \tag{3-9}$$

An electron that has been accelerated through a potential of 1 V has an energy of 1 eV. An electron that has been accelerated through a potential of 10,000 V has an energy of 10,000 eV.

If the electron has an initial velocity of u_o resulting from being accelerated through a potential V_o, then

$$\frac{m}{2}\left(u^2 - u_o^2\right) = e\left(V - V_o\right) \tag{3-10}$$

The total energy of the electron is still eV and the increase in energy is $e(V - V_o)$.

3.2 RELATIVISTIC VELOCITY CORRECTIONS

The relation (3-8) developed in the previous section, showing that electron velocity is proportional to the square root of voltage is, valid only when velocities are low compared with the velocity of light. This is because the mass was assumed to be constant whereas, by the theory of relativity, the mass of a particle changes with velocity. Also, from the theory of relativity, mass and energy are equivalent and are related by the square of the velocity of light, c^2; that is, the energy, w, is related to mass by

$$w = mc^2 \tag{3-11}$$

If a force, F, is applied over a distance, dz, to accelerate an electron, then the energy expended, dw, and the increase in electron mass, dm, are related by

$$F\,dz = dw = c^2\,dm \tag{3-12}$$

Newton's second law is

$$F = \frac{d}{dt}(mu) \tag{3-13}$$

We have been writing

$$\frac{d}{dt}(mu) \qquad \text{as} \qquad m\frac{du}{dt} \tag{3-14}$$

However, at high velocities, where m is variable, this is not true and the force equation must be written as

$$F = \frac{d}{dt}(mu) = m\frac{du}{dt} + u\frac{dm}{dt} \tag{3-15}$$

Using this relation for force in (3-12), then the energy expended, dw, becomes

$$m\frac{du}{dt}dz + u\frac{dm}{dt}dz = m\,u\,du + u^2\,dm = c^2 dm \tag{3-16}$$

or

$$\frac{dm}{m} = \frac{u\,du}{c^2 - u^2} = -\frac{1}{2}\frac{d\left(c^2 - u^2\right)}{c^2 - u^2} \tag{3-17}$$

This may be integrated to obtain

$$\ln m = -\frac{1}{2}\ln\left(c^2 - u^2\right) + \text{constant} \tag{3-18}$$

When the velocity of the electron is zero, the mass is the "rest" mass, m_o, and so

$$\text{constant} = \ln m_o + \frac{1}{2}\ln c^2 \tag{3-19}$$

As a result,

$$\ln\frac{m}{m_o} = \ln\left[\frac{c^2}{c^2 - u^2}\right]^{1/2} \tag{3-20}$$

or

$$m = \frac{m_o}{\left(1 - \frac{u^2}{c^2}\right)^{1/2}} \tag{3-21}$$

The electron velocity, u, in terms of voltage can now be found by reconsidering the energy expended, $c^2\,dm$, in accelerating the electron over distance dz (see (3-12)). If the accelerating force is supplied by an electric field, E, then the energy supplied by that field is $-eEdx$ or edV where dV is the voltage increment through which the electron was accelerated. The energy supplied must equal the increase in electron energy so

$$e\,dV = c^2\,dm \tag{3-22}$$

If the electron starts at rest, where it has "rest" mass, m_o, and is accelerated through a total voltage, V, then by integration

$$eV = c^2\left(m - m_o\right) \tag{3-23}$$

With the substitution of the mass from (3-21),

$$eV = c^2 m_o \left[\frac{1}{\left(1 - \frac{u^2}{c^2}\right)^{1/2}} - 1\right] \tag{3-24}$$

By rearranging terms,

$$u = c\left[1 - \frac{1}{\left(1 + \frac{eV}{c^2 m_o}\right)^2}\right]^{1/2} \tag{3-25}$$

Defining

$$V_n = \frac{m_o c^2}{e} = 5.11 \times 10^5 \;\; \text{V} \qquad (\sim 1/2 \text{ MV}) \tag{3-26}$$

where eV_n is the energy equivalent of the rest mass of the electron, then

$$u = c\left[1 - \frac{1}{\left(1 + \frac{V}{V_n}\right)^2}\right]^{1/2} \tag{3-27}$$

When this expression for velocity is substituted into (3-21) for mass,

$$m = m_o \left[1 + \frac{V}{V_n} \right] \tag{3-28}$$

Electron mass and velocity as functions of voltage are shown in Figure 3-1.

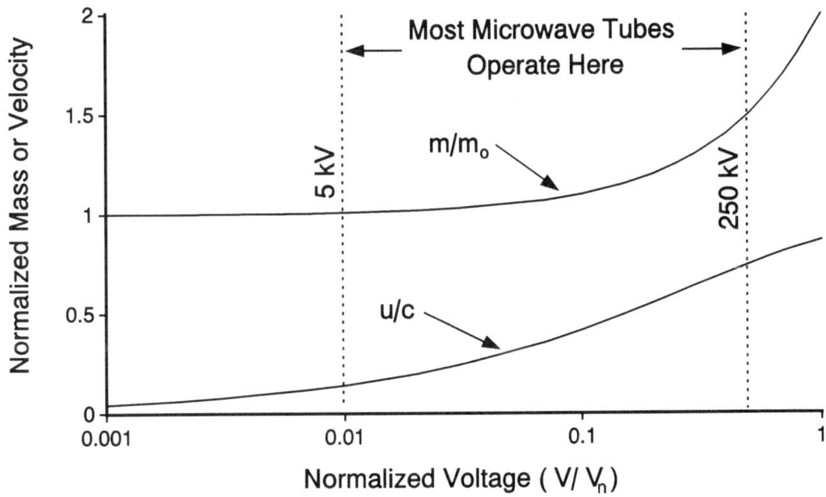

Figure 3-1 Normalized electron mass and velocity as functions of voltage.

As is indicated in Figure 3-1, most microwave tubes operate at voltages between 5 kV and 250 kV. Some low power devices operate below 5 kV and a very few extremely high power devices operate above 250 kV.

3.3 MOTION PERPENDICULAR TO A UNIFORM ELECTRIC FIELD

Consider electron deflection in a cathode ray tube as shown in Figure 3 2. During the transit of the electrons between the deflection plates, the electron is accelerated in the y direction by the field, E_y, between the deflection plates. Thus, the equation of motion of the electron in the y direction is

$$\frac{d^2y}{dt^2} = -\frac{e}{m}E_y = \frac{eV_d}{md} \tag{3-29}$$

where V_d is the voltage applied to the deflection plates. The y directed velocity of the electron after traveling for time, t, in the deflection region is

$$u_y = \frac{eV_d}{md}t \tag{3-30}$$

The time of transit, τ, through the region between the deflection plates is

$$\tau = \frac{\ell}{u_o} \tag{3-31}$$

and so the y directed velocity after the electron leaves the deflection region is

$$u_y = \frac{eV_d}{md}\tau = e\frac{V_d}{md}\frac{\ell}{u_o} \tag{3-32}$$

This velocity is maintained until the electron reaches the screen at time L/u_o, producing a deflection

$$y = \frac{eV_d L\ell}{mdu_o^2} = \frac{V_d L\ell}{2V_o d} \tag{3-33}$$

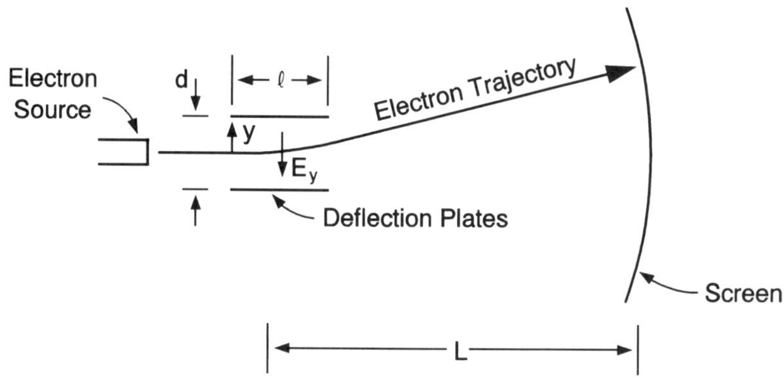

Figure 3-2 Electron deflection in a cathode ray tube.

3.4 ELECTRIC LENSES

In this section, the influence on electron trajectories of electric lenses like those shown in Figure 3-3(a) and (b) is examined. Each electrode set consists of three parallel plates with voltages such that $V_3 > V_2 > V_1$. The center plate at potential V_2 is equally spaced from the other two and has a circular aperture.

As shown in Figure 3-3(a), if the voltage difference between plates 1 and 2 is larger than that between plates 2 and 3, the equipotential contours near the aperture will bulge into the region between 2 and 3. An electron traveling parallel to the axis in region 1–2 will have its trajectory deflected away from the axis in region 2–3.

For the case shown in Figure 3-3(b), the voltage difference between plates 2 and 3 is larger than between 1 and 2 and so the equipotential contours near the aperture

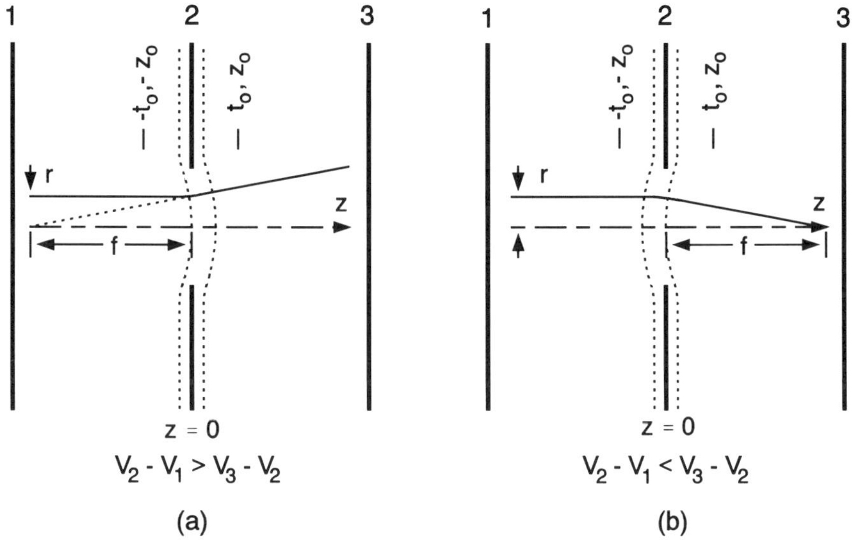

Figure 3-3 Focusing action of single-aperture lens.

bulge into the region 1–2. An electron traveling parallel to the axis in region 1–2 will have its trajectory deflected toward the axis in region 2–1.

The purpose of the analysis in this section is to determine the focal length, f, in terms of the applied voltages. This focal length will be used in the analysis of electron guns in Chapter 6.

Assuming that $V_1 = 0$ and that an electron leaves electrode 1 with negligible velocity, then the velocity of the electron at electrode 2 ($z = 0$) is

$$u_z = \left[2 \frac{e}{m} V_2 \right]^{1/2} \tag{3-34}$$

The approximation made here is that the voltage at $z = 0$ in the aperture is the same as that on electrode 2.

In the vicinity of the aperture in the center electrode, where the equipotential surfaces are curved, there is a radial component to the electric field, E_r. This radial field produces a radial force on the electron and a radial velocity component, u_r, which can be obtained from

$$\frac{d}{dt} m u_r = - e E_r \tag{3-35}$$

and is

$$u_r = -\frac{e}{m}\int_{-t_o}^{t_o} E_r dt \tag{3-36}$$

The time $-t_o$ to t_o is the time spent by the electron in the influence of the radial field. Assuming that u_z is constant during this time period, then the time interval, $-t_o$ to t_o, corresponds to a distance interval, $-z_o$ to z_o and, since $u_z = dz/dt$, u_r may be written as

$$u_r = -\frac{e}{mu_z}\int_{-z_o}^{z_o} E_r dz \tag{3-37}$$

Now, an expression for the integral in (3-37) may be obtained through the use of Gauss's law. Consider a Gaussian surface as shown in Figure 3-4, having a radius r and a length $2z_o$ and located at the center of one of the lenses in Figure 3-3. There is no charge within the surface and so the integral of the normal component of the electric field over the surface is zero. As a result

$$\int_{-z_o}^{z_o} E_r 2\pi r dz + \int_0^r E_1 2\pi r dr - \int_0^r E_2 2\pi r dr = 0 \tag{3-38}$$

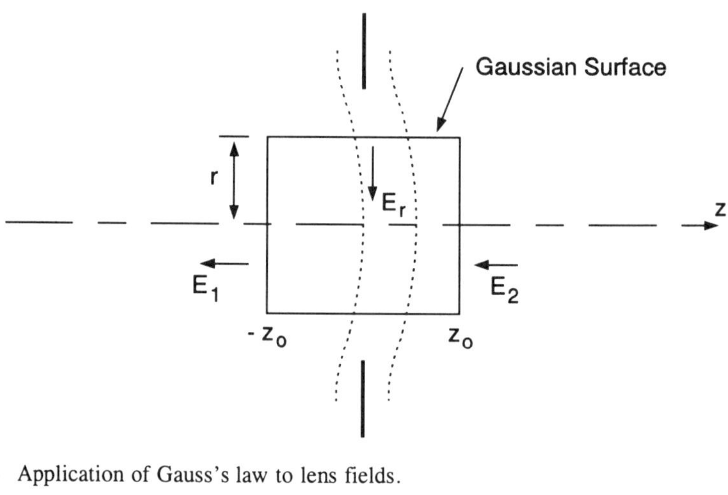

Figure 3-4 Application of Gauss's law to lens fields.

Now, assuming that E_1 and E_2 are constant, then 3-38 becomes

$$\int_{-z_o}^{z_o} E_r 2\pi r dz + \pi r^2 (E_1 - E_2) = 0 \tag{3-39}$$

or

$$\int_{-z_o}^{z_o} E_r dz = -\frac{r}{2}(E_1 - E_2) \tag{3-40}$$

This integral relation for E_r may be used in 3-37 to obtain

$$u_r = \frac{e}{mu_z} \frac{r}{2}(E_1 - E_2) \tag{3-41}$$

Dividing this equation by u_z and using the energy relation (3-34) then

$$\frac{u_r}{u_z} = \frac{r}{4} \frac{(E_1 - E_2)}{V_2} \tag{3-42}$$

But from Figure 3-3

$$\frac{u_r}{u_z} = -\frac{r}{f} \tag{3-43}$$

where the minus sign occurs because the focal position , f, is in the negative z direction when u_r is positive. Now the focal length may be written as

$$f = -\frac{4 V_2}{E_1 - E_2} \tag{3-44}$$

This is the single-aperture lens formula of Davisson and Calbick. This result is valid for $u_r << u_z$ in a thin lens ($z_o <<$ radius of the lens aperture).

3.5 UNIVERSAL BEAM SPREAD CURVE

In the previous sections of this chapter, the motions of individual electrons under the influence of applied fields has been examined. In this section, it will be assumed that there are no applied fields. A beam of electrons is assumed and the only electric fields that are present are those produced by the electrons. The situation that will be analyzed is like that shown in Figure 3-6. The electron beam is converging as it approaches $z = z_m$. But the forces between electrons, which are referred to as *space charge forces*, cause the beam to stop converging and then to diverge.

The electron beam is assumed to be circular with uniform electron density and so cylindrical coordinates will be used. The radial acceleration of the electrons is

$$\frac{d^2 r}{dt^2} = -\eta E_r \tag{3-45}$$

where E_r is the electric field produced by the electrons.

The electric field, E_r, produced by the electron charge in a beam can be found from Gauss's law:

$$\oint_S \mathbf{E}_r \cdot d\mathbf{s} = -\frac{Q}{\epsilon_o} \tag{3-46}$$

Within the length ℓ and radius r of the electron beam shown in Figure 3-5, the charge is

$$Q = \pi r^2 \ell \rho \tag{3-47}$$

and so

$$\oint_S \mathbf{E}_r \cdot d\mathbf{s} = -\frac{\pi r^2 \ell \rho}{\epsilon_o} \tag{3-48}$$

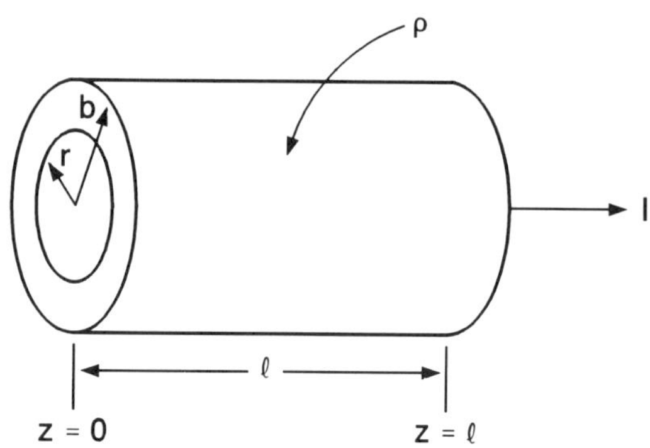

Figure 3-5 Section of electron beam.

The electric field is zero on the ends of the beam section (because adjacent beam sections produce equal and opposite field components) and constant on the cylindrical surface and so the integration produces

$$E_r = -\frac{\pi r^2 \ell \rho}{2\pi r \ell \epsilon_o} = -\frac{r\rho}{2\epsilon_o} \tag{3-49}$$

The charge density, ρ, can be expressed in terms of current and velocity by referring again to Figure 3-5. All of the charge in length ℓ passes position $z = \ell$ in time

$$t = \frac{\ell}{u_o} \tag{3-50}$$

where u_o is the beam velocity. Thus, the current, which is the charge per unit time, is

$$I = \frac{Q}{t} = \frac{\pi b^2 \ell \rho}{\ell / u_o} = \pi b^2 \rho u_o \tag{3-51}$$

As a side note, the current density J is the current per unit area and so

$$J = \rho u_o \tag{3-52}$$

This is, in fact, the definition of current in a beam, and it is a relation that will be used frequently.

The charge density from 3-51 can be combined with 3-49 to obtain the following expression for E_r:

$$E_r = -\frac{rI}{2\pi b^2 \epsilon_o u_o} \tag{3-53}$$

At the beam boundary where $r = b$, the electric field is

$$E_r(b) = -\frac{I}{2\pi b \epsilon_o u_o} \tag{3-54}$$

and so the equation of motion for electrons on the beam boundary is

$$\frac{d^2 b}{dt^2} - \frac{\eta I}{2\pi b \epsilon_o u} = 0 \tag{3-55}$$

Now, assuming, as is shown in Figure 3-6, that the initial beam radius, b_o, occurs at $z = 0$, where $db/dz = db_o/dz$. We want to obtain a relation that defines b as a function of z. Remember that

$$\frac{db}{dt} = \frac{db}{dz}\frac{dz}{dt} \tag{3-56}$$

from which

$$\begin{aligned} \frac{d^2 b}{dt^2} &= \frac{d}{dz}\left[\frac{db}{dz}\frac{dz}{dt}\right]\frac{dz}{dt} \\ &= \left[\frac{d^2 b}{dz^2}\frac{dz}{dt}\right]\frac{dz}{dt} + \left[\frac{db}{dz}\frac{d^2 z}{dt^2}\right]\frac{dz}{dt} \end{aligned} \tag{3-57}$$

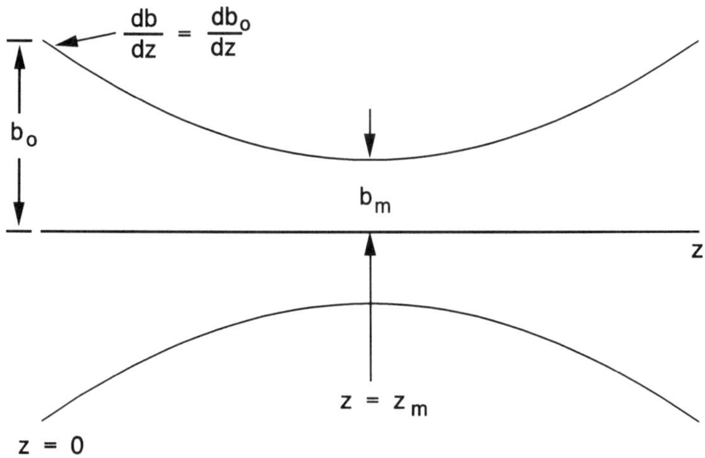

Figure 3-6 Assumed beam shape.

But $dz/dt = u_o$ and $d^2z/dt^2 = 0$ so

$$\frac{d^2b}{dt^2} = u_o^2 \frac{d^2b}{dz^2} \tag{3-58}$$

This means that

$$\frac{d^2b}{dz^2} - \frac{\eta I}{2\pi b \epsilon_o u_o^3} = 0 \tag{3-59}$$

Next, let

$$A^2 = \frac{\eta I}{\pi \epsilon_o u_o^3} = \frac{\eta I}{\pi \epsilon_o (2\eta V)^{3/2}} = \frac{P}{2\pi \epsilon_o \sqrt{2\eta}} = 3.04 \times 10^4 P \tag{3-60}$$

or

$$A = 174\sqrt{P} \tag{3-61}$$

where

$$P = \frac{I}{V^{3/2}} \tag{3-62}$$

In Chapter 5, it will be shown that the relation between current and voltage in most thermionic diodes, including the electron guns on TWTs is given by 3-62. The quantity P is called the *perveance* and depends only on the geometry of the diode or electron gun. Now, by substituting A from 3-60 into 3-59, the following expression for the beam radius is obtained:

$$\frac{d^2b}{dz^2} - \frac{A^2}{2b} = 0 \tag{3-63}$$

This may be rewritten as

$$\frac{d^2\dfrac{b}{b_o}}{d\left[A\dfrac{z}{b_o}\right]^2} - \frac{b_o}{2b} = 0 \tag{3-64}$$

or as

$$\frac{d^2B}{dZ^2} - \frac{1}{2B} = 0 \tag{3-65}$$

where

$$B = \frac{b}{b_o} \quad \text{and} \quad Z = A\frac{z}{b_o} \tag{3-66}$$

After multiplying by dB/dZ, 3-65 may be rewritten and integrated to obtain

$$\left(\frac{dB}{dZ}\right)^2 = \ln B + C \tag{3-67}$$

At $Z = z = 0$, $b = b_o$ so $B = 1$, $db/dz = db_o/dz$ so $dB/dZ = dB_o/dZ$ and so

$$C = -\ln 1 + \left(\frac{dB_o}{dZ}\right)^2 = \left(\frac{dB_o}{dZ}\right)^2 \tag{3-68}$$

which means that

$$\left(\frac{dB}{dZ}\right)^2 = \ln B + \left(\frac{dB_o}{dZ}\right)^2 \tag{3-69}$$

or

$$B = e^{(dB/dZ)^2 - (dB_o/dZ)^2} \tag{3-70}$$

When the beam radius reaches the minimum value, b_m, and $B = B_m$, then $dB/dZ = 0$. Then

$$B_m = e^{-(dB_o/dZ)^2} \tag{3-71}$$

To obtain Z, 3-69 may be rewritten as

$$dZ = \frac{dB}{\left(\ln B + (dB_o/dz)^2\right)^{1/2}} \qquad \text{or} \qquad Z = \int_1^B \frac{dB}{\left(\ln B + (dB_o/dZ)^2\right)^{1/2}} \tag{3-72}$$

This integral may be put in a recognizable form by changing variables and letting

$$u = \frac{dB}{dZ} = \left[\ln B + \left[\frac{dB_o}{dZ}\right]^2\right]^{1/2} \tag{3-73}$$

then

$$B = e^{u^2 - (dB_o/dZ)^2} \qquad \text{and} \qquad dB = 2ue^{u^2 - (dB_o/dZ)^2}\,dz \tag{3-74}$$

The limits of integration now become

$$u = \frac{dB_o}{dZ} \text{ when } B = 1 \text{ and } u = \frac{dB}{dZ} = \pm\left[\ln B + (dB_o/dZ)^2\right]^{1/2} \text{ when } B = B \tag{3-75}$$

and so

$$Z = 2e^{-(dB_o/dZ)^2}\int_{dB_o/dZ}^{dB/dZ} e^{u^2}\,du \tag{3-76}$$

As Z increases from 0, the beam, at first, converges so dB_o/dZ is negative and $B < 1$. In this region, the minus sign in the upper limit of integration is used in finding Z. When $z = z_m$ where $Z = Z_m$, the upper limit becomes zero. For $Z > Z_m$, the positive sign in the upper limit is used.

Plots of B as a function of Z can be generated for various values of dB_o/dZ by selecting values of dB/dZ and then calculating corresponding values of B (3-70) and Z (3-76). The result is shown in Figure 3-7.

The curves in Figure 3-7 are symmetrical about the position at which the minimum beam radius occurs. As dB_o/dZ is made more negative, the minimum first moves to the right (in the $+Z$ direction) and then to the left. Thus, for a given perveance, there is an optimum value of dB_o/dZ that maximizes the length of a tunnel through which a beam can travel without being intercepted by the tunnel walls.

In the design of an electron gun, the position of the minimum beam radius will be of interest. This position, Z_{min}, as a function of dB_o/dZ can be determined by evaluating 3-76 with $dB/dZ = 0$. That is,

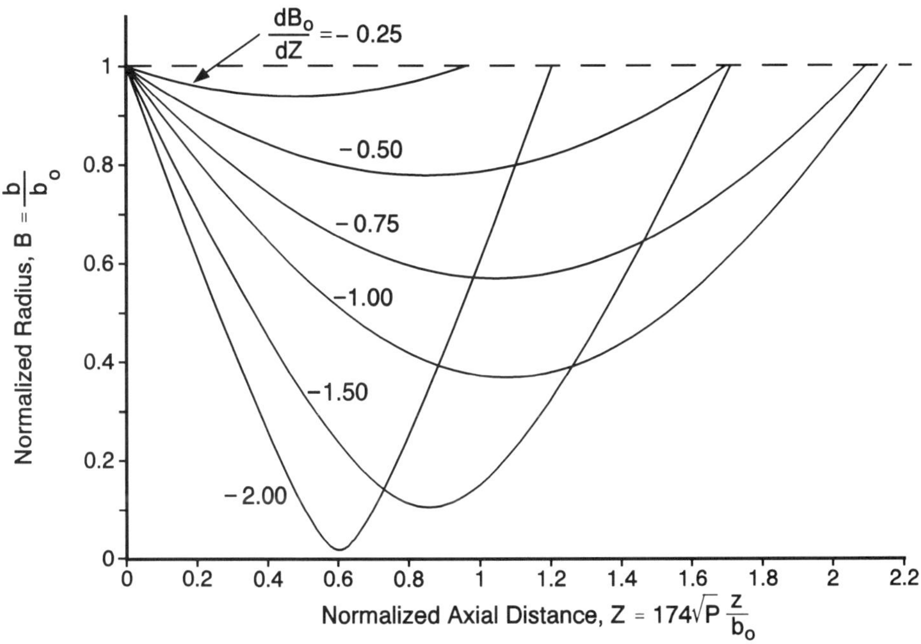

Figure 3-7 Electron trajectories under the influence of space charge.

$$Z_{min} = 2e^{-(dB_o/dZ)^2} \int_{dB_o/dZ}^{0} e^{u^2} du \tag{3-77}$$

The result is shown in Figure 3-8.

A version of this curve that can be obtained by changing the initial conditions to the position where the minimum beam radius occurs and by normalizing the radius and axial distance with respect to the minimum radius, b_m, is the universal beam spread curve shown in Figure 3-9. The curve is symmetrical about $z = z_o$. This curve describes the shape of an electron beam in field-free space. One of its uses is in periodic focusing systems where this curve describes the shape of the beam between lenses.

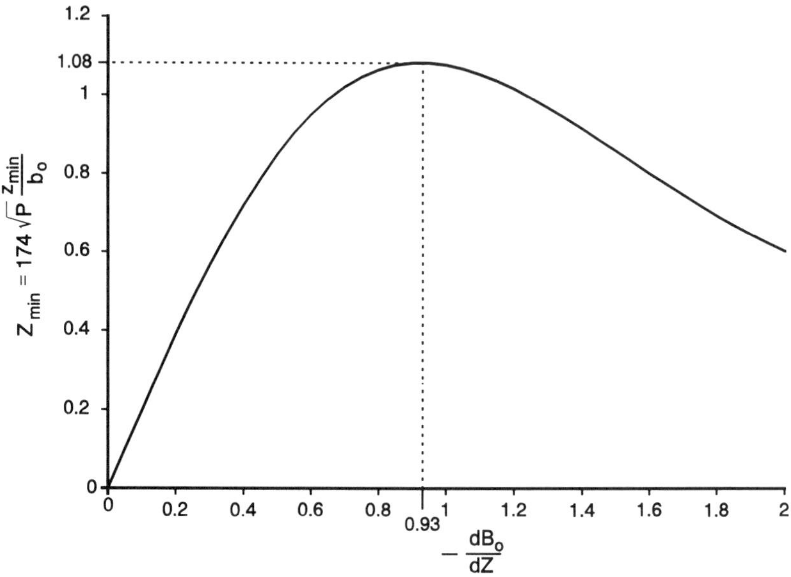

Figure 3-8 Axial position, Z_{min}, of minimum beam radius as a function of dB_o/dZ.

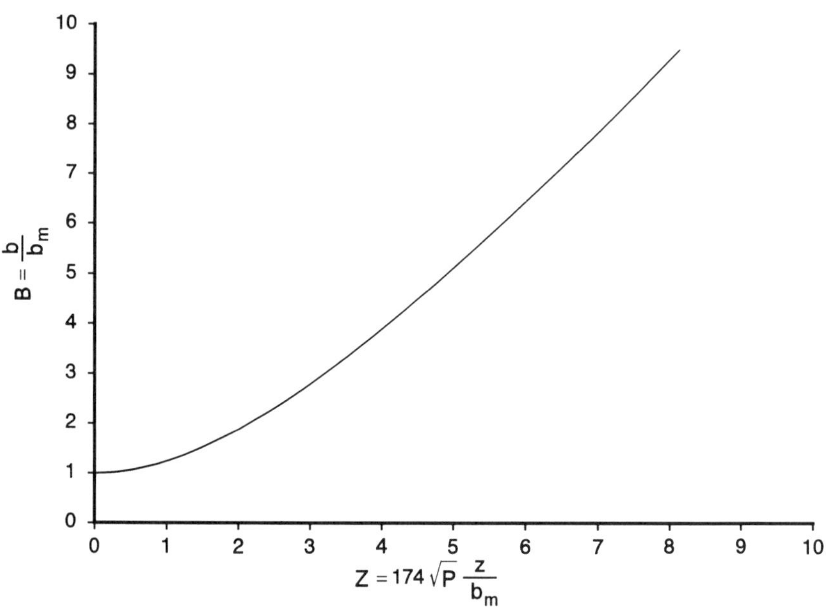

Figure 3-9 Universal beam spread curve.

Chapter 4

Influence of Magnetic Field on Electron Motion

Magnetic fields are used in nearly all microwave tubes for controlling the motions of the electrons. In this chapter, the motion of an electron in a static magnetic field is considered first. Then, the motion in combined electric and magnetic fields is examined. After a general consideration of axially symmetric fields, Busch's theorem is derived. This theorem applies when axial as well as radial electric or magnetic fields are present, and it is extremely important in the analysis of electron beams in all linear-beam tubes.

4.1 ELECTRON MOTION IN A STATIC MAGNETIC FIELD

Before developing the equations of motion for an electron in a magnetic field, it is of value to examine the force equation and see what the influence of the magnetic field on the trajectory is expected to be. Consider an electron traveling with velocity u_o at a right angle to the direction of the flux density, B, which is constant. The force on the electron is eu_oB, and because its direction is perpendicular to the trajectory of the electron, it does not change the speed of the electron. Since the speed is constant, the force is constant.

The force deflects the trajectory of the electron as shown in Figure 4-1 and remains perpendicular to the trajectory. The result is that the path of the electron is circular. The radius of the path may be determined by realizing that there is an outward centrifugal force on the electron as it follows the circular path. The magnitude of this force is mu_o^2/r, where r is the radius of the electron path. This outward force equals the inward force produced by the magnetic field and so

$$eu_oB = \frac{mu_o^2}{r} \tag{4-1}$$

from which

$$r = \frac{mu_o}{eB} \tag{4-2}$$

In terms of the voltage, V_o, through which an electron was accelerated to reach velocity, u_o, the value of r can be written as

$$r = \frac{\left[2\dfrac{e}{m}V_o\right]^{1/2}}{\dfrac{e}{m}B} = 3.37 \times 10^{-6}\,\frac{V_o^{1/2}}{B}\ \text{m} \qquad \text{(4-3)}$$

At V_o = 10,000 V, and B = 1000 G (0.1 T), r = 3.37 mm.

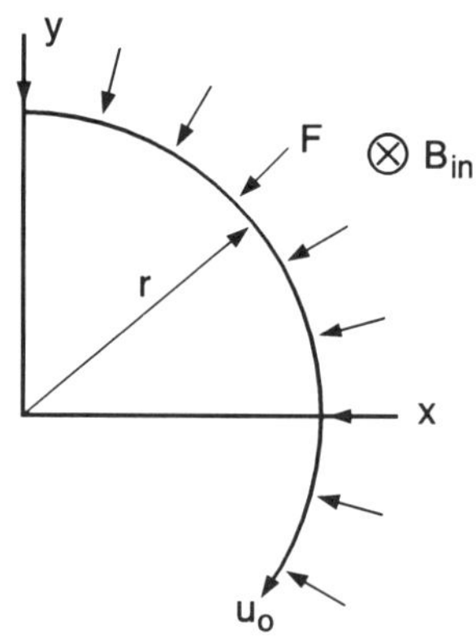

Figure 4-1 Deflection of electron by magnetic field.

The frequency with which the electron rotates is easy to determine because the path length for one rotation is $2\pi r$ and the velocity is u_o. Thus, the time required for each revolution is

$$t = \frac{2\pi r}{u_o} = \frac{2\pi m}{eB} \qquad \text{(4-4)}$$

and the frequency of rotation is

$$f = \frac{1}{t} = \frac{1}{2\pi}\frac{e}{m}B \qquad \text{(4-5)}$$

Note that this frequency depends only on the electron charge and mass and on the magnetic flux density but not on the electron velocity or the radius of the path. The angular frequency with which the electron rotates is called the *cyclotron frequency*, ω_c, and is given by $\omega_c = 2\pi f$ or

$$\omega_c = \frac{e}{m}B \qquad \text{(4-6)}$$

Some values of the frequency of rotation, $f_c = \omega_c/2\pi$, are as follows:

Flux Density (G)	f_c (MHz)
1	2.8
10	28
100	280
1000	2800
10,000	28000

4.2 ELECTRON MOTION IN COMBINED ELECTRIC AND MAGNETIC FIELDS

4.2.1 Perpendicular Fields in Cartesian Coordinates

The general force equation is

$$\boldsymbol{F} = -e(\boldsymbol{E} + \boldsymbol{u} \times \boldsymbol{B}) \tag{4-7}$$

Consider perpendicular electric and magnetic fields in Cartesian coordinates. Assume that the E field is in the $-y$ direction and that B is in the $-z$ direction. Then,

$$\boldsymbol{F} = m\boldsymbol{a} = m(\boldsymbol{i}\ddot{x} + \boldsymbol{j}\ddot{y} + \boldsymbol{k}\ddot{z}) \tag{4-8}$$

where $\boldsymbol{i}$, $\boldsymbol{j}$ and $\boldsymbol{k}$ are unit vectors in the x, y and z directions respectively. The double dots over x, y and z indicate second derivatives with respect to time.

The velocity is

$$\boldsymbol{u} = \boldsymbol{i}\dot{x} + \boldsymbol{j}\dot{y} + \boldsymbol{k}\dot{z} \tag{4-9}$$

and so

$$\boldsymbol{u} \times \boldsymbol{B} = \left(\boldsymbol{i}\dot{x} + \boldsymbol{j}\dot{y} + \boldsymbol{k}\dot{z}\right) \times \left(-\boldsymbol{k}B_z\right) = \boldsymbol{j}\dot{x}B_z - \boldsymbol{i}\dot{y}B_z \tag{4-10}$$

Now, the components of force can be equated to obtain

$$\ddot{x} = -\frac{e}{m}(-\dot{y}B_z) = \omega_c\dot{y} \tag{4-11}$$

$$\ddot{y} = -\frac{e}{m}\left(-E_y + \dot{x}B_z\right) = \eta E_y - \omega_c\dot{x} \tag{4-12}$$

and

$$\ddot{z} = 0 \tag{4-13}$$

With the assumptions that E_y and B_z are constant, these equations can be integrated to obtain

$$\dot{x} = \omega_c y + C_1 \tag{4-14}$$

$$\dot{y} = \eta E_y t - \omega_c x + C_2 \tag{4-15}$$

and

$$\dot{z} = C_3 \tag{4-16}$$

Assume next that, at $t = 0$,

$$\dot{x} = u_o, \quad \dot{y} = 0, \quad \dot{z} = 0, \quad x = 0 \quad \text{and} \quad y = y_o \tag{4-17}$$

that is, the electron enters the coordinate system as shown in Figure 4-2.

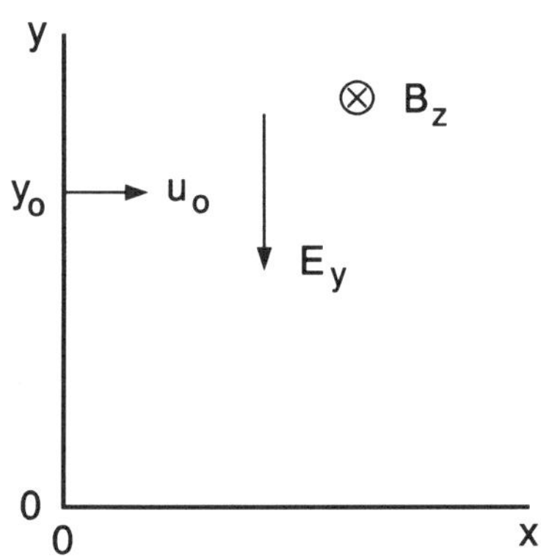

Figure 4-2 Initial conditions for electron motion.

With these initial conditions

$$C_1 = u_o - \omega_c y_o, \quad C_2 = 0 \quad \text{and} \quad C_3 = 0 \tag{4-18}$$

and so

$$\dot{x} = \omega_c (y - y_o) + u_o \tag{4-19}$$

$$\dot{y} = \eta E_y t - \omega_c x \tag{4-20}$$

and

$$\dot{z} = 0 \tag{4-21}$$

Using these relations for velocity, the acceleration equations ((4-11) and (4-12)) become

$$\ddot{x} + \omega_c^2 x = \eta \omega_c E_y t \tag{4-22}$$

and

$$\ddot{y} + \omega_c^2 y = \eta E_y + \omega_c^2 y_o - \omega_c u_o \tag{4-23}$$

The solutions to these equations are

$$x = -r \sin \omega_c t + \frac{\eta E_y}{\omega_c} t = -r \sin \omega_c t + \frac{E_y}{B_z} t \tag{4-24}$$

and

$$y = -r \cos \omega_c t + C \tag{4-25}$$

and the equations for velocity are

$$\dot{x} = -\, r\omega_c \cos \omega_c t + \frac{E_y}{B_z} t \tag{4-26}$$

and

$$\dot{y} = r\omega_c \sin \omega_c t \tag{4-27}$$

The values of r and C can be found using the initial conditions as follows. At $t = 0$

$$\dot{x} = u_o = -\, r \omega_c + \frac{E_y}{B_z} \tag{4-28}$$

so

$$r = \frac{1}{\omega_c} \left[\frac{E_y}{B_z} - u_o \right] \tag{4-29}$$

Also,

$$y = y_o = -\frac{1}{\omega_c} \left[\frac{E_y}{B_z} - u_o \right] + C \tag{4-30}$$

so

$$C = y_o + \frac{1}{\omega_c}\left[\frac{E_y}{B_z} - u_o\right] \tag{4-31}$$

Finally,

$$\begin{aligned} x &= -r \sin \omega_c t + \frac{E_y}{B_z} t \\ &= -\frac{1}{\omega_c}\left[\frac{E_y}{B_z} - u_o\right] \sin \omega_c t + \frac{E_y}{B_z} t \end{aligned} \tag{4-32}$$

and

$$\begin{aligned} y &= -r(\cos \omega_c t - 1) + y_o \\ &= -\frac{1}{\omega_c}\left[\frac{E_y}{B_z} - u_o\right](\cos \omega_c t - 1) + y_o \end{aligned} \tag{4-33}$$

To determine what these equations mean, suppose that we move in the $+x$ direction at the speed E_y/B_z and observe the motion of the electron.

At $\omega_c t = 0$, $x = 0$ and $y = y_o$.
At $\omega_c t = \pi/2$, $x = -r$ and $y = y_o + r$.
At $\omega_c t = \pi$, $x = 0$ and $y = y_o + 2r$.
At $\omega_c t = 3\pi/2$, $x = r$ and $y = y_o + r$.
At $\omega_c t = 2\pi$, $x = 0$ and $y = y_o$.

The electrons appear to rotate as is shown in Figure 4-3; that is, in the clockwise direction at frequency ω_c in a circle with radius, r.

The actual shape of the trajectory depends on the initial velocity of the electron. For example, if $u_o = 0$, then

$$r = \frac{1}{\omega_c}\frac{E_y}{B_z} \tag{4-34}$$

$$x = -r \sin \omega_c t + r \omega_c t \tag{4-35}$$

and

$$y = -r(\cos \omega_c t - 1) + y_o \tag{4-36}$$

and the trajectory is as shown in Figure 4-4.

If $u_o = -E_y/B_z$, then

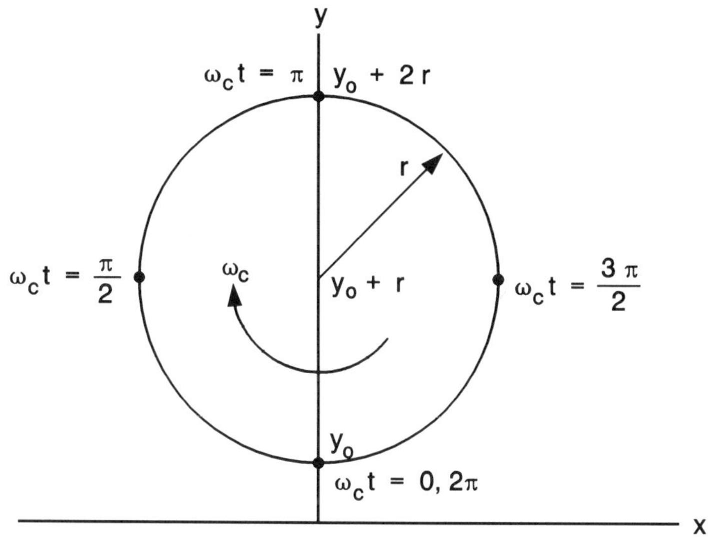

Figure 4-3 Electron motion when frame of reference moves at $\eta E_y/\omega_c$.

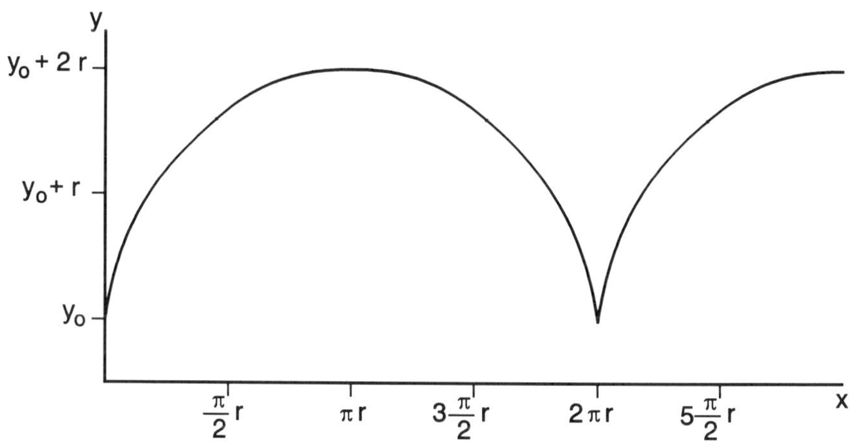

Figure 4-4 Electron trajectory for $u_o = 0$.

$$r = \frac{2}{\omega_c} \frac{E_y}{B_z} \tag{4-37}$$

$$x = -r \sin \omega_c t + \frac{r \omega_c}{2} t \tag{4-38}$$

and

$$y = -r\left(\cos\omega_c t - 1\right) + y_o \tag{4-39}$$

and the trajectory is as shown in Figure 4-5. If $u_o = E_y/B_z$, then $r = 0$ and the electron travels in a straight line in the x direction.

Electron motions similar to these occur in a circular format in a magnetron where the electric field is radial and the magnetic field is parallel to the axis.

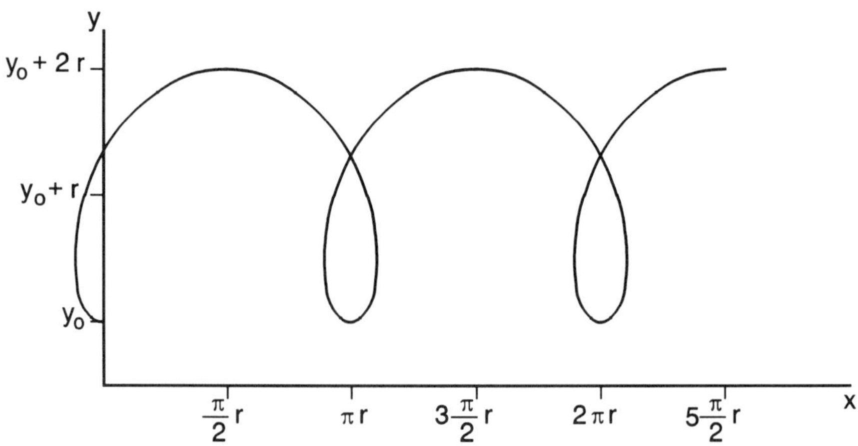

Figure 4-5 Electron trajectory for $u_o = -E_y/B_z$.

4.2.2 Axially Symmetric Fields

Next consider the case of axially symmetric electric and magnetic fields. A common example is that of a radial electric field with no θ variations and an axial magnetic field. For this case it will be easiest to work in cylindrical coordinates. Referring to Figure 4-6, it is seen that the force components F_r, and F_θ in cylindrical coordinates may be written in terms of F_x and F_y as follows:

$$F_r = F_x \cos\theta + F_y \sin\theta \tag{4-40}$$

and

$$F_\theta = F_x \sin\theta - F_y \cos\theta \tag{4-41}$$

Since $x = r\cos\theta$, then

$$\dot{x} = \dot{r}\cos\theta - r\dot{\theta}\sin\theta \tag{4-42}$$

and so

$$F_x = m\ddot{x} = m\left(\ddot{r}\cos\theta - \dot{r}\dot{\theta}\sin\theta - \dot{r}\dot{\theta}\sin\theta - r\ddot{\theta}\sin\theta - r\dot{\theta}^2\cos\theta\right) \quad (4\text{-}43)$$

Also, since $y = r \sin \theta$, then

$$\dot{y} = \dot{r}\sin\theta + r\dot{\theta}\cos\theta \quad (4\text{-}44)$$

and so

$$F_y = m\ddot{y} = m\left(\ddot{r}\sin\theta + \dot{r}\dot{\theta}\cos\theta + \dot{r}\dot{\theta}\cos\theta + r\ddot{\theta}\cos\theta - r\dot{\theta}^2\sin\theta\right) \quad (4\text{-}45)$$

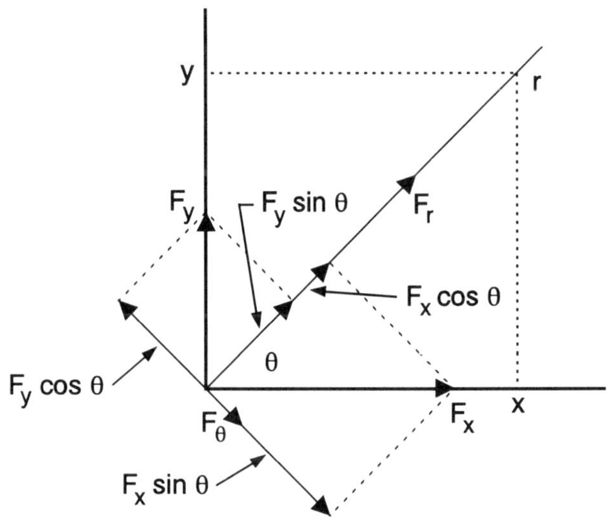

Figure 4-6 Conversion from rectangular to cylindrical coordinates.

Placing these relations into F_r and F_θ, the following are obtained for the force coordinates in cylindrical coordinates:

$$F_r = m(\ddot{r} - r\dot{\theta}^2) \quad (4\text{-}46)$$

and

$$F_\theta = m(2\dot{r}\dot{\theta} + r\ddot{\theta}) \quad (4\text{-}47)$$

Now, consider a practical application. In a smooth-bore magnetron with a voltage applied between the anode and cathode as shown in Figure 4-7, the electric field is in the radial direction and independent of θ. The magnetic field is in the axial direction. As electrons leave the cathode and move toward the anode, they are deflected in the θ direction. In cylindrical coordinates, if r is radially outward from the axis and θ is clockwise about the axis, then z is into the plane of r and θ and the θ directed force is

$$F_\theta = m(2\dot{r}\dot{\theta} + r\ddot{\theta}) = -eu_r B_z = e\dot{r}B_z \tag{4-48}$$

This can be written as

$$2r\dot{r}\dot{\theta} + r^2\ddot{\theta} = \omega_c r\dot{r} \tag{4-49}$$

or

$$\frac{d}{dt}\left(r^2\dot{\theta}\right) = \frac{\omega_c}{2}\frac{d}{dt}r^2 \tag{4-50}$$

which can be integrated to obtain

$$r^2\dot{\theta} = \frac{\omega_c}{2}r^2 + \text{constant} \tag{4-51}$$

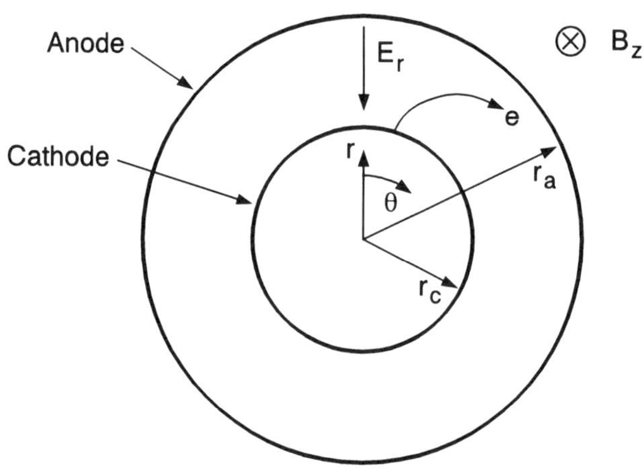

Figure 4-7 Configuration of smooth-bore magnetron.

Next, assume that the initial condition for an electron at the cathode is

$$\dot{\theta} = 0 \quad \text{at} \quad r = r_c \quad \text{so} \quad \text{constant} = -\frac{\omega_c}{2}r_c^2 \tag{4-52}$$

As a result,

$$r^2\dot{\theta} = \frac{\omega_c}{2}\left(r^2 - r_c^2\right)$$

$$\dot{\theta} = \frac{\omega_c}{2}\left[1 - \frac{r_c^2}{r^2}\right] \tag{4-53}$$

At the anode, where $r = r_a$, the angular velocity of the electron is

$$\dot{\theta}_a = \frac{\omega_c}{2}\left[1 - \frac{r_c^2}{r_a^2}\right] \tag{4-54}$$

If the anode is at a potential V_a with respect to the cathode such that an electron just barely grazes the anode (that is, $\dot{r} = 0$) then the velocity of the electron is

$$u_a = \sqrt{2\eta V_a} = r_a\dot{\theta}_a \tag{4-55}$$

This may rewritten and combined with 4-54 to obtain

$$V_a = \frac{\dot{\theta}_a^2 r_a^2}{2\eta} = \frac{\omega_c^2 r_a^2}{8\eta}\left[1 - \frac{r_c^2}{r_a^2}\right]^2 \tag{4-56}$$

Thus, the voltage that causes the electron to just graze the anode depends on the dimensions r_a and r_c and on the flux density (in ω_c) but not on the way in which the voltage varies between the cathode and anode. The electron trajectory is shown in Figure 4-8(a). This condition is known as the *Hull cutoff condition*. When V_a is greater than the value given in 4-56 then the electron strikes the anode as indicated in Figure 4-8(b). When V_a is less than this value, the electron returns to the cathode (Figure 4-8(c)).

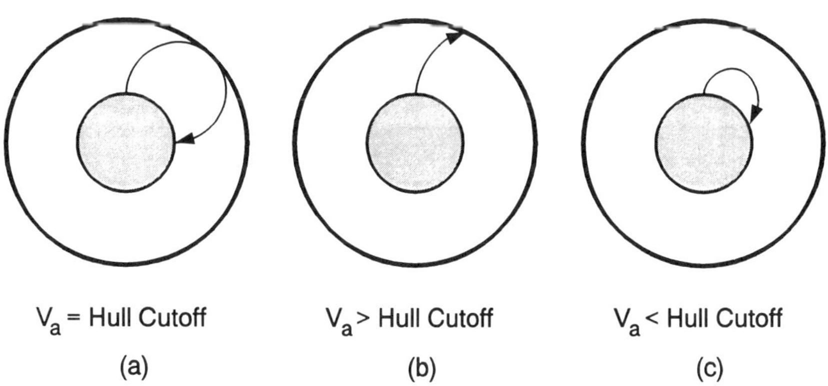

Figure 4-8 Electron trajectories for various anode potentials.

4.2.3 Busch's Theorem

Busch's theorem applies to axially symmetric fields, and it is a general form of the relation for angular velocity in the magnetron (4-53). Busch's theorem is valid when axial as well as radial electric or magnetic fields are present. Thus, it is a powerful tool for the analysis of electron beams in all linear-beam tubes.

As was the case in the magnetron, θ directed forces will be considered. This time, however, forces resulting from axial electron motions, $\dot{z}$, and a radial component, B_r, of the magnetic flux density will be included. Thus, the θ force equation is

$$m\left(2\dot{r}\dot{\theta} + r\ddot{\theta}\right) = e\dot{r}B_z - e\dot{z}B_r \tag{4-57}$$

which may be written as

$$\eta\, r\left(\dot{r}B_z - \dot{z}B_r\right) = \frac{d}{dt}\left(r^2\dot{\theta}\right) \tag{4-58}$$

Multiplying both sides by dt, the following can be obtained:

$$\eta\left(rdrB_z - rdzB_r\right) = d\left(r^2\dot{\theta}\right) \tag{4-59}$$

Next, consider the surface developed by rotating an electron trajectory about the z axis (Figure 4-9). Suppose Ψ is the magnetic flux inside this path at a given value of z; that is,

$$\Psi = \int_0^r B_z 2\pi r dr \tag{4-60}$$

Now, consider the change in Ψ in the distance dz. The flux will diminish by the amount crossing the surface in dz. The amount of flux going outward through the surface is the difference between the radial ($2\pi rdzB_r$) and the axial ($2\pi rdrB_z$) components; that is,

$$2\pi rdzB_r - 2\pi rdrB_z \tag{4-61}$$

and so the change in flux within the surface, $d\Psi$, is

$$d\Psi = -(2\pi rdzB_r - 2\pi rdrB_z) \tag{4-62}$$

Comparing this with the force equation (4-59), it is seen that

$$\frac{\eta}{2\pi}d\Psi = d(r^2\dot{\theta}) \tag{4-63}$$

or

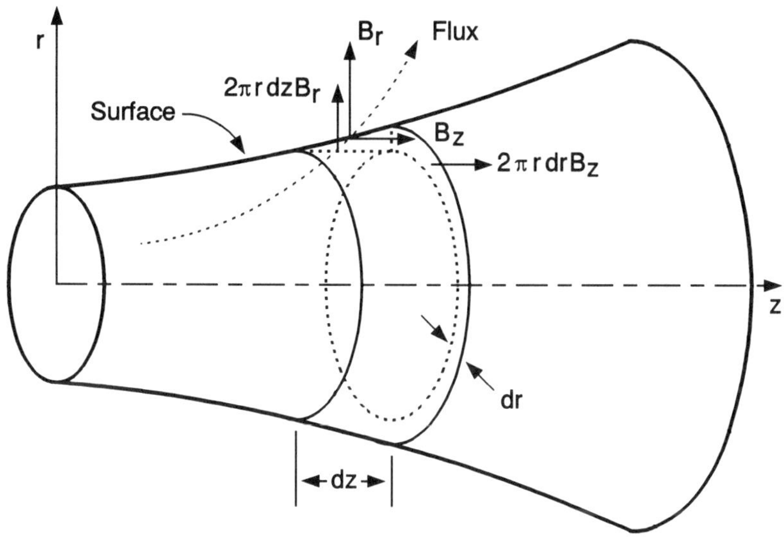

Figure 4-9 Surface developed by rotating electron trajectory about z axis.

$$\dot{\theta} = \frac{\eta}{2\pi r^2}(\Psi - \Psi_o) \tag{4-64}$$

where Ψ_0 is the flux that links the path at the point where $\dot{\theta}$ is zero (at the cathode). Thus, the angular velocity of the electron is directly proportional to the amount of flux that the electron crosses.

In many cases, motion near the axis of a solenoid is of interest. There, $\partial B_z/\partial r \approx 0$ and so

$$\Psi = \pi r^2 B_z \quad \text{and} \quad \Psi_o = \pi r_o^2 B_{zo} \tag{4-65}$$

where B_{zo} is the flux density at the point where $\dot{\theta}$ is zero. Now, Busch's theorem assumes the approximate form

$$\dot{\theta} = \frac{\eta}{2}\left[B_z - B_{zo}\frac{r_o^2}{r^2}\right] \tag{4-66}$$

or

$$\dot{\theta} = \frac{1}{2}\left[\omega_c - \omega_{co}\frac{r_o^2}{r^2}\right] \tag{4-67}$$

Chapter 5

Cathodes

The cathode is the source of electrons for the electron beam in every microwave tube. The current density of electron emission from the cathode ranges from milliamperes to tens of amperes per square centimeter of cathode area.

Two mechanisms for the emission of electrons from the cathode are almost universally used. These are

- thermionic emission
- secondary emission

In cathodes for traveling wave tubes, only thermionic emission is used, and so thermionic cathodes are the primary subject for discussion in this chapter. Secondary emission cathodes, which are used in crossed-field tubes, such as magnetrons and some crossed-field amplifiers, will not be discussed.

J.R. Pierce [1] listed the primary characteristics that an ideal cathode should have:

1. emits electrons freely, without any form of persuasion such as heating or bombardment (electrons would leak off from it into vacuum as easily as they pass from one metal to another);
2. emits copiously, supplying an unlimited current density;
3. lasts forever, its electron emission continuing unimpaired as long as it is needed;
4. emits electrons uniformly, traveling at practically zero velocity.

Of course, real cathodes have none of these ideal characteristics. For example, real cathodes must be heated to a temperature in the vicinity of 1000°C to produce appreciable electron emission. At that temperature, the current density is not unlimited, but instead is on the order of a few tens of amperes per square centimeter at most. Because of the necessity for a high operating temperature, some of the key constituents of a cathode evaporate, which leads to their depletion and, eventually, to the end of the life of the cathode.

The reason for the fourth characteristic of an ideal cathode is not as obvious as the other three. First of all, it should be noted that, in real cathodes, there are random microscopic fluctuations in the rate at which electrons are emitted. Second, electrons are emitted with finite random velocities in random directions relative to the cathode surface. There are at least two important reasons why an ideal cathode "should emit electrons uniformly, traveling at practically zero velocity":

1. Fluctuations in electron emission and variations in electron emission velocities lead to noise currents in the electron beam, which, in turn, produce noise on the output signal from the tube.
2. Variations in electron emission velocities and in the directions with which electrons leave the cathode surface lead to problems in focusing the electrons into a well-defined beam. This problem is particularly severe in very small, high density beams.

While real cathodes have none of the characteristics that Pierce's ideal cathode would have, there has been enormous progress in cathode capabilities since Pierce listed his ideal characteristics. This progress is indicated in Figure 5-1. Note, in particular, the large increase in emission capability that has been achieved in the past decade. As pointed out by Thomas *et al.*, this progress is due largely to the advent of modern surface analytical capabilities that make possible a better understanding of the physics and chemistry of emitter surfaces [2]. In this chapter, we attempt to provide sufficient background material so that, at least qualitatively, the operation of modern cathodes, with the exception of the scandate cathode, can be understood. The operation of the scandate cathode is the subject of active research at the present time.

The chapter starts with the basic theory of thermionic emission. This theory is straightforward and gives a reasonably accurate prediction of the maximum emission that can be expected from a cathode at a given temperature, in the absence of field effects, if the work function is known. Next, the enhancement of thermionic emission that results from the application of an electric field to the surface of the cathode (the Schottky effect) is discussed. The emission regulating effect of the electron cloud adjacent to the surface of the cathode (space charge limited emission) is of extreme importance in the use of thermionic cathodes in TWTs. For the person who must use a cathode in a TWT, the concept of space charge limited emission may be the most important material discussed in this chapter.

To maintain space charge limited emission, the cathode work function must remain low enough so that adequate emission is obtained at a given temperature during the life of a cathode. The work function of the cathode is critically dependent on the materials and condition of the surface of the cathode. Thus, a large portion of this chapter is devoted to a discussion of cathode materials and their impact on life of the cathode and the TWT in which it is used.

The physical arrangement of the cathode and the elements that provide the power necessary to heat it to the required temperature in the time permitted by the application of the TWT are described. Conventional heater assemblies are discussed along with

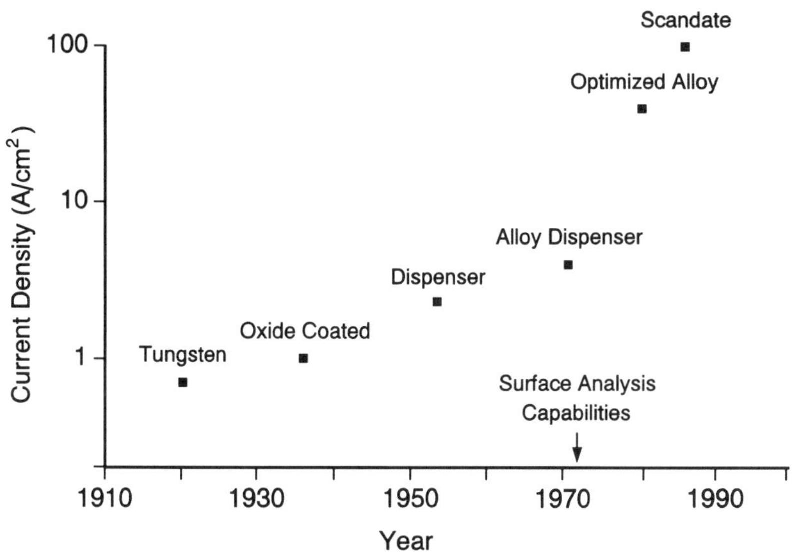

Figure 5-1 Historical perspective of thermionic cathode emission capabilities. From Richard E. Thomas *et al.*, *IEEE Trans. on Electron Devices*, March 1990. © 1990 IEEE.

fast-warm-up assemblies that are required in some special applications of TWTs.

The magnetic field of the filament can have a significant impact on the operation of an electron gun. Electron trajectories near the cathode surface are perturbed by this field with the result that beam generation and focusing are degraded. If an ac filament current is used, the beam shape may be modulated at the ac frequency and this, in turn, modulates the power output of the tube. These effects are discussed at the end of the chapter.

5.1 EMISSION MECHANISMS

5.1.1 Thermionic Emission [3, 4]

At temperatures above absolute zero, some electrons have sufficient energy to escape from the surface of a cathode. As temperature is increased, the number of electrons with sufficient energy to escape increases. In addition to temperature, the nature of the surface of the cathode has an extremely strong effect on the rate at which electrons are emitted. Electron emission resulting from the heating of a surface is referred to as thermionic emission.

The fundamental aspects of thermionic emission may be understood with the aid of Figure 5-2, which is the classical energy level diagram for electrons near the interface between a cathode and vacuum. In the cathode, the parabolic curves represent electron energy levels adjacent to atoms. The energy levels merge to form the conduction band. At a temperature of absolute zero (0 K), no electrons have energy greater than E_o, which is the top of the conduction band and is called the *Fermi level*. The difference in energy between the top of the conduction band in the cathode and the vacuum level adjacent to the cathode is known as the *work function*, commonly denoted by $e\phi$.

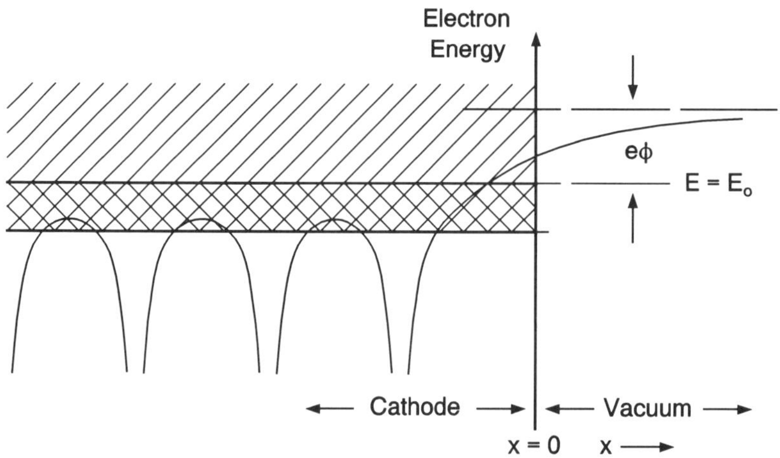

Figure 5-2 Energy level diagram for electrons near the surface of a metal.

At temperatures above absolute zero, some electrons have energy greater than E_o. Emission of the electrons from the cathode can occur if their energy is $E_o + e\phi$ or greater. It is necessary to keep in mind, however, that the electrons move in random directions within the cathode. Those electrons moving toward the surface have the highest probability of being emitted. This is illustrated in terms of the momentum, P, of the electron in Figure 5-3. If the x-directed velocity, v_x (or momentum, P_x), is great enough ($P_x \geq P_{xc}$) the electron will be emitted. If not, the electron will be deflected back into the cathode.

The critical momentum, P_{xc} (that is, the critical component directed toward the surface), required by an electron to overcome $e\phi$ is readily found from

$$\frac{P_{xc}^2}{2m} = \frac{1}{2}mu_x^2 = E_o + e\phi \tag{5-1}$$

Now, the emission current density, J, can be found from

$$J = \rho u_x = en_e u_x \tag{5-2}$$

if the number of electrons per unit volume, n_e, having momentum $P_x \geq P_{xc}$ can be found. The density of electrons with $P_x > P_{xc}$ can be computed from density of energy

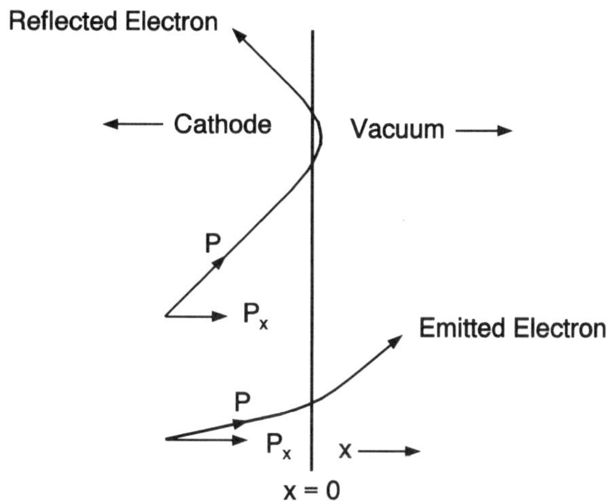

Figure 5-3 Two electron trajectories with the same initial energy ($> E_o + e\phi$) but different directions.

states and the Fermi-Dirac distribution function, which gives the probability that the states are occupied. From the exclusion principle, the density of states with momentum in the increment dP_x, dP_y and dP_z is

$$\left[\frac{2}{h^3}\right] dP_x dP_y dP_z \tag{5-3}$$

where h is Planck's constant (6.625×10^{-34} J/s).

The fraction of the states occupied is given by the Fermi function, f, which is

$$f = \frac{1}{e^{(E - E_o)/kT} + 1} \tag{5-4}$$

where k is Boltzmann's constant (1.38×10^{-23} J/deg). This function is shown in Figure 5-4. Notice that at $T = 0$ K, $f = 0$ for $E > E_o$. This is in agreement with the previous statement that no electrons exist with energies greater than the top of the conduction band. As T is raised above 0 K, f becomes > 0 for $E > E_o$ and so electrons exist that have energy greater than the top of the conduction band.

Because $E - E_o \sim e\,\phi$, which is on the order of 1 eV or greater and $kT \sim 1/10$ eV, then

$$e^{(E - E_o)/kT} >> 1 \tag{5-5}$$

and so

$$f = e^{-(E - E_o)/kT} \tag{5-6}$$

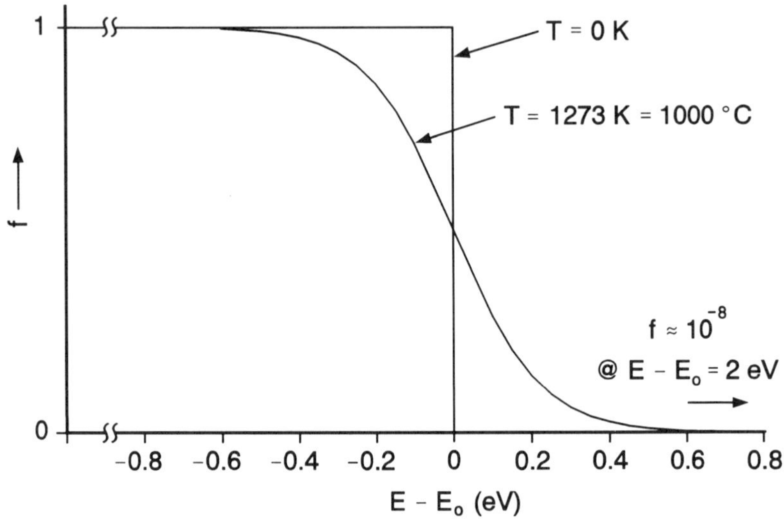

Figure 5-4 Fermi-Dirac distribution function.

The energy, E, of an electron is related to the total electron momentum by

$$E = \frac{1}{2m}\left(P_x^2 + P_y^2 + P_z^2\right) \tag{5-7}$$

and so the Fermi function can be written as

$$f = e^{-\left(P_x^2 + P_y^2 + P_z^2 - 2mE_o\right)/2mkT} \tag{5-8}$$

Now the number of electrons, dn_e, with momentum in the range dP_x, dP_y, dP_z is

$$dn_e = \frac{2}{h^3}dP_x dP_y dP_z f = \frac{2}{h^3}dP_x dP_y dP_z e^{-\left(P_x^2 + P_y^2 + P_z^2 - 2mE_o\right)/2mkT} \tag{5-9}$$

The increment of current density, dJ, produced by these electrons is

$$dJ = e\,dn_e u_x = e\,dn_e \frac{P_x}{m} \tag{5-10}$$

Thus, the total current produced by all electrons with $P_{xc} \le P_x < \infty$, $-\infty < P_y < \infty$, $-\infty < P_z < \infty$ is

$$J = \frac{2e}{mh^3}\int_{P_y=-\infty}^{\infty}\int_{P_z=-\infty}^{\infty}\int_{P_x=P_{xc}}^{\infty} P_x dP_x dP_y dP_z e^{-\left(P_x^2 + P_y^2 + P_z^2 - 2mE_o\right)/2mkT} \tag{5-11}$$

Using the value for the definite integral given by

$$\int_{-\infty}^{\infty} e^{-ax^2} dx = \left[\frac{\pi}{a} \right]^{1/2} \tag{5-12}$$

the current density can be written as

$$J = \frac{2e}{mh^3} (2\pi mkT)^{1/2} (2\pi mkT)^{1/2} \int_{P_{xc}}^{\infty} e^{-\left(P_x^2 - 2mE_o\right)/2mkT} P_x \, dP_x \tag{5-13}$$

Next, the third integral may be evaluated by, first, changing variables. Define u as follows:

$$u = \frac{P_x^2 - 2mE_o}{2mkT} \tag{5-14}$$

so, when $P_x = P_{xc}$ (see (5-1)),

$$u = \frac{e\phi}{kT} \tag{5-15}$$

Next, taking the derivative of (5-14) and rearranging terms,

$$P_x \, dP_x = mkT \, du \tag{5-16}$$

and so the third integral may be easily evaluated as follows:

$$\int_{\frac{e\phi}{kT}}^{\infty} e^{-u} mkT \, du = mkT \, e^{-e\phi/kT} \tag{5-17}$$

When this result is placed in the expression for current density (5-13), the result is

$$J = \frac{2e}{mh^3} (2\pi mkT) mkT \, e^{-e\phi/kT} \tag{5-18}$$

$$= \left[\frac{4\pi mek^2}{h^3} \right] T^2 e^{-e\phi/kT} \tag{5-19}$$

or

$$J = A_o T^2 e^{-e\phi/kT} \tag{5-20}$$

which is the Richardson-Dushman equation for thermionic emission. A_o is a universal constant and has the value

$$A_o = 1.20 \times 10^6 \quad \text{A/m}^2\text{deg}^2 \tag{5-21}$$

The most significant aspect of the Richardson-Dushman equation is the exponential variation of current density with work function and temperature. The variation of current density with T^2 is negligible by comparison. This can be readily demonstrated by considering the relative effects of a small change in temperature. For example, for a typical operating temperature of 1000°C, a 1% change in T produces roughly a 70% change in the value of the exponential term, but only a 2% change in T^2.

To compare theory with experiment, the Richardson-Dushman equation is first written in the following form:

$$\ln \frac{J}{T^2} = \ln A_o - \frac{e\phi}{kT} \tag{5-22}$$

This suggests that a plot of J/T^2 versus $1/T$ should be a straight line. The work function, $e\phi$, should be the slope of the line and ln A_o should be the intercept of the extrapolation of the line to infinite temperature ($1/T = 0$). The data are plotted in the form shown in Figure 5-5, which is known as a *Richardson plot*.

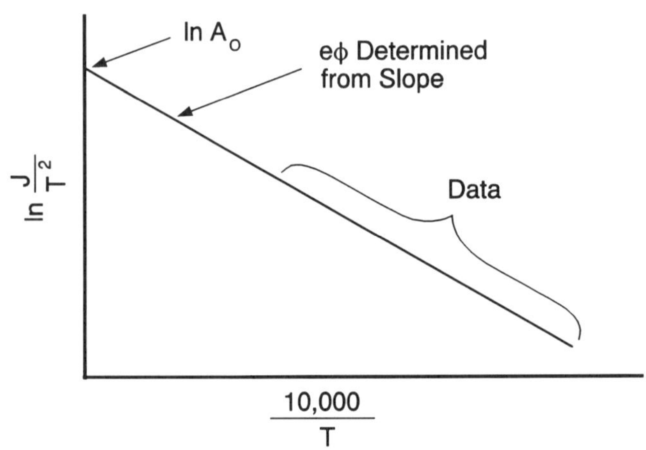

Figure 5-5 Richardson plot for determining work function and emission constant A_o.

In practice, the experimental value of A_o turns out to be on the order of one quarter to one half of the predicted value. Two reasons for this are:

1. The work function has been assumed to be independent of temperature. In fact, $e\phi$ may vary with temperature. A first order approximation sometimes made is that the incremental change in $e\phi$ varies linearly with temperature. Thus,

$$\phi = \phi_R + \alpha T \tag{5-23}$$

where $e\phi_R$ is the Richardson work function and α is the temperature coefficient of change in work function. Now, the Richardson-Dushman equation may be rewritten as

$$J = \left(A_o e^{-e\alpha T/kT}\right) T^2 e^{-e\phi_R/kT} \tag{5-24}$$

where the quantity in parentheses is the corrected value of A_o.

2. The same value of $e\phi$ has been assumed for all parts of the emitting surface. In fact, $e\phi$ may be different for different crystal surfaces. Suppose, for example, that half of the emitter area has work function $e\phi_1$ and the other half has work function $e\phi_2$. In addition, suppose that $e\phi_2 > e\phi_1$. Then, because of the exponential dependence of emission on work function, practically all of the emission will be from the $e\phi_1$ areas. The emitting area is, therefore, only half of the nominal cathode area, and the measured value of A_o is only half of the theoretical value.

From the Richardson-Dushman equation, it is clear that to obtain high values of emission, the work function must be low and the temperature must be high. Unfortunately, as shown in Table 5-1, when pure metals are considered, those materials with low work functions also have low melting temperatures. Cesium, with a work function of 2.1 eV and a melting temperature close to room temperature (28°C), is an extreme example. Barium also has a reasonably low work function, but has a moderate melting temperature (725°C). It turns out that barium is commonly used in thermionic cathodes, but not because it has a low work function. Instead, barium is used because its capability (accompanied by other elements including oxygen) to reduce the work function of the underlying metal surface. Thus, even though they have high work functions, it is possible to use a high melting temperature metal such as tungsten and osmium by introducing barium and other elements to reduce the work function of the surface.

Figure 5-6 shows theoretical emission densities as functions of temperature and work function (dashed lines). Also shown are measured emission densities for several cathodes, most of which contain metals listed in Table 5-1. For W and ThW, the experimental curves are located where expected, based on the values of $e\phi$ in Table 5-1. (For ThW, thorium diffuses to the surface of the tungsten and so the work function is approximately that of thorium.) Notice that, for the cathodes containing barium (BaW, BaNi, Type B, Alloy, Scandate and Oxide) the locations of the experimental curves indicate work functions well below that of barium ($\sim$ 2.7 eV). As noted in the previous paragraph, in most cases it is the interaction of the barium and other elements with the underlying metal that produces the low work function. This will be discussed later in this chapter when the theory of cathode operation is examined.

TABLE 5-1
Room Temperature Work Functions for Various Metals

Metal	*Work Function** eV	*Melting Temperature* °C
Barium	2.7	725
Calcium	2.9	839
Carbon	5.0	~3550
Cesium	2.1	28
Hafnium	3.9	2227
Iridium	5.2	2410
Molybdenum	4.5	2620
Nickel	5.2	1455
Osmium	5.4	3045
Platinum	5.3	1773
Rhenium	5.1	3180
Scandium	3.5	1539
Sodium	2.7	97
Strontium	2.6	769
Tantalum	4.2	2996
Thorium	3.4	1750
Tungsten	4.6	3410
Zirconium	4.1	1852

*Work function is dependent on crystalline orientation. The values listed are estimates of average values.

5.1.2 Schottky Effect

In the derivation of the Richardson-Dushman equation, the effect of an electric field at the cathode surface was not taken into consideration. As a result, neither voltage (from an anode adjacent to the cathode) nor electric field appear in the Richardson-Dushman equation. In fact, when an electric field is applied to the surface of a cathode, it is found that the emission increases slightly. To understand this, we must look closely at the nature of the forces on emitted electrons as a function of distance from the emitting surface.

Figure 5-7(a) shows the electric field lines for an electron near the surface of a cathode. It would be possible, but tedious, to calculate the positive charge on the surface of the cathode as a function of position and then to calculate the force exerted on the electron by this charge distribution. There is a much easier way to solve this problem, and this is based on the "method of images." Consider the field pattern

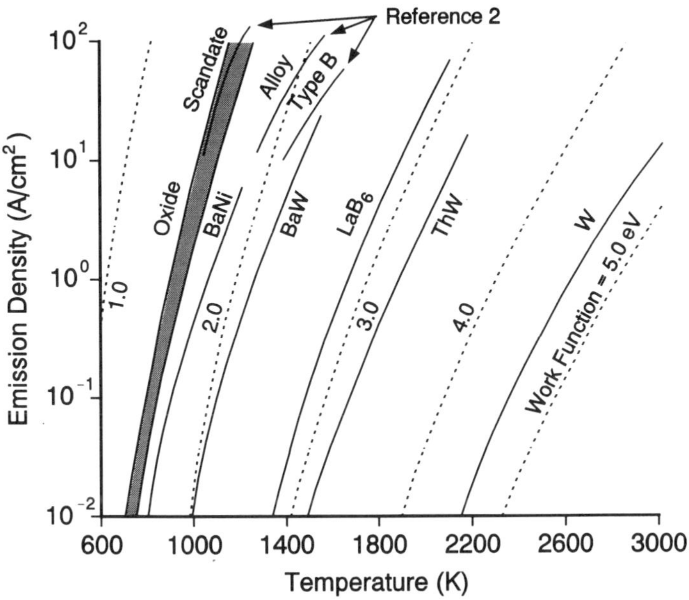

Figure 5-6 Emission characteristics of several cathodes. Adapted from [2] and from *Methods of Experimental Physics*, Vol. 4, Part A, Section 1.1.1 by G.A. Haas, copyright 1967 by Academic Press.

established by an electron and another positive "image" having the same charge as the electron and positioned as shown in Figure 5-7(b). The field to the right of $x = 0$ is the same as that in (a). This means that the force on the electron at a distance x from the surface is the same as if the cathode surface were replaced by a charge $+e$ at $-x$.

From Coulomb's law, the magnitude of the force of attraction between two equal charges of opposite polarity is

$$F = \frac{e^2}{16\pi\epsilon_o x^2} \tag{5-25}$$

and this is also the force of attraction between the electron and the surface of the metal. If the electron is removed to infinity, then, of course, the force goes to zero. If it is assumed that the electron energy at infinity is zero, then as the electron is moved from infinity to position x, it is accelerated and its kinetic energy increases as its potential energy decreases. At position x, the potential energy, P_{Es}, is

$$P_{Es} = \int_{\infty}^{x} F dx = \int_{\infty}^{x} \frac{e^2}{16\pi\epsilon_o x^2} dx \tag{5-26}$$

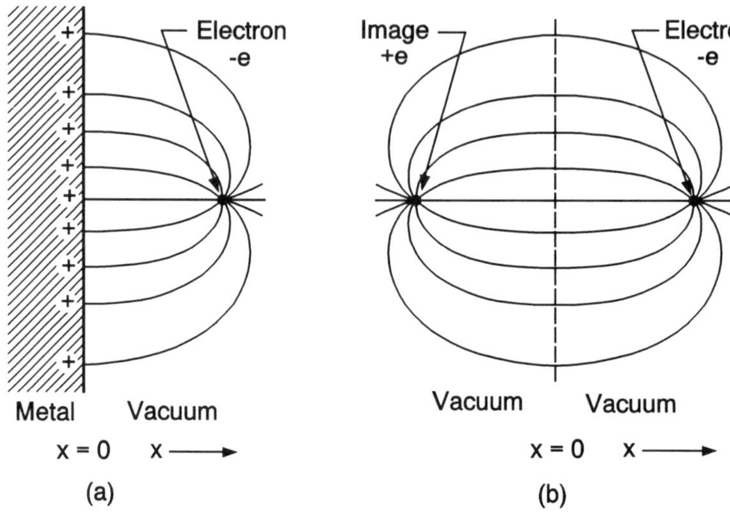

Figure 5-7 Electric field lines for an electron near the surface of the cathode.

or

$$P_{Es} = -\frac{e^2}{16\pi\epsilon_o x} \tag{5-27}$$

This potential energy is shown in Figure 5-8(a).

When a voltage, V_a, exists on an anode near the cathode, then an electric field, E_a, is applied in the region between the cathode and anode. The potential energy of an electron in that field would be zero at the cathode and would decrease while the kinetic energy increased as the electron moved toward the anode. The increase in kinetic energy at position x would be $-eE_ax$, and this would be the same as the decrease in potential energy as indicated in Figure 5-8(b). The total field experienced by an electron would be the sum of the surface and applied fields, and so the total potential energy would be the sum of the potential energies produced by the two fields. The potential energy profile for the electron is modified as shown in Figure 5-8(b), and it is seen that the electron needs somewhat less energy ($e\Delta\phi$) than the energy $E_o + e\phi$ to be emitted and so more electrons are emitted. The resulting increase in J is called the *Schottky effect*.

To calculate the increase in current, first the position x_o at which the maximum potential energy occurs is found. Then, $e\Delta\phi$ can be found and, from that, the change in current can be found. The potential energy of the electron in the presence of the surface and applied fields is

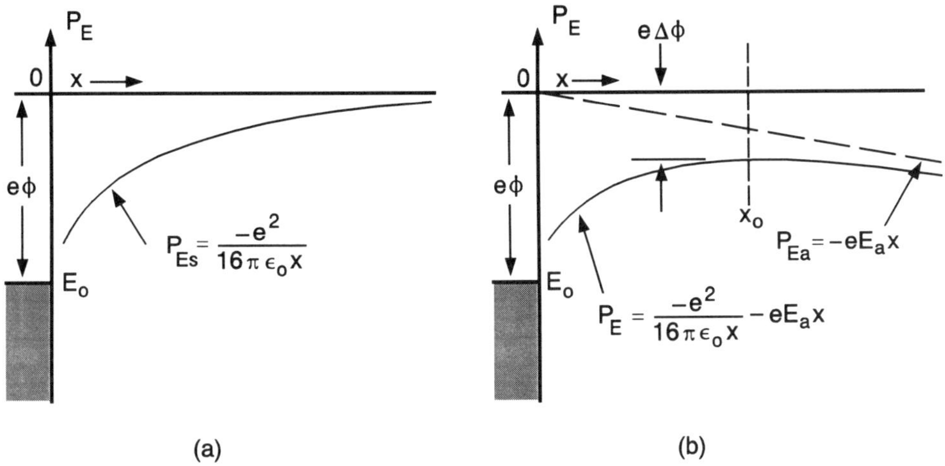

Figure 5-8 Potential energy of an electron as a function of distance from the surface of the cathode.

$$P_E = P_{Es} + P_{Ea} = \frac{-e^2}{16\pi\epsilon_o x} - eE_a x \tag{5-28}$$

The maximum potential energy occurs at $dP_E/dx = 0$. There,

$$\frac{e^2}{16\pi\epsilon_o x_o^2} - eE_a = 0 \tag{5-29}$$

or

$$x_o = \left[\frac{e}{16\pi\epsilon_o E_a}\right]^{1/2} \tag{5-30}$$

The potential energy of the electron at x_o is $-e\,\Delta\phi$ and so

$$-e\Delta\phi = \frac{-e^2}{16\pi\epsilon_o x_o} - eE_a x_o \tag{5-31}$$

Now, substituting in the value of x_o, $\Delta\phi$ is found to be

$$\Delta\phi = \left[\frac{eE_a}{4\pi\epsilon_o}\right]^{1/2} \tag{5-32}$$

Using this $\Delta\phi$, the Richardson-Dushman equation may be modified as follows:

$$J = A_o T^2 e^{-\frac{e}{kT}(\phi - \Delta\phi)} = A_o T^2 e^{-\frac{e}{kT}\left[\phi - \left(\frac{eE_a}{4\pi\epsilon_o}\right)^{1/2}\right]} \tag{5-33}$$

This is sometimes written as

$$J = J_o e^{\frac{e}{kT}\left(\frac{eE_a}{4\pi\epsilon_o}\right)^{1/2}} \tag{5-34}$$

where J_o is the Richardson-Dushman current density and is sometimes referred to as the *zero-field* current density [5]. J_o is a useful parameter for comparing performances of different cathodes.

In practice, to determine J_o experimentally, it is necessary to extrapolate from values measured at relatively high electric fields. This is because, at low electric field levels, electrons near the cathode surface alter the applied electric field. This is known as the *electron space charge effect* and will be discussed in detail in subsequent paragraphs.

To determine J_o, it is first noted that (5-34) can be written as

$$\ln J = \ln J_o + \frac{e}{kT}\left[\frac{eE}{4\pi\epsilon_o}\right]^{1/2} \tag{5-35}$$

Thus, a plot of ln J *versus* E should be a straight line with an intercept, when extrapolated to $E = 0$, of the value of ln J_o. This is illustrated in Figure 5-9. Note that, at low values of electric field, the measured value of current density falls below the extrapolated value because of the effect of the charge of the electrons in the region near the cathode surface. Some error in J_o may result from the change of the true work function by the applied electric field.

5.1.3 Field Emission

As the electric field applied to the surface of a cathode is increased to the 10^9–10^{10} V/m (0.1–1.0 V/Å) level, it is found that electron emission increases very rapidly. Furthermore, the increase in emission is almost independent of cathode temperature.

The electron energy profile is like that for the Schottky effect (Figure 5-8), except that the applied field is much larger. As a result, as indicated in Figure 5-10, the width of the potential barrier at the surface of the cathode is very narrow. It is so narrow, in fact, that some electrons can pass through the barrier even though they do not have sufficient kinetic energy to overcome the barrier. This effect is known as the *tunneling effect* and results from the quantum mechanical nature of the electron.

The resulting electron emission is called *field emission*. Field emission increases exponentially with applied voltage. Because extremely large emission densities are possible at all temperatures, including room temperature, there have been many

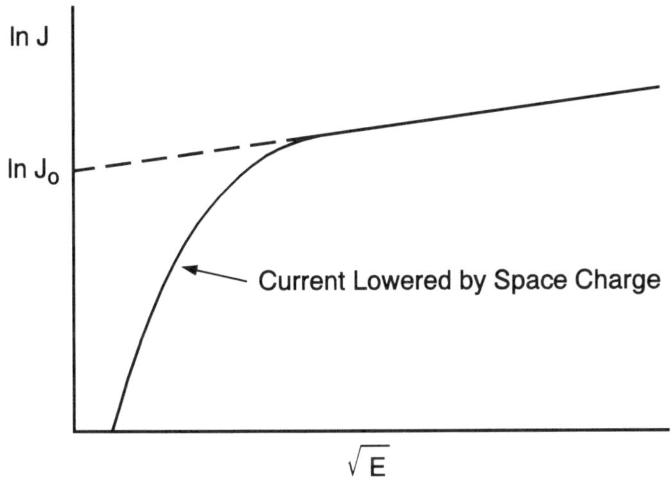

Figure 5-9 Illustration of extrapolation to find zero-field current density.

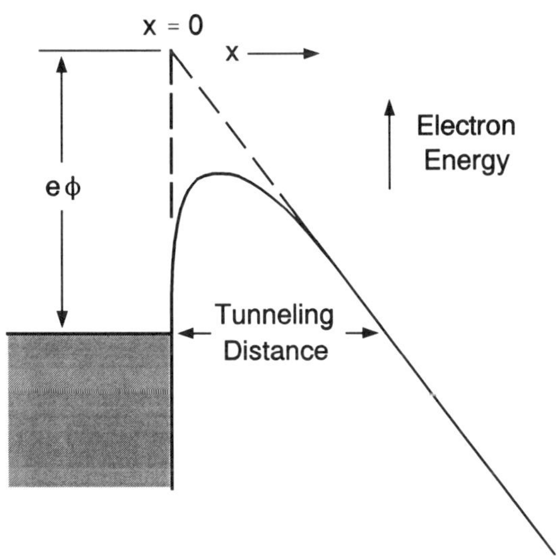

Figure 5-10 Electron potential energy profile with a very strong electric field.

attempts to develop field emission cathodes for use in microwave tubes. Unfortunately, these attempts have not been successful. The primary reason is that, to achieve the high electric field levels ncccssary for substantial emission, it has been necessary to fabricate the cathodes from arrays of very small needle-shaped emitters. The needle shape is used to enhance the electron emission so that emission occurs at the tip of each emitter, as illustrated in Figure 5-11. Emitted electrons ionize gas molecules to form positive ions, some of which impinge on and damage the tips of the emitter. As a result, emission is degraded.

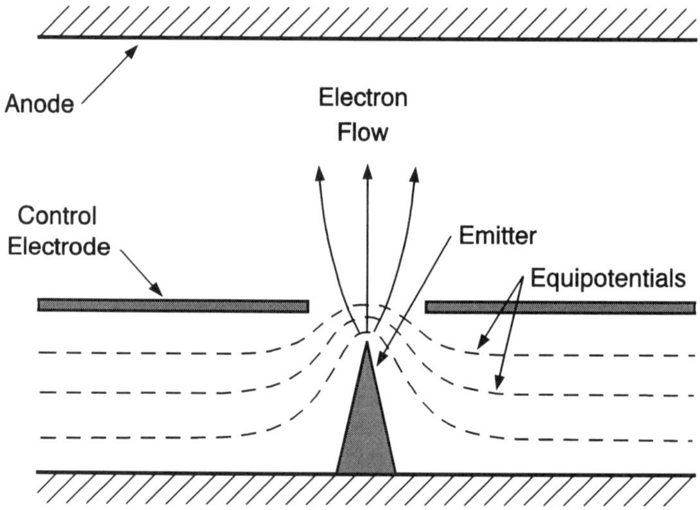

Figure 5-11 Field enhancement and electron emission from a needle-shaped emitter.

5.1.4 Space Charge Limitation [6, 7]

The effect of the negative charge of an electron is, of course, to reduce the potential that is present in the absence of the electron. In the region near an emitting cathode where many electrons are present, the reduction in potential can be appreciable. This is shown for a parallel-plane diode in Figure 5-12. In the absence of electrons, the potential increases linearly from the cathode to the anode. When electrons are present, the potential profile is depressed as indicated by the dashed line.

As the electron emission rate is increased (by increasing temperature, for example) the potential is further decreased. As indicated by Figure 5-13, there is a limit to how far the electron emission rate can be increased. This limit occurs when the density of electrons adjacent to the cathode surface becomes so large that it tends to depress the potential below zero. If the potential were to become negative, electrons would be forced back to the cathode.

For (a) in Figure 5-13, the electric field at the cathode surface causes all emitted electrons to leave the cathode, thereby depressing the potential near the surface. For (b) the electric field at the cathode surface forces electrons back to the cathode surface and increases the potential. An equilibrium condition occurs when, as shown in (c) the potential adjacent to the cathode surface is zero; that is, when the electric field at the cathode surface is zero. If the potential tends to become positive, more electrons flow from the cathode, thereby depressing the potential. If the potential tends to become negative, fewer electrons flow from the cathode, thereby raising the potential.

When the electric field at the cathode surface is forced to zero by the electron cloud near the cathode surface, the emission is said to be space charge limited. When a sufficient number of electrons is present so that emission is space charge limited, then cathode temperature, surface condition *et cetera* have no effect on emission. This is indeed a very fortunate situation because it eliminates the necessity for extreme uniformity of temperature across the surface of a cathode. It also eliminates the necessity for precise control of the voltage and current to the heater in the cathode. When the voltage is increased in a diode with a space charge limited cathode, the potential at all points between the cathode and anode tends to increase. As a result, additional electrons flow from the cathode to cause the electric field at the cathode surface to remain at zero. It is evident, therefore, that diode current is dependent on anode voltage.

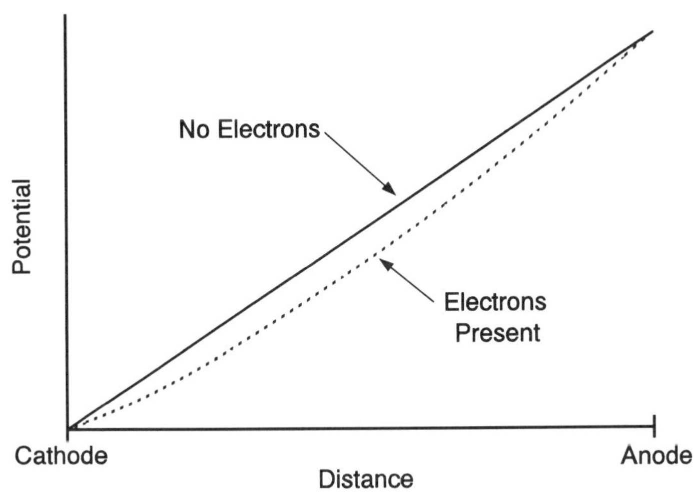

Figure 5-12 Potential distribution with and without electrons in a parallel-plane diode.

The relation between voltage and current in a parallel-plane, space charge limited diode can be determined in a straightforward manner. It is assumed that the electric field at the cathode surface is zero, as is indicated in Figure 5-13. Poisson's law applies, and the potential, V, at any distance, x, from the cathode is

$$\nabla^2 V = \frac{\rho}{\epsilon_o} \tag{5-36}$$

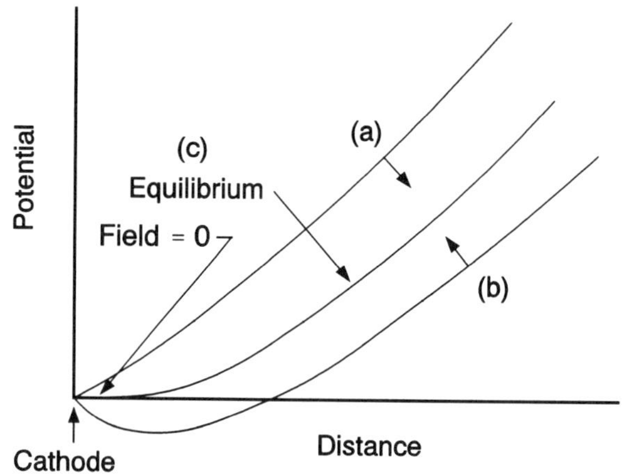

Figure 5-13 Potential profiles near the surface of the cathode.

where ρ is the electron charge density and ϵ_0 is the dielectric constant of free space. Assuming that the parallel plane diode extends to infinity in the transverse directions, then there are variations only in the x direction, so

$$\frac{d^2 V}{dx^2} = \frac{\rho}{\epsilon_o} \tag{5-37}$$

The charge density, ρ, is related to the current density, J, and the electron velocity, u, by

$$J = \rho u \tag{5-38}$$

Thus,

$$\frac{d^2 V}{dx^2} = \frac{J}{u \epsilon_o} \tag{5-39}$$

From the conservation of energy,

$$u = \sqrt{2 \eta V} \tag{5-40}$$

where η is the electron charge-to-mass ratio. Now,

$$\frac{d^2 V}{dx^2} = \frac{J}{\sqrt{2 \eta V}\, \epsilon_o} \tag{5-41}$$

Multiplying both sides by 2 dV/dx gives

$$2\frac{dV}{dx}\frac{d^2V}{dx^2} = \frac{2J}{\sqrt{2\eta}\,\epsilon_o}V^{-1/2}\frac{dV}{dx} \tag{5-42}$$

This equation may be integrated directly to obtain

$$\left(\frac{dV}{dx}\right)^2 = \frac{4JV^{1/2}}{\epsilon_o\sqrt{2\eta}} + C_1 \tag{5-43}$$

where C_1 is the constant of integration.

At $x = 0$, $V = 0$ and for space charge limited emission with no initial electron velocity, $dV/dx = 0$. Thus $C_1 = 0$. As a result,

$$\frac{dV}{dx} = 2\left(\frac{J}{\epsilon_o}\right)^{1/2}\left(\frac{V}{2\eta}\right)^{1/4} \tag{5-44}$$

This equation may be integrated to obtain

$$\frac{4}{3}V^{3/4} = 2\left(\frac{J}{\epsilon_o}\right)^{1/2}\left(\frac{1}{2\eta}\right)^{1/4}x + C_2 \tag{5-45}$$

Because $V = 0$ at $x = 0$, $C_2 = 0$. Thus,

$$J = \frac{4}{9}\epsilon_o(2\eta)^{1/2}\frac{V^{3/2}}{x^2} \tag{5-46}$$

When the values for η and ϵ_0 are used, the result is

$$J = 2.33 \times 10^{-6}\frac{V^{3/2}}{x^2} \tag{5-47}$$

This is the Child-Langmuir law for the flow of electrons in a parallel-plane, space charge limited diode. For a diode with a cathode-to-anode spacing $x = d$, this equation is commonly written as

$$I = PV^{3/2} \tag{5-48}$$

where $I = JA$, with A being the area of the cathode. P, known as the *perveance* of the diode, is

$$P = \frac{4}{9}\frac{A\epsilon_o}{d^2}(2\eta)^{1/2} \tag{5-49}$$

or

$$P = 2.33 \times 10^{-6}\frac{A}{d^2} \tag{5-50}$$

for a parallel-plane diode.

It is important to note that perveance is a function only of the geometry of the diode. Space charge limited diodes with geometries other than planar (cylindrical, spherical, *et cetera*) all obey $I = PV^{3/2}$, where the expression for perveance is the one appropriate for the geometry used.

In the preceding discussion and analysis, it has been assumed that the initial velocities of the electrons are zero when they are emitted. In fact, the electrons are emitted with small initial velocities. If the electric field was zero at the cathode surface, then all of the emitted electrons would be able to leave the surface and the saturation emission, J_o (total thermionic emission capability of the cathode), would flow. In practice, the space charge limited emission is very much less than the saturated emission. This is because the presence of large quantities of electrons causes a potential minimum to be formed just outside the cathode surface. The voltage at the minimum potential is a small fraction of a volt below the cathode potential. Only the most energetic electrons can overcome this potential minimum and pass on to the anode. The less energetic electrons are returned to the cathode. The effect of increasing the anode voltage so that more current is drawn is to decrease the depth of the potential minimum and move it closer to the cathode. As J approaches the saturation current density, J_o, dV/dx approaches 0 at the cathode surface. If the voltage is increased further, the current will be limited to the saturation current density and space charge limited operation will cease.

In practice, space charge limited diodes behave as predicted. This is readily demonstrated by plotting current as a function of voltage using two-thirds power graph paper. As shown in Figure 5-14 the resulting plot is a straight line for space charge limited operation. The slope of the line is the perveance.

When the voltage is increased to a high enough value that the cathode cannot supply sufficient current to maintain the space charge limited condition, the cathode is said to be *TL* (temperature limited). The TL region is readily apparent in the two-thirds power *I-V* graph because the plotted results depart from a straight line.

In Figure 5-15, current is shown plotted as a function of temperature for a thermionic cathode. In the region of TL operation, current increases with temperature. At high temperatures, current is independent of temperature, indicating space charge limited operation. As voltage is increased, the temperature at which the transition from TL to space charge limited operation occurs is increased. As indicated in Figure 5-16, the TL portion of a current *versus* temperature curve is translated to higher temperatures as a function of time. The reason for this will be discussed later in this chapter. It is necessary to select an operating temperature for a cathode that remains above the knee of the curve during the life of the cathode. Of course, the rate at which the TL curve moves increases with increasing temperature, so an excessive temperature will reduce cathode life.

In Figure 5-16, there is a slight upward slope with increasing temperature in the space charge limited region. This is commonly observed and results from movement of the cathode toward the anode as a result of thermal expansion of elements of the

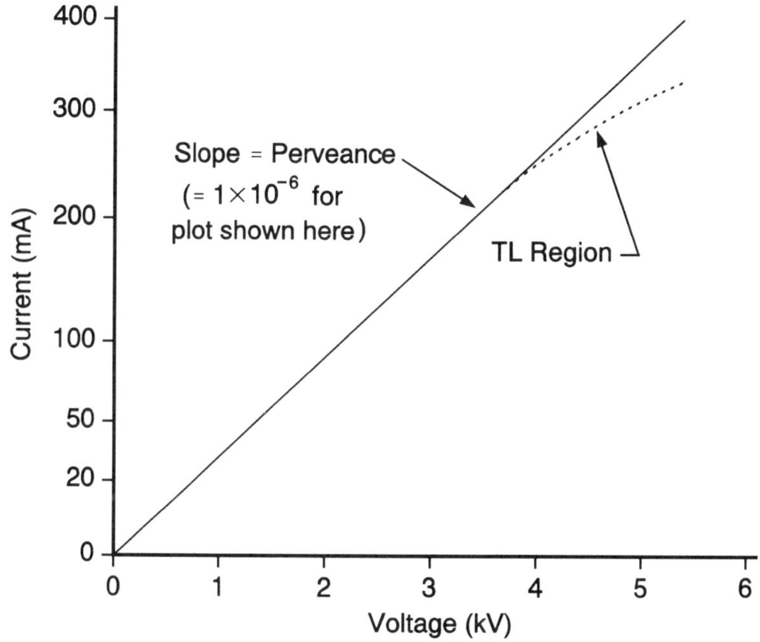

Figure 5-14 Current *versus* voltage plotted on two-thirds power graph paper.

cathode support structure.

The mechanisms dominating current flow in a thermionic diode are summarized in Figure 5-17. At low voltages, current is space charge limited. As voltage is increased, the emission limit of the cathode is reached and current becomes dependent on cathode temperature, T_c. Current continues to increase slowly with voltage because of the Schottky effect. At high voltages, field emission causes current to increase rapidly.

5.2 EVOLUTION OF THERMIONIC CATHODES

The first thermionic cathodes used in quantity were those used in early radio tubes. They were made from pure tungsten, although other cathode materials were known that were capable of producing adequate emission at lower temperatures than tungsten. Because of the high work function ($\sim$4.6 eV), the operating temperature was high. For example, 2200°C yields about 0.3 A/cm^2. Apparently, tungsten was used in early tubes rather than other cathodes because it was foolproof. A tungsten cathode always had plenty of emission. Other cathodes were temperamental and were sometimes "poisoned" by mysterious agents.

For some applications, a small percentage of thoria was mixed with the tungsten. Thorium would diffuse to the surface of the tungsten and reduce the work function,

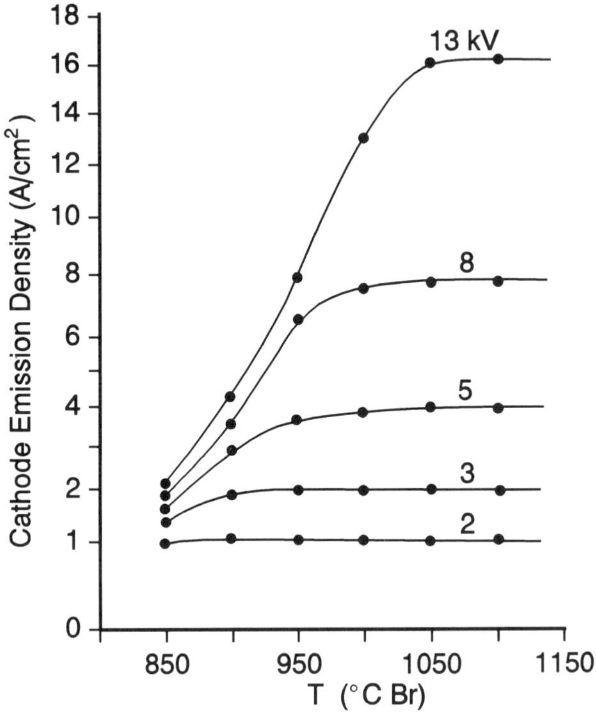

Figure 5-15 Emission as a function of temperature and anode voltage.

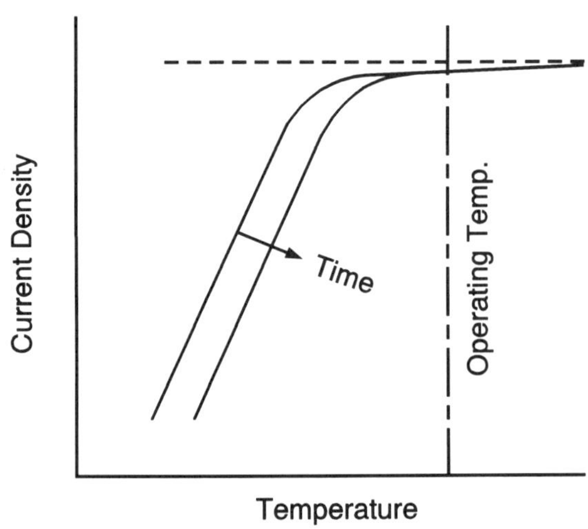

Figure 5-16 Movement of the TL curve with time.

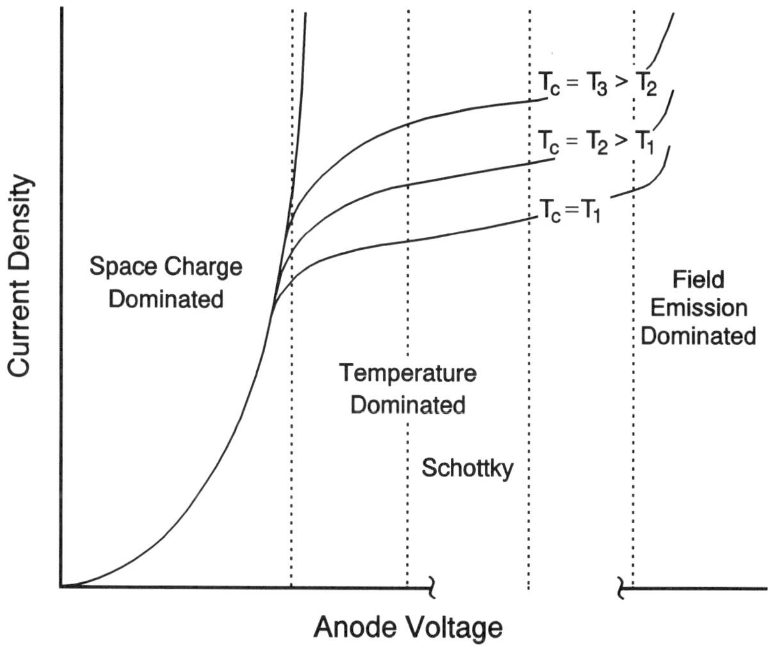

Figure 5-17 Summary of mechanisms dominating current flow through a thermionic diode.

thereby permitting the use of a lower operating temperature. Thoriated tungsten (ThW) cathodes are still in common use in high-power switch tubes, transmitter tubes, and microwave oven magnetrons. The cathode surface is sometimes carburized to reduce the evaporation rate of the thorium. The life expectancy of high-power switch tubes containing ThW cathodes is plotted in Figure 5-18.

Early cathodes were all directly heated; that is, the filament, which was heated by passing a current through it, was also the cathode. Thus, as a result of the heater current, there was a voltage variation along the surface of the cathode. Because batteries were used as power sources, the filament voltage was dc, and so there was no modulation of the emission current.

The development of ac radios and the need to heat the filaments with ac current brought about the need for a cathode, the surface of which was all at the same potential. The indirectly heated cathode resulted, whereby the heater was separate from the electron emitting surface. Directly and indirectly heated cathodes are illustrated in Figure 5-19.

In the indirectly heated cathode, tungsten is used for the filament. Because of heat losses and the relatively large cathode area, the cathode surface had to operate at a temperature substantially below the heater temperature. This provided the requirement for a low work function emitting surface.

The oxide-coated cathode, which had been discovered by Wehnelt in 1904, long

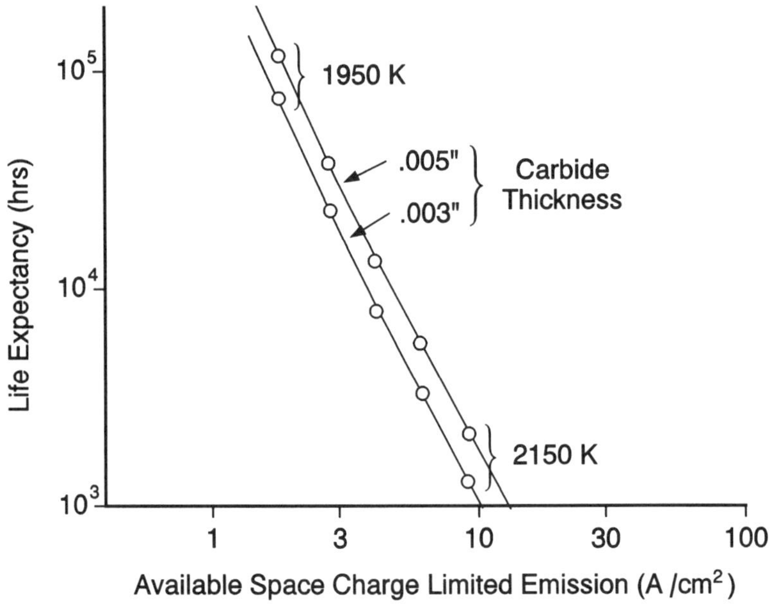

Figure 5-18 Life expectancy for a tube with a thoriated tungsten cathode (actual life ≈ 1/2 of life expectancy). From T.E. Yingst *et al.*, *Proc. IEEE,* March 1973. © 1973 IEEE.

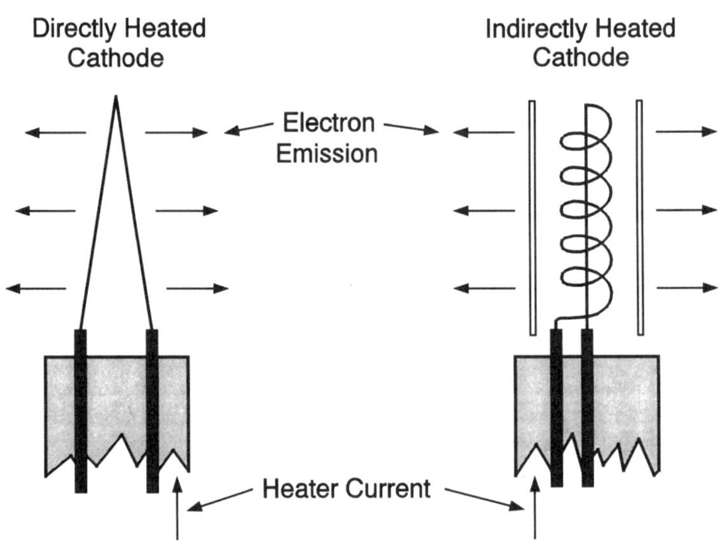

Figure 5-19 Comparison of directly and indirectly heated cathodes.

before radio tubes, provided the solution to the problem because of its very low work (≈ 1.5 eV) function and resulting low operating temperature. Oxide-coated cathodes served the industry well from the 1920s through World War II and into the early 1950s. They are still prevalent today in applications requiring relatively modest cathode emission and long life. Early oxide-coated cathodes were described by J.R. Pierce [1] as follows:

> The early history of oxide coated cathodes seemed without rhyme or reason. It was found that if a wire intended to emit electrons was heated and rubbed with sealing wax and the sealing wax then burned off a truly phenomenal cathode was produced. Further work traced the emission to alkaline earth oxides, and the compound from which they came was humorously described as a barium resinate. The sealing wax technique died hard, and for several years the process of coating filaments consisted in stringing fine platinum ribbon out in long lengths, heating it with an electric current, and rubbing a resinous compound containing barium and strontium over it repeatedly. The narrow ribbon was then cut into lengths, put into a crude vacuum tube of the day, the tube was pumped and baked until a "hard vacuum" was achieved, and the filament was activated. This activation was a wishful process consisting of glowing the filament brightly, applying voltage, and drawing current to the plate until, in happy circumstances the plate got red-hot, giving off enough gas to produce a blue glow and further increase the activity and the current. After a lucky period of this treatment the filament became active; that is, it would emit electrons at a lower and reasonable temperature, at which it would last for a number of thousands of hours.

The art and science of oxide cathodes is now highly developed. A mixture of barium, strontium, and calcium carbonates is commonly used. These are applied to a metal base (usually nickel) and then activated in vacuum. Activation consists of heating and reducing the carbonates to oxides.

As oxide-coated cathodes were improved, dc emission increased from mA/cm^2 to hundreds of mA/cm^2. Pulsed emission densities are dependent on pulse length and reportedly can be as high as tens of A/cm^2. Oxide-coated cathodes are not capable of high dc emission densities because of the high resistivity of the coating and the resulting resistive heating that occurs at high currents. They are also easily poisoned.

The search for better cathodes than oxide-coated cathodes resulted from their low dc emission and susceptibility to damage. The L cathode, illustrated in Figure 5-20, was developed by Philips about 1950. In this cathode, the emissive material is sandwiched between the heater and a porous tungsten pellet. During operation, emissive material diffuses through the porous tungsten and forms a low work function emitting surface.

Cathodes like the L cathode are referred to as *reservoir cathodes* for the obvious reason that a reservoir of emissive material is used. Some problems with early L

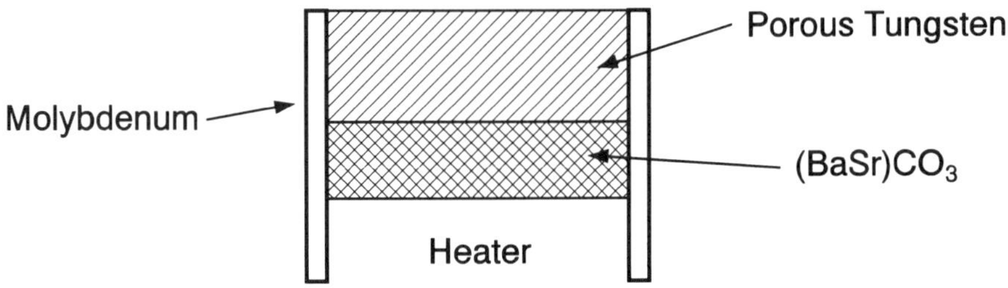

Figure 5-20 Essential features of Philips L cathode.

cathodes were

1. Leakage of barium vapor between the tungsten and molybdenum;
2. Nonuniformity of the temperature of the cathode surface because of poor heat transfer through the cavity containing the emissive mix;
3. Difficult outgassing and activation.

These problems have been overcome and commercial cathodes with a structure like the L cathode are now made by Siemens. These are referred to as *MK cathodes*.

A reservoir cathode developed by the Naval Research Laboratory is known as the *controlled porosity dispenser* (CPD) cathode. The reservoir contains a mixture of BaO, CaO and Al_2O_3 as well as particles of tungsten, a reducing agent that frees barium from the BaO. Instead of the porous tungsten emitting surface that the L and MK cathodes have, the CPD cathode uses a thin sheet of tungsten or alloy (to be discussed in following paragraphs) in which pores have been etched or drilled with a laser. The CPD approach to cathode construction offers advantages of improved emission uniformity and improved reproducibility and performance. At present, the difficulties of fabricating CPD cathodes, especially for concave cathodes, limit their usefulness.

From 1950 to 1955, the evolution of the impregnated dispenser cathode illustrated in Figure 5-21 occurred. The Philips B-type cathode developed by Levi in 1953 is still used in the microwave tube industry. The pores of the porous tungsten pellet are impregnated with a compound of BaO, CaO and Al_2O_3. Barium is released when the impregnant reacts with tungsten. The freed barium migrates to the surface of the porous tungsten to form the emitting surface. The most common mix, which is used in the B-type cathode, consists of five parts BaO, three parts CaO and two parts Al_2O_3 and is frequently referred to simply as 5:3:2. Another mix is 4:1:1, and when this is used, the cathode is referred to as an *S-type cathode*.

The B- and S-type cathodes are rugged, relatively impervious to damage and can readily provide an emission density of several A/cm^2. At these high emission densities, the operating temperature must be 1100°C or higher because of the relatively high work functions of these cathodes. Thus, life is limited to a few thousand hours.

The search for a cathode capable of an emission density comparable to or higher

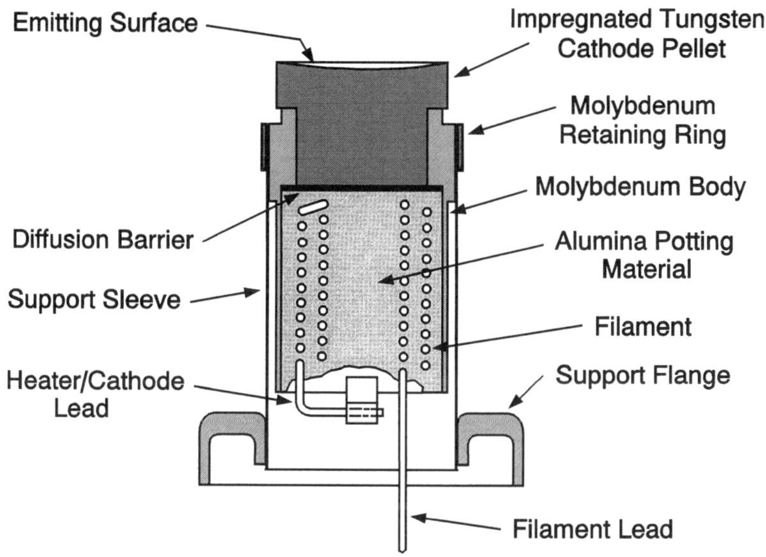

Figure 5-21 Typical B-type impregnated dispenser cathode.

than the B- and S-type cathodes, but with a lower operating temperature and longer life, led to the development of the M-type cathode in the 1960s. In its simplest form, the M-type cathode is a B- or S-type with a thin film several thousand Angstroms thick of osmium, iridium or rhenium applied to the surface. The effect of the thin film is to reduce the work function about 0.2 eV and the cathode operating temperature about 90°C (dependent on film metal) for the same emission density as is obtained from a B or S cathode.

A problem observed early in the development of the M-type cathode was degradation of the thin enhancing film. At normal operating temperatures, the tungsten substrate and the coating metal would interdiffuse, and so an alloy surface on the cathode was formed with a composition that changed with time. The lowest work function for a cathode with this alloy surface occurs when the alloy is about 50/50 base metal and coating metal. Thus, during the life of one of these early M cathodes, emission would actually increase at first and later degrade and would eventually revert to that of a B cathode as the surface became, primarily, tungsten.

Several fabrication techniques have been examined in attempts to control the properties of the alloy formed on the surface of an M-type cathode. In one cathode, called the *mixed metal matrix* (MMM) cathode, the entire matrix of the cathode pellet contains particles of the enhancing metal (usually iridium). The fabrication is difficult because of the need to carefully control the size distribution of the tungsten and iridium particles.

In the CD (controlled doping) cathode developed by Varian, multiple layers of alloy material, each having a different composition, are deposited. The composition of the layers is adjusted to minimize interdiffusion of the tungsten and the enhancing metal. A few percent of the enhancing metal can be added to the tungsten matrix prior to coating and this reduces the diffusion of tungsten into the coatings.

The performance characteristics of the cathodes described in the preceding paragraphs are summarized in Figure 5-22. The very significant increase in emission capability over B and S cathodes of cathodes containing osmium, iridium or rhenium is clearly shown.

In recent years, there has been considerable interest in scandate cathodes. These cathodes have a very low work function (≈ 1.5 eV) and are capable of emission densities on the order of 100 A/cm^2 for thousands of hours. In one version developed by Philips, the emitting surface is composed of a layer of tungsten mixed with Sc_2O_3 ($\approx 5\%$ by weight). The entire matrix of the cathode is impregnated with BaO, CaO and Al_2O_3. Limitations at present are poor reproducibility, nonuniform emission and degradation due to ion bombardment. Research on the scandate cathode is very active at the present time.

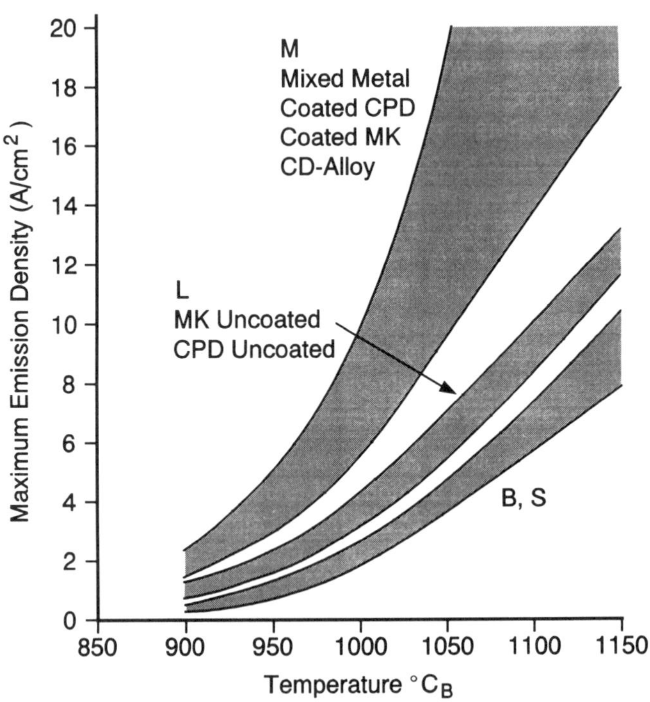

Figure 5-22 Performance comparison for various dispenser cathodes. Adapted from L.R. Falce, *Technical Record*, 1983 IEDM. © 1983 IEEE.

5.3 IMPREGNATED DISPENSER CATHODE OPERATION

Figure 5-23 shows the scale of the features of the surface of a typical B-type cathode [8–10]. The cathode is made up of small grains of tungsten that have been pressed together under high pressure and temperature (sintered) to form a porous pellet. A coated cathode (M type, for example) has a very thin film of enhancing metal on the surfaces of the tungsten grains. Note that all dimensions are typically a few microns.

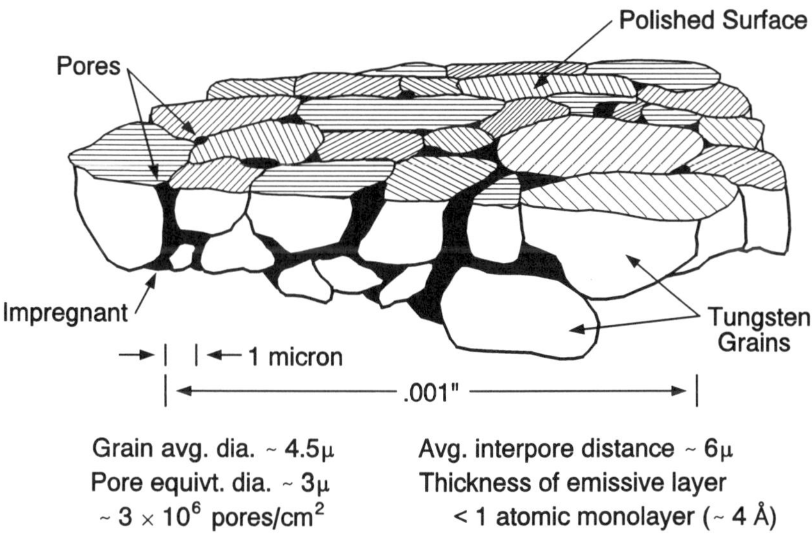

Figure 5-23 Scale of features of an impregnated cathode. From M.C. Green, Technical Report RADC-TR-81-211, July 1981.

When the cathode is heated to the operating temperature, barium is released from the interior of the tungsten matrix by reactions of the emissive mix with the tungsten. The freed barium then migrates to the surface of the emitter. Some of the barium spreads out over the emitter surface and eventually evaporates. The fraction of the surface that is covered depends on the rates of diffusion of barium to the surface and the rate of evaporation.

The work function of the cathode is not a single valued quantity but instead has a distribution of values. One reason is that each tungsten grain forms an emission patch that may have a unique crystalline orientation. The B cathode can have twice as many crystalline surface orientations as a coated cathode, causing a broader work function distribution [11]. Work function distributions for B and M cathodes are shown in Figure 5-24.

The next question to be addressed is how the distributed nature of the work function affects the overall emission characteristic of the cathode. An important consideration here is the thickness of the electron space charge cloud at the surface of

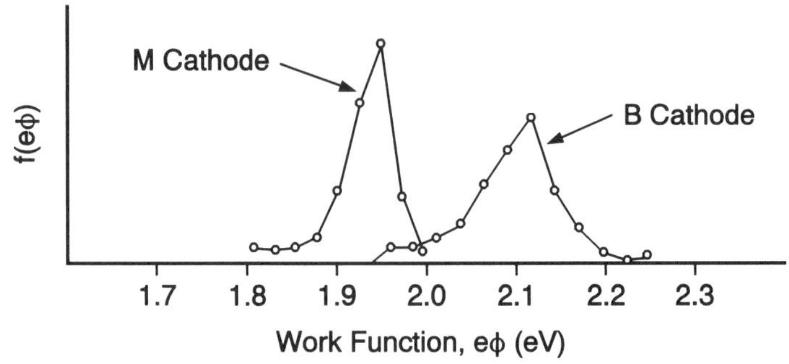

Figure 5-24 Work function distributions for a B cathode and an M cathode early in life. From Thomas J. Grant, *Technical Digest*, 1986 IEDM. © 1986 IEEE.

the cathode. Because of the finite velocities of emission of electrons, there is a small depression in the potential (~ 0.2 V) near the surface of the cathode. (This was ignored when we analyzed space charge limited emission earlier in this chapter. This potential depression will be discussed in detail in Chapter 15 when noise is considered.) The location of the position where the potential is a minimum is a measure of the thickness of the space charge cloud near the cathode surface. Figure 5-25 shows the distance between the potential minimum and the cathode surface as functions of temperature and emission density. Note that, at emission densities of a few A/cm^2, which are typical for coated dispenser cathodes, the position of the potential minimum is comparable to the tungsten grain size (see Figure 5-23) and thus the emission patch size. As a result, we can conclude that the various patches emit independently at rates determined by their individual work functions [12].

Because the various patches operate independently, the surface of the cathode behaves like an array of diodes with varying work functions and connected together in parallel. Thus, the emission characteristic for the cathode can be modeled as a parallel combination of the emission characteristics of the diodes (patches).

In the temperature limited (TL) region of operation of each patch, the Richardson-Dushman equation, corrected for the Schottky effect applies. In the space charge limited region (indicated by *FSCL* to show that the calculation is based on operation that is *fully* space charge limited, that is, no part of a patch is operating temperature limited), the Child-Langmuir law applies. Figure 5-26 shows, conceptually, how the TL and FSCL curves can be combined to obtain the overall emission characteristic of the cathode. In this figure, all currents are plotted as percentages of the FSCL values. It has been assumed that all emission patches operate at the same temperature and that the electric field applied by the anode is the same for all emission patches.

The composite plot in Figure 5-26 gives the total emission from the cathode for one FSCL value. Miram *et al.* have developed a technique for presenting emission data which is an extension of that in Figure 5-26 and which provides a powerful tool for diagnosing the performance of dispenser cathodes and for helping to understand their

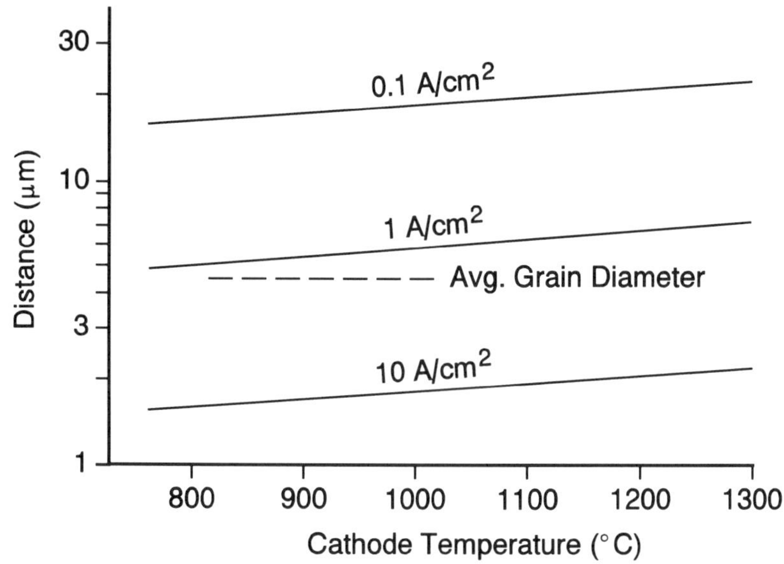

Figure 5-25 Distance between potential minimum and cathode surface.

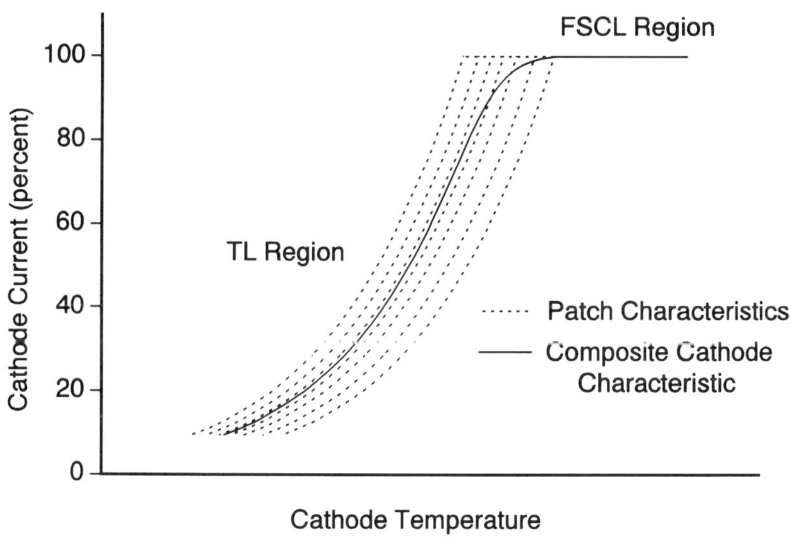

Figure 5-26 Combination of emission curves for several patches to form overall emission characteristic for cathode.

operation [13, 14]. The basic concepts of the Miram technique may be understood with the aid of Figure 5-27. The dashed curves represent emission current calculated for a cathode work function having a single value. The temperature of all parts of the cathode is assumed to be the same and the electric field is assumed to be uniform across the cathode surface.

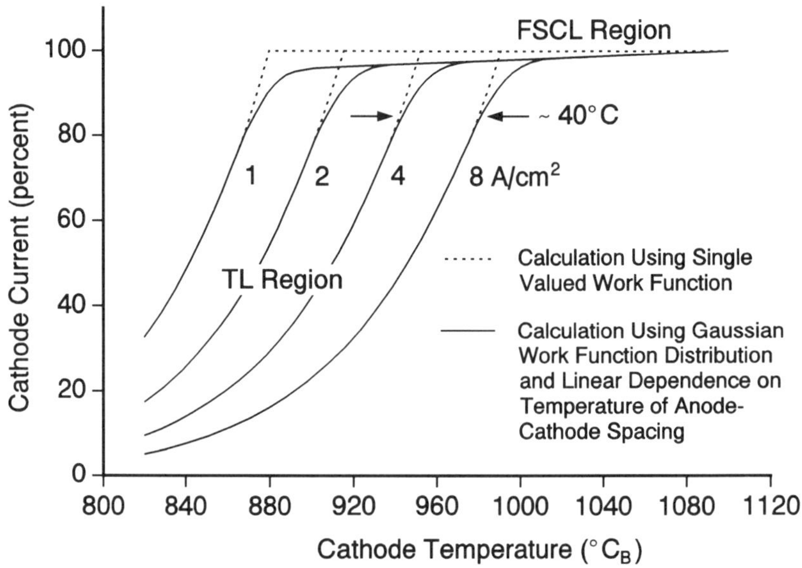

Figure 5-27 Calculated Miram curves based on ideal cathode emission models. From M.J. Cattelino *et al.*, 1982 IEDM. © 1982 IEEE.

Several curves are shown for TL operation. With the Miram technique, the curves are plotted for FSCL currents that are separated by a factor of two. Thus, for example, if the lefthand curve is for a FSCL current of 1 A, the others would be at 2 A, 4 A, 8 A, *et cetera*. From the Richardson-Dushman equation, the temperature difference required to produce an increase in current by a factor of two is about 40°C so the TL curves are separated by about 40°C.

The solid curves in Figure 5-27 were calculated using the following assumptions:

1. The work function has a distribution of values rather than a single value. A Gaussian distribution was assumed.
2. The spacing between the cathode and anode decreased slowly as temperature increased because of thermal expansion of the cathode support structure. Thus, the perveance was a function of temperature and so the FSCL characteristic has a small slope rather than being a horizontal line.

Shown in Figure 5-28 is a Miram plot for a good cathode. The characteristics of this plot are very much like those given by the solid curves in Figure 5-27, which lends support to the validity of the approach used in calculating those curves.

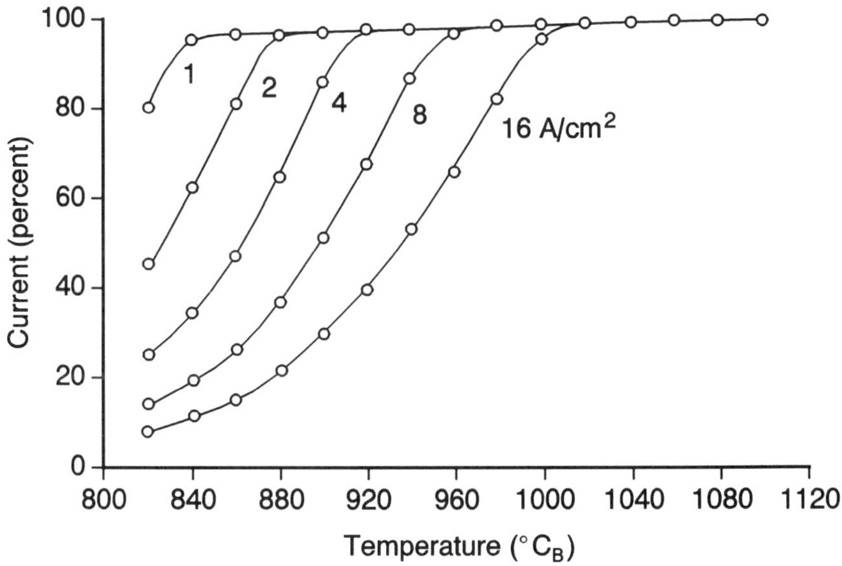

Figure 5-28 Miram curves for a good cathode. From T.J. Grant, *Technical Digest*, 1984 IEDM. © 1984 IEEE.

The use of Miram curves provides a technique, not only for evaluating cathode performance as functions of temperature and loading, but also for revealing defects in cathode fabrication and operation. Examples of Miram curves for defective cathodes are given in Figure 5-29. The curves in (a) were obtained from a cathode with chipped edges. The chipped regions had a higher effective work function than the central portion of the cathode. As a result, the emission plots exhibited two types of knees. The ones for the central portion were much like the knees for a good cathode. The other knees were very broad and occurred at higher temperatures.

Figure 5-29(b) shows results for a cathode for which the effective work function increased as temperature was raised above 940°C. Note that, above 940°C, the separation between the 8 A/cm^2 curve and the 4 A/cm^2 curve is greater than the separations between the 4 and 2 A/cm^2 curves and between the 2 and 1 A/cm^2 curves. This large change in separation indicates an increase in work function.

In the plots shown in Figure 5-29(c), the knees of the curves have changed as is indicated by the dashed lines. The change may have occurred as a function of time, or the change may represent the difference between a cathode assembly that is known to be operating properly and another assembly that is supposed to be identical to the first.

If only the knees of the curves have changed as shown in (c), then it is likely that something has caused a nonuniformity in the electric field near the cathode surface. With a gridded electron gun, this could be movement or misalignment of one of the grids.

If both the spacing and the shapes of the knees have changed as shown in Figure 5-29(d), then there is a cathode emission problem and possibly a misalignment of electron gun electrodes.

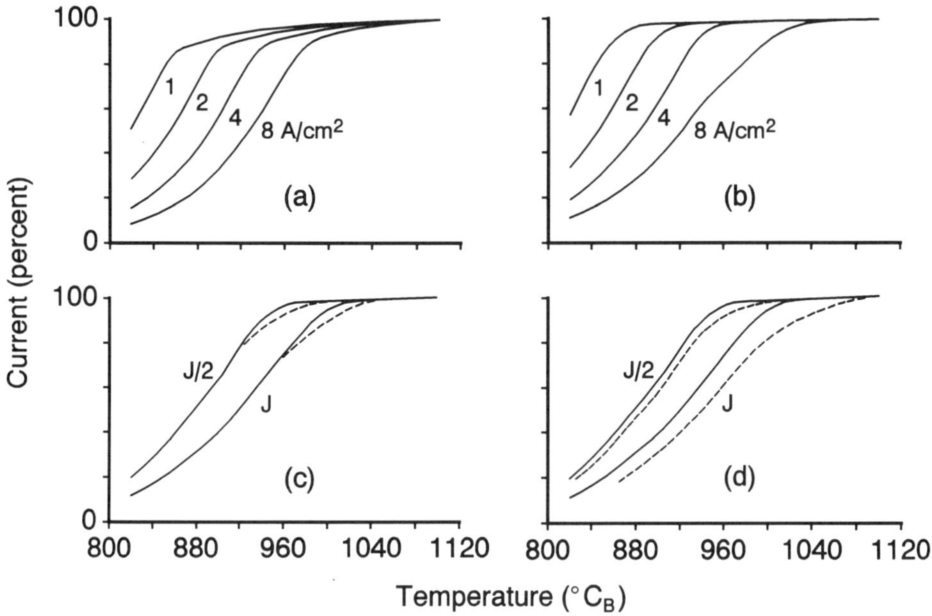

Figure 5-29 Miram plots showing evidence of cathode and grid defects. Adapted from M.J. Cattelino *et al.* and T.J. Grant, *Technical Digests,* 1982 and 1984 IEDMs. © 1982 and 1984 IEEE.

5.4 LIFE CONSIDERATIONS

In the great majority of cathode applications, long life is essential. Long life normally implies that the life should be as long as possible to ensure reliability, to help minimize tube replacement costs, and to help limit redundancy costs (in satellite applications for example).

In a very limited number of applications (decoys and space defense, for example) cathode life can be very short. Except for the very fast warm-up times and long shelf lives that are also usually required, there does not appear to be a cathode problem in these applications.

Long-life requirements commonly range from 10,000 hrs to over 100,000 hrs. While life testing for these long periods is sometimes carried out, it is much more common to predict cathode life on the basis of a knowledge of cathode degradation mechanisms. Grant has identified three distinguishable types of emission degradation mechanisms [12]:

- Type I gradual decrease in perveance, that is, FSCL emission decreases with time;
- Type II change in work function distribution with time and without dependence on changes in cathode temperature;
- Type III change in work function distribution with temperature as well as with time.

The type II mechanism results from a change in the base metal work function with time. This mechanism applies to coated cathodes and is attributed to the diffusion of tungsten through the coating. Grant points out that the work function distribution can move to lower as well as higher work functions early in life depending on coating composition, thickness, and interdiffusion rate.

When the type III mechanism occurs, the work function distribution changes in a reversible manner as temperature changes. That is, if temperature is increased or decreased, the distribution increases or decreases and the measurements are repeatable. This phenomena can occur early in life and is attributed to an insufficient supply of barium, a change in sticking coefficient of the barium to the cathode surface or to external poisoning.

The reason given for the Type I degradation is usually the gradual depletion of the barium supply in the impregnated tungsten pellet [2, 15–18]. Thus, the rate at which barium diffuses to the cathode surface decreases and so the barium coverage of the surface decreases and work function increases.

A characteristic feature of Miram curves that sometimes results from barium depletion is the movement of the entire set of curves toward higher temperatures as a function of time [13]. An example is given in Figure 5-30. By observing the rate at which the curves shift to higher temperatures, life projections can be made. End of life is predicted to occur when the knee of the current-density curve of interest reaches the operating temperature and emission current drops by a prescribed amount such as 10%. For the curves shown in Figure 5-30, for example, the movement of the curves over a 2,500 hr period is shown. Assuming that the curves continue to move at the same rate, then the time for the 1 A/cm^2 or the 2 A/cm^2 curves to reach the operating temperature (1050°C in this case) can be estimated. At 4 A/cm^2 the knee of the curve has already reached the operating temperature and current has dropped about 5%. For this current density, a 10% drop and end of life would soon occur.

Grant notes that the Miram curve does not always support the conclusion that barium coverage of the entire cathode depletes with time [12]. In some cases, a reduction in current persists even if cathode temperature is increased 150°C above the

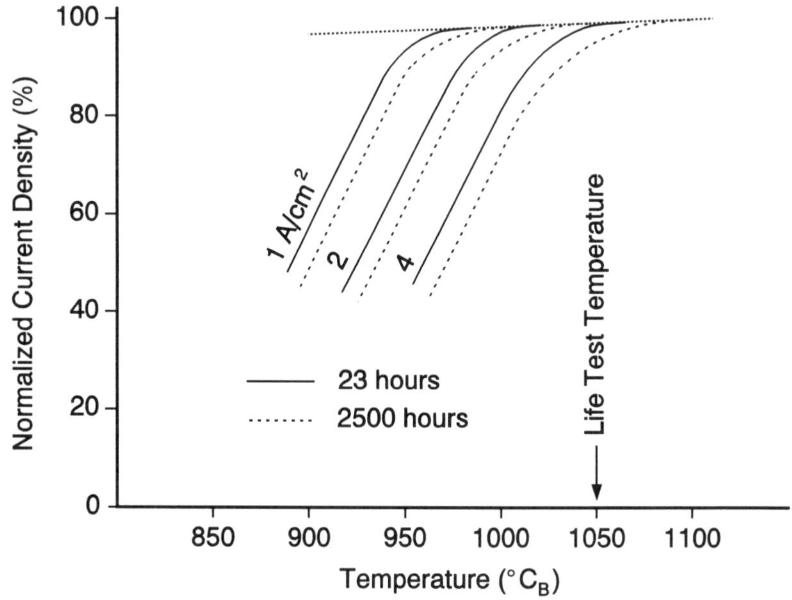

Figure 5-30 Miram plot for B-type cathode. From M.C. Green, Technical Report RADC-TR-81-211, July 1981.

point where the cathode is 99.5% FSCL. If depletion of the entire cathode had been the reason for the reduced current, the increase in temperature would have increased the current. Grant states that "the preferred reason for the change in perveance is the complete depletion of barium from the edges of the cathode surface." Depletion from the cathode edge is rapid because, if there is no barrier to diffusion on the outer diameter of the cathode pellet, barium can diffuse and evaporate radially outward as well as from the front surface of the cathode.

Another example of life prediction based on the depletion model is by Longo [15–17]. He observed that the following simple relation described the emission from an impregnated cathode over a wide range of voltages and temperatures:

$$\frac{1}{J_{OP}} = \frac{1}{J_{FSCL}} + \frac{1}{J_{TL}} \tag{5-51}$$

In this equation, J_{OP} is the observed current density, J_{FSCL} is the current density that would be observed if the entire cathode was operating space charge limited, and J_{TL} is the current density that would be observed if the entire cathode was operating temperature limited. Shown in Figure 5-31 is a comparison of (5-51) with measured results.

Before proceeding with Longo's analysis, it should be noted that the knee of the

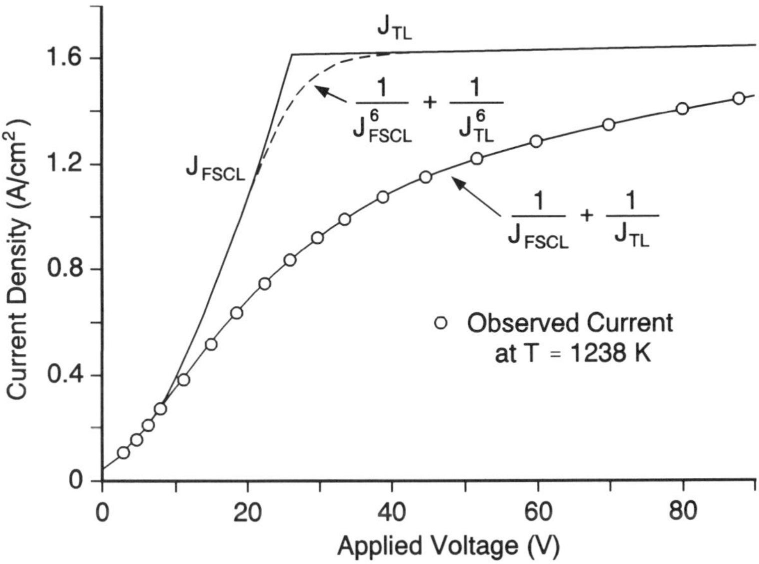

Figure 5-31 Empirical relations for emission current. Adapted from R.T. Longo, *Technical Digest,* 1978 IEDM. © 1978 IEEE.

transition curve from TL to FSCL operation described by Longo's equation is much broader than that found for most cathodes. Vaughan proposed the relation

$$\frac{1}{J_{OP}^{n}} = \frac{1}{J_{FSCL}^{n}} + \frac{1}{J_{TL}^{n}} \tag{5-52}$$

where n can be adjusted to provide the best fit to data [19]. A value for n of 6 to 10 was given as being appropriate for good electron guns. The dashed curve in Figure 5-31 shows Vaughan's expression plotted for $n = 6$.

Longo uses (5-51) to predict emission current density as a function of time by assuming that the work function in J_{TL} (the Richardson-Dushman current) varies with time. The time dependence of work function is assumed to result from the variation with time of the fraction of the cathode surface that is covered by barium and by how much barium is on the covered surface. Surface coverage by barium is determined by considering losses due to evaporation and increases due to the evolution of active material in the porous tungsten and the migration of this material to the surface by gas flow under conditions known as *Knudsen flow*.

To use the prediction model for an M-type cathode, Longo takes into account the diffusion of tungsten into the osmium coating and the resulting variation with time of the base metal work function and of the minimum cathode work function. He calibrates his model against the results of accelerated life tests as shown in Figure 5-32(a) and then uses the model to predict operating lifetime at normal operating temperatures as shown in Figure 5-32(b). In this figure, $J(0)$ is the initial cathode current density and $J(t)$ is the current density at a later time. Note that lifetimes of over 100,000 hrs are predicted with only small decreases in emission current.

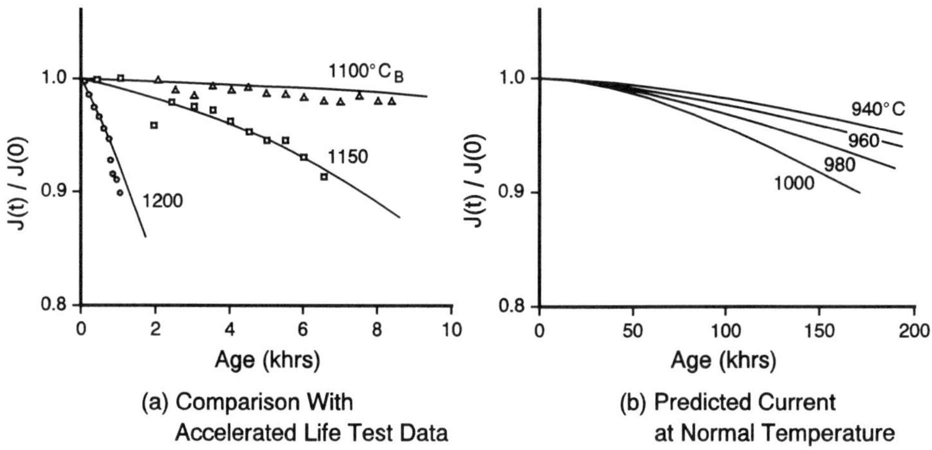

Figure 5-32 Results of Longo's life prediction model. From R.T. Longo *et al., Technical Digest*, 1984 IEDM. © 1984 IEEE.

Since the first use of cathodes many decades ago, there has been a continuing requirement for increased emission current capability along with long life. At the present time, for microwave tubes, these requirements are driven primarily by mm-wave tubes where small, high current density cathodes are required for acceptable beam formation. There are also requirements for high current density cathodes for technologies other than microwave tubes. One of these is for free electron lasers where a high current density cathode (commonly referred to in this case as a *high brightness cathode*) is needed to help limit the divergence of the beam generated by the electron gun.

We should point out that there is a practical limit to the current density that can be drawn from a cathode, regardless of its emission capability. This limit results from the electrical breakdown characteristics of vacuum, which will be discussed in Chapter 17. For a 1 cm gap with a uniform field, the breakdown field is about 10^5 V/cm and so breakdown across the 1 cm gap would be expected for a voltage of about 100 kV. For a gap with hot electrodes (such as a thermionically emitting cathode) and filled with a high electron density, the breakdown voltage would be below 100 kV.

For a planar space charge limited diode with an anode-cathode separation of 1 cm and an emission current density of 100 A/cm^2, an anode voltage of about 120 kV is required. While the average field would be 120 kV/cm, we must remember that, for space charge limited flow, the field at the cathode is zero and at the anode is above the average field. Even if breakdown did not occur within the current flow region of the diode, there is very high probability that it would occur between the nearby focus electrode and the anode where the spacing would be less than 1 cm. Thus, it appears that the practical upper limit to emission current densities is less than 100 A/cm^2. Higher current densities are possible at very short pulse lengths because the breakdown field strength increases as pulse length becomes shorter.

By optimizing the alloy composition on coated cathodes, Varian has been able to demonstrate emission densities of over 100 A/cm^2 [20]. The Miram curves in Figure 5-33 show this emission capability and also show that the cathode was well behaved up to the highest temperatures and currents tested.

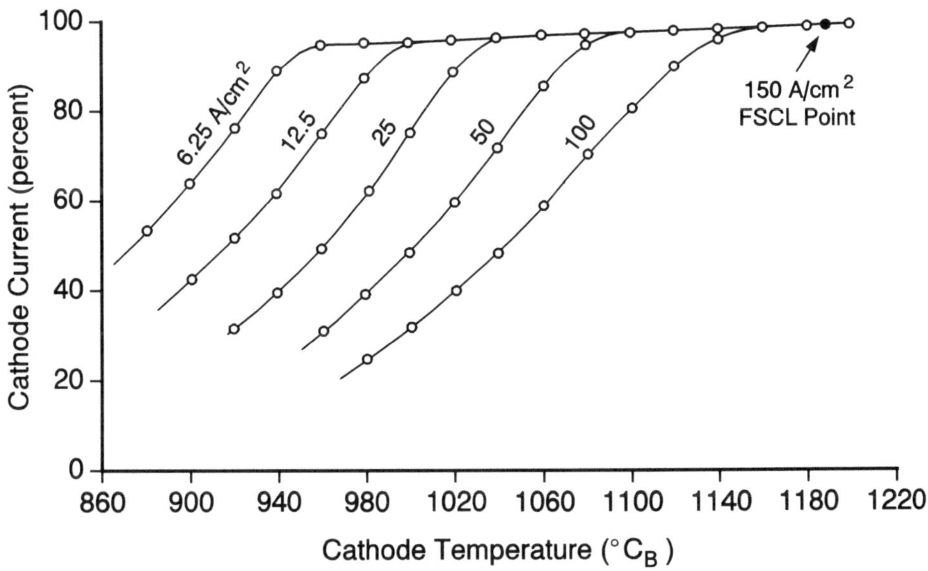

Figure 5-33 100 A/cm^2 dispenser cathode test data. From George Miram *et al.*, Vacuum Electronics and Display Conference, 1989.

Thomas *et al.* have used barium depletion calculations to predict cathode life at high emission densities [2]. The criteria used for end of life was when the temperature limited emission dropped to half the initial emission level. Evaporation rate data for B-type cathodes was used to predict life as a function of temperature for each cathode type. Thomas *et al.* note that the approximation is crude because of differences in impregnant composition, pore structure and surface composition, all of which affect the net rate of evaporation. Still, a B-type cathode tested at LLNL and an M type tested at NRL, both at about 40 A/cm^2, are in reasonable agreement with predicted life. Thus, the results as shown in Figure 5-34 appear to be valid within a factor of two or three uncertainty.

The scandate cathode life was also assumed to result from barium depletion.

Thomas *et al.* note that, at the time the paper was written, it was not clear what other mechanisms may limit the life of this cathode. Also, for all of the cathodes shown in Figure 5-34, it is often found in practice that a higher operating temperature than expected (with resulting shortened life) is required to maintain emission because of the presence of poisoning gases.

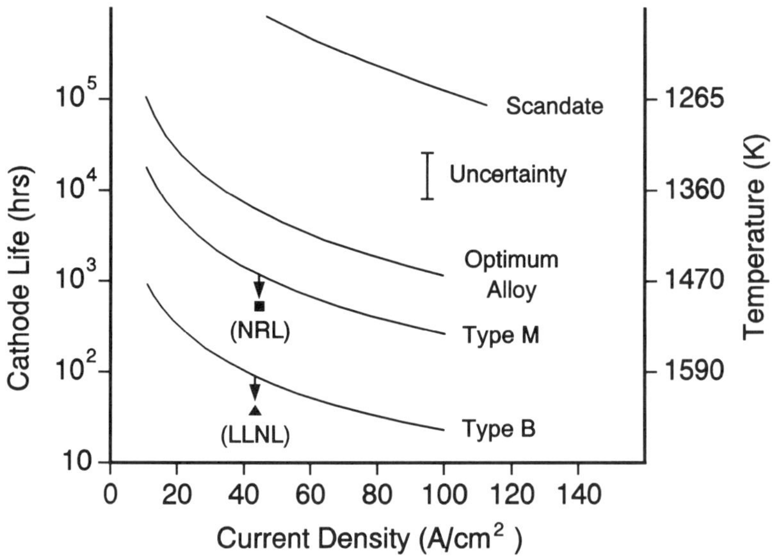

Figure 5-34 Predicted cathode life *versus* current density. From R.E. Thomas *et al.*, *IEEE Trans. on Electron Devices*, March 1990. © 1990 IEEE.

5.5 DISPENSER CATHODE SURFACE PHYSICS

The basic theory of thermionic emission discussed at the beginning of this chapter has been well established and understood for several decades. What has not been well understood until very recently has been the work functions of surfaces with adsorbed atoms of various materials. As a result, cathodes have historically been developed more by trying various combinations of materials and techniques, rather than by design based on a thorough knowledge of surface physics. Fortunately, however, the modern state of cathode theories is significantly advanced from that given by Pierce in 1946 [1].

> It is clear that as far as science goes, the state of the oxide coated cathode is confused indeed. Not only does theory fail to help us make good cathodes; it seems not to fit all the evidence. Maybe cathodes work sometimes one way and sometimes another, which isn't as ridiculous as it may sound. At any rate,

while some physicists are uninterested and others confused, cathode making continues to be an art practiced, it seems, by artists with a good deal of witch doctor in their make-up. When something goes wrong, this or that demon is exorcised, and eventually things go right again. And, unlikely as it seems, cathodes continue to improve.

While it is not yet possible in theory to design a new type of cathode having a predetermined emission density, operating temperature, and life, present cathode theories do explain very well how most dispenser cathodes operate. In the following section, an attempt is made to provide an overview of the present theories of dispenser cathode operation.

In the basic theory of thermionic emission discussed early in this chapter, an energy level diagram for electrons in and near the metal surface, like that shown in Figure 5-35(a), was used. In dispenser cathodes, the emitting surface is far more complex than that shown in Figure 5-35(a). This is because of the diffusion of barium, oxygen and perhaps other elements to the surface of the cathode. For the B-type cathode, for example, measurements show that the surface consists of layers of barium and oxygen on the metal (tungsten in this case) base as indicated in Figure 5-35(b). The layers of barium and oxygen are very thin, and each consists of approximately a monolayer of atoms. The atoms in these layers are charged. On average, the barium atoms have lost electronic charge and are positively charged. The oxygen atoms have acquired electronic charge and are negatively charged. Thus, the positive barium atoms and negative oxygen atoms, along with negative image charges on the surface of the metal (not indicated in Figure 5-35(b)), form electric dipoles. These dipoles, in turn, affect the potential distribution at the surface of the metal in some way as indicated by the curved potential profile followed by a question mark. The resulting work function that electrons must overcome in order to be emitted from the cathode is therefore changed.

The positive charge of the barium atoms also results in electrostatic forces that attract the atoms to the surface of the metal. This causes the rate of evaporation of the barium atoms from the tungsten surface to be far below (by over an order of magnitude) the rate of evaporation of barium atoms from a barium surface.

Before attempting to explain the effect of two (or more) layers of charged atoms of different elements on the work function of a metal surface, it is helpful to consider a single layer of atoms of one element. Following the work of Zalm and van Stratum [21], who were early developers of the M-type cathode, consider again the energy level diagram for a metal surface as shown in Figure 5-36(a). With no foreign atoms in the vicinity of the surface, the energy levels are as have been previously discussed. If a small number (far fewer than required to form a monolayer) of atoms that can easily give up electrons (have a low ionization potential) are brought into the electric field near the surface of the metal, it is probable that some of the atoms will become ionized and give up electrons to the metal surface (see Figure 5-36(b)). The presence of the resulting positive charge on the surface of the metal produces a reduction, $\Delta e\phi$, in the

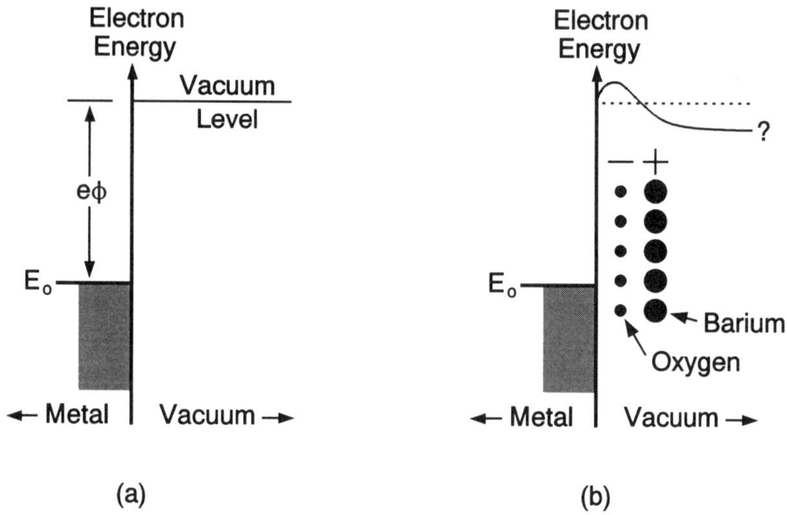

Figure 5-35 Effect of oxygen and barium on energy level diagram.

work function, which, in turn, increases the probability of electron emission. The influence of the positive ions near the surface of the metal is similar to that of a microscopic grid at a positive potential. Such a grid would reduce the potential barrier at the surface of the metal and permit increased electron emission.

As additional atoms are brought near the surface, some become ionized and further reduce the work function (Figure 5-36(c)). For small numbers of atoms (much fewer than required for a monolayer) the amount by which the work function is reduced, $\Delta e\phi$, increases linearly with the density of atoms on the metal surface.

As more and more atoms are brought near the surface, and more and more ions are formed, two primary effects reduce the rate of ion formation:

1. The reduction in electric field near the surface that the ions produce results in a reduced probability of ionization. Eventually, the field is reduced enough that the probability of ionization approaches zero. The addition of more atoms will not produce additional reductions in the work function.
2. As the number of ions increases, the distance between ions decreases. As this distance decreases, the repulsion forces between ions increases, and this causes more and more ions to be ejected from the metal surface.

As still more atoms are brought near the surface, and the film thickness grows to over a monolayer, the work function gradually changes to that of the atoms on the surface.

Using the concepts outlined in the preceding discussion, it is possible to understand the reasoning that led Zalm and van Stratum to the use of osmium and other high work

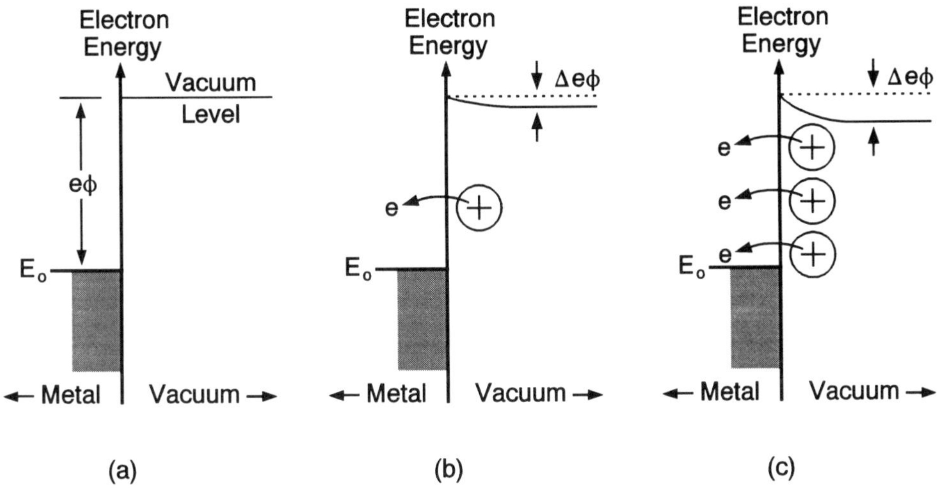

Figure 5-36 Energy level diagrams for a metal without (a) and with (b) and (c) ions near the surface.

function metals in cathodes. Consider two metals, one having a higher work function than the other. For example, consider osmium with a work function of about 5.4 eV and tungsten with a work function of about 4.6 eV. The energy level diagram shown in Figure 5-37 indicates why, with an adsorbed layer of ions, Zalm and van Stratum expected the work function for osmium to be lower than for tungsten. In each case, as more and more atoms are brought near the surface, ionization will occur until the electric field has dropped to the point where the probability of ionization approaches zero ($P_i \rightarrow 0$). This field will be approximately the same for the two metals. Because the work function for osmium is higher than for tungsten, the field at the surface is higher, and more ions can be produced than for tungsten. Thus, the resulting reduction in work function for osmium is higher than for tungsten. In fact, the reduction for osmium is high enough that the overall work function of the metal plus adsorbed ion layer is lower than for tungsten. This results, of course, in higher electron emission from osmium (with an adsorbed layer) than for tungsten.

The explanation given in the previous paragraph for the improved performance of an osmium cathode fails when some other metals are considered. For example, as indicated in Figure 5-38, the work function for platinum (which is a 5d transition metal in the same family as osmium) is comparable to that of osmium. However, the emission of a platinum coated cathode is significantly below that of osmium and is even below that of tungsten. As another example, the emission capability of a cathode coated with a tungsten-osmium alloy is superior to that of an osmium coated cathode even though the work function of the tungsten-osmium alloy is below that of osmium.

To understand the differences in performance between cathodes coated with a tungsten-osmium alloy, osmium or platinum it is necessary to examine the atomic

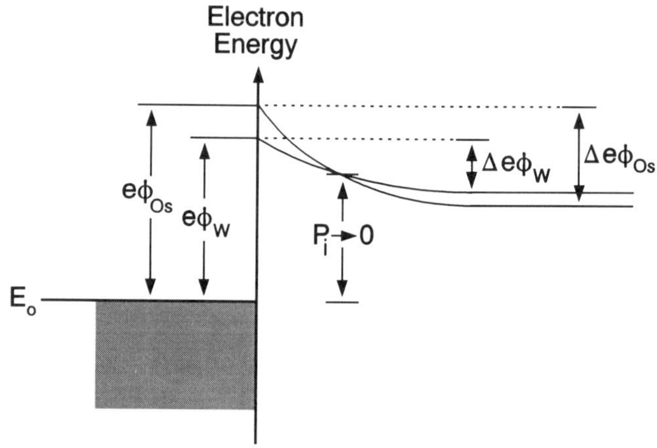

Figure 5-37 Effect of work function of metal on energy profiles.

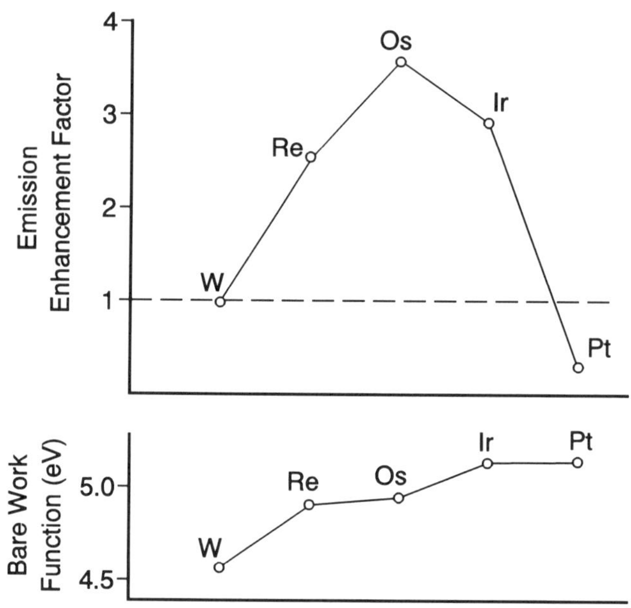

Figure 5-38 Emission enhancement and work function for dispenser cathode metals. From M.C. Green, Technical Report RADC-TR-81-211, July 1981.

structure of the metal cathode surface and its relation to adsorbed oxygen and barium atoms. The work of Müller [22–24] has led to calculations of work functions for dispenser cathodes that are in excellent agreement with experimental results. For this reason, the following discussion relies heavily on the work of Müller.

In Müller's work, a microscopic portion of the surface of a cathode is modeled as a cluster of atoms. The models are chosen on the basis of crystal structures found on actual cathodes. An example is shown in Figure 5-39. Here, the B-type cathode is modeled as a cluster consisting of 19 tungsten atoms in a body centered cubic configuration and 3 groups of BaO molecules. The tungsten substrate in an actual B-type cathode is polycrystalline with many crystal faces present. The (100) orientation was chosen because of its simple structure. The clusters used for modeling the osmium and osmium-tungsten alloy substrates represented the hexagonal-close-packed lattice. For platinum, the face-centered cubic lattice was used.

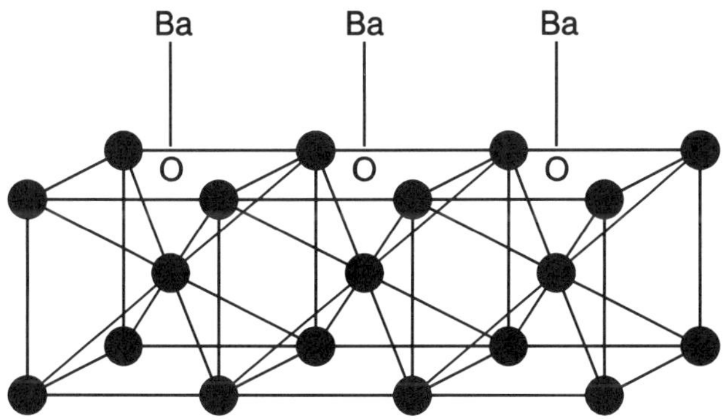

Figure 5-39 Body-centered cubic cluster (100 face) assumed for BaO/W computations. From W. Müller, *Technical Record*, 1991 IEDM. © 1991 IEEE.

The calculations were performed using very large computer codes capable of analyzing the interactions between the relatively large numbers of atoms in the clusters. Calculated properties included electronic energies, densities of states, electron densities, charge distributions, surface electric dipoles and barium binding energies on different substrates.

Only the work function change that occurs with the adsorption of barium and oxygen on the metal surface is obtained from the calculated properties. Thus, the minimum work functions depend critically on the bare-metal work functions of the various crystal faces considered. Because accurate values of the bare surface work functions are not known in most cases, some assumptions and approximations had to be made. In addition, empirical temperature dependencies of polycrystalline work functions for pure metals were used to estimate work functions at 1200 K. The initial, bare-metal work functions affect the results by their magnitude and temperature dependence. A high value leads to a large charge transfer from barium to oxygen and the metal substrate. Alloying with W, in general, reduces the work functions for Re, Os, Ir and Pt.

The work-function change that is induced by the adsorption of BaO is found to be the largest for the hexagonal structures (Os, Ir, Os-W). The larger change more than compensates for the initially higher bare-metal work functions of the alloy substrates and, as shown in Figure 5-40, leads to minimum work functions that are up to 0.2 eV lower than for BaO on tungsten.

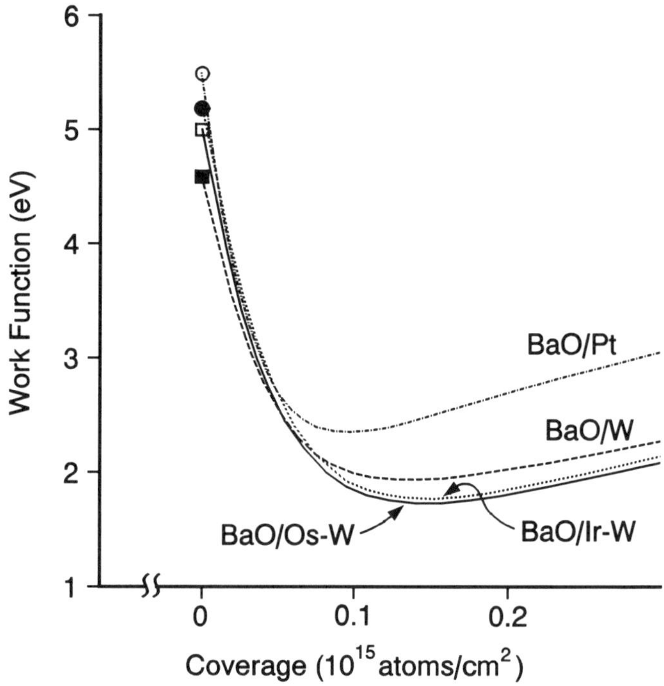

Figure 5-40 Calculated work functions for B- and M-type cathodes. From W. Müller, 1992 Tri-Service/NASA Cathode Workshop.

The calculated barium binding energies for high surface coverages are 5.4 eV for BaO/W and ~6.0 eV for the alloy substrates. This increase of 0.6 eV is in excellent agreement with the observed increase in the energy of desorption for Ba from B- and M-type cathodes. This, along with a reduced operating temperature for a given emission current, is the reason for the long life of the M-type cathode. For BaO on platinum, the calculated binding energy drops to 3.6 eV. Thus, barium evaporates at a higher rate from platinum than from the other metals considered, and the resulting reduced coverage of the surface leads to a higher minimum work function.

5.6 HEATERS

5.6.1 Conventional Heater Assemblies [25]

Typical oxide coated cathodes operate near 850°C and B-type cathodes operate near 1100°C. To raise a cathode to this temperature, the heater must be very hot. The materials of choice for heaters are pure tungsten or tungsten with a small amount of rhenium added. The rhenium makes the tungsten easier to work.

After forming, the heaters are coated with an insulating layer .002–.003 in thick of aluminum oxide. The oxide can be applied by spraying or by cataphoretic coating (an electrolytic process). The aluminum oxide insulation is sintered at high temperature to drive off organic binders and to form a hard, scuff-resistant coating.

Heaters for oxide coated cathodes generally operate at a temperature near 1100°C. This is not excessive for long life in tungsten or tungsten-rhenium heaters. Oxide cathode heaters are generally mounted in a fairly simple cathode structure with no special steps taken to improve the transfer of heat from the heater to the cathode. Two typical structures are shown in Figure 5-41.

One side of the heater is often connected electrically to the cathode to make certain that no large voltage difference exists between the cathode and any part of the heater. The aluminum oxide insulation on the heater is thin and will not withstand much voltage without electrical breakdown.

Heaters for B-type dispenser cathodes present a more severe problem because of the 1100–1150°C operating temperature of the cathode. The usual solution is to embed the heater in an aluminum oxide pellet at the back of the cathode as shown in Figure 5-42. Heat transfer between heater and cathode is much improved by this embedding or potting and so the heater operating temperature needs to be only 150–200°C higher than the cathode. Also, the potting gives the heater mechanical support, making it resistant to damage from vibration and shock. The heater potting material is added as a powder, possibly compacted under pressure and then sintered at high temperature.

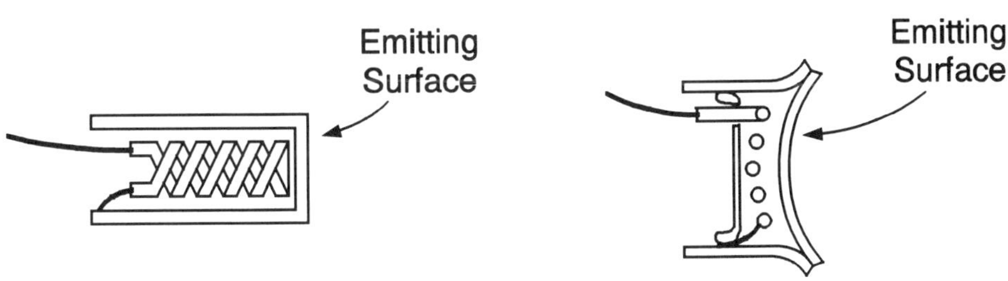

Figure 5-41 Typical heater assemblies for oxide coated cathodes.

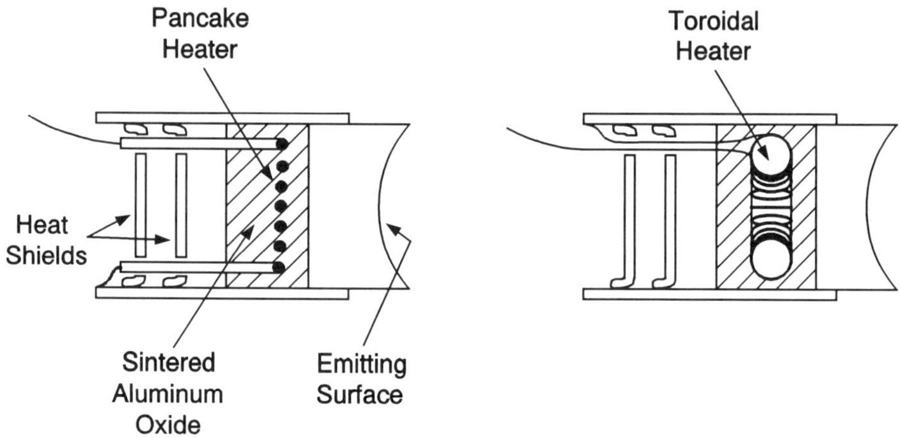

Figure 5-42 Heater assemblies for dispenser cathodes.

Potted heaters have one disadvantage. The extra mass of the potting material increases the warm-up time over an unpotted version in a similar cathode. In most applications this warm-up time penalty (typically 2–3 min) is accepted as the price for improving heater reliability. Some applications, in missiles, for example, require fast warm up. (Fast warm-up heaters are discussed in the following section.) Fortunately, such applications do not require long life, so the heater can be designed for low mass and high temperature operation.

Dispenser cathodes require special steps to conserve heater power. Heat shields are used extensively to minimize radiation losses to the side and back. Support members are made with as thin a cross section as possible to minimize conduction losses. Finally, considerable care must be taken by the manufacturer to ensure consistency in the heater potting operation. Variations in contact between the heater, the potting material and the cathode can lead to dangerous heater "hot spots" and cathode "cold spots."

Heaters can be operated from either ac or dc supplies. When an ac supply is used the heater may cause "hum" to appear as modulation of the signal. The source of this hum is often magnetic flux generated by the heater current. This flux extends to the cathode surface and perturbs the trajectories of electrons leaving the cathode. The toroidal heater on the right in Figure 5-42 is designed to minimize the generation of stray magnetic flux. The perturbation of the beam in a tube by magnetic flux from the cathode is discussed further in the following section.

The dc heater supplies require extra circuitry, but they are sometimes used. Of course, a dc heater will eliminate hum but stray magnetic flux from the heater can still perturb the beam emerging from the gun and cause increased interception current. The rectifier circuitry for a dc heater supply adapts easily to voltage regulation. In some systems the ac line voltage varies too much to keep the heater at a uniform enough temperature. Voltage regulation is used to reduce this variation.

When a dc supply is used, polarity of the supply can be important. If one side of the heater is connected to the cathode, that side should go to the positive side of the heater supply. The heater is then negative with respect to the cathode. This precaution is taken to avoid emission current from the back of the cathode to the heater. If the heater were positive with respect to the cathode it would attract any electrons unintentionally emitted from the back of the cathode. The voltage and current involved are small, but over a long period of time such a current can damage the heater.

5.6.2 Fast Warm-up Heaters

Fast warm-up heaters are needed for cathodes in expendable microwave tubes for missiles and may be needed for some space defense applications. Full emission is required from these cathodes within a few seconds after heater power is applied. This emission requirement must be met even though the cathodes may have been dormant for years prior to turn on. Also, most tubes with this fast turn-on requirement must meet severe shock and vibration specifications.

Before describing fast warm-up assemblies, we should comment on the reactivation of cathodes after periods of shelf storage. The problems of rapid reactivation after storage have been discussed in detail by Haas *et al.* [26]. The factors involved include cathode type, surface composition, pore spacing, Ba impregnant concentration, Ba-O bonding, exposure to poisoning gases, and heating time and temperature. Haas *et al.* point out that it is the elimination of the poisoning products of ambient gases from the surface along with the arrival of new Ba surface atoms via the impregnant diffusing out of the pores that constitutes the reactivation process of the cathode surface. Some turn-on tests indicate that essentially no delay occurs. This could have been due to temperature overshoot as the cathode is very quickly brought to the operating temperature or to the passivating effect of other gases such as hydrogen. Other tests have correlated turn-on delays with measured pressures of oxidizing gases.

Shown in Figure 5-43 are the temperature characteristics of some fast warm-up heater-cathode assemblies [20]. The cathode diameter in each case was 0.2 in (0.79 mm). In all cases, the fast warm-up times were achieved by using low thermal mass components. The three-second design used an integral tungsten-3% rhenium heater cataphoretically coated with alumina. A voltage surge as high as twice the normal operating voltage was used.

The bombarder-type assembly, which is shown in Figure 5-44, achieved a warm-up time of one second [27]. The heater is made of thoriated tungsten and designed to be heated to about 1700°C in a fraction of a second by a large overvoltage. At that temperature the heater is also an emitter. Several hundred volts are applied between the heater and the cathode. Bombardment of the back of the cathode by electrons from the heater raises the cathode to the operating temperature.

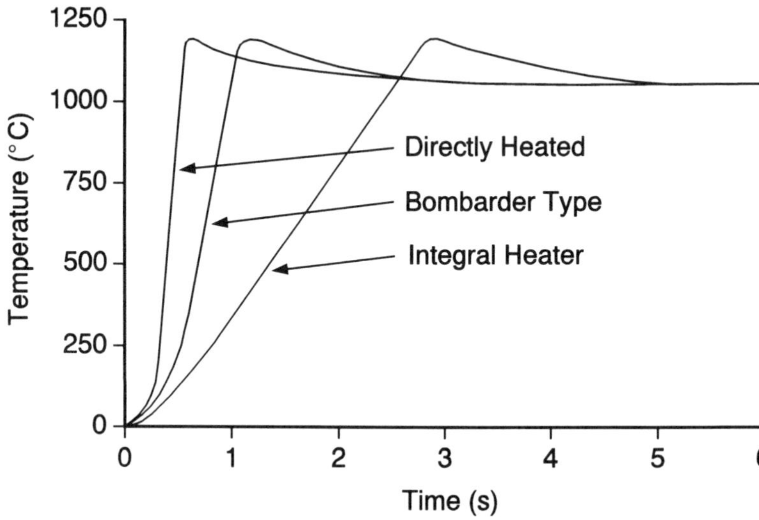

Figure 5-43 Temperature characteristics for fast warm-up heater-cathode assemblies. From George Miram *et al.*, Vacuum Electronics and Display Conference, 1989.

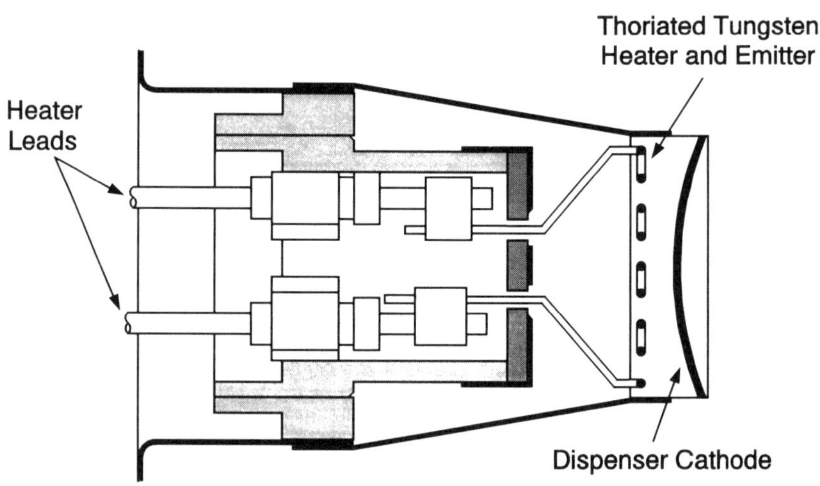

Figure 5-44 Bombarder-type fast warm-up heater. From R.C. Treseder *et al., Technical Digest,* 1983 IEDM. © 1983 IEEE.

As shown in Figure 5-43, a directly heated cathode (such as the thoriated tungsten heater in the bombarder design) can achieve warm-up times of less than one second. The current flow through and voltage across a directly heated cathode as well as the geometry of the cathode have significant affects on the trajectories of electron leaving the cathode. Thus directly heated cathodes are used only in applications where beam formation is not critical (for example, in magnetrons such as those used in microwave ovens).

Cattelino *et al.* have described fast warm-up heaters for large diameter cathodes [28]. Anisotropic pyrolytic graphite deposited onto anisotropic boron nitride is used to fabricate heater wafers. Both planar and spherical heaters can be made. Warm-up times of 1 to 2 seconds for 4 to 6.4 mm-diameter cathodes and less than 20 seconds for 60 mm-diameter cathodes have been demonstrated.

5.6.3 Heater Testing

By examining heater current and voltage it is possible to determine if a heater is operating properly and, usually, if the cathode is at or near the proper operating temperature. It is important to realize that heater resistance varies rapidly with heater temperature. For the tungsten-3% rhenium wire often used for heaters, resistance increases by a factor of six as the wire is heated from room temperature to approximately 1550°C. Thus, a plot of heater current *versus* heater resistance, as shown in Figure 5-45 can reveal malfunctions of the heater assembly.

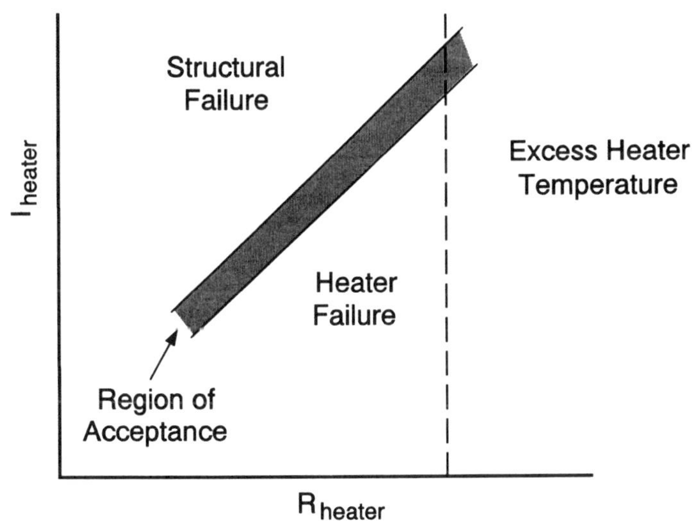

Figure 5-45 Heater performance. Adapted from T.J. Grant, *Technical Digest*, 1984 IEDM. © 1984 IEEE.

To provide meaningful information, the "region of acceptance" in Figure 5-45 must be determined for a particular class of heater assemblies that are known to operate at the desired temperature. Then, Figure 5-45 becomes the benchmark against which other heater assemblies can be compared. Excessive heater current (operation above the region of acceptance) implies that the heater assembly is at a lower temperature than it should be (for that current, the heater resistance is too low and so the temperature is too low). Possible problems are that the tube is gassy (the gas cools the heater assembly) or that a heat shield is touching the heater or cathode assembly and carrying heat away. Operation below the "region of acceptance" indicates excessive heater wire temperature and possible heater failure.

5.6.4 Effect of Filament Magnetic Field

Detailed measurements made on the electron beam from a magnetically shielded Pierce gun (to be discussed in following chapters) illustrate the effect that an ac heater can have on the beam. In Figure 5-46, the beam-scanning device is illustrated that was used for current density measurements [29]. An ac current was used in the cathode heater, and the voltage to the cathode was pulsed on at various times during the heater current cycle [30].

When the heater current was zero, the electron beam cross section of Figure 5-47 was obtained. When the heater current was at a maximum value, cross sections like that of Figure 5-48 were obtained. Under very specialized conditions, an image of the spiral-wound heater could be detected in the beam, as shown in Figure 5-49.

The perturbations shown in Figures 5-48 and 5-49 resulted from the linking through the cathode surface of magnetic flux lines generated by the heater current. The variations in the shape of the beam as the filament current changes result in modulation of the RF output of the tube at the frequency of the filament current.

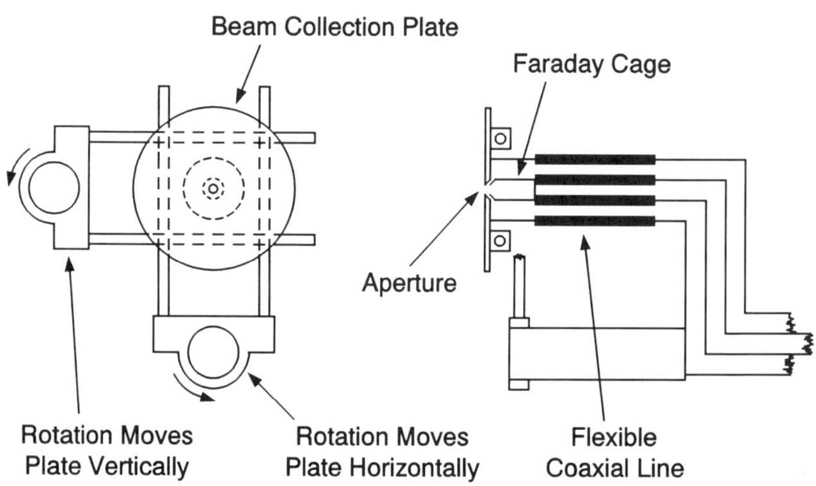

Figure 5-46 Beam scanning device used for measurements of beam current density. From A.S. Gilmour, Jr., *Proc. High Power Microwave Tube Symposium*, The Hexagon, 1962.

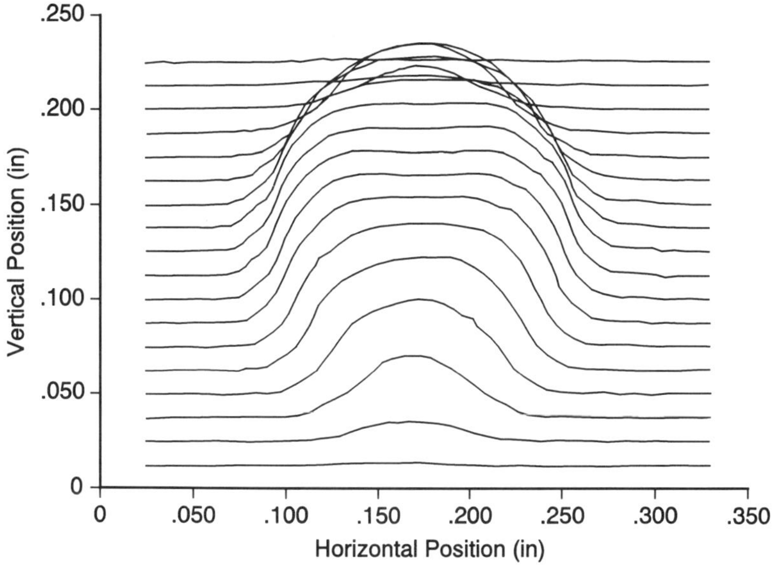

Figure 5-47 Electron beam cross section obtained for no current through cathode heater. From A.S. Gilmour, Jr., *Proc. IRE*, vol. 49, 1961. © 1961 IRE (now IEEE).

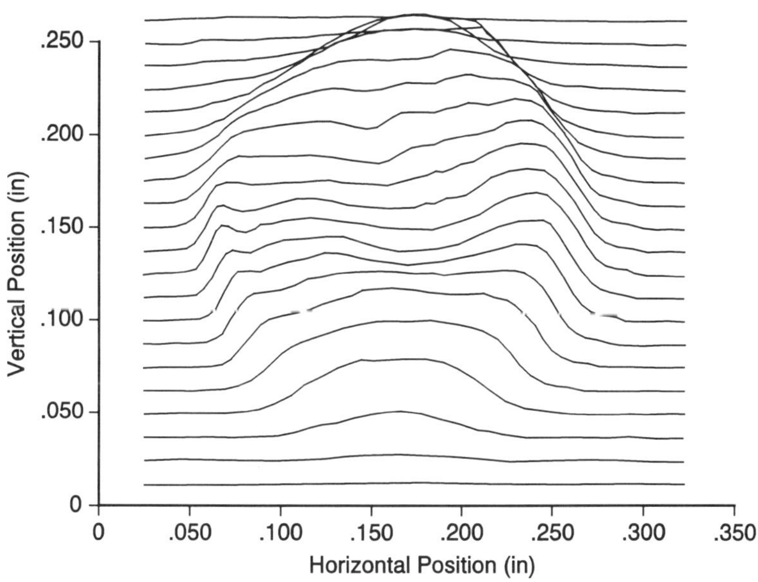

Figure 5-48 Electron beam cross section obtained with the cathode heater at a maximum value. From A.S. Gilmour, Jr., *Proc. IRE*, vol. 49, 1961. © 1961 IRE (now IEEE).

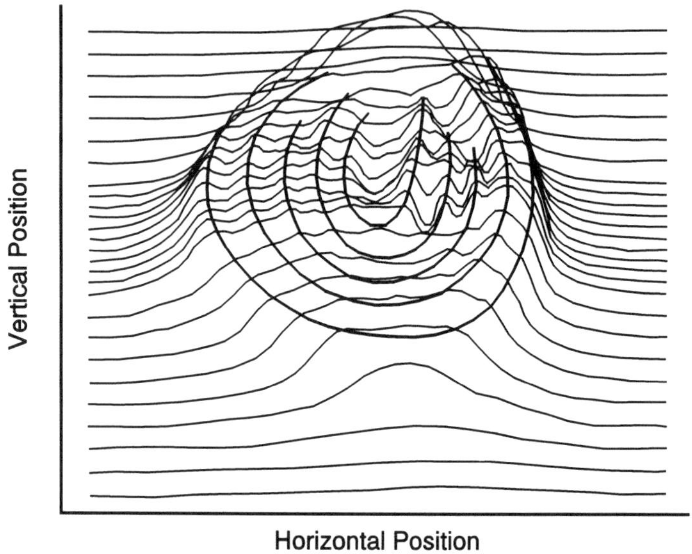

Figure 5-49 Image of spiral-wound cathode heater in electron beam cross section. From A.S. Gilmour, Jr., *Microwave Tubes*. Copyright 1986 by Artech House. Reproduced with permission.

REFERENCES

[1] Material extracted from an article by J.J. Coupling, a pseudonym of J.R. Pierce, which appeared in the November 1946 edition of *Science Fiction*, a Street and Smith publication.

[2] Richard E. Thomas, John W. Gibson, George A. Haas and R.H. Abrams, Jr., "Thermionic sources for high-brightness electron beams," *IEEE Trans. on Electron Devices*, vol. 37, no. 3, March 1990, pp. 850–861.

[3] Additional suggested reading: G. Herrman and P.S. Wagener, *The Oxide-Coated Cathode*. London: Chapman and Hall, 1951.

[4] Additional suggested reading: R.L. Sproull, *Modern Physics*. New York: John Wiley and Sons, 1956.

[5] J.L. Cronin, "Modern dispenser cathodes," *IEE Proc.* vol. 128, pt. 1, no. 1, Feb. 1981, pp. 19–32.

[6] Additional suggested reading: C.L. Hemenway, R.W. Henry, and M. Caulton, *Physical Electronics*. New York: John Wiley and Sons, 1967.

[7] Additional suggested reading: M. Chodorow and C. Susskind, *Fundamentals of Microwave Electronics*. New York: McGraw-Hill, 1964.

[8] M.C. Green, "Dispenser cathode physics," Final Technical Report, RADC-TR-81-211, July 1981.

[9] M.C. Green, "Dispenser cathode design," Final Technical Report, RADC-TR-83-99, May 1983.

[10] M.C. Green, "Modern thermionic cathodes," *Technical Digest*, 1987 International Electron Devices Meeting, pp. 925–928.

[11] R.A. Tuck, H.B. Skinner, T.M. Gardiner and P.J. Wadsworth, "Surface studies of dispenser cathodes using surface extended absorption fine structure," Tri-Service Cathode Workshop, 1986.
[12] Thomas J. Grant, "Emission degradation characteristics of coated dispenser cathodes," *Technical Digest*, 1986 International Electron Devices Meeting, pp. 700–703.
[13] M.J. Cattelino, G.V. Miram and W.R. Ayers, "A diagnostic technique for evaluation of cathode emission performance and defects in vehicle assembly," *Technical Digest*, 1982 International Electron Devices Meeting, pp. 36–39.
[14] T.J. Grant, "A powerful quality assurance technique for dispenser cathodes and electron guns," *Technical Digest*, 1984 International Electron Devices Meeting, pp. 334–337.
[15] R.T. Longo, "Long life, high current density cathodes," *Technical Digest*, 1978 International Electron Devices Meeting, pp. 152–158.
[16] R.T. Longo, "A study of thermionic emitters in the regime of practical operation," *Technical Digest*, 1980 International Electron Devices Meeting, pp. 467–470.
[17] R.T. Longo, E.A. Adler and L.R. Falce, "Dispenser cathode life prediction model," *Technical Digest*, 1984 International Electron Devices Meeting, pp. 318-321.
[18] A.M. Shroff and G. Firmain, "Long-life travelling-wave tubes. Use of M-type cathode. Life prediction model," *Technical Digest*, 1985 International Electron Devices Meeting, pp. 346–349.
[19] Rodney Vaughan, "A synthesis of the Longo and Eng cathode emission models," *IEEE Trans. on Electron Devices*, vol. ED-33, no. 11, November 1986, pp. 1925–1927.
[20] George Miram, Michael Green and Mark Cattelino, "Gridded gun design considerations for pulse and CW applications," presented at Vacuum Electronics and Display Conference, Garmisch Partenkirchen, Austria, May 9, 1989.
[21] P. Zalm and A.J.A. van Stratum, "Osmium dispenser cathodes," *Philips Technical Review*, vol. 27, no. 3/4, 1966, pp. 69–75.
[22] W. Müller, "Theoretical study of cathode surface," Final Technical Report, NASA CR-182166, July 1988.
[23] W. Müller, "Computational modeling of dispenser cathode emission properties," *Technical Digest*, 1991 International Electron Devices Meeting, Dec. 1991, pp. 399–402.
[24] W. Müller, "Mechanism of emission enhancement in barium dispenser cathodes," *Proc. 1992 Tri-Service/NASA Cathode Workshop*, Greenbelt, MD, March 17–19, 1992.
[25] A.T. Raczinski, "TWT principles," Notes from TWT training course presented at NSWC, Crane, IN, March–May, 1982.
[26] G.A. Haas, R.E. Thomas, Christie R.K. Marrian and Arnold Shih, "Rapid turn-on of shelf-stored tubes: An update," *IEEE Trans. on Electron Devices*, vol. ED-38, no. 10, October 1991, pp. 2244–2251.
[27] R.C. Treseder, T.J. Grant and G.V. Miram, "Design of quick start, high current density electron guns," *Technical Record*, 1983 International Electron Devices Meeting, pp. 453–455.
[28] M.J. Cattelino, G.V. Miram and Bernard Smith, "Fast-and super-fast-warm-up cathodes using novel APG/APBN heaters," *IEEE Trans. on Electron Devices*, vol. ED-38, no. 10, October 1991, pp. 2239–2243.
[29] A.S. Gilmour, Jr., and D. Hallock, "A demountable beam analyzer for studying magnetically confined electron beams," *Advances in Electron Tube Techniques*, vol. 2, pp. 153–157, New York: Pergammon Press, 1963.
[30] A.S. Gilmour, Jr., "Effect of filament magnetic field on the electron beam from a Pierce gun," *Proc. IRE*, vol. 49, 1961, p. 976.

Chapter 6

Electron Guns

The electron guns for all traveling wave tubes and, in fact, for all linear beam tubes have operating principles that are essentially the same. The only major differences are in size and operating voltage and current, which result from differences in power requirements for the tubes. This chapter contains a discussion of the operating principles of the electron guns used on nearly all linear beam tubes. These guns are named after J.R. Pierce and are known as Pierce guns [1].

6.1 PIERCE GUNS

The electron gun in a microwave tube is used to form the electrons from the cathode into a beam suitable for interaction with a microwave circuit. Electron gun designers must cope with the following two basic problems:

1. The electrostatic repulsion forces between electrons tend to cause the beam to diverge.
2. The current density required in the electron beam is normally far greater than the emission density that the cathode can supply with an acceptable life cxpcctancy.

The approach to solving these problems, described by Pierce, is summarized in Figure 6-1. It is assumed that there is no magnetic field in the gun. Thermal velocities of electrons are ignored and laminar electron flow is assumed. The gun is divided into three regions. First of all, the shape of the cathode disc (often referred to as the cathode *button*) is spherical. The focus electrode is designed to produce equipotential surfaces in region 1 that are nearly spherical, with the same center of curvature as the cathode. As a result, electrons flow toward the center of curvature of the cathode.

In region 2, because the anode must contain a hole to let the electron beam pass through, equipotential surfaces bow into the anode aperture. As a result, a divergent electrostatic lens exists that produces a defocusing action on the electron beam.

Finally, in region 3, the electrons have escaped from the accelerating field of the

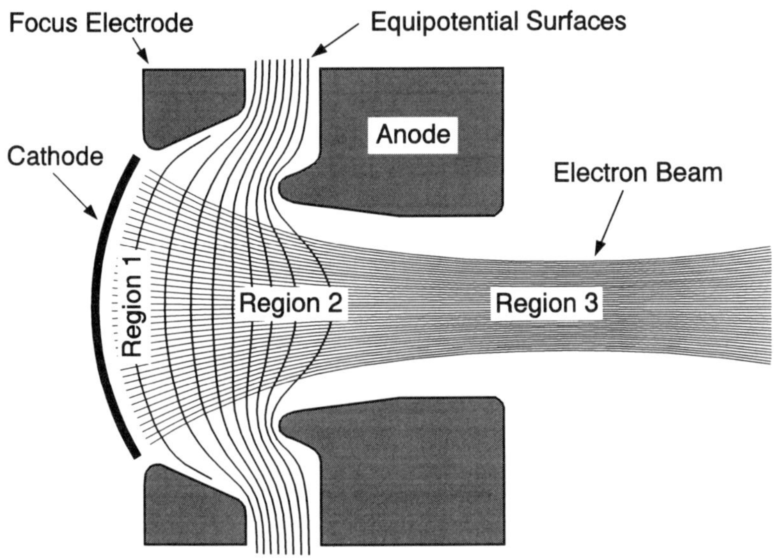

Figure 6-1 Overview of Pierce gun.

cathode-to-anode regions and is drifting under the influence of space charge forces. Thus, the electrons in the beam follow universal beam-spread trajectories.

In the following sections, we will discuss each of the three regions of the Pierce gun. Then, using the equations developed for each of the regions, a procedure by Vaughan [2] for designing a Pierce gun will be described.

6.1.1 Focus Electrodes for Parallel Flow

Dealing first with the repulsion force problem as electrons leave the cathode and move toward the anode, it is instructive to examine the problem of attempting to cause electrons to travel in parallel paths as they leave a cathode as indicated in Figure 6-2.

When no electrons are present, the equipotential profiles in a parallel-plane diode are parallel and spaced equal distances apart for equal voltage increments. This is shown by the dashed lines in Figure 6-3. When electrons are present, the equipotential profiles are deflected to the right as indicated by the solid lines. Electrons tend to move perpendicular to the equipotential profiles, so the electron trajectories diverge as the electrons move away from the cathode. This is simply another way of saying that the electrons are pushing on each other and this causes their trajectories to diverge. If an electrode arrangement could be devised that would straighten the equipotential profiles even though electron space charge forces were present, then the electrons would travel in parallel paths.

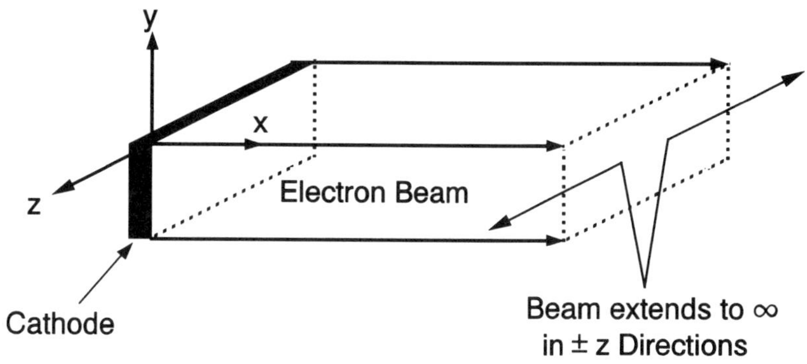

Figure 6-2 Parallel electron flow from a planar cathode.

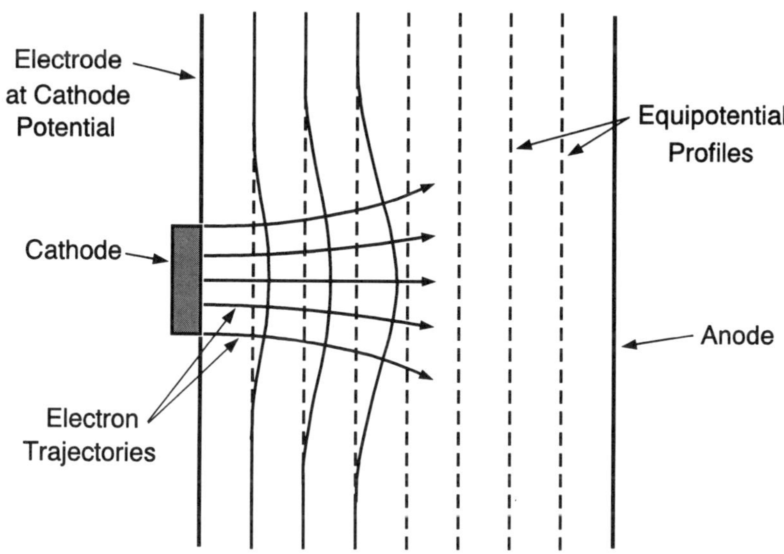

Figure 6-3 Equipotential profiles without (dashed lines) and with (solid lines) electrons.

Intuitively, it can be seen that, if the electrodes adjacent to the cathode are tipped toward the anode, as indicated in Figure 6-4, then the equipotential profiles will be straightened. In the absence of electrons, the profiles curve toward the cathode. Electron charge causes them to move toward the anode.

The mathematical solution that leads to a definition of the electrode configuration for parallel electron flow is similar to the solution for the problem of spacc charge limited flow discussed in Chapter 5. The main difference now is that the equations must be solved inside and outside the electron stream, and then the solutions must

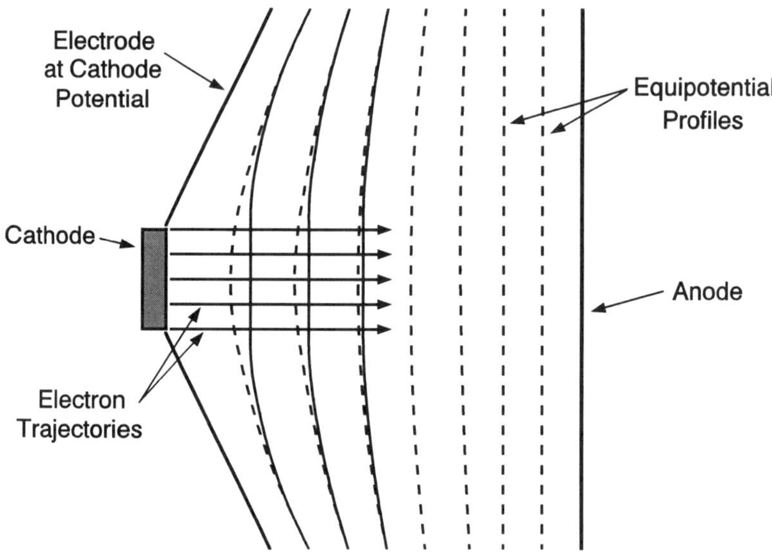

Figure 6-4 Straightening of equipotential profiles in electron beam.

match at the edge of the beam.

Inside the electron beam, Poisson's equation applies and so

$$\nabla^2 V = \frac{\rho}{\epsilon_o} = \frac{J}{u\epsilon_o} = \frac{J}{\epsilon_o\sqrt{2\eta V}} \tag{6-1}$$

If the electrons are to flow parallel to one another, then there can be no electric fields in the y and z directions because these would deflect the electrons in those directions. The potential, V, can vary only in the x direction; that is, $V = f(x)$. As a result,

$$\frac{\partial V}{\partial y} = 0, \quad \frac{\partial V}{\partial z} = 0 \tag{6-2}$$

and so Poisson's equation becomes

$$\frac{\partial^2 V}{\partial x^2} = \frac{J}{\epsilon_o\sqrt{2\eta V}} \tag{6-3}$$

Outside the beam, that is, for $y > 0$, Laplace's equation applies and so

$$\nabla^2 V = 0 \tag{6-4}$$

The boundary conditions that must be satisfied at $y = 0$ are

$$\frac{\partial V}{\partial y} = \frac{\partial V}{\partial z} = 0, \quad V = \mathrm{f}(x) \tag{6-5}$$

Next, it will be assumed that the cathode is space charge limited. The Child-Langmuir law for a parallel-plane diode can be rewritten as

$$V = \left[\frac{J}{2.33 \times 10^{-6}}\right]^{2/3} x^{4/3} \tag{6-6}$$

and so the f(x) that we are looking for in $V = \mathrm{f}(x)$ inside the beam and at $y = 0$ is

$$V = \mathrm{f}(x) = Ax^{4/3} \tag{6-7}$$

Now, make an educated guess and assume that the potential outside the beam (that is, for $y > 0$) may be expressed by the following complex function:

$$V(x,y) = \mathrm{Re}\ \mathrm{f}(x + jy) = \mathrm{Re}\ A(x + jy)^{4/3} \tag{6-8}$$

where only the real part of the complex function is to be used. If it can be shown that this function satisfies Laplace's equation and the boundary conditions, then it is the correct solution to the problem.

First, we will show that this function satisfies Laplace's equation. The function f($x + jy$) may be separated into real and imaginary parts as follows:

$$\mathrm{f}(x + jy) = \mathrm{u}(x,y) + j\mathrm{v}(x,y) \tag{6-9}$$

Then

$$\nabla^2 \mathrm{f} = \nabla^2(\mathrm{u} + j\mathrm{v}) = \frac{\partial^2 \mathrm{u}}{\partial x^2} + \frac{\partial^2 \mathrm{u}}{\partial y^2} + j\left[\frac{\partial^2 \mathrm{v}}{\partial x^2} + \frac{\partial^2 \mathrm{v}}{\partial y^2}\right] \tag{6-10}$$

and it must be shown that the real part is zero. The only functions of practical interest for solving Laplace's equation are analytic functions and these have the characteristic that

$$\frac{\partial \mathrm{u}}{\partial x} = \frac{\partial \mathrm{v}}{\partial y} \quad \text{and} \quad \frac{\partial \mathrm{v}}{\partial x} = -\frac{\partial \mathrm{u}}{\partial y} \tag{6-11}$$

This means that

$$\frac{\partial^2 \mathrm{u}}{\partial x^2} + \frac{\partial^2 \mathrm{u}}{\partial y^2} = \frac{\partial^2 \mathrm{v}}{\partial x\,\partial y} - \frac{\partial^2 \mathrm{v}}{\partial x\,\partial y} = 0 \tag{6-12}$$

and so the real part satisfies Laplace's equation. (The imaginary part also does.)

Considering the boundary conditions, at $y = 0$, f($x + jy$) = f(x) and so this boundary condition is satisfied. Also,

$$\mathrm{Re}\ \frac{\partial}{\partial y}\mathrm{f}(x + jy) = \mathrm{Re}\ \frac{\partial}{\partial y}(x + jy)^{4/3} = \mathrm{Re}\ j\frac{4}{3}(x + jy)^{1/3} = \mathrm{Re}\ j\frac{4}{3}(x)^{1/3} = 0 \tag{6-13}$$

and so $\partial V/\partial y = 0$ at $y = 0$ is satisfied. Thus, the required solution is

$$V = \text{Re}\, A(x + jy)^{4/3} \tag{6-14}$$

To understand what this means, it is convenient to change to polar coordinates. That is, let

$$x + jy = |r|\, e^{j\theta} \tag{6-15}$$

Then,

$$\text{Re}\,(x + jy)^{4/3} = \text{Re}\, |r|^{4/3} e^{j(4/3)\theta} = |r|^{4/3} \cos \frac{4}{3}\theta \tag{6-16}$$

and so

$$V(r,\theta) = A\,|r|^{4/3} \cos \frac{4}{3}\theta \tag{6-17}$$

Now, notice that, if we let $4/3\ \theta = \pi/2$, or $\theta = 3/8\pi$ (67.5°), $V = 0$. This means that planar electrodes with $V = 0$ placed at an angle of 67.5° with respect to the edge of the beam, as shown in Figure 6-5 (along with a curved electrode at the anode potential, V_a), will cause parallel electron flow. These electrodes at 67.5° are called *Pierce electrodes* and the 67.5° angle is referred to as the *Pierce angle*.

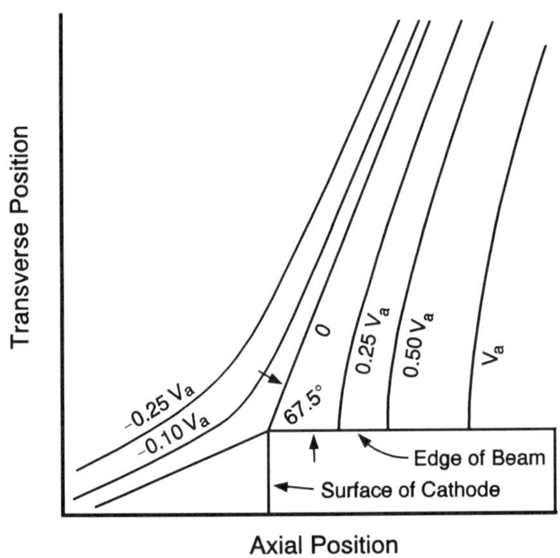

Figure 6-5 Equipotential profiles for parallel electron flow in a planar two-dimensional diode. Adapted from *Theory and Design of Electron Beams* by J.R. Pierce, copyright 1954 by D. Van Nostrand, Inc.

Figure 6-6 shows an electron gun containing 67.5° Pierce electrodes for producing parallel (rectilinear) electron flow. Note that grid wires are shown across the aperture in the positive electrode. This is to maintain the electrode potential within the beam and, thus, prevent divergence of the beam after it passes through the positive electrode. Beam divergence caused by an anode aperture will be discussed in subsequent paragraphs.

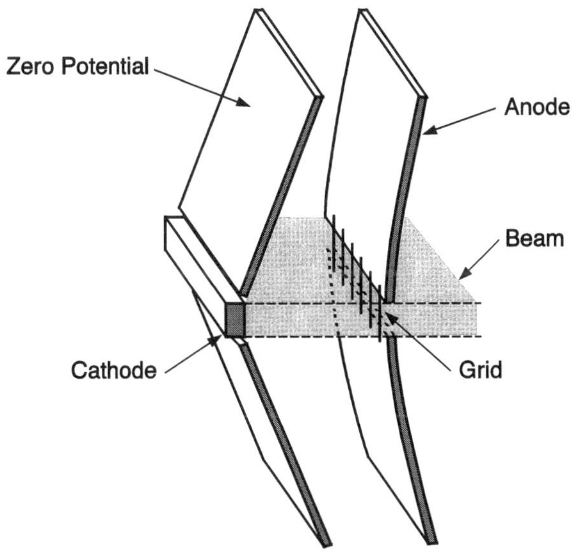

Figure 6-6 Electron gun for producing rectilinear (parallel) electron flow. Adapted from *Theory and Design of Electron Beams* by J.R. Pierce, copyright 1954 by D. Van Nostrand, Inc.

6.1.2 Focus Electrodes for Convergent Flow

The current density required in the electron beam in a microwave tube is normally far greater than the emission density capability of the cathode. As a result, it is necessary to use a cathode having a relatively large area and then to compress the flow of electrons as the beam is formed. The starting point in the analysis is a spherical diode. The inner surface of the outer sphere is the cathode and the inner sphere is the anode. The configuration is shown in Figure 6-7. Space charge limited emission is assumed.

The procedure used in analyzing the spherical diode is like that for the planar diode. The only difference is that the mathematics is more complicated. Poisson's equation in spherical coordinates, when there are no angular variations in any quantities, is

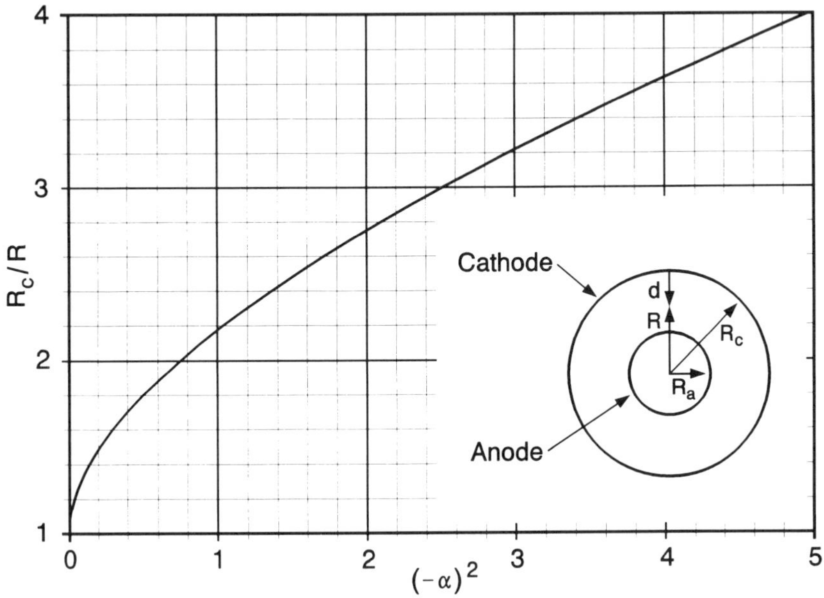

Figure 6-7 Langmuir and Blodgett's $(-\alpha)^2$.

$$\nabla^2 V = \frac{1}{R^2}\frac{d}{dR}\left[R^2\frac{dV}{dR}\right] = \frac{\rho}{\epsilon_o} \tag{6-18}$$

Here, R is used for radial position in the spherical configuration to avoid confusion with the radial position, r, in the electron beam in cylindrical coordinates.

The charge density, ρ, is

$$\rho = \frac{J}{u} = \frac{J}{\sqrt{2\eta V}} = \frac{I_s}{4\pi R^2\sqrt{2\eta V}} \tag{6-19}$$

where I_s is the total current in the spherical diode. Thus, Poisson's equation becomes

$$\frac{d}{dR}\left[R^2\frac{dV}{dR}\right] = \frac{I_s}{4\pi\sqrt{2\eta V}} \tag{6-20}$$

The solution, as worked out by Langmuir and Blodgett [3], is as follows. First, let

$$\gamma = \ln\left[\frac{R_c}{R}\right] = \ln\left[\frac{1}{1-\dfrac{d}{R_c}}\right] \tag{6-21}$$

and

$$\left(-\alpha\right) = \gamma + 0.3\,\gamma^2 + 0.075\,\gamma^3 + 0.01432\,\gamma^4 + 0.00216\,\gamma^5 + 0.00035\,\gamma^6 \tag{6-22}$$

where the coefficient of the last term has been changed from 0.00027 to 0.00035 for increased accuracy [2].

Then the current is

$$I_s = \frac{16\,\pi\,\epsilon_o}{9}\sqrt{2\,\eta}\,\frac{V^{3/2}}{\left(-\alpha\right)^2} \tag{6-23}$$

or

$$I_s = 29.33 \times 10^{-6}\,\frac{V^{3/2}}{\left(-\alpha\right)^2} \tag{6-24}$$

The quantity $(-\alpha)^2$ is shown in Figure 6-7. Of course, the current varies as the three-halves power of voltage as it did for the planar diode.

To form an electron gun, a conical portion, as shown in Figure 6-8, of the spherical diode is used. For a cone of half-angle θ, the current I is related to the total current in the spherical diode by

$$I = \frac{1 - \cos\theta}{2}\,I_s \tag{6-25}$$

so

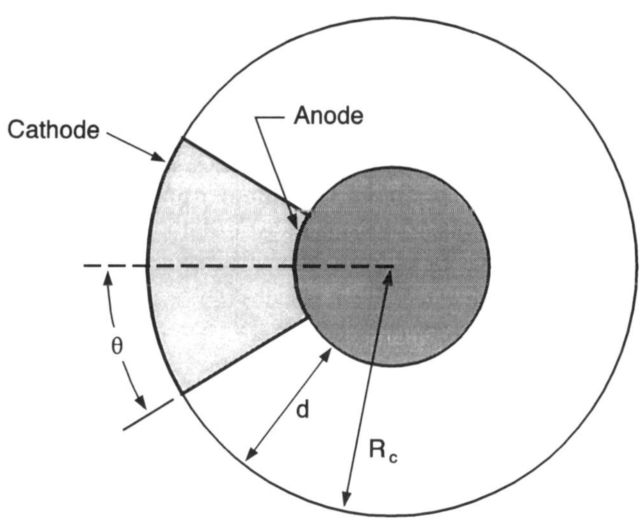

Figure 6-8 Conical diode with half angle θ.

$$I = 14.67 \times 10^{-6} \frac{(1 - \cos\theta)}{(-\alpha)^2} V^{3/2} \tag{6-26}$$

Figure 6-9 shows perveance as a function of cone angle and normalized cathode-anode spacing. It will be shown in subsequent paragraphs that the extremely high values of perveance indicated in Figure 6-9 cannot be achieved in real electron guns when anode effects are considered.

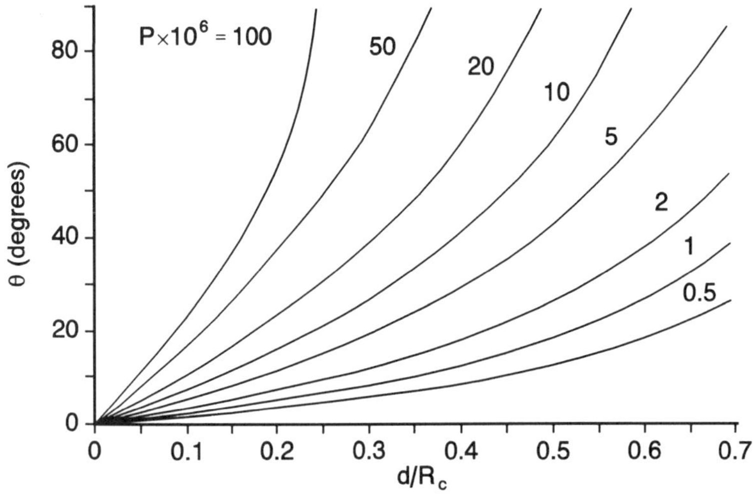

Figure 6-9 Perveance as a function of cone half angle and cathode-anode spacing.

The problem with the conical diode is that space charge forces will cause the beam to diverge just as they did for the planar diode. Equipotential contours, which should be spherical in shape, are deflected by the electron space charge potential as indicated in Figure 6-10. To retain radial electron flow when the remainder of the sphere is removed, it is necessary, as was the case for rectilinear flow in a parallel plane diode, to use electrodes that cause the equipotential profiles to have the shape which they had in the absence of the electrons (spherical in this case).

Conceptually, as was the case for the planar diode, it can be seen that, by tipping the electrodes adjacent to the cathode forward, the equipotential contours can be made more nearly spherical. In fact, as indicated in Figure 6-11(a), close to edge of the cathode, electron flow appears to originate from a planar portion of the cathode with electron flow perpendicular to that portion. Thus, an electrode placed at a 67.5° angle with respect to the edge of the beam will have the desired effect on the equipotential profiles near the edge of the cathode. For small cone half angles (small θ), this effect extends to the center of the cathode and so no further consideration of the shape of the focus electrode is necessary.

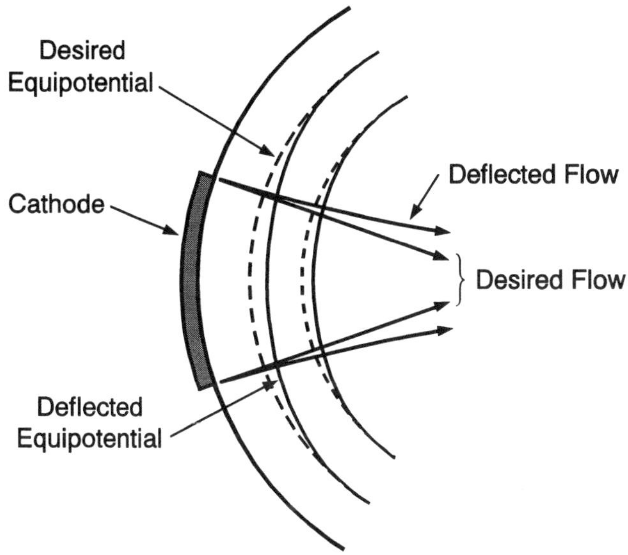

Figure 6-10 Deflection of equipotential contours in conical diode by electron space charge potential.

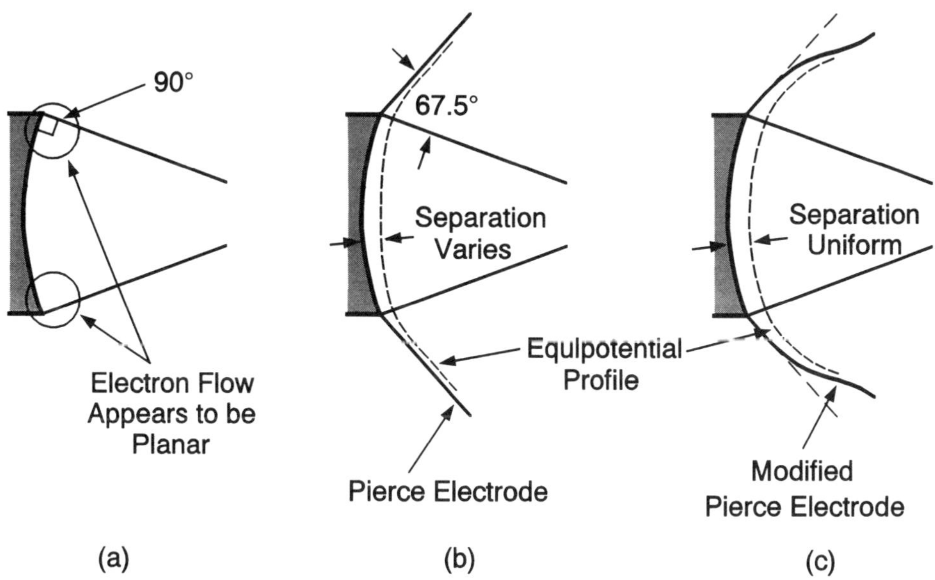

Figure 6-11 Modification of Pierce electrodes to produce spherical equipotential profiles.

For large cone angles, as indicated in Figure 6-11(b), the potential profiles will still not be spherical. This results in nonuniform current flow from the cathode (the current density drawn from the outer regions of the cathode is greater than that at the center). Also, the electron trajectories, as the electrons leave the cathode region, will not be focused toward the center of curvature of the cathode. To correct these problems the shape of the focusing electrode must be modified so that the equipotential profiles near the cathode surface become spherical. Researchers have experimented with many possible shapes, one of which is shown in Figure 6-11(c) [4]. In effect, the modified focus electrodes lift the equipotential profiles away from the edges of the cathode. There is little effect on the shapes of the profiles in the center region of the cathode because the electron space charge fields dominate in controlling the potentials there.

6.1.3 Defocusing Effect of Anode Aperture

To remove electrons from the conical diode, it is necessary to cut a hole in the anode. The resulting distortion of the equipotential profiles near the anode aperture is illustrated in Figure 6-12 [5]. Notice that, at a low perveance as in (a) the distortion of the profiles is limited to the region very close to the anode aperture. As perveance is increased as in (b) and (c) the size of the anode aperture increases and the separation between the anode and cathode decreases. At a perveance of 3×10^{-6} the distortion of the profiles extends to the cathode surface. As a result, the electron trajectories in the cathode-to-anode region are seriously perturbed and emission from the cathode is nonuniform (much higher nearer the cathode edge than near the center).

In [5], Brewer describes the modification of the focus electrode as shown in Figure 6-12(d) to reduce distortion of the equipotentials and improve electron focusing and cathode emission uniformity. The modified focus electrode has the same effect as that shown in Figure 6-11(c). That is, the equipotentials are lifted away from the regions near the edge of the cathode. Although focusing and emission uniformity are improved, perveance decreases because of the reduced electric field near the cathode edges.

The distortion of the equipotential profiles near the anode aperture results in the formation of a divergent lens, which changes the focus of the electron flow to a point beyond the center of the sphere, as shown in Figure 6-13. A comprehensive treatment of the effect of the anode aperture was given by Amboss [6]. We will limit our discussion to the much simpler analysis given by Pierce [1] and then to subsequent corrections to this analysis. Pierce's treatment used the Davisson-Calbick equation for thin lenses (derived in Chapter 3). A thin lens is one in which the axial length of the lens is much less than the radial dimension. In Figures 6-1 and 6-12, it is apparent that the axial length of the distortion of the equipotential profiles near the anode aperture is comparable to the radius of the aperture. This means, of course, that the validity of the thin lens approximation is questionable. Still, it is of value to use the Davisson-Calbick equation as a starting point in the analysis. Then, corrections to this equation may be made to improve the accuracy of the gun design.

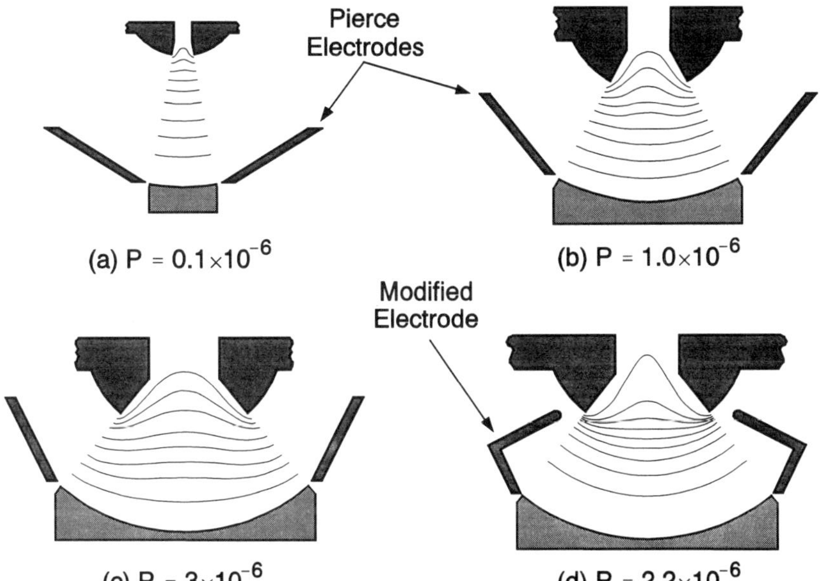

Figure 6-12 Effect of anode aperture on equipotential profiles. Adapted from G.R. Brewer, *Jour. Appl. Phys.*, January 1957 and *Focusing of Charged Particles*, copyright 1967 by Academic Press.

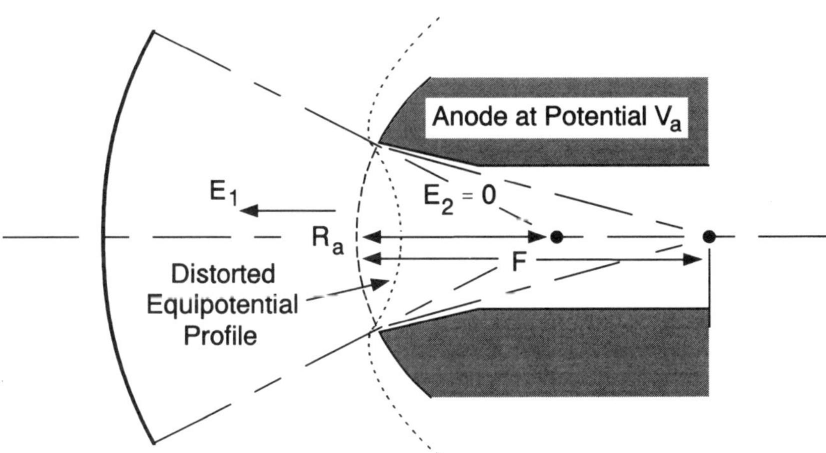

Figure 6-13 Defocusing effect of anode aperture.

First, assume, temporarily, that the electrons are moving parallel to the axis in the region to the left of the aperture. Then, from the Davisson-Calbick equation, the focal length, f, is

$$f = 4\frac{V_a}{E_2 - E_1} \tag{6-27}$$

In the electron gun being considered, the accelerating voltage is zero after the electrons pass through the aperture so $E_2 = 0$. This means that

$$f = -\frac{4V_a}{E_1} \tag{6-28}$$

Now, we remember from optics that, for a pair of lenses with two focal lengths, f_1 and f_2, the focal length of the combination, F, is

$$\frac{1}{F} = \frac{1}{f_1} + \frac{1}{f_2} \tag{6-29}$$

In the electron gun being considered, f_1 is the anode radius, R_a, and f_2 is $-4V_a/E_1$, so

$$\frac{1}{F} = \frac{1}{R_a} - \frac{E_1}{4V_a} \tag{6-30}$$

The value of E_1 can be found from the current-voltage relation for the conical diode; that is,

$$I = 14.67 \times 10^{-6}\frac{(1 - \cos\theta)}{(-\alpha)^2}V^{3/2} \tag{6-31}$$

In this expression, V and α are functions of the radius of curvature of the cathode, R_c, and the radius of curvature, R, of the spherical surface on which V is to be determined. As shown in Figure 6-14, that radius is the negative of the position z that will be used in this analysis. Thus, $E_1 = -dV/dz$ can be found by finding dV/dR from (6-31) as follows:

$$V = \left[\frac{I}{14.67 \times 10^{-6}(1 - \cos\theta)}\right]^{2/3}(-\alpha)^{4/3} \tag{6-32}$$

so

$$\frac{dV}{dR} = \left[\frac{I}{14.67 \times 10^{-6}(1 - \cos\theta)}\right]^{2/3}\frac{d(-\alpha)^{4/3}}{dR} \tag{6-33}$$

But

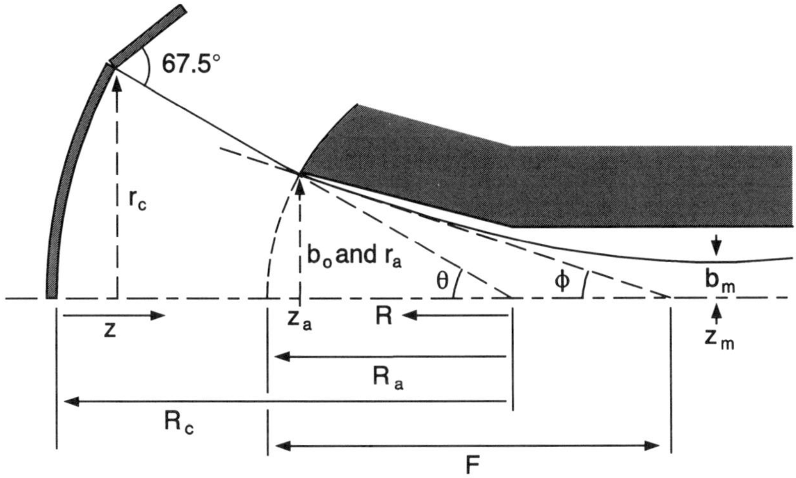

Figure 6-14 Quantities used in analysis of effect of anode aperture.

$$\frac{d(-\alpha)^{4/3}}{dR} = \frac{d(-\alpha)^{4/3}}{d(R_c/R)}\frac{d(R_c/R)}{dR} = -\frac{R_c}{R^2}\frac{d(-\alpha)^{4/3}}{d(R_c/R)} \tag{6-34}$$

and so, with the substitution of I from (6-31),

$$\frac{dV}{dR} = -\frac{V}{(-\alpha)^{4/3}}\frac{R_c}{R^2}\frac{d(-\alpha)^{4/3}}{d(R_c/R)} = -\frac{4V}{3(-\alpha)}\frac{R_c}{R^2}\frac{d(-\alpha)}{d(R_c/R)} \tag{6-35}$$

At the anode, $V = V_a$, $R = R_a$ and $dV/dR = -E_1$. As a result,

$$\frac{E_1}{4V_a} = \frac{1}{3(-\alpha)}\frac{R_c}{R_a^2}\frac{d(-\alpha)}{d(R_c/R_a)} \tag{6-36}$$

and so the focal length, F, becomes

$$\frac{1}{F} = \frac{1}{R_a}\left[1 - \frac{1}{3(-\alpha)}\frac{R_c}{R_a}\frac{d(-\alpha)}{d(R_c/R_a)}\right] \tag{6-37}$$

This relation may be simplified somewhat using $(-\alpha) = f(\gamma)$ (from (6-22)). Thus,

$$\frac{d(-\alpha)}{d(R_c/R_a)} = \frac{df(\gamma)}{d\gamma}\frac{d(\gamma)}{d(R_c/R_a)} \tag{6-38}$$

But from (6-21),

$$\frac{d(\gamma)}{d(R_c/R_a)} = \frac{d\ln(R_c/R_a)}{d(R_c/R_a)} = \frac{R_a}{R_c} \tag{6-39}$$

and so the relation for focal length becomes

$$\frac{1}{F} = \frac{1}{R_a}\left[1 - \frac{1}{3(-\alpha)}\frac{d\mathrm{f}(\gamma)}{(d\gamma)}\right] \tag{6-40}$$

The corresponding slope of the trajectory at the edge of the beam is very nearly

$$\tan\phi = \frac{b_o}{F} = \frac{b_o}{R_a}\left[1 - \frac{1}{3(-\alpha)}\frac{d\mathrm{f}(\gamma)}{(d\gamma)}\right] \tag{6-41}$$

From (6-22), the derivative in these expressions for F and $\tan\phi$ is

$$\frac{d\mathrm{f}(\gamma)}{d\gamma} = 1 + 0.6\,\gamma + 0.225\,\gamma^2 + 0.0573\,\gamma^3 + 0.0108\,\gamma^4 + 0.0021\,\gamma^5 \tag{6-42}$$

Shown in Figure 6-15 is a plot (the $\Gamma = 1.0$ curve) from (6-40) of F/R_a as a function of R_c/R_a. This plot shows what would be expected. That is, as the spacing between the anode and cathode decreases (so that R_c/R_a decreases) the focal length increases (because of increased field distortion). As shown by the dashed line to the right in the figure, for $R_c/R_a < 1.46$, the beam diverges after passing through the anode lens (the focal length becomes negative).

As noted at the beginning of this section, the use of the Davisson-Calbick thin-lens formula is in question in treating the defocusing effect of the anode aperture. This is because, in most cases of interest, the axial length of the defocusing region is comparable to the radius. Thus, a thick lens is present and the defocusing effect is more severe than is predicted by the thin-lens approximation. Danielson, Rosenfeld and Saloom [7] made a more detailed analysis of the effect of the anode aperture and applied a correction factor, Γ, to the Davisson-Calbick result. Their formula for ϕ is

$$\tan\phi = \frac{b_o}{F} = \frac{b_o}{R_a}\left\{1 - \frac{\Gamma}{3(-\alpha)}\frac{d\mathrm{f}(\gamma)}{(d\gamma)}\right\} \tag{6-43}$$

The magnitude of Γ depends on the size of the anode aperture relative to the spacing between the cathode and anode as shown by Wilson and Brewer [8] in Figure 6-16. The data in this figure are from Harris [9] and Buckey [10]. Harris found that, because of the shift in position of the equivalent lens, the lens strength passed through a maximum as spacing is changed. For most of the calculations presented by Danielson *et al.*, a value for Γ of 1.1 was used and shown to produce good agreement with experiment. Vaughan [2] used a value of 1.25, which along with a correction for spherical aberration produced good agreement with practice. The increase in focal length for values of Γ of 1.1 and 1.25 are shown in Figure 6-15.

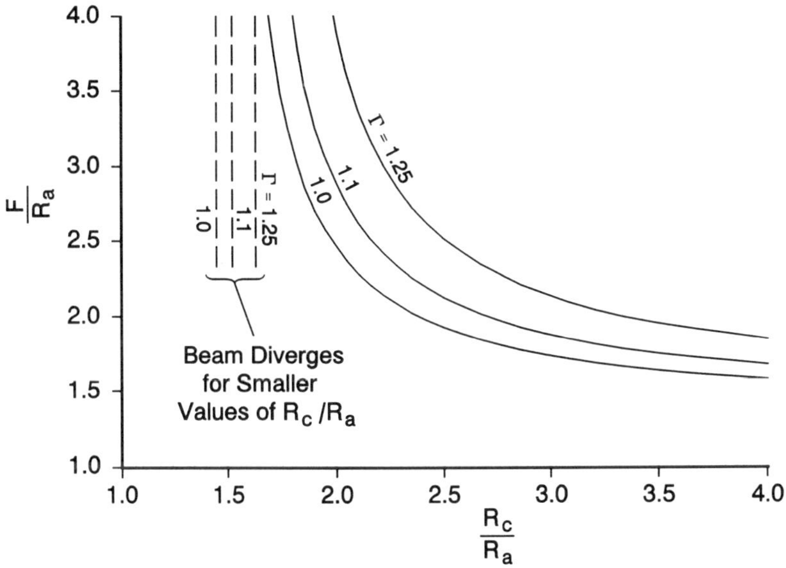

Figure 6-15 Relation of focal length to anode and cathode radii. Γ is a correction factor to the Davisson-Calbick lens equation.

6.1.4 Formation of Minimum Beam Diameter

After the beam leaves the anode aperture, the axial electric field is nearly zero and the only forces on the electrons are those resulting from the space charge. As a result, the shape of the beam is defined by the universal beam spread curve, shown in Figure 6-17. From the analysis for this curve (Chapter 3), the minimum beam radius, b_m, and the axial position of the minimum radius, z_m, can be found as functions of the entrance conditions into region 3. It is essential to have accurate knowledge of these quantities to properly inject the beam into the magnetic focusing structure.

From the universal beam-spread analysis (3-71) the relation between b_m and tan ϕ can be found as follows:

$$B_m = \frac{b_m}{b_o} = e^{-(dB_o/dZ)^2} = e^{-(db_o/A\,dz)^2} = e^{-(\tan\phi/A)^2} \tag{6-44}$$

so

$$b_m = b_o e^{-(\tan\phi/A)^2} \tag{6-45}$$

where $A = 174 \times \text{perveance}^{1/2}$.

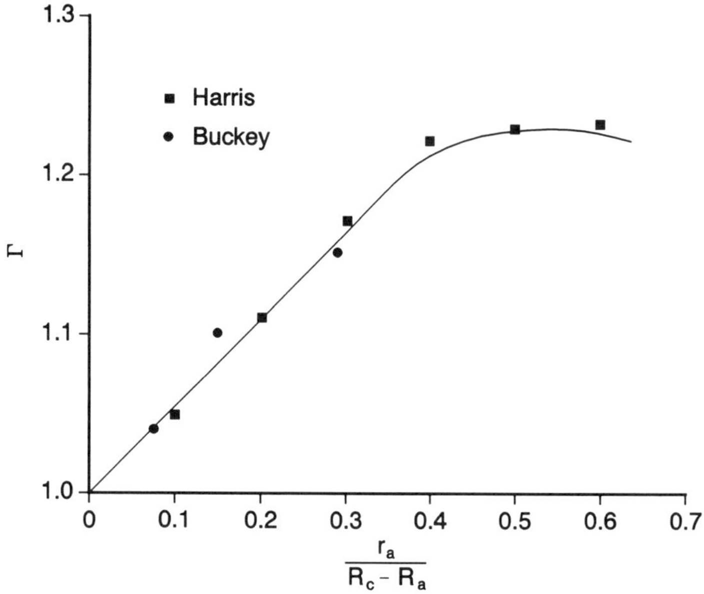

Figure 6-16 Correction factor for the Davisson-Calbick equation. Adapted from R.G. Wilson and G.R. Brewer, *Ion Beams*, copyright 1973 by John Wiley and Sons, Inc.

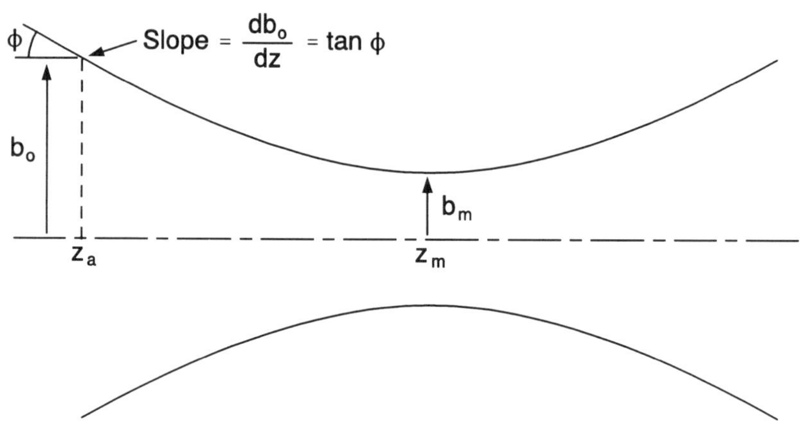

Figure 6-17 Beam shape in region 3 (drift region).

The universal beam spread curve (Figure 3-9) is usually given as normalized b *versus* normalized z. To complete the basic gun design, we need z as a function of b, since b_m is known from (6-45). Vaughan [2] has inverted the relation for the universal beam spread curve, and in terms of the quantities used here, the result is

$$z_m = z_a + \frac{b_m}{A}\left[1.914\left(\frac{b_o}{b_m} - 1\right)^{1/2} + 0.230\left(\frac{b_o}{b_m} - 1\right)\right.$$

$$\left. + 0.0107\left(\frac{b_o}{b_m} - 1\right)^2 - 0.000291\left(\frac{b_o}{b_m} - 1\right)^3\right] \tag{6-46}$$

6.1.5 Example of Gun Design

Consider a requirement for a 10,000 V, 1 A ($P = 1 \times 10^{-6}$) beam with a radius of 1 mm. Assume also, that the cathode emission density should be limited to about 2 A/cm^2 to ensure adequate life. Thus, the disc radius of the cathode, r_c, should be

$$r_c = \left(\frac{A}{\pi}\right)^{1/2} = \left(\frac{I}{\pi J}\right)^{1/2} = 0.4 \text{ cm} \tag{6-47}$$

The procedure that we will follow in illustrating the steps in designing a Pierce gun meeting these requirements is that given by Vaughan [2]. With this procedure, tan ϕ is calculated in two ways:

1. By assuming an initial value of θ and using the equations for regions 1 and 2 (the cathode and lens regions),
2. From the universal beam-spread curve.

The value of θ is adjusted so that the two values of tan ϕ are nearly the same. In this way, the beam will leave the lens region with the characteristics required at the entrance to the drift region for the proper beam radius to be obtained. The calculations are simple enough to be programmed into a handheld calculator. The calculations take a few seconds at most.

Step 1. Assume $\theta = 30°$. Then, from (6-26),

$$(-\alpha) = \left[\frac{14.67 \times 10^{-6}(1 - \cos\theta)}{P}\right]^{1/2} \tag{6-48}$$

so

$$(-\alpha) = 1.402 \tag{6-49}$$

Step 2. Determine γ. To do this, Vaughan inverts (6-22) for $(-\alpha)$ in terms of γ to obtain

$$\gamma = (-\alpha) - 0.275(-\alpha)^2 + 0.06(-\alpha)^3 - 0.006(-\alpha)^4 \tag{6-50}$$

and so the value of γ is

$$\gamma = 1.004 \tag{6-51}$$

Step 3. Determine tan ϕ_2 (subscript 2 for region 2) from (6-43). For b_o/R_a, sin θ is used and, as noted earlier, $\Gamma = 1.25$ is used. Vaughan found that this value of Γ along with the correction for spherical aberration to be discussed in the following section produced good agreement with practice. $df(\gamma)/d\gamma$ is calculated from (6-42) and is

$$\frac{d\,\mathrm{f}(\gamma)}{d\gamma} = 1.900 \tag{6-52}$$

As a result,

$$\tan\phi_2 = 0.218 \tag{6-53}$$

Step 4. Determine tan ϕ_3 (subscript 3 for region 3) from (6-45). That is,

$$\tan\phi_3 = A\left[\ln\frac{b_o}{b_m}\right]^{1/2} \tag{6-54}$$

The value of b_o may be determined with the aid of Figure 6-18 and (6-21); that is,

$$\frac{b_o}{r_c} = \frac{R_a}{R_c} = e^{-\gamma} \tag{6-55}$$

and so

$$b_o = 0.147 \text{ cm} \tag{6-56}$$

Now,

$$\tan\phi_3 = 0.108 \tag{6-57}$$

Step 5. Revise the value of θ. We note that tan ϕ_3 is smaller than tan ϕ_2 and so θ must be reduced to bring these values closer together. Vaughan obtains the new value of θ (θ(new)) in terms of the old value (θ(old)) as follows:

$$\theta(\text{new}) = \theta(\text{old})\left[\frac{\tan\phi_3}{\tan\phi_2}\right]^{1/2} \tag{6-58}$$

From this, the new value of θ is

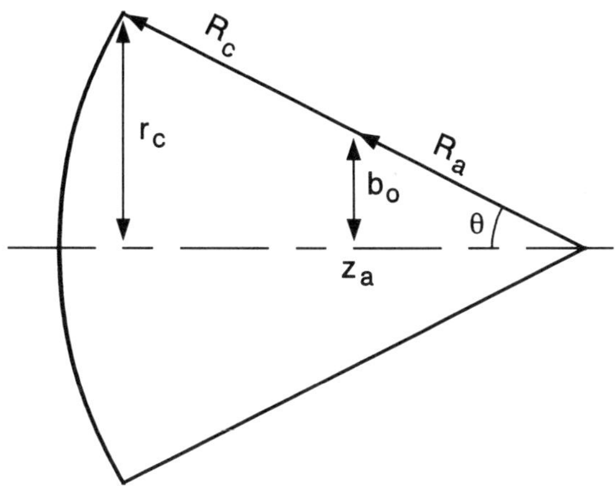

Figure 6-18 Relation of beam radius at anode aperture to cathode and anode radii.

$$\theta(\text{new}) = 21.1° \tag{6-59}$$

Step 6. Iterate the procedure until a final value of θ is determined. Three more iterations are more than adequate and the final value, θ(final), is

$$\theta(\text{final}) = 22.6° \tag{6-60}$$

With the correction for spherical aberration to be discussed in the next section, this final value of θ will increase about 10%.

Step 7. Complete the design by calculating the remaining gun parameters. Thus,

$$R_c = \frac{r_c}{\sin\theta} = 1.039 \text{ cm} \tag{6-61}$$

and

$$R_a = R_c e^{-\gamma} = 0.461 \text{ cm} \tag{6-62}$$

The value of r_a was shown in Figure 6-14 to be the same as b_o. Clearly, r_a must be larger than b_o to avoid excessive beam interception on the anode. Vaughan uses a factor of 1.2; that is,

$$r_a = 1.2\, b_o = 0.213 \text{ cm} \tag{6-63}$$

With r_a, z_a can be found:

$$z_a = \left(R_c^2 - r_a^2\right)^{1/2} = 0.630 \text{ cm} \tag{6-64}$$

Finally, using (6-46), the axial position of the beam minimum is

$$z_m = 1.708 \text{ cm} \tag{6-65}$$

A scale drawing of the gun is shown in Figure 6-19.

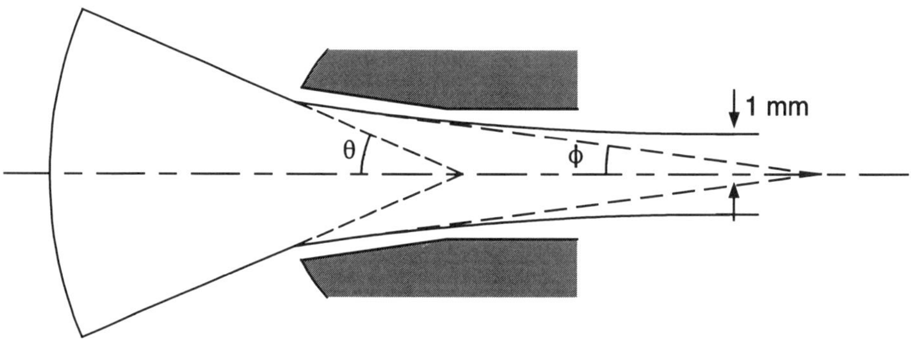

Figure 6-19 Scale drawing of 10 kV, 1 A, Pierce gun for producing a beam with a radius of 1 mm from a cathode with an emission density limited to about 2 A/cm^2.

It is important to realize that the values of b_m and z_m are very sensitive to R_c/R_a. This is illustrated in Figure 6-20, where beam profiles are shown for values of R_c/R_a that differ by 25%. As R_c/R_a increases from 1.6 to 2.0, significant changes occur in both the minimum beam diameter and its axial position. As R_c/R_a increases to 2.5, the minimum beam diameter is again reduced by a significant amount. This figure clearly illustrates the importance of precisely locating the cathode with respect to the anode.

6.1.6 Spherical Aberration

Lens aberrations result in a focal length that varies with radial position in a lens. Frost, Purl and Johnson [11] found that the measured beam from a high perveance gun was larger in diameter and less uniform in current density than expected from theory. By modifying the spherical radius of the cathode, as shown in Figure 6-21, they were able to correct for the lens aberration. The resulting improvements in beam characteristics are shown in Figure 6-22. The beam-current profiles at the top left of the figure were obtained before the correction was made. Note the severe nonuniformity in current as a function of radial position in the beam.

After modifying the cathode radius, the profiles at the bottom left of Figure 6-22 were obtained. Not only was the uniformity of the current improved, but the radius of

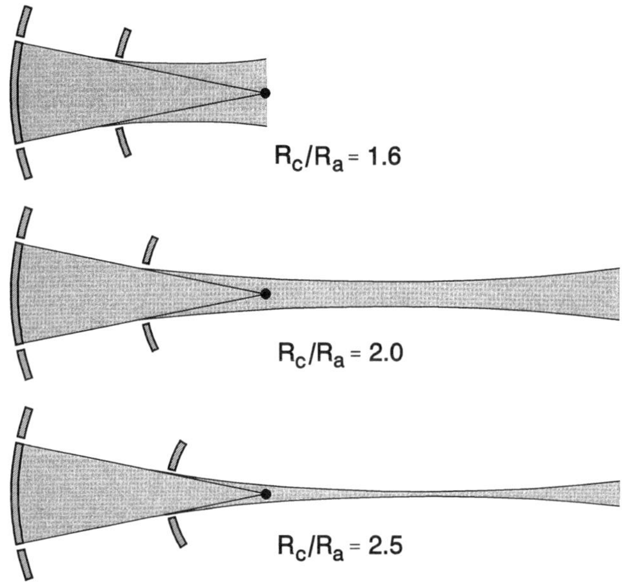

Figure 6-20 Emerging electron stream for various ratios of cathode radius to anode radius. Adapted from *Theory and Design of Electron Beams* by J.R. Pierce, copyright 1954 by D. Van Nostrand, Inc.

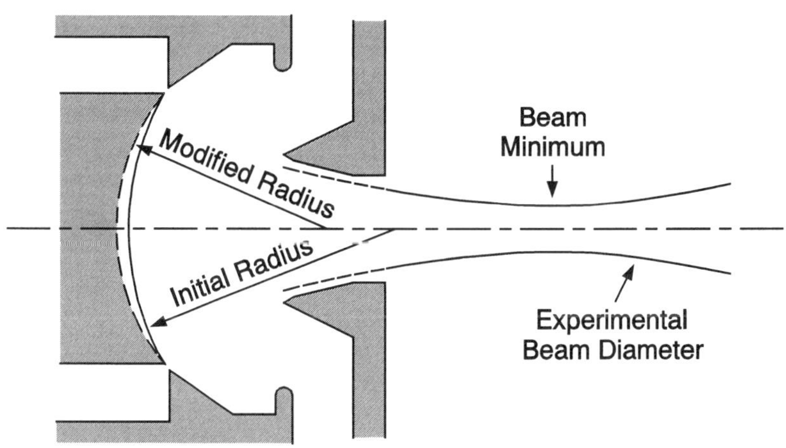

Figure 6-21 Modification to cathode radius used by Frost, Purl and Johnson [11] to correct for spherical aberration.

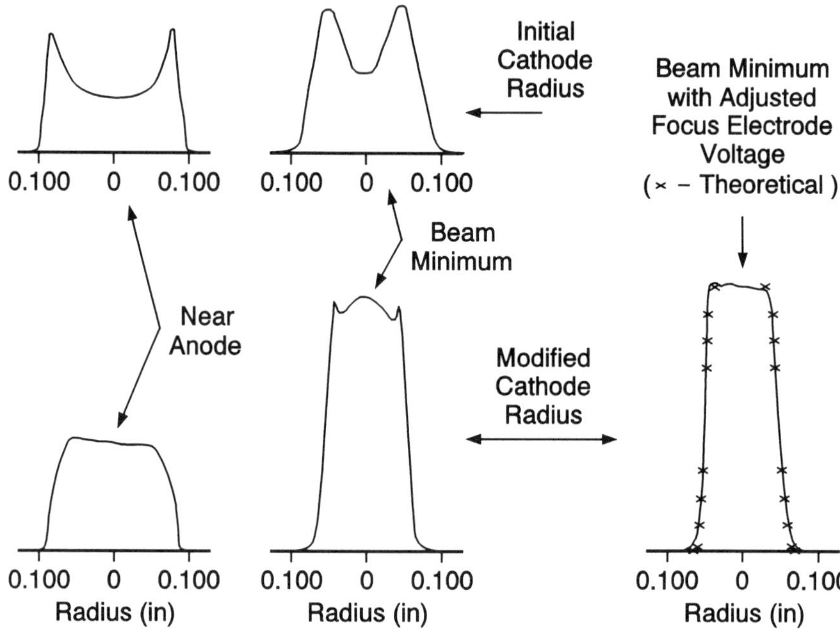

Figure 6-22 Beam-current profiles measured before and after correcting for spherical aberration. Adapted from R.D. Frost *et.al.*, *Proc. IRE*, August 1962. © 1962 IRE (now IEEE).

the beam at the position of the beam minimum was reduced. After an additional small change in the focus electrode voltage, the profile to the right was obtained. As indicated by the ×s, the shape of this profile was very close to that predicted by theory (the theories of Pierce and of Cutler and Hines [12], who predicted thermal velocity effects. The Cutler-Hines theory is discussed in the following section.)

As reported by Vaughan [2], R.B. True in an unpublished work has defined values of θ_T and R_{cT} corrected for spherical aberration by

$$\frac{\sin\theta}{\sin\theta_T} = k \tag{6-66}$$

and

$$\frac{R_{cT}}{R_c} = k \tag{6-67}$$

where k is an empirical constant less than unity. Vaughan reports that, for diode guns, the value

$$k = 0.905 \tag{6-68}$$

yields results that agree well with practice. For the gun design in the previous section, θ(final) increases to 25.1° and R_c decreases to 0.940 cm for $k = .905$.

6.1.7 Thermal Velocity Effects

In this section we will consider the effects of the thermal velocities of electrons on the beam produced by an electron gun. It is assumed that the magnetic field is zero. The nature of electron emission from a thermionic cathode is discussed early in the chapter on cathodes (Chapter 5) and also in the chapter on noise (Chapter 15). The Maxwellian velocity distribution of the emitted electrons is of importance here.

Consider a small spot on the cathode of an electron gun. Electrons are emitted in all directions, and the beamlet of electrons from the spot grows in diameter and the current density decreases as the beamlet is accelerated toward the anode. The current has a Maxwellian distribution with a normal deviation, σ, as shown in Figure 6-23.

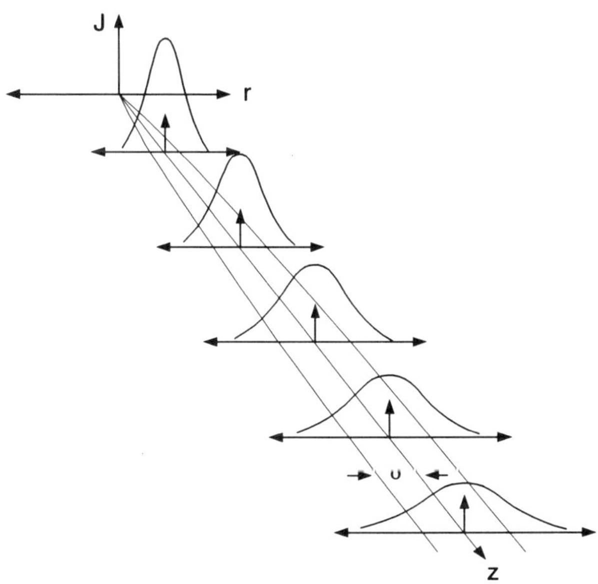

Figure 6-23 Effect of thermal velocities on a beamlet emitted by a spot on a cathode.

Cutler and Hines [12] analyzed the effects of thermal velocities on electron flow in a Pierce gun. They define the normal deviation by

$$\sigma = \frac{\Delta r}{(dr/dt)_c}\left[\frac{kT}{m}\right]^{1/2} \tag{6-69}$$

where Δr is the trajectory deviation of an electron with an initial transverse velocity at the cathode $(dr/dt)_c$. Figure 6-24 contains plots from which σ can be obtained at the anode and at the position of the minimum beam diameter. As an example of the application of these plots, consider the 10 kV, 1 A gun design given previously in the chapter (before correction for spherical aberration). The value of R_c/R_a was 2.25 and so

$$\frac{\sigma}{R_c\left[\dfrac{kT}{2eV_a}\right]^{1/2}} \approx 3.9 \tag{6-70}$$

The value of R_c was 1.039 cm and so, using a cathode temperature of 1000°C (1273 K),

$$\sigma \approx 0.01 \text{ cm} \tag{6-71}$$

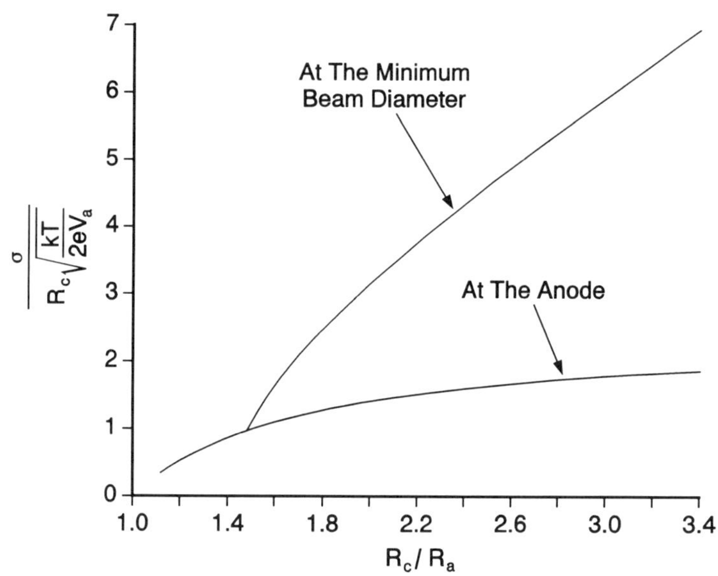

Figure 6-24 Curves from which the normal deviation, σ, can be determined. From C.C. Cutler and M.E. Hines, *Proc. IRE*, March 1955. © 1955 IRE (now IEEE).

Cutler and Hines sum the currents in beamlets from the entire cathode surface to find an expression for the total current distribution in the beam leaving the gun. The result is shown in Figure 6-25. Here, r is the radial position in the beam, b is the nominal beam radius (the radius in the absence of thermal effects) and J_0 is the current density in a beam having the nominal radius.

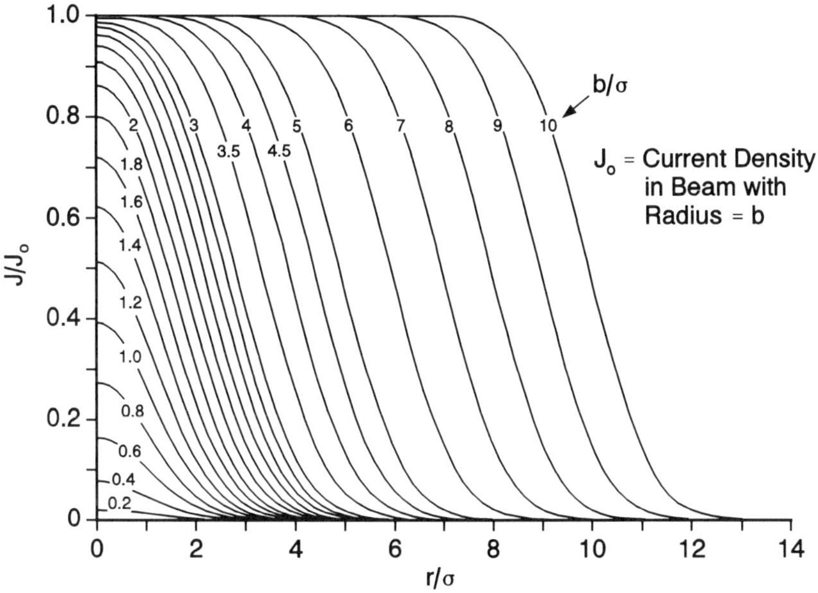

Figure 6-25 Current density *versus* radial position; *b* is the radius without thermal effects. From C.C. Cutler and M.E. Hines, *Proc. IRE*, March 1955. © 1955 IRE (now IEEE).

For the 10 kV, 1A gun design given previously, the nominal beam radius was 1 mm. Thus, for $\sigma = 0.01$, $b/\sigma = 10$, and so the current profile at the beam minimum would be like the $b/\sigma = 10$ curve in Figure 6-25.

Figure 6-26 shows the percent of the total beam current to be found within any given radius of the beam. For the 10 kV, 1 A gun design, about 92% of the beam current is contained within the nominal beam radius of 1 mm. Thus, if the radius of the tunnel at the beam minimum were the same as the nominal beam radius, 8% of the current would be intercepted. To reduce the interception to 0.5%, the tunnel radius would have to be increased to about 1.17 mm.

When very small beams are required, thermal effects are far more severe than they are in the 10 kV, 1 A gun example. An example, from Danielson *et al.* [7] is given in Figure 6-27. Ignoring thermal effects, the minimum beam radius was calculated to be less than 2×10^{-3} in. The experimental radius enclosing 95% of the beam current (r_{95}) was over four times larger (about 8×10^{-3} in). The minimum r_{95} radius calculated with the Cutler-Hines theory agreed will with experiment but the variation of radius with distance was quite different.

Danielson *et al.* corrected the Cutler-Hines theory by including the correction to the Davisson-Calbick equation and by using the space charge field in the thermal beam rather than that in the nominal beam. The minimum radius that they predict also agrees well with experiment, and the variation of radius with distance is closer to experiment

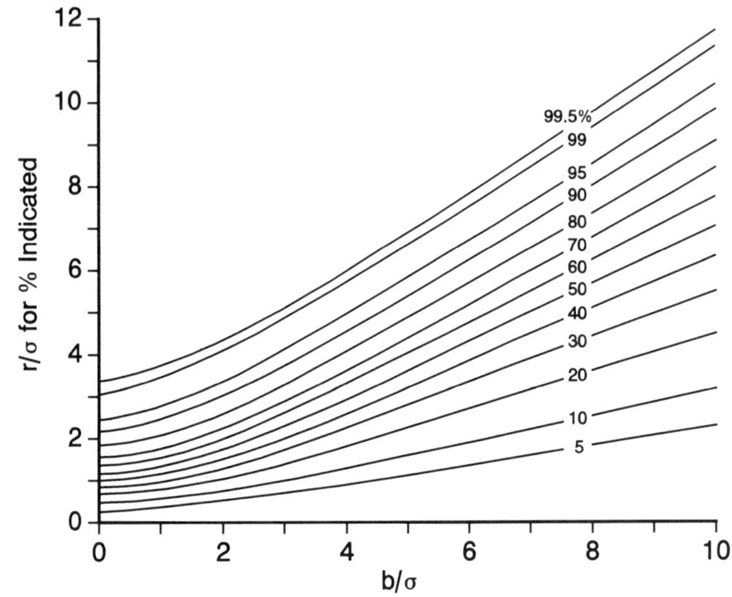

Figure 6-26 Percent of total beam current to be found within any given radius. From Cutler and Hines, *Proc. IRE*, March 1955. © 1955 IRE (now IEEE).

than that predicted by Cutler and Hines.

Before leaving the discussion of thermal velocities, it is important to consider the effect of beam convergence in the electron gun on the transverse velocity distribution of the electrons. Figure 6-28 shows the trajectories of several nonthermal electrons, which are convergent on point F. The trajectory of one thermal electron is shown crossing the trajectories of the nonthermal electrons. Electric fields are assumed to be negligible. It is clear that the angle that the trajectory of the thermal electron makes with the trajectories of the nonthermal electrons increases as the beam converges. Thus, the component of thermal-electron velocity perpendicular to the nonthermal-electron trajectories increases as the beam converges.

Using Figure 6-28, it is easy to calculate the increase in the transverse component of the velocity of the thermal electron. If v is the velocity of the thermal electron, then at the positions where the radius is r_1 and r_2, the transverse velocity components are

$$v_{t1} = v \sin\alpha_1 \qquad \text{and} \qquad v_{t2} = v \sin\alpha_2 \tag{6-72}$$

so

$$\frac{v_{t2}}{v_{t1}} = \frac{\sin\alpha_2}{\sin\alpha_1} \tag{6-73}$$

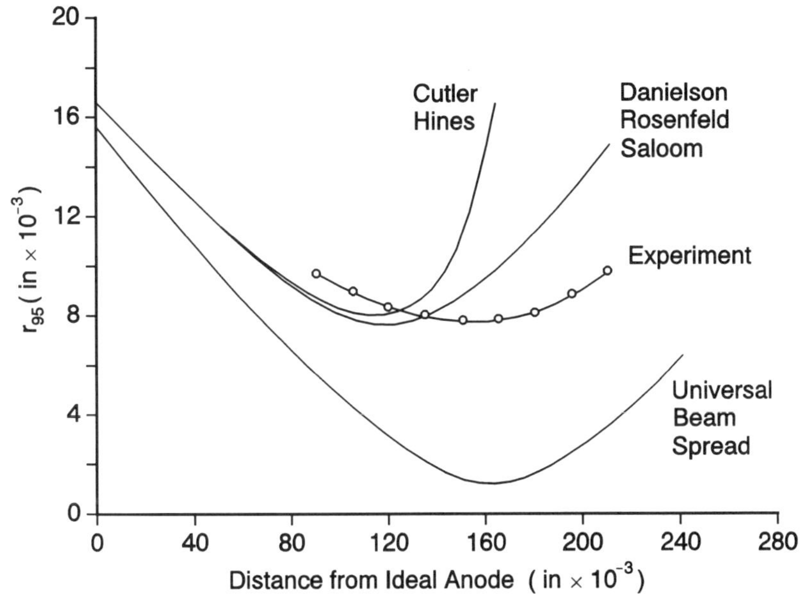

Figure 6-27 Computed and experimental beam profiles for a very small beam dispersed by thermal velocities. Adapted from Danielson *et.al.*, *Bell Sys. Tech. Jour.*, March 1956.

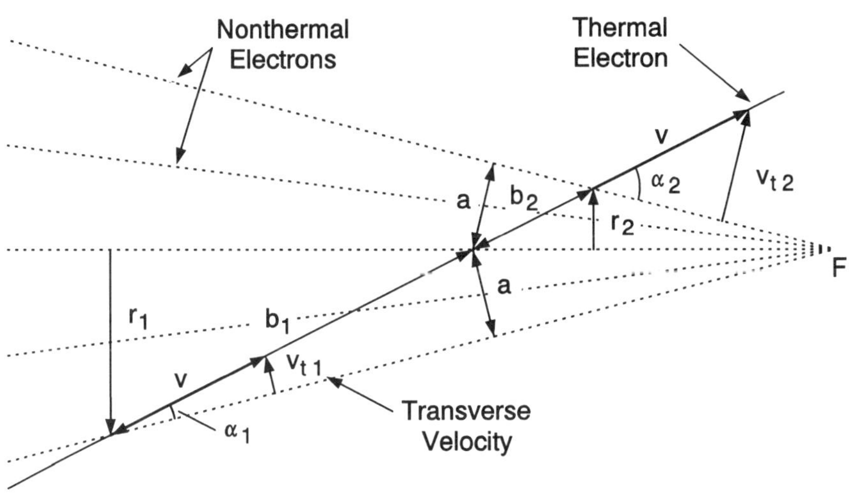

Figure 6-28 Illustration of increase in transverse velocity of thermal electron in convergent flow.

but

$$\frac{\sin \alpha_2}{\sin \alpha_1} = \frac{a / b_2}{a / b_1} = \frac{b_1}{b_2} \tag{6-74}$$

and

$$\frac{r_1}{b_1} = \frac{r_2}{b_2} \quad \text{so} \quad \frac{r_1}{r_2} = \frac{b_1}{b_2} \tag{6-75}$$

Thus,

$$\frac{v_{t2}}{v_{t1}} = \frac{r_1}{r_2} \tag{6-76}$$

That is, the transverse velocity increase is inversely proportional to the radius compression ratio. Thus, to limit the growth of transverse velocities in an electron gun, the amount by which the beam is compressed should be held to a minimum. A cathode with a high emission capability helps in this regard because, for a given beam size and current from a gun, the cathode can be smaller and so the radius compression ratio is smaller than for a cathode with a low emission capability.

6.1.8 Effects of Patchy Emission and Cathode Roughness

The thermal velocity effects on beam shape discussed in the previous section are frequently overshadowed by the relatively large transverse electron velocities produced by patchy emission and cathode roughness. The patchy characteristic of the emission from a cathode was discussed in Chapter 5. The surface of a dispenser cathode is made up of grains of tungsten that have various crystalline orientations and that may have various work functions. Because the thickness of the space charge cloud at the cathode surface is comparable to the dimensions of the grains and the pores between grains, each grain emits independently of the others and the emission is said to be patchy. For those grains that are temperature limited, emission varies with grain work function and most of the emission comes from the grains with low work functions.

There are at least two ways by which the patchy nature of the emission produces transverse velocity components to the electrons leaving the cathode surface. These are indicated in Figure 6-29(a) and (b). In (a) we note that the space charge forces within the beamlet emitted from a patch causes spreading of the beamlet. In Figure 6-29(b) we show that the equipotential surfaces near the cathode surface are rippled because of the varying space charge density near the patches. This produces spreading of the beamlets. As was pointed out in Chapter 5, coated dispenser cathodes have a narrower distribution of work functions than uncoated cathodes. As a result, patchy emission is less pronounced for coated than for uncoated cathodes. If the work function distribution of a cathode increases with time, then the effects of patchy emission become worse.

This may cause the beam diameter to increase so that the current intercepted by the circuit increases.

If the surface of the cathode is rough, as indicated by the protuberance in Figure 6-29(c), then local transverse electric fields are present, which produce transverse electron velocities. A problem sometimes encountered with coated dispenser cathodes is that the coating may develop microscopic cracks and peel away from the tungsten matrix. Thus, a cathode that was originally very smooth can become rougher as it is operated. If this happens, transverse velocities increase, beam size increases and interception current may increase.

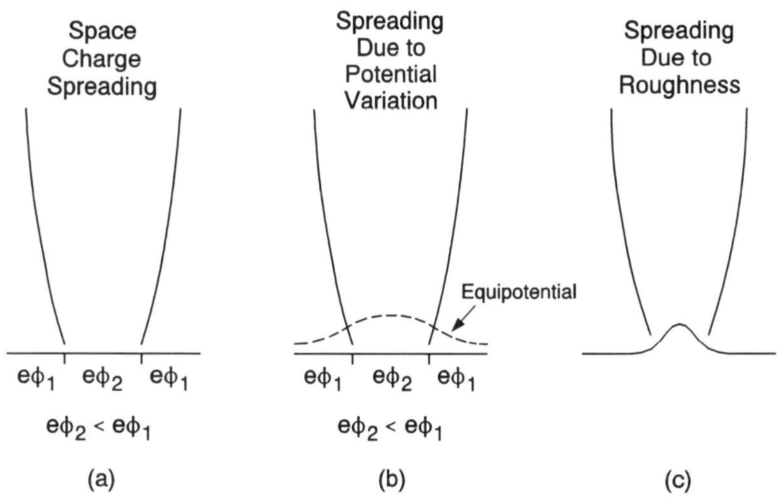

Figure 6-29 Effects on electron trajectories of patchy emission and cathode roughness.

6.2 BEAM CONTROL TECHNIQUES

In all applications of microwave tubes, beam current must be controlled. Usually, the beam current must be switched on and off. Sometimes, continuous control is required. The "brute force" way to control beam current is with the cathode-to-anode voltage as indicated in Figure 6-30(a). A second anode (called a *modulating anode*) may be placed between the cathode and final anode for current control as shown in Figure 6-30(b). In some guns (usually low perveance guns) current can be controlled with the focus electrode as shown in Figure 6-30(c). Finally, in the majority of applications, one or more grids is placed near the cathode surface, as indicated in Figure 6-30(d), for current control. In this section these control techniques are discussed. The primary emphasis is on grid control because of its wide use in TWTs.

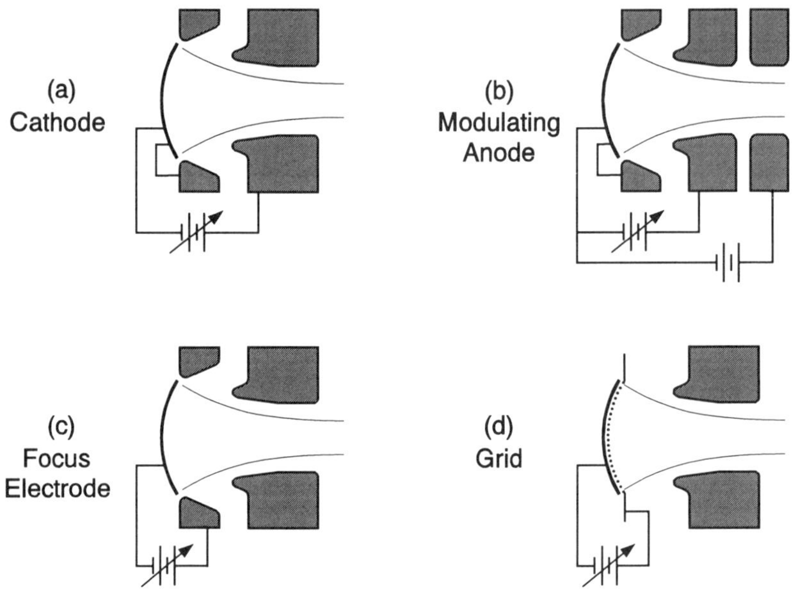

Figure 6-30 Techniques for current control in an electron gun.

6.2.1 Cathode Pulsing

Conceptually, cathode pulsing is the simplest technique for switching a microwave tube on and off. In virtually all applications, the anode of the gun and the body of the tube are at ground potential. To turn the beam on, the cathode is pulsed negative. To turn the beam off, the cathode is pulsed back to ground potential. The primary problem with cathode pulsing is that the power supply (the power modulator) must switch *all* of the beam power so the power modulator must be large, heavy and, usually, inefficient.

6.2.2 Control Focus Electrodes

If a sufficiently large negative voltage is applied to the control electrode of a Pierce gun, the beam current can be reduced to zero. The voltage required to cut off the beam current can be reduced by locating an additional electrode in the center of the cathode as shown in Figure 6-31.

The control focus electrodes can be used to turn on the beam by setting the voltage at or very close to the cathode potential. The beam is turned off by setting the voltage significantly below cathode potential. This system is not useful for the intermediate control of current because of the severe defocusing that occurs in the electron gun.

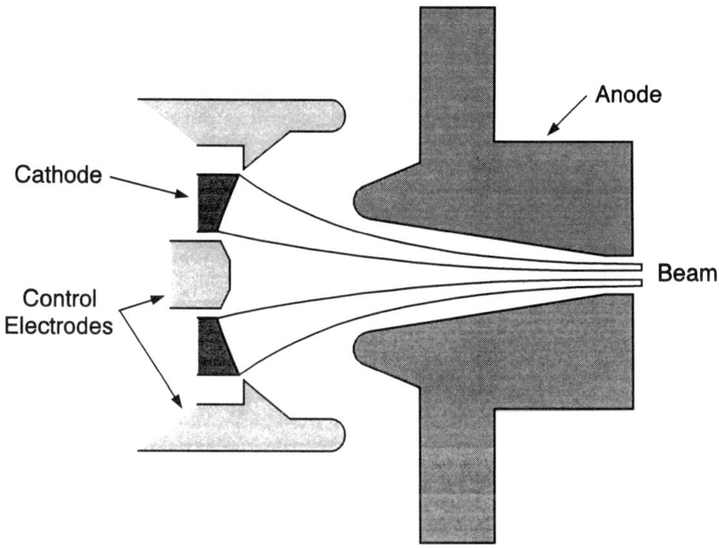

Figure 6-31 Control focus electrode gun. From Armand Staprans *et al., Proc. IEEE,* March 1973. © 1973 IEEE.

6.2.3 Modulating Anode

In the modulating anode gun, the anode of the electron gun (the modulating anode) is separated from the body of the tube, a part of which is a second (final) anode. The configuration of a modulating anode gun is shown in Figure 6-32.

Although the voltage range over which the modulating anode must be varied in order to vary beam current is large, the power required to drive the modulating anode is small because the current intercepted by this anode is very small. At intermediate voltage levels, perveance may be controlled with the modulating anode without severely affecting beam focusing.

6.2.4 Grids

Under normal operating conditions, the guns with modulating anodes and those with control electrodes operate as Pierce-designed guns. When grids are used, however, electron trajectories near the cathode are affected, and so beam focusing is degraded and beam dynamics are affected. Still, grid control is very widely used because of the small modulator size that it makes possible.

Several types of grid structures are illustrated in Figure 6-33. The simplest structure has a single grid element. A computer simulation of electron flow in a Pierce

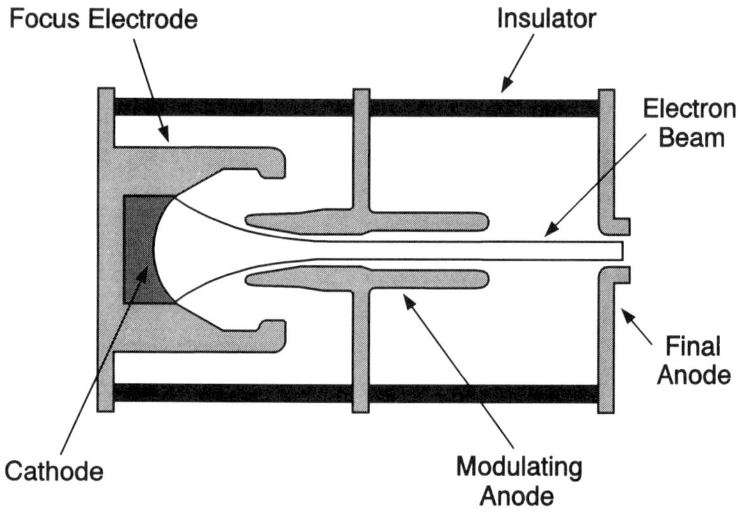

Figure 6-32 Modulating anode gun. From Armand Staprans *et al., Proc. IEEE,* March 1973. © 1973 IEEE.

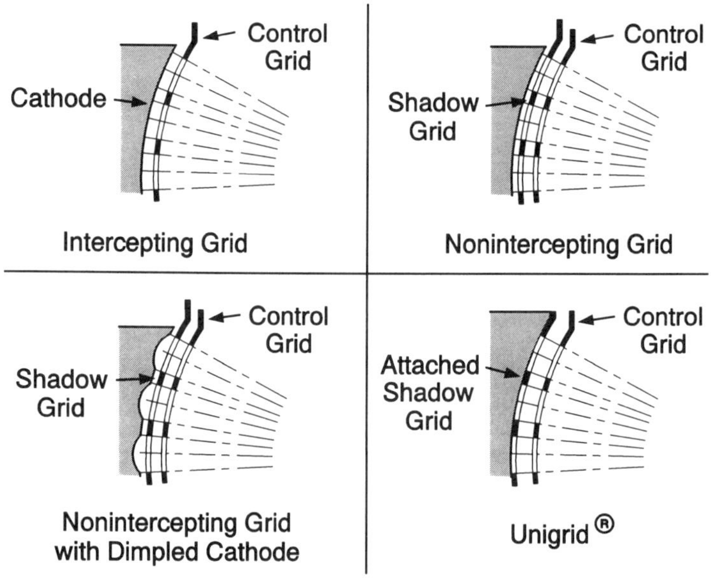

Figure 6-33 Various grid structures used in Pierce guns for current control. From *Microwave Tube Manual* by Varian, Air Force Publication No. T.O. 00-25-251, October 1979.

gun containing a single grid is shown in Figure 6-34. Thermal velocities are ignored here, and electron flow in seen to be very laminar. The problem with a gun containing a single grid is that the cathode current intercepted by the grid can be very high. In some cases as much as 15% or more of the cathode current is intercepted. The results of the interception are excessive heating of the grid at high duty cycles and the need for a relatively powerful grid modulator.

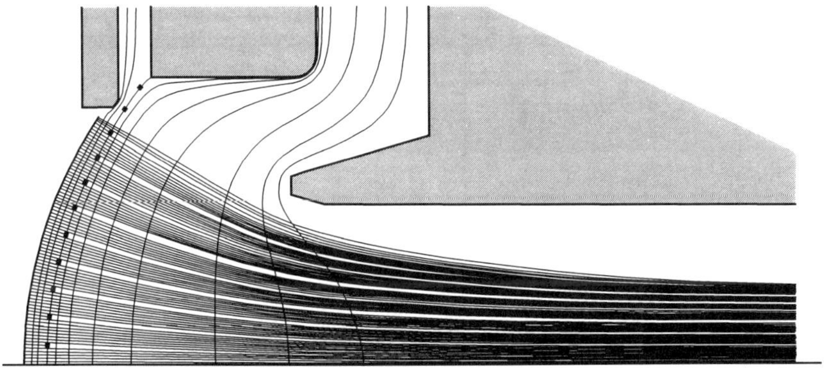

Figure 6-34 Computer simulation of electron flow in a Pierce gun with a single grid. Adapted from R. True, *Technical Digest*, 1985 IEDM. © 1985 IEEE.

Most grid structures contain a shadow grid between the control grid and the cathode surface. The shadow grid operates at or near cathode potential so that heating by intercepted current is small. The elements of the shadow grid are aligned with the elements of the control grid so that the control grid is shielded and intercepts very little current. Several shadow-grid structures are shown in Figure 6-33.

Grid control of current in electron tubes has been used since the invention of the vacuum tube. In 1960, Wolkstein discussed methods of grid placement and associated design considerations for convergent-flow Pierce guns [13]. The basic concept is to form the grid so that its shape is identical to that of an equipotential contour near the cathode surface (usually spherical). Then, if the grid is at the potential of that contour and intercepts negligible current, electron flow will not be disturbed. To interrupt current flow, the grid voltage is switched negative with respect to the cathode.

A simple relationship for grid voltage as a function of grid position can be obtained by approximating the grid-cathode region as a parallel-plane diode. This approximation is excellent for small grid-cathode spacings. The current flow from the cathode to grid is then given by the Child-Langmuir equation for parallel-plane diodes (Equations 5-47 through 5-49). The cathode area, A, to be used is

$$A = 4\pi R_c^2 \left[\frac{1 - \cos\theta}{2} \right] \tag{6-77}$$

and the electrode spacing, d (grid-cathode spacing), is d_{gc} as shown in Figure 6-35.

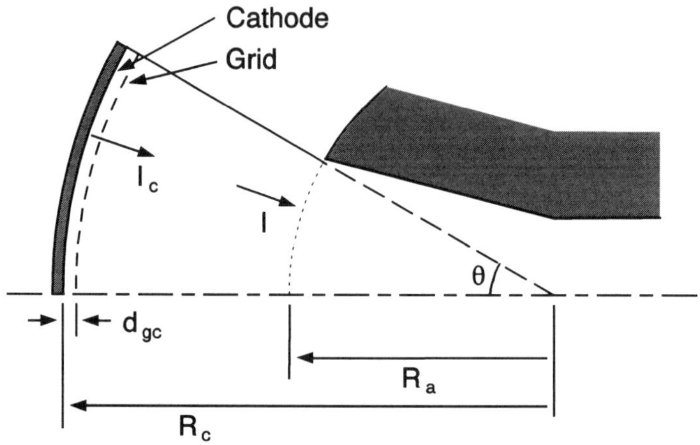

Figure 6-35 Gun parameters for determining grid-cathode spacing.

With negligible grid interception, the cathode current, I_c, must equal the current flow, I, in the gun, that is,

$$I_c = I \tag{6-78}$$

Now, substituting for I_c and for I (from Equation 6-26),

$$\frac{2.335\times 10^{-6}\, V_g^{3/2}}{d_{gc}^2}\, 4\pi R_c^2 \left[\frac{1-\cos\theta}{2}\right] = \frac{29.34\times 10^{-6}\, V_a^{3/2}}{(-\alpha_a)^2}\left[\frac{1-\cos\theta}{2}\right] \tag{6-79}$$

or

$$\frac{V_a}{V_g} = \left[\frac{(-\alpha_a)}{d_{gc}/R_c}\right]^{4/3} \tag{6-80}$$

where V_g is the grid-to-cathode voltage, V_a is the anode-to-cathode voltage and $(-\alpha_a)$ is Langmuir and Blodgett's $(-\alpha)$ evaluated at the anode. This equation is plotted in Figure 6-36. As an example of its use, consider the 10 kV, 1 A gun design given earlier in this chapter. The value of R_c/R_a was 2.25 so, for a grid voltage of 100 V,

$$\frac{d_{gc}}{R_c} \approx .032 \tag{6-81}$$

Since $R_c = 1.039$ cm, then

$$d_{gc} \approx 0.033 \text{ cm} \approx 0.013 \text{ in} \tag{6-82}$$

Thus, the grid must be very close to the cathode if the grid voltage is to be low.

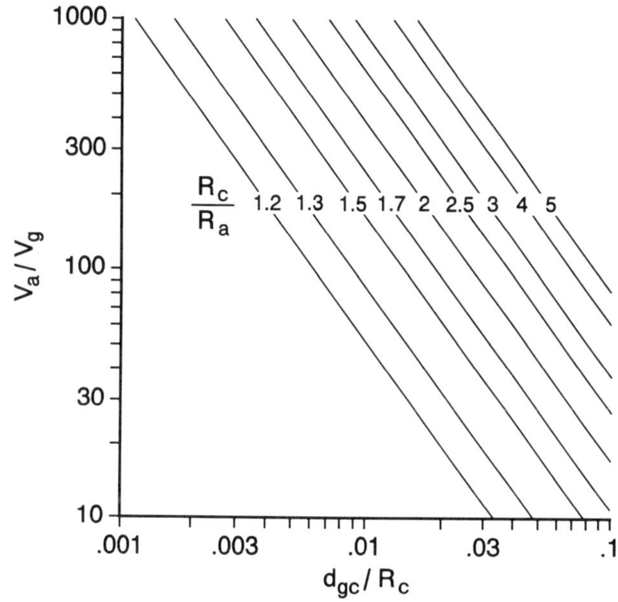

Figure 6-36 Relation of grid and anode voltages to grid, cathode and anode positions. Adapted from H.J. Wolkstein, *RCA Review*, September 1960.

Next, we will examine the configuration of the grid elements. Examples of control grids for Pierce guns are shown in Figure 6-37. The grid configuration on the left is often used in guns with intercepting grids and that on the right is used in shadow gridded guns. To stop (cutoff) electron flow from the cathode, as shown in Figure 6-38, a negative voltage (with respect to the cathode) is applied to the grid. In Figure 6-38, the dashed curves show the potential variation between the cathode and anode for the 10 kV, 1 A electron gun with the grid 100 V above cathode potential. The solid curves show the potential variation when the grid is sufficiently negative with respect to the cathode to cut off electron flow. Note in the expanded drawing of the grid region that two solid curves are shown. One is the potential variation through a grid wire and the other is the potential variation midway between grid wires. The net effect of these two potential variations must be that the potential near the cathode is zero or negative with respect to the cathode if current flow is to be prevented.

To avoid electrical breakdown between the grid and the cathode, the grid voltage used for cutoff must not be excessive. The cutoff grid voltage is controlled, largely, by the spacing between grid wires. Clearly, for large spacings, the cutoff grid voltage must be large because of the strong tendency for the anode field to penetrate the regions between grid wires. Thus, for relatively small cutoff grid voltages (and low probability of grid-cathode breakdown) the grid wire spacing must be small.

In addition to small grid wire spacing, the area of the grid wires (and, thus, the

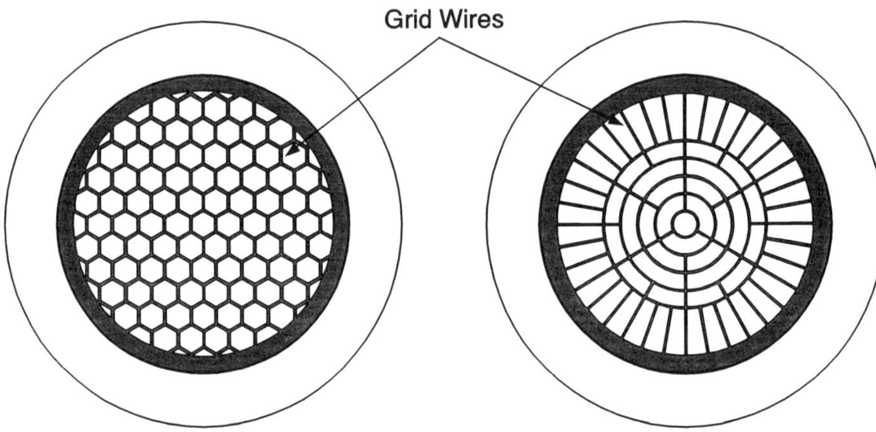

Figure 6-37 Examples of control grids for Pierce guns.

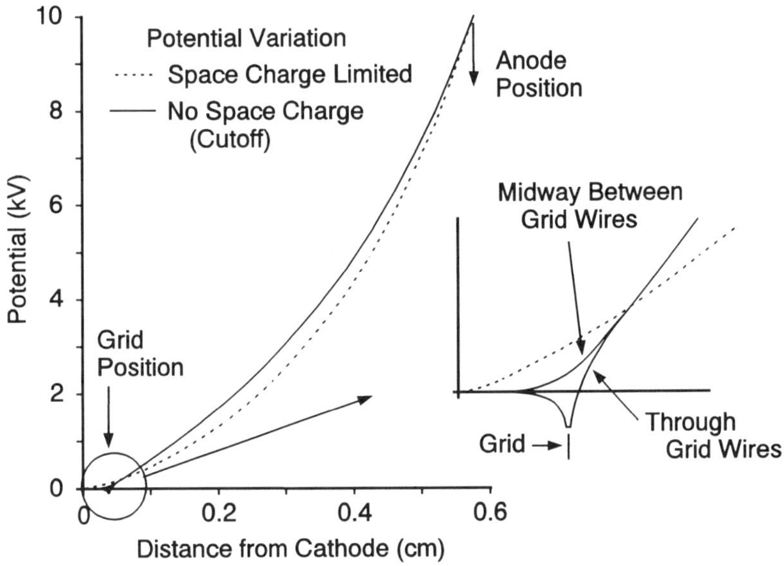

Figure 6-38 Potential variations in a gridded 10 kV Pierce gun.

width of the grid wires) must be small compared with the cathode area if the current intercepted by the grid is to be small. The ratio of the grid area to the cathode area is the screening fraction and this is the fraction of the cathode current intercepted by the grid. Thus, for small interception, the screening fraction must be small.

The spacing between grid wires and grid wire width can be estimated using the procedure given for planar triodes by Spangenberg [14]. First, the voltage distribution in a cylindrical triode (cathode in center surrounded by a grid and an anode) is determined. Then, this distribution is transformed to a planar configuration. The grid voltage required for cutoff and the cutoff amplification factor, μ_c, which is the ratio of anode voltage to the cutoff grid voltage, is determined as functions of grid dimensions. The result is given in Figure 6-39.

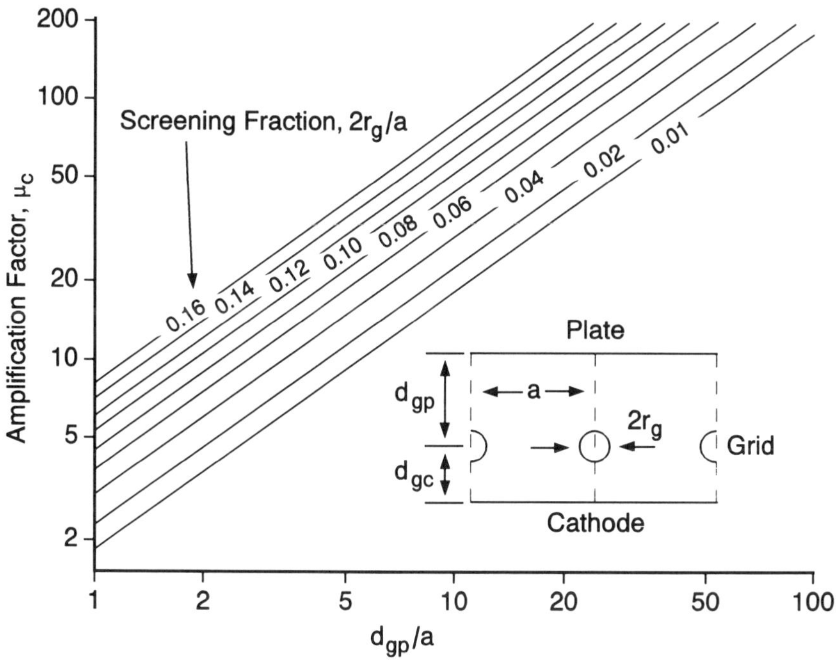

Figure 6-39 Amplification factor for a planar triode. Adapted from *Vacuum Tubes* by Karl R. Spangenberg, copyright 1948 by McGraw Hill.

As an example of the use of Figure 6-39, assume that the grid voltage for cutoff of the 10 kV, 1 A gun should be -100 V with respect to the cathode. Thus, $\mu_c = 100$. For a screening fraction of 0.10,

$$\frac{d_{gp}}{a} \approx 20 \tag{6-83}$$

where d_{gp} would be the grid-to-plate distance in a planar triode. To apply this result to the spherical configuration of a Pierce gun, we must determine the anode position in a planar triode that would produce the same electric field near the grid as that in the Pierce gun. For our example, that field is 8545 V/cm. (This can be determined from the slope of the potential variation near the grid). Thus, for 10 kV on the plate,

$$d_{gp} = \frac{10{,}000}{8{,}545} = 1.17 \text{ cm} = 0.47 \text{ in} \tag{6-84}$$

and so the spacing between grid wires, a, is

$$a = 0.059 \text{ cm} = 0.023 \text{ in} \tag{6-85}$$

Since the screening fraction is 0.10, the width of the grid wires (long parallel wires are assumed) is

$$2\,r_g = 0.0059 \text{ cm} = 0.0023 \text{ in} \tag{6-86}$$

In summary, for an anode voltage of 10 kV, control grid voltages (with respect to the cathode) of 100 V and -100 V, d_{gc} = 0.013 in, grid wire spacing = 0.023 in and grid wire width = 0.0023 in. These results are in reasonably good agreement with those reported by True [15] for the control grid in a shadow gridded gun similar to that on the right in Figure 6-37. For a 9.6 kV gun with control grid voltages (with respect to the cathode) of 102 V and -86 V, d_{gc} = 0.015 in, effective grid wire spacing = 0.023 in and grid wire width = 0.002 in.

At this point, we can start to appreciate the care that must be taken in designing and fabricating a gridded Pierce gun. Not only are dimensions small but also, proper grid-to-cathode spacing must be maintained as the cathode is heated to a temperature of 1000°C or higher. The problem of maintaining spacings and alignments is even more critical in shadow gridded guns, which are discussed in the following paragraphs.

The main problem with a single grid is that the cathode current intercepted by it can be very high. To alleviate this problem, many gridded Pierce guns use a shadow grid between the control grid and the cathode. The shadow grid was invented by Drees [16], who described its configuration and operation as follows: "In accordance with the present invention, the novel grid construction comprises a first control grid and a second grid spaced from and aligned with the first grid. The second grid is disposed between the control grid and the cathode. In this arrangement, with the second grid preferably being held at cathode potential, the second grid acts as a (shadow) grid in that it shadows electrons from the control grid so that they are not intercepted by the control grid."

The effects on electron trajectories of the shadow grid, its location and the voltage applied to it are illustrated in Figure 6-40 [17]. Each of these plots shows computer generated electron trajectories from a cathode for the region from midway between grid wires to the grid wires. In an ideal shadow gridded gun, both the shadow grid and the control grid would be at positions and voltages that coincide with the equipotential surfaces that exist in the absence of the grids. Thus, except for interception and the finite sizes of the grid wires, electron flow would not be disturbed. The plot in Figure 6-40(d) shows trajectories with grid voltages adjusted to approximate the ideal shadow gridded gun voltages.

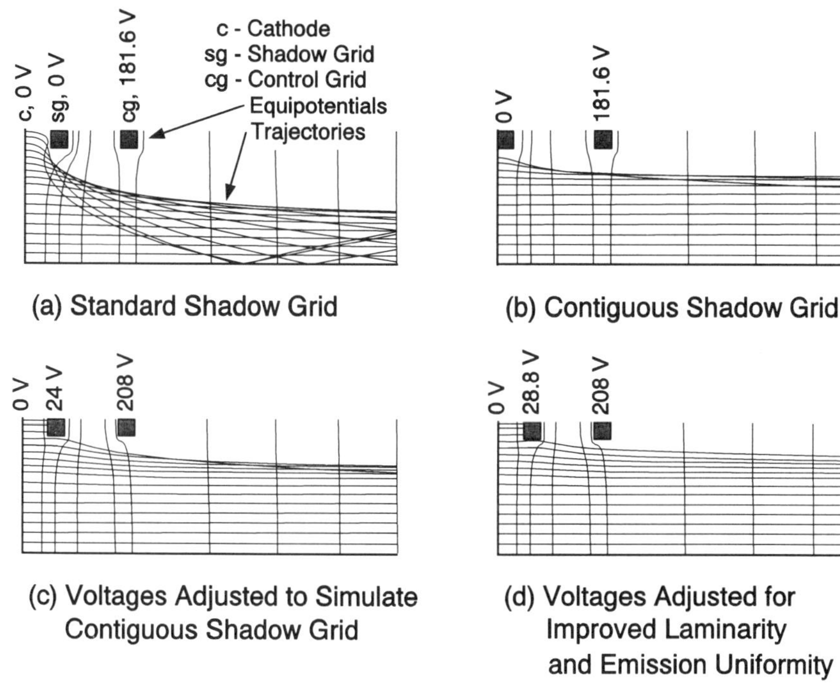

Figure 6-40 Electron trajectories for shadow-gridded structures (simplified planar model). Adapted from R. True, *Technical Digest*, 1979 IEDM. © 1979 IEEE.

For the standard shadow gridded gun (Figure 6-40(a)), the shadow grid is at cathode potential but is placed a short distance (usually a few thousandths of an inch) from the cathode surface. Thus, a lens action occurs near each wire of the shadow grid. This is evident from the bending of the equipotential surfaces near the shadow grid. As a result of the lens action, electron trajectories are severely distorted.

When the shadow grid is attached to or placed on the cathode surface, as in Figure 6-40(b), the distortion of the trajectories is significantly reduced. Unfortunately, of course, the shadow grid must operate at cathode temperature, which can cause difficulties with grid emission and grid alignment.

An advantage of placing the shadow grid at cathode potential, as in Figure 6-40(a) and (b), is that only one grid power supply is required. If a separate power supply for the shadow grid can be used, then, as shown in (c) and (d), the beam laminarity of the contiguous shadow grid can be equaled as in (c) or improved upon as in (d).

Shown in Figure 6-41 are computer generated plots of electron trajectories in a shadow gridded gun with the standard shadow grid [18]. The beam radius containing 95% of the current, r_{95}, is indicated by dots. If the electron trajectories outside the r_{95} radius are followed back to the cathode, it is seen that they are the trajectories of electrons originating close to the grid wires.

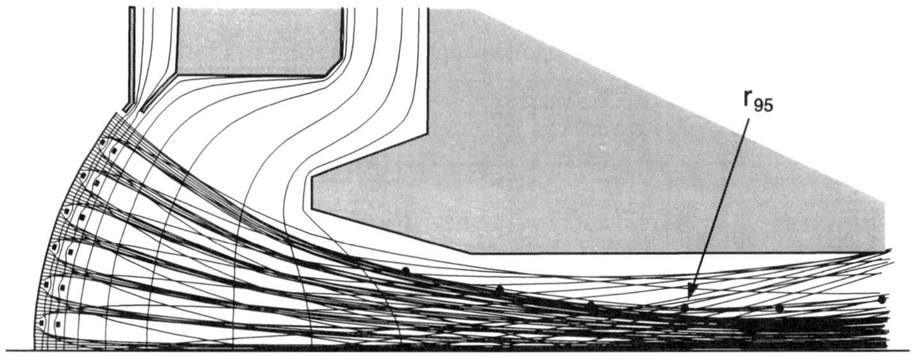

Figure 6-41 Electron trajectories in a shadow gridded gun. Adapted from R. True, *IEEE Trans. on Electron Devices*, March 1984. © 1984 IEEE.

If we carefully examine Figure 6-41 we note that the grid-to-cathode spacing is somewhat smaller near the edge of the cathode than near the center. This was also evident in Figure 6-34. The reason for the reduced spacing at the cathode edge is, of course, because the equipotentials were spaced that way. As a result of the nonuniform spacing, the electric field and the emission density at the edge of the cathode are somewhat larger than at the center.

To cutoff current flow, the control grid is pulsed negative with respect to the cathode. The resulting voltage distribution in the region of the grids is shown in Figure 6-42. Here, the grid voltage was -160 V and the inner most potential contour from grid wire to grid wire was at -10 V, so the grid voltage was slightly below that required for cutoff.

We have just noted that the electric field is somewhat higher at the edge of the cathode than at the center. Also, when grids like that shown on the right in Figure 6-37 are used, the spacing between grid wires increases with increasing radius. Thus, for negative control grid voltages, the anode field has a greater tendency to penetrate the regions between grid wires near the edge of the cathode than near the center. This means that, as control grid voltage is made negative with respect to the cathode, the last part of the cathode to be cut off is that near the outer edge. To compensate for this effect, True and Laycock tapered the widths of the grid wires so that they increased in width near the cathode edge [19]. As a result, cut-off voltage was reduced by a factor of almost two.

As the control grid voltage is varied in a shadow-gridded electron gun, cathode current varies as shown in Figure 6-43. In addition, the beam current intercepted (the body current), mostly by the RF structure, varies. This is because, as control grid voltage and gun current are changed, focusing conditions in the gun change. As a result, the entrance conditions of the beam from the gun into the focusing magnetic field vary. This causes beam shape and body current to vary.

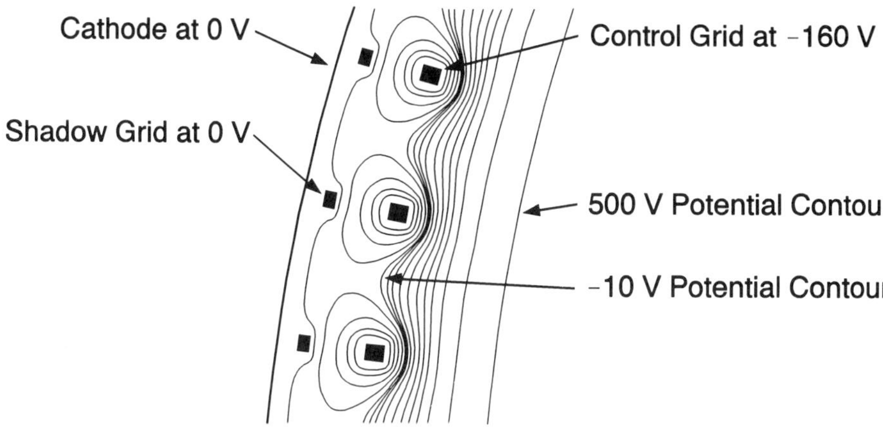

Figure 6-42 Voltage distribution in region of grids at a control grid voltage slightly below cutoff. Adapted from R. True, *Technical Digest*, 1989 IEDM. © 1989 IEEE.

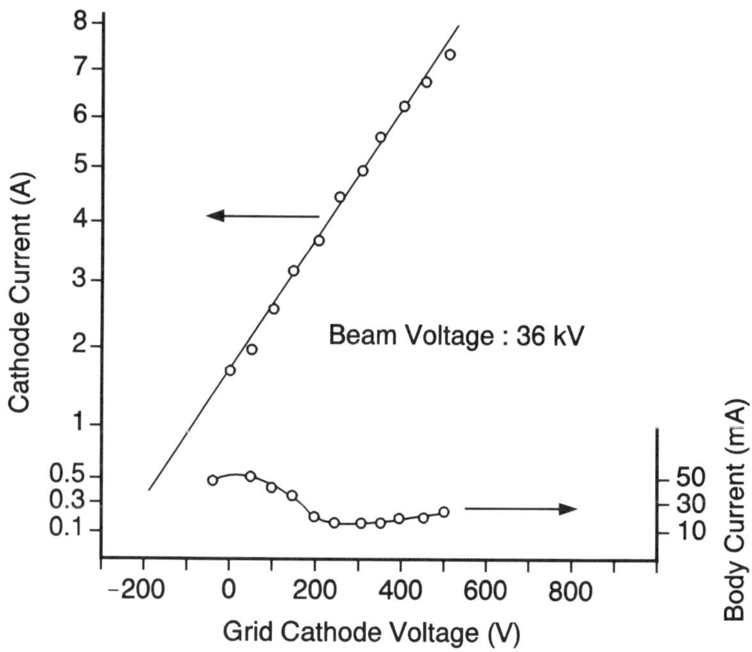

Figure 6-43 Shadow gridded electron gun characteristics. From Armand Staprans *et al.*, *Proc. IEEE*, March 1973. © 1973 IEEE.

When the control grid is pulsed negative to cutoff beam current, complete cutoff may not occur if the grid is emitting electrons. The resulting small beam current through the TWT can result in unwanted signal generation and interpulse noise.

Grids are usually fabricated from molybdenum, which has a high work function. Thus, grid emission from bare molybdenum is insignificant even at high grid temperatures. However, when barium and barium oxide vapors from the cathode collect on the molybdenum grid, the work function can decrease to the point where grid emission occurs.

To prevent grid emission, it is necessary to coat the molybdenum with a material that prevents the work function from decreasing in the presence of evaporation products from the cathode. Zirconium and hafnium are metals that serve this purpose. Haas and Shih [20] reported that zirconium acts as a getter by reducing barium oxide molecules and absorbing the oxygen. The freed barium then immediately vaporizes from the zirconium surface so that the work function is not reduced.

Unfortunately, as the oxygen content of the zirconium increases, its effectiveness decreases and so eventually excessive grid emission occurs. Green and Goeser [21] have reported that a thin layer of tungsten deposited on the molybdenum prior to zirconium deposition serves to extend the life of the zirconium film. In fact, the zirconium/tungsten/molybdenum combination is found to outperform pure zirconium grids. The role of the tungsten layer is not known but its use results in grids with acceptably low emission for lifetimes in excess of 10,000 hours.

The fabrication of shadow grid structures is difficult, especially in high-frequency tubes, where dimensions are small. Alignment of the elements of the control and shadow grids is critical. Alignment must be maintained despite the fact that shadow grid operates at a higher temperature than the control grid because it is closer to or perhaps in contact with the cathode.

An example of the grid motions that occur as a cathode warms up is given in Figure 6-44. These results were computed using the thermal expansion characteristics of the grids and their support structures [22]. Note that, at one point, the shadow grid moved to within about 0.001 in of the cathode surface. This small spacing exceeded the total tolerance on the position of the grid. Thus, the shadow grid could touch the cathode, which, in turn, could lead to deformation of the shadow grid [23]. The author of [22] states that this phenomena had also been noticed experimentally and corrective action was undertaken.

In an effort to overcome the problems of maintaining proper grid spacings in shadow gridded guns, Miram *et al.* developed the bonded grid structure shown in Figure 6-45 [24–26]. The grid elements were pyrolytic graphite, and these were deposited on a thin boron nitride spacer. To suppress grid emission, the control grid was overcoated with a thin boron nitride film. Technical feasibility was demonstrated in a 2,000 hr life test in which grid leakage and emission were shown to be very small and stable. The bonded grid had not yet found application because boron nitride having the required characteristics is not available and the cost of establishing a facility to produce it is, at present, prohibitive.

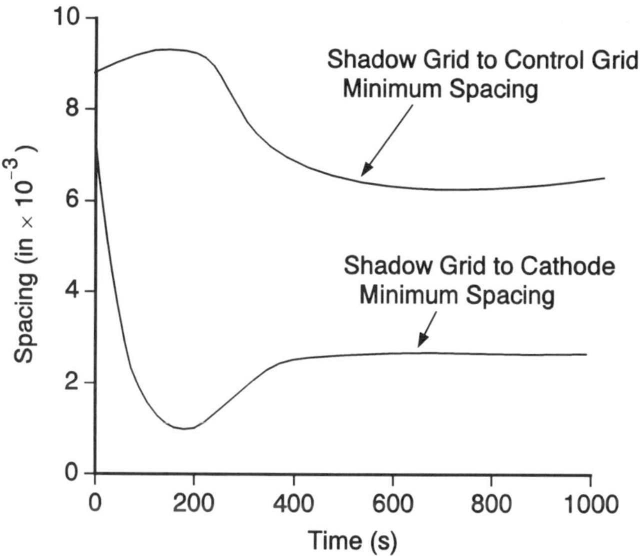

Figure 6-44 Relative spacing between shadow grid, control grid and cathode during gun warm-up. From R. Sawicki, *Technical Digest*, 1978 IEDM. © 1978 IEEE.

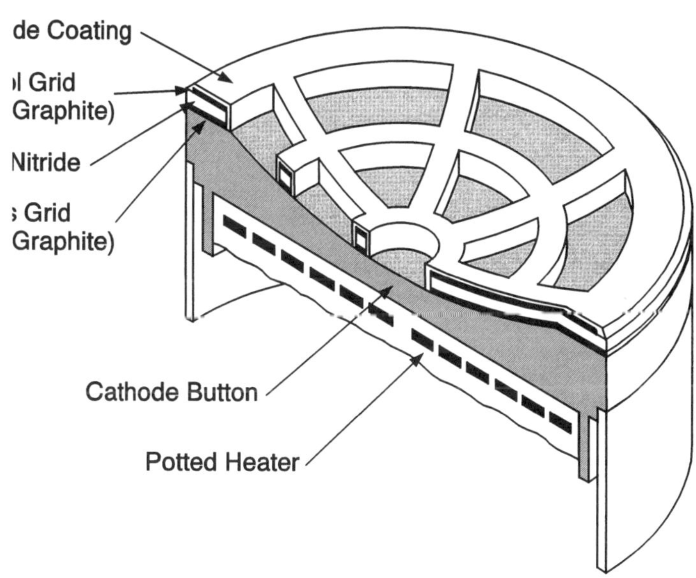

Figure 6-45 Bonded grid and cathode assembly. From T.J. Grant *et al.*, *Technical Digest*, 1983 IEDM. © 1983 IEEE.

Very sophisticated grid structures have been developed for some microwave tube applications. In the example shown in Figure 6-46, two control grids are used along with a shadow grid. This particular grid structure was developed to make possible a tube that could operate efficiently at two different power levels. This is referred to as *dual-mode operation*. The grid closest to the anode controls emission from the central part of the cathode. The next grid controls emission from the outer portion of the cathode. Thus, beam current and diameter can be switched to change the beam power level while maintaining constant beam voltage and proper focusing conditions.

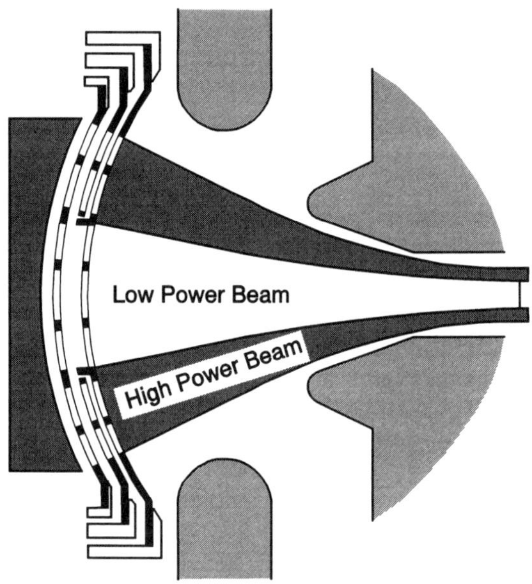

Figure 6-46 Dual-mode gun. From *Microwave Tube Manual* by Varian, Air Force Publication No. T.O. 00-25-251, October 1979.

In [27], True describes a dual-mode grid structure using two grids rather than three. The grid farthest from the cathode is a control grid like that shown on the right in Figure 6-37. The grid next to the cathode is similar to a shadow grid except that the outer openings contain closely spaced grid wires. When this grid is pulsed positive with respect to the cathode (+36 V in [27]) current flows from the entire cathode surface. When the grid is negative (-36 V in [27]) the outer portion of the cathode is cut off and current flows only from the center portion.

6.2.5 Summary of Beam Control Electrode Characteristics

Table 6-1 contains a summary of the more important characteristics of beam control electrodes. Two amplification factors are given. The total amplification factor, μ, is the ratio of beam to control electrode voltage required for beam control. The cutoff amplification factor is the ratio of beam voltage to the negative control electrode voltage (with respect to the cathode) required to reduce beam current to zero.

TABLE 6-1
Typical Characteristics of Control Electrodes
From Armand Staprans *et al.*, *Proc. IEEE,* March 1973. © 1973 IEEE.

Type	μ	μ_c	*Capacitance*	*Grid Interception*	*Focusing at Low Voltage*
Modulating Anode	1 to 3		50 pF	0	Good
Control Focus Electrode	2 to 10	2 to 10	100 pF	0	Poor
Intercepting Grid	50	100	50 pF	15%	Fair
Shadow Grid	30	300	50 pF	0.1%	Fair

As would be expected, the gridded guns have by far the highest amplification factors and the modulating anode gun has the lowest amplification factor. The capacitances of all of the control electrode structures to the surrounding electrodes are essentially the same. Capacitance is important, of course, in the design of the modulator that must be used to drive the control electrode structure. Finally, as would be expected, the interception of the intercepting grid structure is much higher than the other structures. Only the modulating anode gun has good focusing because the design of that gun is essentially the design of an ideal Pierce gun.

REFERENCES

[1] J.R. Pierce, *Theory and Design of Electron Guns*, 2nd ed. New York: Van Nostrand, 1954.

[2] J. Rodney M. Vaughan, "Synthesis of the Pierce gun," *IEEE Trans. Electron Devices*, vol. ED-28, January 1981, pp. 37–41.

[3] I.L. Langmuir and K. Blodgett, "Currents limited by space charge between concentric spheres," *Phys. Rev.*, vol. 24, July 1924, p. 53.

[4] R. Helm, K. Spangenberg and L.M. Field, "Cathode-design procedure for electron-beam tubes," *Elec. Comm.*, vol. 24, March 1947, pp. 101–107.

[5] G.R. Brewer, "Formation of high-density electron beams," *Jour. Appl. Phys.*, vol. 28, no. 1, January 1957, pp. 7-15.

[6] K. Amboss, "The effect of the anode aperture in conical flow Pierce guns," *Jour. Electronics and Control*, vol. 13, 1962, pp. 545–572.

[7] W.E. Danielson, J.L. Rosenfeld and J.A. Saloom, "A detailed analysis of beam formation with electron guns of the Pierce type," *Bell Sys. Tech. Jour.*, March 1956, pp. 375–420.

[8] R.G. Wilson and G.R. Brewer, *Ion Beams With Application to Ion Implantation*, New York: John Wiley and Sons, 1973.

[9] L.A. Harris, *Jour. Electronics*, vol. 8, 1960, p. 241.

[10] C.R. Buckey, Hughes Research Laboratories, unpublished data.

[11] R.D. Frost, O.T. Purl and H.R. Johnson, "Electron guns for forming solid beams of high perveance and high convergence," *Proc. IRE*, August 1962, pp. 1800–1807.

[12] C.C. Cutler and M.E. Hines, "Thermal velocity effects in electron guns," *Proc. IRE*, March 1955, pp. 307–315.

[13] H.J. Wolkstein, "Design considerations for grid-controlled electron guns for pulsed traveling-wave tubes," *RCA Review*, September 1960, pp. 389–413.

[14] Karl R. Spangenberg, *Vacuum Tubes*, New York: McGraw-Hill, 1948.

[15] R. True, "Calculation and design of grids in Pierce guns," *Technical Digest*, 1989 International Electron Devices Meeting, pp. 215–218.

[16] J.M. Drees, "Non-intercepting grid structure for an electron tube," U.S. Patent 3,484,645, filed March 6, 1967, issued December 16, 1969.

[17] R. True, "An ultra-laminar tetrode gun for high duty cycle applications," *Technical Digest*, 1979 International Electron Devices Meeting, pp. 286–289.

[18] R. True, "A theory for coupling gridded gun design with PPM focusing," *IEEE Trans. on Electron Devices*, vol. ED-31, no. 3, March 1984, pp. 353–362.

[19] Richard B. True and Donald E. Laycock, "Gridded electron gun," U.S. Patent 4,737,680, Filed April 10, 1986, Issued April 12, 1988.

[20] G.A. Haas and A. Shih, "Study of high work function materials needed for close-spaced grid applications," *Applications of Surface Science*, vol. 4, 1980, pp. 104–126.

[21] M. Green and G. Goeser, "Deposition techniques for suppressed emission," Final Report, Contract No. N00014-84-C-2270, October 1988.

[22] R. Sawicki, "Analytical determination of the thermal/mechanical performance of traveling wave tube electron guns," *Technical Digest*, 1978 International Electron Devices Meeting, pp. 160–163.

[23] Hsien-Yang Yeh, "The role of structural stability in the reliability of microwave vacuum devices," *IEEE Trans. on Electron Devices*, vol. ED-34, no. 2, February 1987, pp. 494–497.

[24] G.V. Miram and E.L. Lien, "Convergent electron gun with bonded nonintercepting control grid," *Technical Digest*, 1978 International Electron Devices Meeting, pp. 164–167.

[25] G.V. Miram and E.L. Lien, "Convergent electron gun with bonded nonintercepting control grid," *Technical Digest*, 1979 International Electron Devices Meeting, pp. 290–292.

[26] T.J. Grant, R. Garcia and G.V. Miram, "Bonded grid electron gun for 95 GHz extended interaction amplifier (EIA)," *Technical Digest*, 1983 International Electron Devices Meeting, pp. 141–143.

[27] R. True, "Dual-mode electron gun," U.S. Patent 4,593,230, Filed March 29, 1982, Issued June 3, 1986.

Chapter 7

Electron Beams

In Chapter 6, Pierce electron guns used in TWTs were discussed. In this chapter, the magnetic focusing of a beam from a Pierce gun is examined [1]. Before proceeding with the discussion of magnetically focused beams, it is important to point out that numerous electrostatic focusing techniques have been examined in the more than 50 years that microwave tubes have been in existence. With the exception of the very few cases where the tube is designed to accommodate a beam following the universal beam-spread curve, in which case no focusing is required, only magnetic focusing is used at the present time.

The primary problem in focusing, as is illustrated in simplified fashion in Figure 7-1, is that space charge forces are so high for the typical beams used in microwave tubes that rapid divergence of the beam occurs in the absence of focusing forces. With equal and opposite constraining (focusing) forces, beam divergence can be prevented.

In TWTs, the focusing forces are provided by a magnetic field aligned with the axis of the electron beam. Frequently, a solenoid like that illustrated in Figure 7-2, is used to generate the magnetic field. Before developing the equations of beam motion in the presence of a magnetic focusing field, it is of value to discuss, qualitatively, how magnetic focusing works.

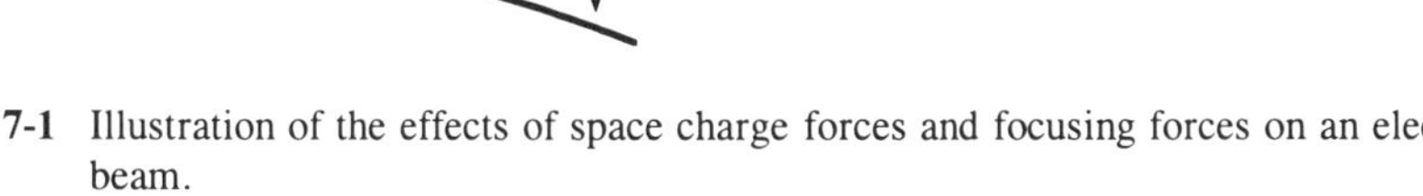

Figure 7-1 Illustration of the effects of space charge forces and focusing forces on an electron beam.

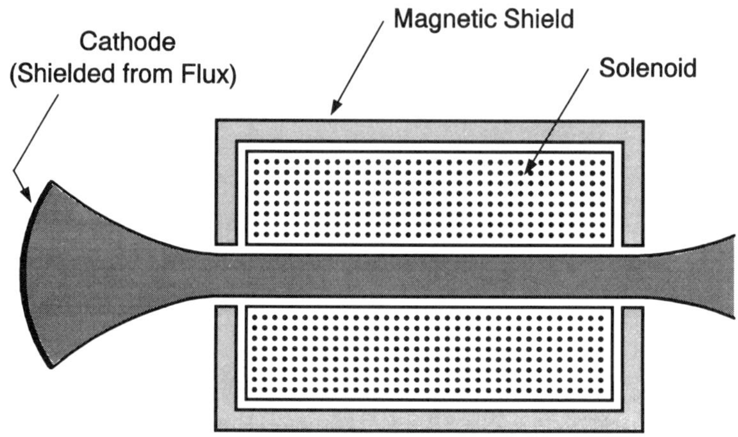

Figure 7-2 Illustration of use of solenoid to focus the beam in a TWT.

7.1 OVERVIEW OF UNIFORM-FIELD FOCUSING

To simplify the description of the focusing process, it is initially assumed that there is no magnetic flux through the cathode and that the beam first encounters magnetic flux lines as it enters the solenoid. It is also assumed that there are no transverse velocity components (resulting from thermal effects, nonuniform emission from the cathode or grid effects, *et cetera*).

Magnetic focusing of the beam is initiated as the electron beam crosses radially directed components of magnetic flux lines (illustrated in Figure 7-3). Consider, for example, an electron above the axis of the beam (and solenoid). That electron encounters a component of magnetic flux that is radially directed downward. The magnetic force on the electron is out of the paper (the cross product of the velocity, u, and the flux density, B, is into the paper and the force on an electron is in the reverse direction from the $u \times B$ product).

Consider, next, an electron below the axis. That electron encounters a component of magnetic flux that is radially directed upward. The magnetic force on that electron is into the paper.

When the magnetic forces on all electrons are considered, the net effect is that the electron beam starts rotating in the clockwise direction as it enters the field with the flux lines in the directions shown in Figure 7-3. If the directions of the flux lines were reversed, the direction of rotation of the beam would be counterclockwise.

As the electron beam moves further into the magnetic field, it encounters stronger and stronger axial components of magnetic flux. Consider again an electron above the beam axis. As a result of crossing radial components of flux, this electron has a velocity component out of the paper. The cross product of that velocity component with

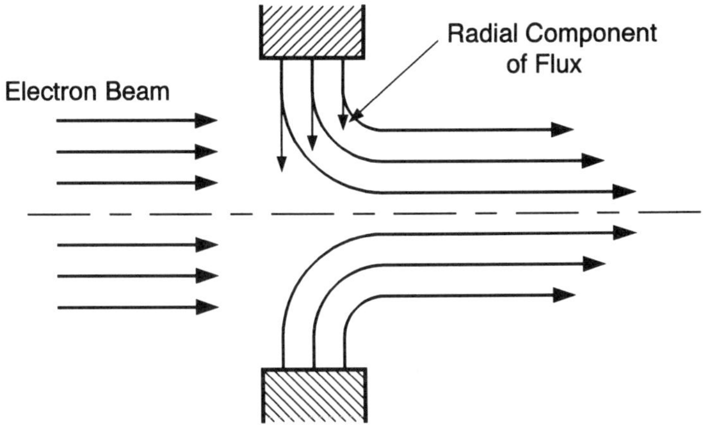

Figure 7-3 Configuration of magnetic flux lines at entrance of electron beam into solenoid.

the axial magnetic flux is a vector away from the axis so that the force on the electron is toward the axis. Similar considerations for other electrons show that they all experience a magnetic force toward the axis.

Thus, it is the rotation of the beam (established as the beam crosses radially directed flux lines upon entering the field) interacting with the axial components of flux that produces the magnetic focusing force. As each electron in the beam is followed, it is found that a helical trajectory about the axis of the solenoid is traversed as indicated in Figure 7-4.

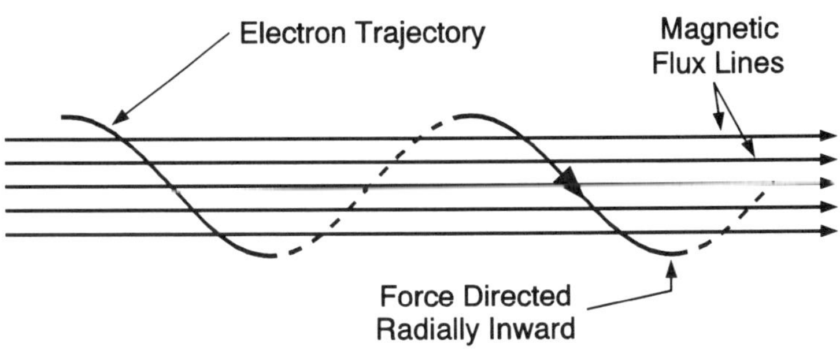

Figure 7-4 Electron trajectory in axial field.

7.1.1 Brillouin Flow

If the space charge forces (plus centrifugal forces resulting from beam rotation) tending to cause the beam to diverge are just balanced by equal and opposite magnetic forces tending to compress the beam, and if the electrons enter the magnetic field with no radial velocity components, then the beam diameter will remain constant as the beam rotates and moves through the magnetic field.

For a beam with uniform charge density throughout and no magnetic flux through the cathode, the magnetic flux level that produces a magnetic force that exactly balances the space charge and centrifugal forces is called the *Brillouin flux level*, commonly denoted by B_B. The resulting electron motion in the form of a rotating beam with constant diameter is called *Brillouin flow*. The configuration of the electron gun and magnetic pole piece for launching Brillouin flow are shown in Figure 7-5.

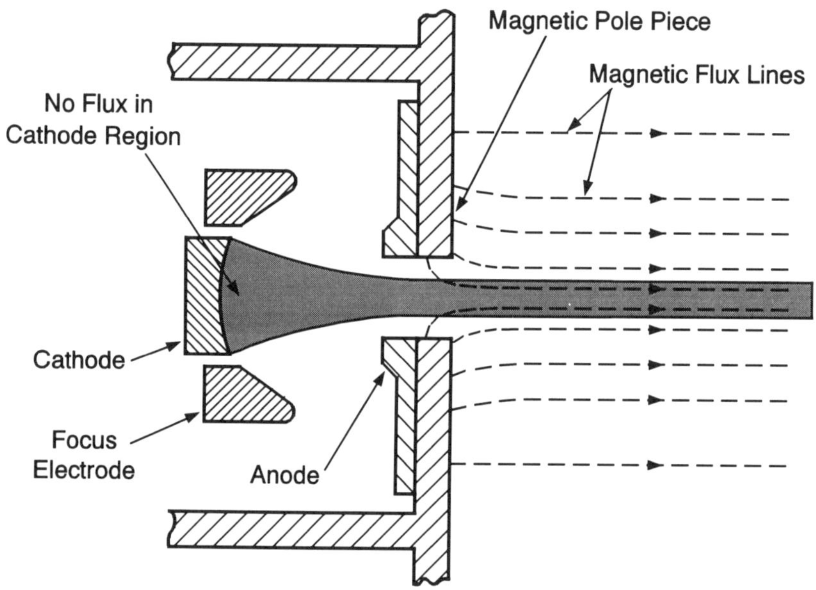

Figure 7-5 Gun and pole piece configuration for launching Brillouin flow.

When the beam exits from the magnetic field at the collector end of the tube, the sequence of interactions is just the opposite of that at the entrance of the beam into the field. That is, as the beam leaves the magnetic field, the radial compressive force disappears and rotation of the beam stops. As a result, the beam expands under the influence of space charge forces as it enters the collector.

Now, as we have said, for Brillouin flow to occur, the beam must have uniform charge density and must enter the magnetic field on axis with no radial electron motion and with a radius, b_o, equal to the equilibrium radius, a (the radius at which space charge and centrifugal forces are balanced by magnetic forces). This situation is illustrated in Figure 7-6. In the analysis given later in this chapter, it will be shown that, for Brillouin flow, the equilibrium radius, a, of the beam is

$$a = \frac{1}{B_z}\left[\frac{2I}{\pi \eta \epsilon_o u_o}\right]^{1/2} \tag{7-1}$$

where

B_z = axial flux density
I = beam current
u_o = axial beam velocity
η = electron charge to mass ratio
ϵ_o = dielectric constant of free space

There are no surprises in (7-1). As B_z increases, the magnetic forces increase, and so the equilibrium radius must decrease. As current increases, the charge density in the beam increases and so the space charge forces and equilibrium radius must increase. As velocity increases, the space charge density and the space charge forces decrease so the equilibrium radius must decrease.

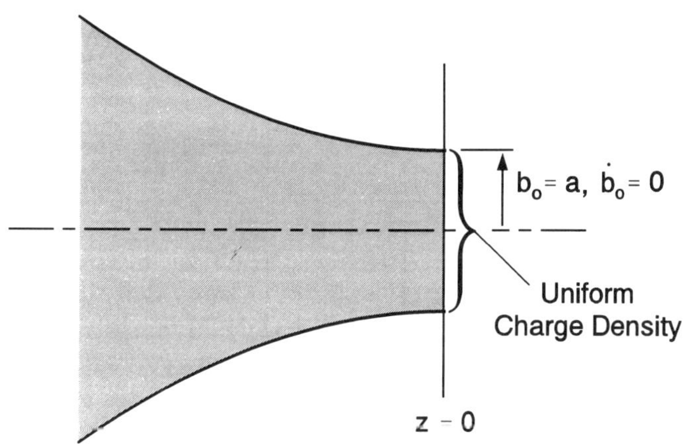

Figure 7-6 Entrance conditions for Brillouin flow. (The initial beam radius, b_o, is the equilibrium radius, a, and no radial velocities exist so $\dot{b}_o = 0$.)

7.1.2 Scalloping

In practice, when the electron gun is shielded from magnetic flux, it is extremely difficult to achieve the perfectly balanced force condition that is required for Brillouin flow. For example, consider the case when the electron beam enters the magnetic field with a radius smaller than the equilibrium radius as shown in Figure 7-7.

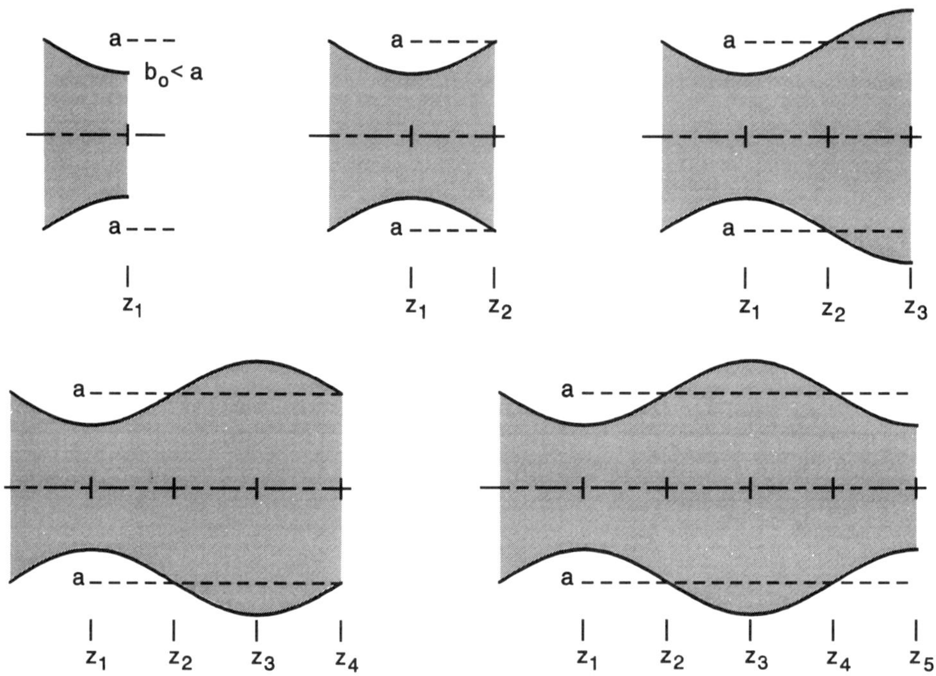

Figure 7-7 Beam dynamics resulting from beam entrance into magnetic field at less than the equilibrium radius.

At plane z_1, where the beam enters the magnetic field, the space charge force is larger than the magnetic focusing force, so the beam starts to expand. When the beam reaches plane z_2, the radial forces are equal but the electrons are moving radially outward. As a result, the electrons overshoot the equilibrium radius. By the time the electrons have reached plane z_3, the radial motion of the electrons has been stopped because the space charge and centrifugal forces have become smaller than the magnetic forces. In fact, this force imbalance causes the beam to become smaller again and, by plane z_4, the electrons are again crossing the equilibrium radius. Because the electrons are moving radially inward at plane z_4, they overshoot the equilibrium radius. By the time the beam reaches plane z_5, the inward radial motion has been stopped because the space charge and centrifugal forces exceed the magnetic forces.

In a uniform axial field, the beam radius at plane z_5 is the same as it was at plane z_1 and the entire process that took place between planes z_1 and z_5 starts over again. As a result, the beam radius continues to oscillate as the beam moves in the axial direction. This periodic variation of beam radius is referred to as *scalloping*.

For the preceding consideration (electrons moving parallel to axis as the beam enters the magnetic field) it is possible to adjust the magnetic field level so that the beam radius is the equilibrium radius. Then, scalloping can be eliminated.

A similar situation occurs when a beam traveling parallel to the axis but having a radius larger that the equilibrium radius enters the magnetic focusing field. Then the beam scallops inward because the focusing forces are larger that the space charge and centrifugal forces. Again, it is possible to adjust the magnetic field level to eliminate scalloping.

If the electron gun is at the wrong axial position so that the beam is either diverging or converging as it enters the magnetic field, then no adjustment of the magnetic field can be made to eliminate the resulting scalloping.

At this point it is appropriate to comment on the beam behavior that results if the axis of the electron beam is not aligned with the axis of the magnetic field. This could happen for any one of several reasons. For example, the electron gun could be misaligned during tube fabrication. The gun axis could be either off center or tilted with respect to the axis of the tube. Other causes for misalignment include errors in cathode, grid or control electrode positioning or nonuniformities in magnetic pole piece material. The result of beam misalignment is that the entire beam spirals about the magnetic field axis at the cyclotron frequency.

Figure 7-8 contains a summary of the problems that could be encountered in attempting to form a Brillouin beam flowing parallel to the axis. These problems are

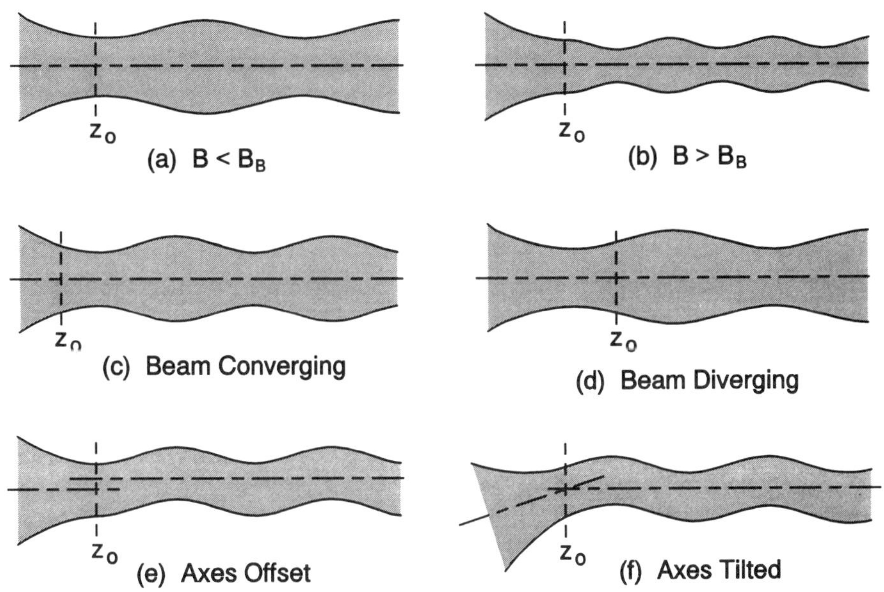

Figure 7-8 Problems in forming a Brillouin beam flowing parallel to the axis.

a. Flux density is too small.
b. Flux density is too large.
c. Beam is converging at the entrance.
d. Beam is diverging at the entrance.
e. Beam axis is offset from the magnetic field axis.
f. Beam axis is tilted with respect to the magnetic field axis.

Brillouin flow has the advantage that the magnetic field required is the lower than for any other focusing technique. The primary disadvantage is that the beam is extremely sensitive to misalignments and perturbations (such as those produced by RF modulation). The effect of a change in beam current, such as that resulting from RF bunching of electrons, is shown in Figure 7-9. Note that an increase in current of 50% results in an increase in beam diameter of over 40%.

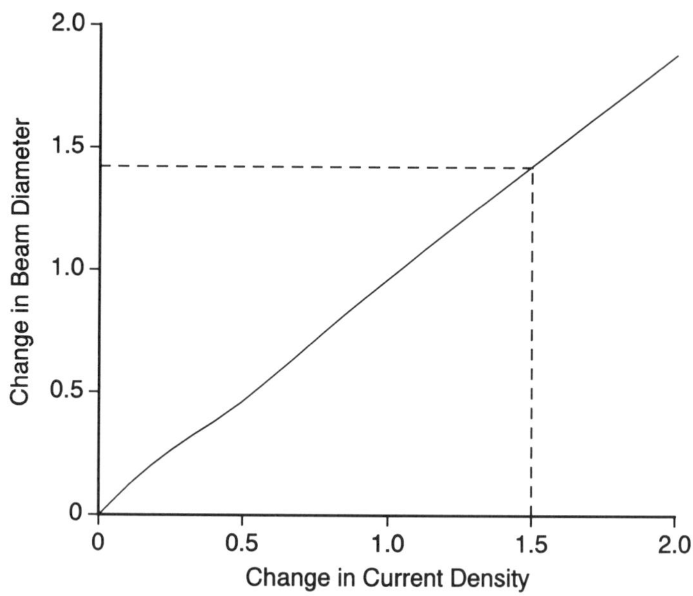

Figure 7-9 Change in beam diameter resulting from a change in current density for a Brillouin beam. From *Power Travelling Wave Tubes* by J.F. Gittins, published 1965 by American Elsevier Publishing, Inc.

We will show later in this chapter that Brillouin flow can be very nearly achieved under laboratory conditions. In practice, the magnetic focusing field that is used, even when the cathode is shielded, is greater than the Brillouin value. Reasons for this include transverse electron velocities (from thermal and other effects in the gun) and the increase in beam size that results from RF modulation. These and other reasons are discussed later in this chapter.

7.1.3 Confined (Immersed) Flow

In practice, the use of Brillouin focusing is avoided because of the high sensitivity of the beam to perturbations. Instead, as shown in Figure 7-10, the magnetic circuit providing the focusing field for a tube is purposely designed so that magnetic flux lines pass through the cathode. This makes possible the use of a much larger magnetic field than the Brillouin value for focusing the beam. In addition, this can be accomplished without causing scalloping of the beam. This type of focusing is referred to as *confined flow* or *immersed flow focusing*.

It is helpful in understanding the effect of flux through the cathode in confined flow to assume that the electron trajectories follow the magnetic flux lines in the electron gun. With this assumption, the magnetic field has no affect on the electrons in the gun. It should be pointed out that electrons do, in fact, cross flux lines in the gun (as was indicated in Figure 7-10). The resulting behavior of the beam is essentially the same as it is if it is assumed that the electrons do not cross flux lines.

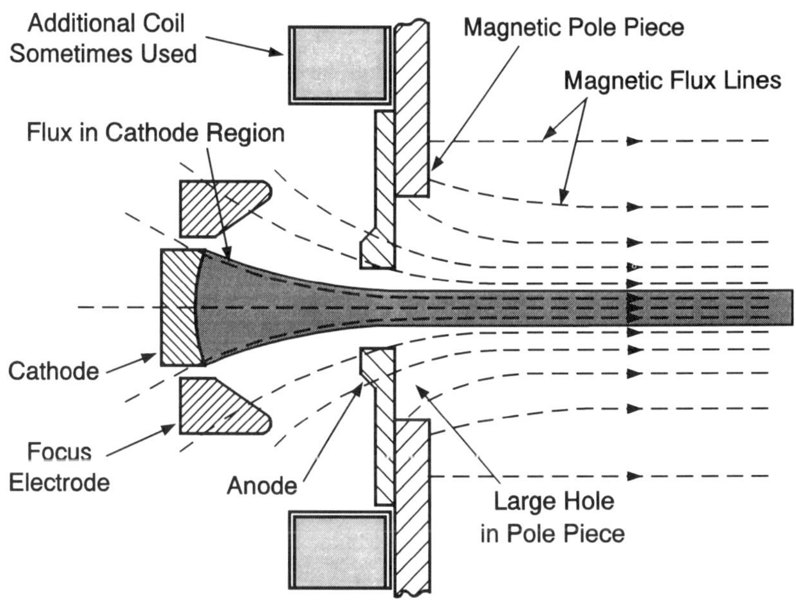

Figure 7-10 Magnetic circuit for launching confined (immersed) flow.

As the electron beam leaves the gun and enters the main part of the focusing field, the *increase* in flux density encountered must be sufficient to produce a magnetic focusing force that counterbalances the space charge and centrifugal forces of the beam. The focusing force results, just as it did for Brillouin focusing, from the interaction of the beam rotation with the axial magnetic field. In this case, however, the axial field is larger than it was for the Brillouin flow and so the beam rotates more slowly than for the Brillouin case. (Note, also, that, since the beam rotates more slowly, the centrifugal force is reduced and this further reduces the rotational rate required for equilibrium.)

Next, consider a perturbation to the beam that causes the electrons to move radially outward. This could result, for example, from electron bunching caused by interaction of the electrons with an RF field. As the outward moving electrons cross flux lines of the very large confined flow focusing field (often two to three times the Brillouin value) the rate of rotation of the beam increases rapidly. The interaction of this increased rotation with the large focusing field produces a focusing force far larger than that which occurs with Brillouin focusing. As a result, the radial expansion of the beam is much less than for Brillouin focusing.

The primary benefit of confined flow focusing is that beam control is significantly better than for Brillouin focusing. This is illustrated in Figure 7-11, which is like Figure 7-9 except that curves for field levels two times ($m = 2$) and three times ($m = 3$) the Brillouin value are shown. For an increase in beam current of 50%, the change in beam diameter is decreased from over 40% for Brillouin focusing to about 8% for $m = 2$ and less than 4% for $m = 3$. The primary disadvantage of confined flow focusing is that the magnet(s) must be larger and heavier than for Brillouin focusing to produce the large field that is required. Also, if electromagnets are used, the power and cooling requirements are larger than for Brillouin flow.

7.2 UNIFORM-FIELD FOCUSING AND LAMINAR FLOW

7.2.1 The Beam Equation

In the electron beams to be analyzed, the fields are axially symmetric. The radial electric field, E_r, is produced by the electron beam and the magnetic flux density, B, used to focus the beam is in the axial direction. It is assumed that the charge density, ρ, and the axial velocity, u_o, are uniform in the radial direction (but not necessarily in the axial direction). The radial force equation, which will be used to find the motion of the electrons, is

$$\ddot{r} - r\dot{\theta}^2 = -\eta(E_r + Br\dot{\theta}) \tag{7-2}$$

where $\dot{\theta}$ is given by Busch's theorem:

$$\dot{\theta} = \frac{\eta}{2}\left[B - B_c\frac{r_c^2}{r^2}\right] \tag{7-3}$$

Here, it has been assumed that the electrons originate at the cathode at a position, r_c,

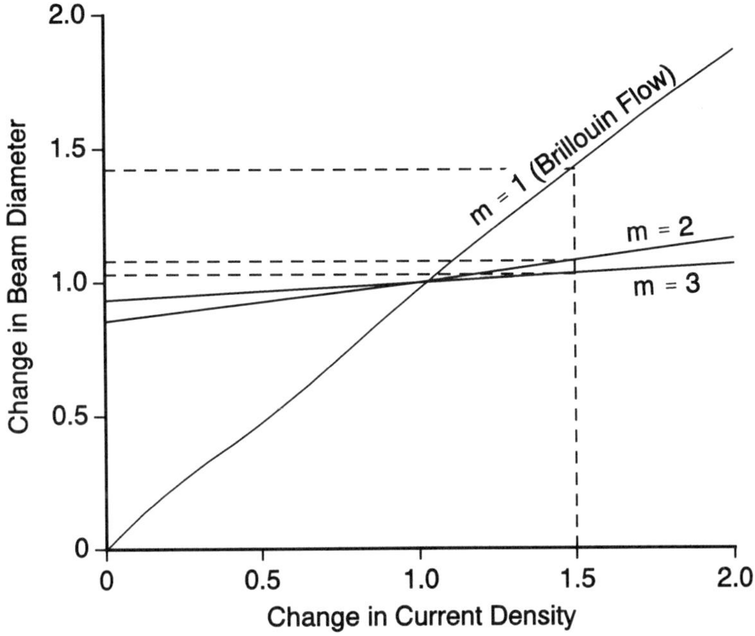

Figure 7-11 Effect of current density on beam diameter for fields greater than Brillouin. From *Power Travelling Wave Tubes* by J.F. Gittins, published 1965 by American Elsevier Publishing, Inc.

where the flux density is B_c. At a given axial position, it should be noted that, with the assumption of uniform charge density in the radial direction and an additional assumption that electron flow is laminar in the gun (electron trajectories do not cross), then r_c/r is a constant. Thus, at any axial position, $\dot{\theta}$ is constant throughout the beam.

The electric field, which was found using Gauss's law in the universal beam spread analysis, is

$$E_r = -\frac{rI}{2\pi b^2 \epsilon_o u_o} \tag{7-4}$$

Now it is important to realize that the space charge force, the centrifugal force and the magnetic force are all proportional to radial position. As a result, the radial acceleration and radial motions of the electrons are proportional to r and so electron flow in the beam is laminar.

At the beam boundary where $r = b$ (and $r_o = b_o$), the electric field, $E_r(b)$, is

$$E_r(b) = -\frac{I}{2\pi b \epsilon_o u_o} \tag{7-5}$$

Now, by combining the force equation with Busch's theorem and Gauss's law, the following equation of motion may be obtained for the electrons on the outer edge of the beam:

$$\ddot{b} - b\left[\frac{\eta}{2}\right]^2\left[B - B_c\frac{b_c^2}{b^2}\right]^2 = -\eta\left[-\frac{I}{2\pi b\epsilon_o u_o} + Bb\frac{\eta}{2}\left[B - B_c\frac{b_c^2}{b^2}\right]\right] \tag{7-6}$$

By rearranging and combining terms this may be written as

$$\ddot{b} + b\omega_L^2\left[1 - \left[\frac{B_c}{B}\frac{b_c^2}{b^2}\right]^2\right] - \frac{\eta I}{2\pi b\epsilon_o u_o} = 0 \tag{7-7}$$

where $\omega_L = \eta B/2$ is the Larmor frequency, which is half of the cyclotron frequency. This equation is sometimes referred to as the beam equation.

It is not possible to obtain an exact solution to the beam equation. However, for the cases of interest in linear beam tubes, the radius of the beam is nearly constant. For these cases, it is possible to make approximations that permit solutions to be obtained.

Before proceeding with the analysis of beams, it is of value to determine the significance of the Larmor frequency. This may be done by neglecting space charge forces (i.e., letting $I = 0$) and assuming that $B_c = 0$ and so the beam equation reduces to

$$\ddot{b} + b\omega_L^2 = 0 \tag{7-8}$$

This is the equation of simple harmonic motion for which the general solution is

$$b = A\sin\omega_L t + B\cos\omega_L t \tag{7-9}$$

By assuming that, at $t = 0$, the beam has radius b_o and no radial velocity component (that is $\dot{b}_o = 0$ so the beam is not expanding or contracting) then $B = b_o$ and $A = 0$. Thus,

$$b = b_o\cos\omega_L t \tag{7-10}$$

As indicated by the plot in cylindrical coordinates in Figure 7-12, there is one complete rotation of an electron as $\omega_L t$ goes through π rad. The actual frequency of rotation of the electron is $\omega_c = 2\omega_L$, as it should be. Thus, the Larmor frequency arises from the coordinate system used for describing the beam behavior.

As a function of axial distance, z, the beam appears as is shown in Figure 7-13. The radius of the beam periodically goes to zero and then increases to the original value of b_o. Poorly focused beams having low charge density actually behave in a manner similar to this.

The first step in obtaining a general solution to the beam equation is to examine the

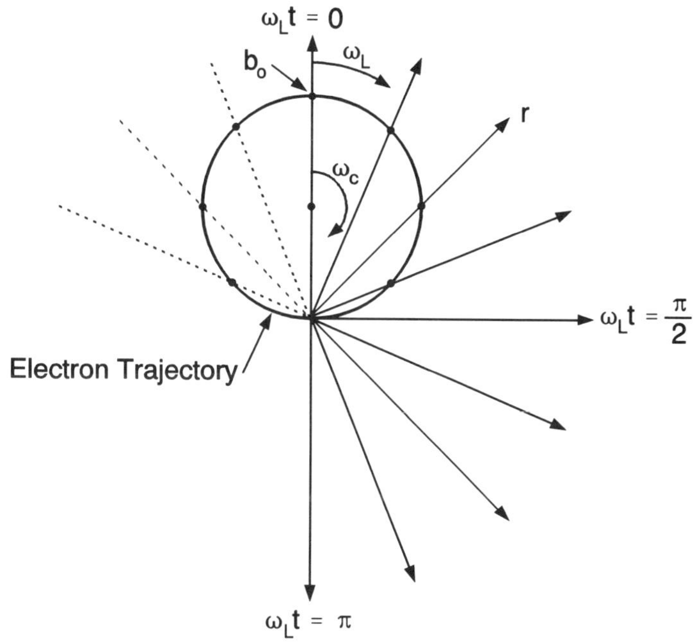

Figure 7-12 Electron trajectory with no space charge and $B_c = 0$.

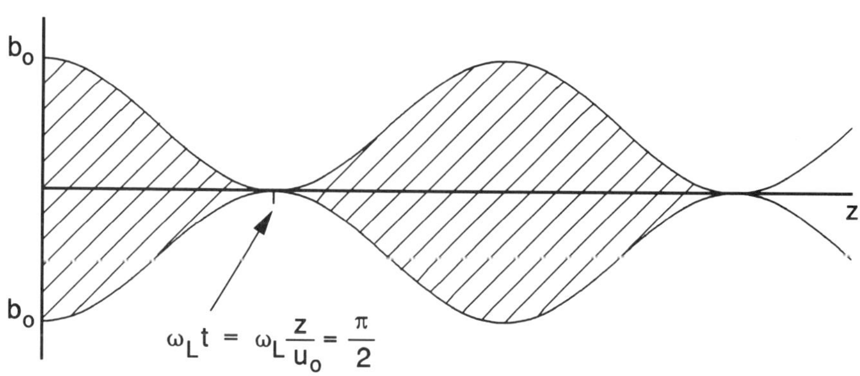

Figure 7-13 Beam shape as a function of distance.

following quantity:

$$\left[\frac{B_c}{B}\,\frac{b_c^2}{b^2}\right]^2 \tag{7-11}$$

The total flux through the cathode surface is $\pi b_c^{\,2} B_c$. This is the same as the total flux through the area defined by radius g (shown in Figure 7-14) which is $\pi g^2 B$. Note that when $g = 0$, there is no flux through the cathode. When $g = b$, the total flux through the cathode is the same as the total flux contained in the electron beam at position z.

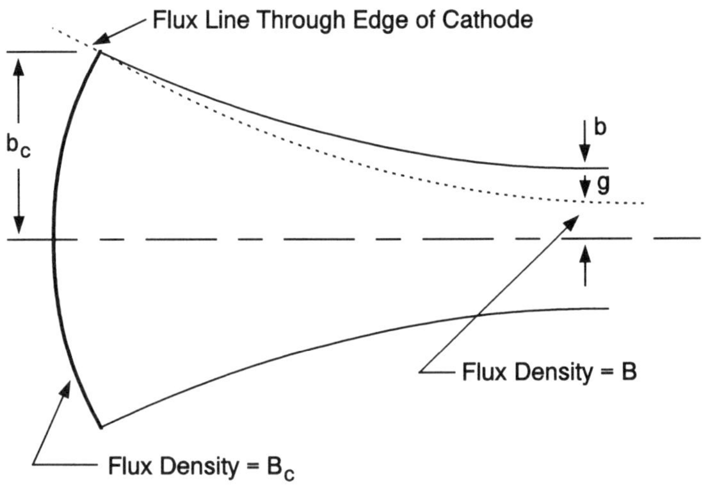

Figure 7-14 Definition of quantity g.

As a result,

$$\pi b_c^2 B_c = \pi g^2 B \tag{7-12}$$

or

$$\frac{B_c}{B}\frac{b_c^2}{b^2} = \frac{g^2}{b^2} \tag{7-13}$$

Using this result and letting

$$a = \frac{1}{\omega_L}\left(\frac{\eta I}{2\pi\epsilon_o u_o}\right)^{1/2} \tag{7-14}$$

the beam equation may be written as

$$\ddot{b} + \omega_L^2\left[b\left(1 - \frac{g^4}{b^4}\right) - \frac{a^2}{b}\right] = 0 \tag{7-15}$$

Next, with the normalizations

$$R = \frac{b}{a} \quad \text{and} \quad R_g = \frac{g}{a} \tag{7-16}$$

the beam equation becomes

$$\ddot{R} + \omega_L^2 \left[R \left(1 - \frac{R_g^4}{R^4} \right) - \frac{1}{R} \right] = 0 \tag{7-17}$$

To understand what this equation means and to obtain beam information from it, it is useful to write the equation as

$$\frac{\ddot{R}}{\omega_L^2} = - \left[R \left(1 - \frac{R_g^4}{R^4} \right) - \frac{1}{R} \right] \tag{7-18}$$

and then to examine the right-hand side (RHS) as a function of R.

First of all, it should be noted that the left-hand side is proportional to the radial acceleration and, therefore, the radial force on the electrons. As a result, when the RHS is zero, the net radial force on the electrons is zero. (This occurs because the space charge and centrifugal forces causing the beam to expand are just equal and opposite to the magnetic focusing forces that compress the beam.) When the radial forces are zero, the beam is in equilibrium and there are no forces causing it to expand or contract. If this condition can be sustained, then the beam diameter remains constant and a cylindrical beam has been formed. This is normally the desired beam shape in a microwave tube.

The RHS of (7-18) is plotted in Figure 7-15 as a function of R for several values of R_g. Note that, for $R \approx 1$, the RHS is close to zero. If $R < \approx 1$, the force is positive and so the beam expands. If $R > \approx 1$, the force is negative and so the beam contracts. Thus, we would expect the beam to remain close to $R = 0$ where the forces are zero. If the normalized radius at which the forces are zero is called the *equilibrium radius*, R_e, then the RHS of (7-18) can be written as

$$R_e \left(1 - \frac{R_g^4}{R_e^4} \right) - \frac{1}{R_e} = 0 \quad \text{or} \quad R_e^4 - R_e^2 - R_g^4 = 0 \tag{7-19}$$

As a result,

$$R_e = \pm \left[\frac{1}{2} \pm \frac{1}{2} \sqrt{1 + 4R_g^4} \right]^{1/2} \tag{7-20}$$

Only the positive signs have meaning because the radius cannot be negative or imaginary. When $R_g = 0$, $R_e = 1$ and so the beam radius $b = a$. Thus, the value of a given in (7-14) is the equilibrium radius when no flux links the cathode ($R_g = 0$).

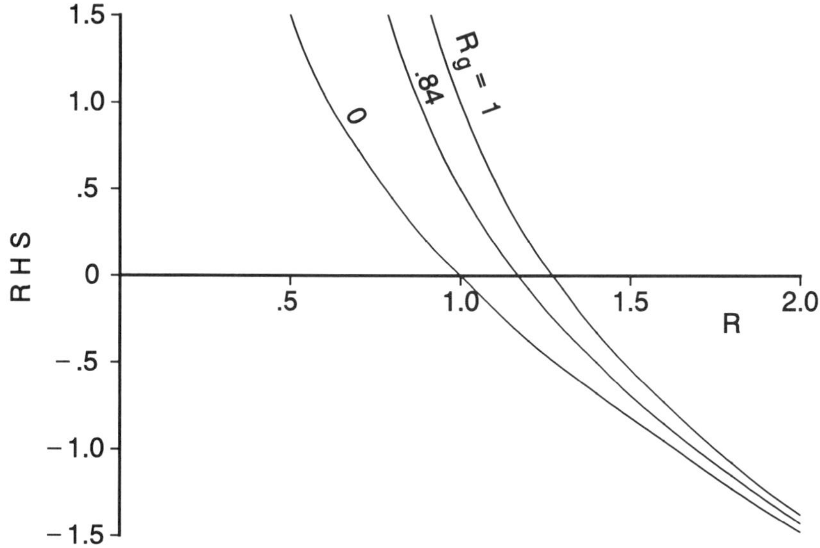

Figure 7-15 Right-hand side of (7-18) as a function of R for several values of R_g.

If a beam is properly focused, then R will normally be close to R_e. As a result, it is useful to look for solutions to the beam equation for $R \approx R_e$. To do this, let

$$R = R_e(1 + \delta) \tag{7-21}$$

Then, the beam equation may be written as

$$R_e\ddot{\delta} + \omega_L^2 \left[R_e(1 + \delta) \left\{ 1 - \frac{R_g^4}{R_e^4(1 + \delta)^4} \right\} - \frac{1}{R_e(1 + \delta)} \right] = 0 \tag{7-22}$$

Now, assume δ is small enough that

$$\frac{1}{1 + \delta} \approx 1 - \delta \quad \text{and} \quad \frac{1}{(1 + \delta)^3} \approx 1 - 3\delta \tag{7-23}$$

Then

$$\ddot{\delta} + \omega_L^2 \left[(1 + \delta) - \frac{R_g^4}{R_e^4}(1 - 3\delta) - \frac{1}{R_e^2}(1 - \delta) \right] = 0 \tag{7-24}$$

or

$$\ddot{\delta} + \omega_L^2 \delta \left[1 + \frac{3R_g^4}{R_e^4} + \frac{1}{R_e^2} \right] = \omega_L^2 \left[-1 + \frac{R_g^4}{R_e^4} + \frac{1}{R_e^2} \right] \tag{7-25}$$

but the right-hand side of this equation is the same as (7-19) and is, therefore, zero. Making the following substitution

$$1 + \frac{3R_g^4}{R_e^4} + \frac{1}{R_e^2} = 2\Omega \tag{7-26}$$

where, with the aid of (7-19)

$$\Omega = 2 - \frac{1}{R_e^2} \tag{7-27}$$

and has a value in the range $1 \leq \Omega \leq 2$ as R_e varies from unity to infinity. With the substitution of (7-26) into (7-25),

$$\ddot{\delta} + 2\Omega\omega_L^2\delta = 0 \tag{7-28}$$

This is the equation for simple harmonic motion and has the general solution

$$\delta = A \sin\sqrt{2\Omega}\,\omega_L t + B\cos\sqrt{2\Omega}\,\omega_L t \tag{7-29}$$

but $t = z/u_o$ and $R = R_e(1 + \delta)$ so

$$R = R_e \left\{ 1 + A\sin\sqrt{2\Omega}\,\frac{\omega_L}{u_o} z + B\cos\sqrt{2\Omega}\,\frac{\omega_L}{u_o} z \right\} \tag{7-30}$$

The beam radius, b, is

$$b = Ra = R_e a + A'\sin\sqrt{2\Omega}\,\frac{\omega_L}{u_o} z + B'\cos\sqrt{2\Omega}\,\frac{\omega_L}{u_o} z \tag{7-31}$$

or

$$b = R_e a + C\sin\left\{ \sqrt{2\Omega}\,\frac{\omega_L}{u_o} z + \phi \right\} \tag{7-32}$$

where

$$C = \sqrt{A'^2 + B'^2} \tag{7-33}$$

and

$$\phi = \tan^{-1}\frac{B'}{A'} \tag{7-34}$$

and where $A' = R_e aA$ and $B' = R_e aB$.

The values of A' and B' may be determined by examining the conditions of entrance of the beam at $z = 0$ into the main magnetic focusing field. There, the initial conditions are as shown in Figure 7-16 and are

$$b = b_o \quad \text{and} \quad \frac{db_o}{dz} = \tan\gamma_o \tag{7-35}$$

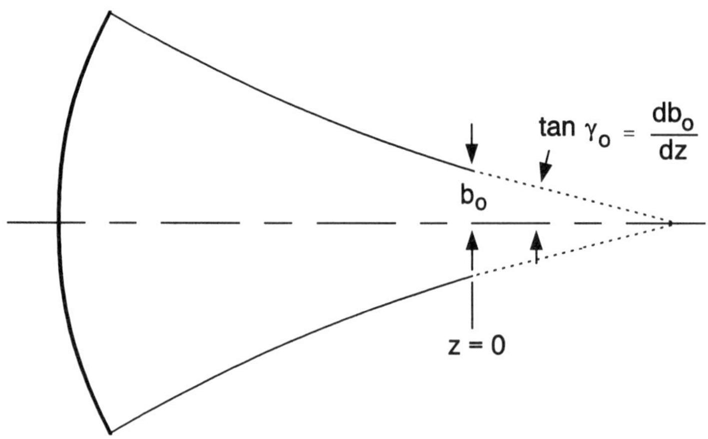

Figure 7-16 Initial conditions for entrance of beam into focusing field.

As a result, from (7-31) and d/dz of (7-31) evaluated at $z = 0$,

$$b_o = R_e a + B' \quad \text{and} \quad \frac{db_o}{dz} = \tan\gamma_o = A'\sqrt{2\Omega}\,\frac{\omega_L}{u_o} \tag{7-36}$$

so

$$B' = b_o - R_e a \tag{7-37}$$

and

$$A' = \frac{\tan\gamma_o}{\sqrt{2\Omega}\,\dfrac{\omega_L}{u_o}} \tag{7-38}$$

To determine the meaning of (7-32), consider typical focusing techniques used in linear beam tubes.

7.2.2 Brillouin Flow

When Brillouin flow focusing is used, no magnetic flux exists in the cathode region, $R_g = 0$, and so the configuration of the magnetic field at the entrance to the focusing structure is as shown in Figure 7-17.

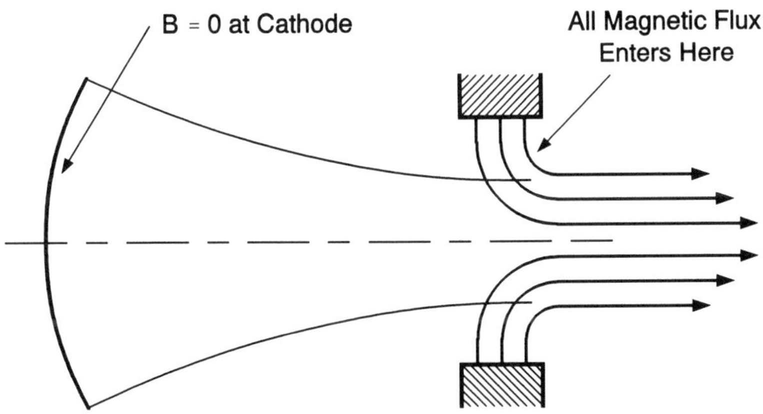

Figure 7-17 Configuration of magnetic field at entrance to focusing structure for Brillouin flow ($R_g = 0$).

The value of $R_e = 1$ and so $\Omega = 1$. Thus, (7-32) becomes

$$b = a + C \sin \left[\sqrt{2} \, \frac{\omega_L}{u_o} z + \phi \right] \tag{7-39}$$

where

$$C = \left[\left(\frac{\tan \gamma_o}{\sqrt{2} \, \frac{\omega_L}{u_o}} \right)^2 + (b_o - a)^2 \right]^{1/2} \tag{7-40}$$

and

$$\phi = \tan^{-1} \left[(b_o - a) \frac{\sqrt{2} \, \frac{\omega_L}{u_o}}{\tan \gamma_o} \right] \tag{7-41}$$

First, let the electron beam enter the focusing structure at the equilibrium radius traveling parallel to the axis. Then, $b_o = a$, $\gamma_o = 0$ and $\tan \gamma_o = 0$. Thus, $C = 0$ and so (7-39) reduces to $b = a$. As a result, the beam radius remains constant as the beam travels through the magnetic focusing field.

The magnetic flux density that produces a beam of constant radius when no magnetic flux links (passes through) the cathode is called the *Brillouin flux density*, usually denoted by B_B. The resulting beam is referred to as a *Brillouin beam*, and the resulting electron motion is referred to as *Brillouin flow*. The magnetic flux required for Brillouin flow is smaller than for any other magnetic focusing technique. The Brillouin flux density can be determined as follows from (7-14):

$$a = \frac{1}{\omega_L}\left[\frac{\eta I}{2\pi\epsilon_o u_o}\right]^{1/2} = \frac{2}{\eta B}\left[\frac{\eta I}{2\pi\epsilon_o u_o}\right]^{1/2} \tag{7-42}$$

so for Brillouin flow, the flux density is

$$B_B = \frac{2}{a}\left[\frac{I}{2\pi\epsilon_o \eta u_o}\right]^{1/2} = 0.83\times 10^{-3}\,\frac{I^{1/2}}{aV^{1/4}}\quad \mathrm{T} \tag{7-43}$$

Because the flux density in minimized, the size of the magnet structure (permanent magnet or solenoid) is minimized with Brillouin flow. The penalty for using Brillouin flow, as will be shown in the following discussion, is that the beam is very sensitive to misalignments, incorrect magnetic fields, RF drive, *et cetera*.

If the magnetic flux density is changed so that it is less than the Brillouin value, that is, $B < B_B$, then, from (7-42), the value of the equilibrium radius, a, increases and so the beam diameter increases as the beam enters the focusing field. Subsequently, the beam diameter oscillates about the equilibrium diameter as shown in Figure 7-18.

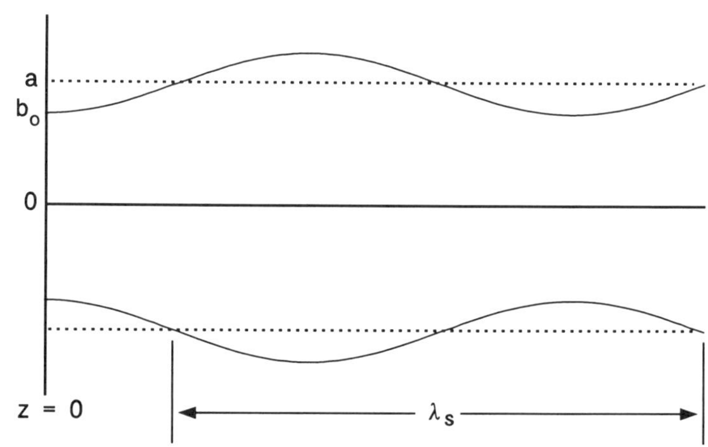

Figure 7-18 Beam shape for $B < B_B$.

The condition in which the beam radius varies periodically with distance is referred to as *scalloping*. The wavelength, λ_s, of the scallop can be found from 7-39 by letting

$$\sqrt{2}\,\frac{\omega_L}{u_o}z = \sqrt{2}\,\frac{\omega_L}{u_o}\lambda_s = 2\pi \tag{7-44}$$

from which

$$\lambda_s = \frac{2\pi u_o}{\sqrt{2}\,\omega_L} = \frac{4\pi u_o}{\sqrt{2}\,\eta B} = .030 \times 10^{-3}\frac{V_o^{1/2}}{B}\ \text{m} \tag{7-45}$$

so when $B < B_B$, $\lambda_s > \lambda_{sB}$, where λ_{sB} is the Brillouin scallop wavelength (although, with Brillouin flow, the amplitude of the radius fluctuation is zero, and so no scalloping occurs).

When $B > B_B$, $a < b_o$, and so the beam scallops inward from the initial entrance radius, b_o, rather than outward. Also, $\lambda_s < \lambda_{sB}$. In summary, the way in which beam shape varies with B is shown in Figure 7-19.

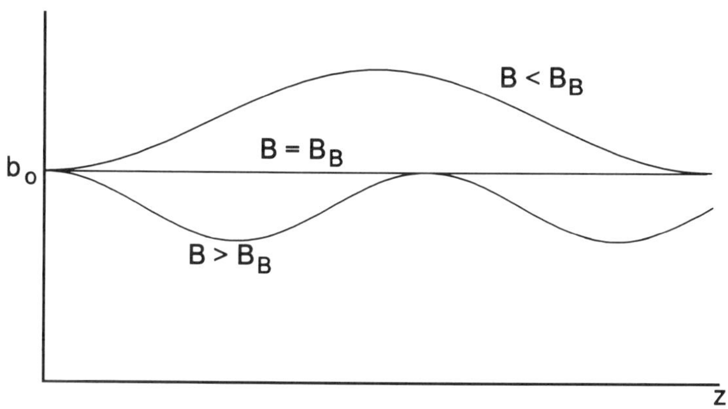

Figure 7-19 Beam shape as a function of magnetic flux density.

Next, consider the case when the beam enters the focusing field at $b_o = a$ but γ_o is not zero. That is, the beam is either expanding or contracting as it enters the magnetic field, as is shown in Figure 7-20. In this case,

$$b = a + \frac{\tan\gamma_o}{\sqrt{2}\,\dfrac{\omega_L}{u_o}}\sin\sqrt{2}\,\frac{\omega_L}{u_o}z \tag{7-46}$$

and so, for a positive γ_o, the beam radius varies with B as shown in Figure 7-21.

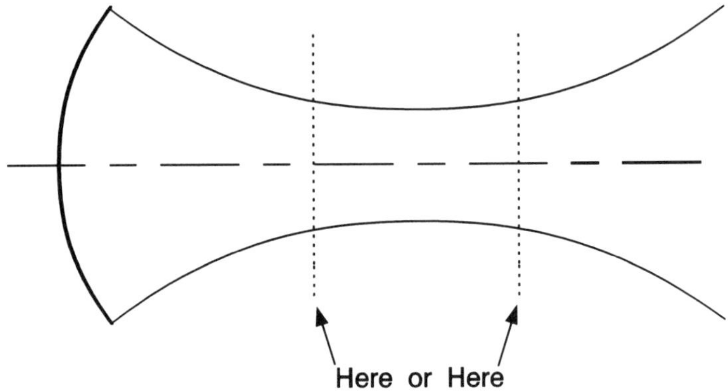

Figure 7-20 Entrance conditions when γ_o is not zero.

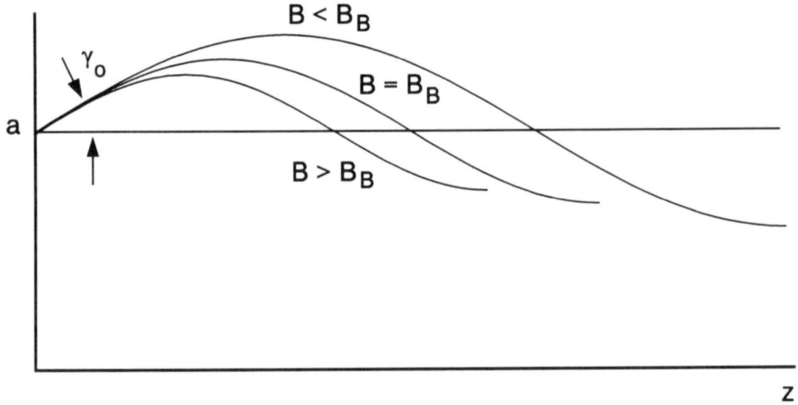

Figure 7-21 Beam shapes for $\gamma_o > 0$.

Detailed measurements made on Brillouin focused electron beams show that, in fact, these beams can behave very nearly as predicted by analysis. Results obtained by scanning a current probe transversely across the center of a beam at equally spaced axial positions are shown in Figure 7-22 [2, 3]. The inward scalloping of the beam when $B > B_B$ is clearly shown. When $B = B_B$, the beam diameter was essentially constant.

Ideally, each of the current profiles shown in Figure 7-22 should be rectangular in shape. At least three factors prevented the current from rising abruptly from zero to a constant value at the edges of the beam profiles. One was the finite diameter (0.010 in (0.25 mm)) of the aperture in the current probe used for current sampling. The second factor was the thermal velocities of the electrons, and the third was the patchiness and roughness of the oxide coated cathode.

From data of the type shown in Figure 7-22, beam radius as functions of axial position and magnetic flux density were determined and these are given in Figure 7-23. At the predicted Brillouin flux density of 285 G, the variation in beam diameter was less than 2%.

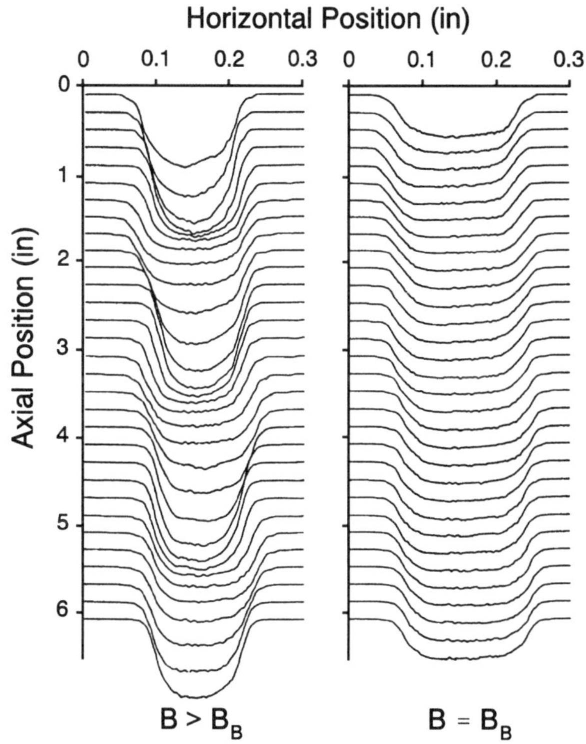

Figure 7-22 Current density profiles for a beam from a shielded gun. From A.S. Gilmour, Jr., *Proc. High Power Microwave tube Symposium*, The Hexagon, 1962.

Brewer reported detailed measurements on beams from several electron guns with a wide range of focusing conditions [4]. He observed translaminar streams of electrons that originated from aberrations in the electron gun but these constituted a small percentage of the current in the beam. He concluded: "The close agreement between the experimental data and the laminar-flow theory would appear to indicate that the bulk of the beam is in a very nearly laminar-flow situation. This does not mean that no electrons cross other electron paths in the beam, since many anomalies have certainly been observed which could result from electron-trajectory crossings. However, the author believes that most of the electrons in a well-formed beam do behave as in laminar flow for magnetic field values near the Brillouin-flow condition."

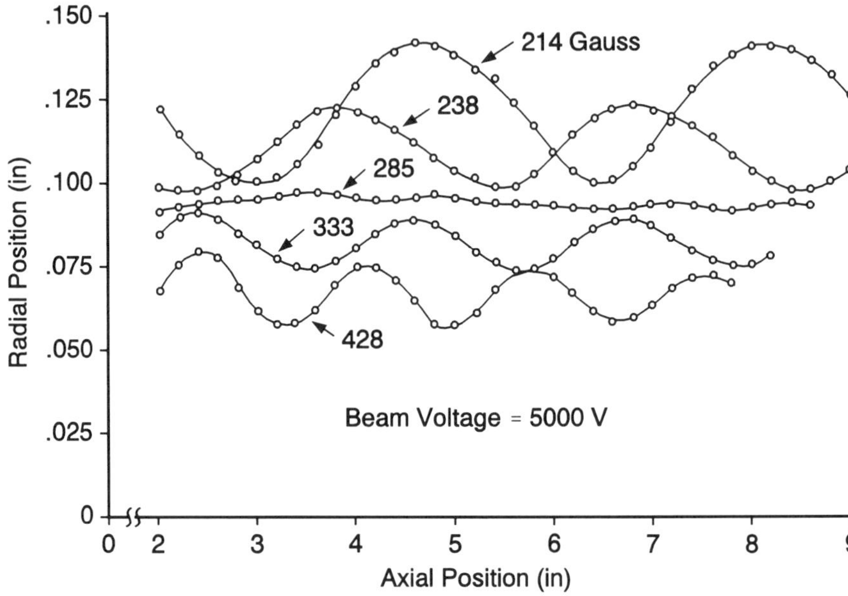

Figure 7-23 Measured beam radius *versus* axial position and magnetic flux density. From A.S. Gilmour, Jr., *Proc. High Power Microwave Tube Symposium,* The Hexagon, 1962.

7.2.3 Confined (Immersed) Flow

When part of the magnetic flux used for focusing the beam passes through the cathode, the beam is said to be partially confined or confined (or sometimes partially immersed or immersed) and the behavior of the beam is quite different than for Brillouin flow. First, the scalloping of the beam is reduced when the beam is expanding or contracting as it enters the focusing field. Second, when changes in beam conditions occur, such as bunching of the electrons by an RF field, the resulting perturbations of the beam radius are significantly reduced.

To illustrate these effects, it will be assumed that the beam radius, b, at the entrance into the focusing field is the same in all cases considered and is the equilibrium value. Thus, $b = b_o = R_e a$ (see (7-32)). As g and R_g increase, then R_e must increase (7-20) and so a must decrease to keep b_o constant. Because $a \propto 1/\omega_L$, and $\omega_L \propto B$ (the focusing flux density), then B must increase to keep b_o constant.

The value of B in terms of the flux through the cathode and B_B is readily determined from the force equation. At equilibrium, the radial acceleration is zero and so, for Brillouin flow,

$$-b_o\dot{\theta}_B^2 = -\eta\left(E_r + B_B b_o \dot{\theta}_B\right) \tag{7-47}$$

or

$$\eta E_r = -\eta B_B b_o \dot{\theta}_B + b_o \dot{\theta}_B^2 \tag{7-48}$$

where $\dot{\theta}_B$ is the value of $\dot{\theta}$ for Brillouin flow.

For confined flow,

$$-b_o\dot{\theta}^2 = -\eta\left(E_r + B b_o \dot{\theta}\right) \tag{7-49}$$

or

$$\eta E_r = -\eta B b_o \dot{\theta} + b_o \dot{\theta}^2 \tag{7-50}$$

But E_r is the same for both cases because the current, velocity and beam radii are equal. Thus,

$$\eta B_B \dot{\theta}_B - \dot{\theta}_B^2 = \eta B \dot{\theta} - \dot{\theta}^2 \tag{7-51}$$

From Busch's theorem,

$$\dot{\theta}_B = \frac{\eta}{2} B_B \qquad \text{and} \qquad \dot{\theta} = \frac{\eta}{2}\left[B - B_c \frac{b_c^2}{b_o^2}\right] \tag{7-52}$$

and so

$$B^2 = B_B^2 + B_c^2 \frac{b_c^4}{b_o^4} = B_B^2 + B_c^2 \frac{A_c^2}{A_o^2} \tag{7-53}$$

where A_c is the area of the cathode and A_o is the cross-sectional area of the beam at the entrance into the focusing field. The ratio A_c/A_o is referred to as the *area compression ratio* of the electron gun. This result is significant because A_c/A_o is often as high as 25 or more and so a flux density of only a few tens of gauss at the cathode surface has an effect equivalent to a Brillouin flux density of 500 to 1000 G or more.

It is often convenient to reference the actual flux density used for focusing, B, to the Brillouin value, B_B, by the confinement factor m, which is defined by

$$B = m B_B \tag{7-54}$$

The relation between m and g can be found, with the aid of (7-12) and (7-53), as follows:

$$b_c^2 B_c = g^2 B = g^2 m B_B \tag{7-55}$$

and so

$$B^2 = m^2 B_B^2 = B_B^2 + \frac{g^4 m^2 B_B^2}{b_o^4} \tag{7-56}$$

or

$$g = \left[1 - \frac{1}{m^2}\right]^{1/4} b_o \tag{7-57}$$

In terms of normalized radii, $R_g = g/a$ and $R_e = b_o/a$,

$$R_g = \left[1 - \frac{1}{m^2}\right]^{1/4} R_e \tag{7-58}$$

But, by combining this with (7-19),

$$R_e = m \tag{7-59}$$

and so

$$R_g = \left[m^2 (m^2 - 1)\right]^{1/4} \tag{7-60}$$

It is interesting to note that, for reasonable values of m, g (and R_g) is very close to b_o (and R_e). For example, for $m = 2$, $g = 0.93\ b_o$. Thus, the flux that causes rotation of the beam is equivalent to that contained in the outer 7% of the beam.

With the aid of (7-51) it is possible to more clearly understand the details of confined flow focusing. For example, the rate of rotation of the beam, $\dot{\theta}$, compared to the rate of rotation of the Brillouin focused beam, $\dot{\theta}_B$, is

$$\frac{\dot{\theta}}{\dot{\theta}_B} = \frac{\frac{\eta}{2}\left[B - B_c \frac{b_c^2}{b_o^2}\right]}{\frac{\eta}{2} B_B} = \frac{B - \left(B^2 - B_B^2\right)^{1/2}}{B_B} \tag{7-61}$$

or, because $B = mB_B$,

$$\frac{\dot{\theta}}{\dot{\theta}_B} = m - \left(m^2 - 1\right)^{1/2} \tag{7-62}$$

Equation 7-62 is plotted in Figure 7-24. Note that, as m increase from unity (the Brillouin value) to the range of two to three commonly used for confined flow focusing, the rate of rotation of the beam drops by a factor of four to over five from the Brillouin value.

Because the beam rotates so slowly with confined flow focusing, the centrifugal force (which is equal to the space charge force for Brillouin focusing) almost vanishes.

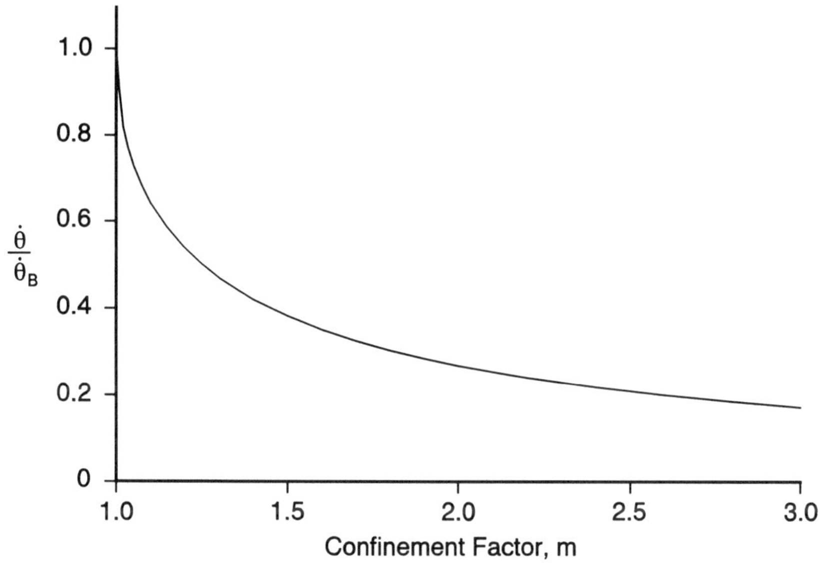

Figure 7-24 Ratio of rotation rates of confined and Brillouin beams as a function of the confinement factor, m.

Thus, nearly all of the focusing force produced by the slowly rotating beam in the strong focusing field is used to offset the space charge forces.

It is of value to compare the beam dynamics with confined focusing to those that occur with Brillouin focusing. First, assume that the beam enters the focusing field at the equilibrium radius but with a radial component of velocity. Then, $\gamma_o > 0$ and the amplitude of the resulting scallop, from (7-33), (7-37) and (7-38), is

$$C = \frac{\tan\gamma_o}{\sqrt{2\Omega}\,\dfrac{\omega_L}{u_o}} \tag{7-63}$$

When this amplitude is compared with the amplitude of the scallop on a Brillouin beam, C_B, for the same entrance conditions, the result is

$$\frac{C}{C_B} = \frac{\dfrac{\tan\gamma_o}{\sqrt{2\Omega}\,\dfrac{\omega_L}{u_o}}}{\dfrac{\tan\gamma_o}{\sqrt{2}\,\dfrac{\omega_L}{u_o}}} = \frac{\sqrt{2}\,\omega_{LB}}{\sqrt{2\Omega}\,\omega_L} \tag{7-64}$$

where ω_{LB} is the Brillouin Larmor frequency. Since $B = mB_B$, $\omega_L = \omega_{LB}/m$ and so

$$\frac{C}{C_B} = \frac{m}{\sqrt{\Omega}} \tag{7-65}$$

Now, from the definition of Ω in (7-27) and, because $R_e = m$,

$$\Omega = 2 - \frac{1}{R_e^2} = 2 - \frac{1}{m^2} \tag{7-66}$$

so

$$\frac{C}{C_B} = \frac{1}{\left(2m^2 - 1\right)^{1/2}} \tag{7-67}$$

This result is shown in Figure 7-25. Note that a significant decrease in scallop amplitude occurs as flux density is increased.

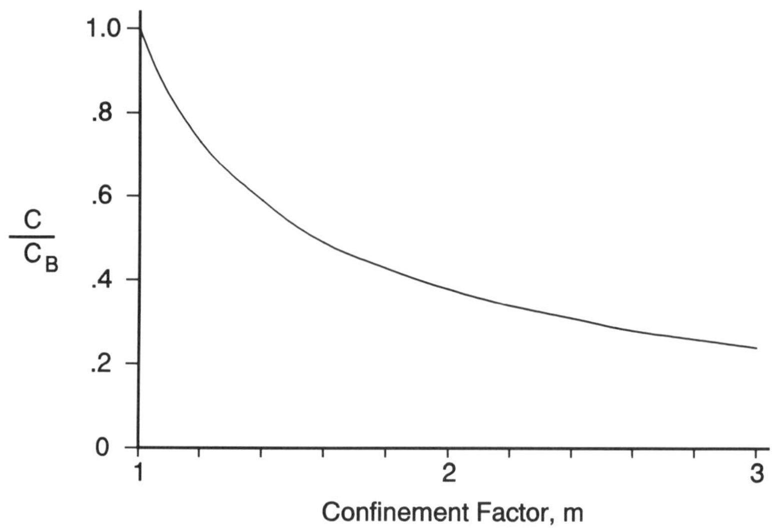

Figure 7-25 Variation of scallop amplitude with flux density for tan $\gamma_o > 0$.

Next, consider the effect on beam radius of a change in current in the beam. Here it will be assumed that the beam velocity remains constant. Current may change because of the use of a control electrode, such as a grid, in the electron gun to modulate electron flow. Also, current changes in localized regions when the beam interacts with the RF field on a nearby circuit to produce bunching of the electrons.

It will be assumed that, before the change in current, the beam is in equilibrium so that there are no radial motions of electrons. As a result, from the beam equation in the form given in (7-15) (or from (7-19)),

$$0 = -b_o + \frac{g^4}{b_o^3} + \frac{a^2}{b_o} \tag{7-68}$$

or

$$b_o^2 - \frac{g^4}{b_o^2} = a^2 \tag{7-69}$$

Now, the quantity a contains the beam current; that is,

$$a \propto I^{1/2} \quad \text{or} \quad a^2 \propto I \tag{7-70}$$

When the current changes an amount Δ to I_2 such that $I_2 = I(1 + \Delta)$ and $a_2^2 = a^2(1 + \Delta)$ then the beam radius changes to $b_2 = b_o(1 + \delta)$. At equilibrium

$$b_2^2 - \frac{g^4}{b_2^2} = a_2^2 \tag{7-71}$$

or

$$b_o^2(1 + \delta)^2 - \frac{g^4}{b_o^2(1 + \delta)^2} = a^2(1 + \Delta) \tag{7-72}$$

By dividing by b_o^2 and substituting for g/b_o and a/b_o from (7-57) and (7-59) (where it is remembered that $R_e = b_o/a$), then

$$(1 + \delta)^2 - \left[1 - \frac{1}{m^2}\right]\frac{1}{(1 + \delta)^2} = \frac{1}{m^2}(1 + \Delta) \tag{7-73}$$

For small values of δ the following approximations may be made

$$(1 + \delta)^2 \simeq 1 + 2\delta \quad \text{and} \quad \frac{1}{(1 + \delta)^2} \simeq 1 - 2\delta \tag{7-74}$$

and so

$$1 + 2\delta - \left[1 - \frac{1}{m^2}\right](1 - 2\delta) = \frac{1}{m^2}(1 + \Delta) \tag{7-75}$$

or

$$\delta = \frac{\Delta}{4m^2 - 2} \tag{7-76}$$

For Brillouin flow when $m = 1$, the value of δ is

$$\delta_B = \frac{\Delta}{2} \tag{7-77}$$

and so the ratio of δ for $m > 1$ to δ_B is

$$\frac{\delta}{\delta_B} = \frac{1}{2m^2 - 1} \tag{7-78}$$

The ratio, δ/δ_B, is plotted in Figure 7-26 as a function of the confinement factor m. It is clear that the effect of confined flow on beam perturbations resulting from current changes is significant and that the use of confined flow is very desirable to reduce these perturbations. The penalty is that the increase in flux density required for confined flow necessitates an increase in the size of the magnets used and, if solenoid focusing is employed, an increase in solenoid power.

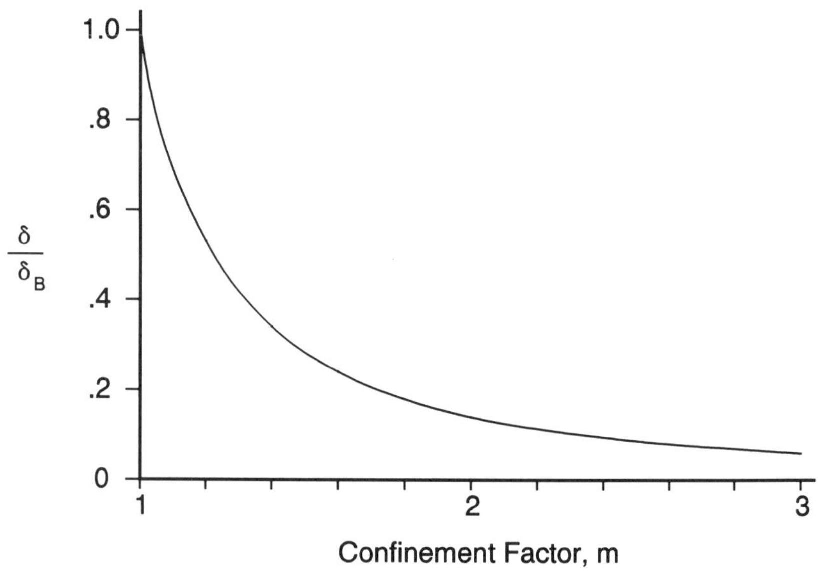

Figure 7-26 Change in beam radius for confined flow ($m > 1$) normalized to the Brillouin value ($m = 1$) resulting from a change in beam current.

In his paper on magnetically focused electron beams, Brewer clearly showed the confining effect of magnetic flux through the cathode [4]. As the cathode flux density was increased, beam scalloping decreased and variations in current density with distance were reduced. He also noted that cathode flux prevents translaminar electrons from crossing or coming close to the beam axis. Thus, cathode flux improves the laminarity of the beam.

7.3 UNIFORM-FIELD FOCUSING AND NONLAMINAR FLOW

In the chapters on cathodes and guns, several mechanisms have been identified that produce transverse velocity components to electron trajectories:

1. Thermal effects;
2. Patchy emission;
3. Cathode roughness;
4. Filament magnetic field;
5. Field distortion by grids;
6. Gun aberrations.

As a result of the transverse velocity components, electron trajectories cross and the electron flow is said to be nonlaminar.

There is no single theory for treating all forms of nonlaminar flow. In fact, the only technique for determining, with reasonable accuracy, electron behavior in nonlaminar flow is by tracing electron trajectories by numerical techniques. Some results of this approach will be shown in a following section, which is concerned with electron flow from gridded guns in PPM focused structures.

Several theories exist for special cases of nonlaminar flow. One of the first was that by Pierce and Walker for the effects of thermal velocities on Brillouin flow [5]. This theory was based on the assumption of a Maxwellian distribution of charge density within the beam. The beam was assumed to be in thermal equilibrium with a temperature that was suggested to be the cathode temperature multiplied by the area compression ratio of the gun. Pierce and Walker computed the percentage of the beam current to be found outside the nominal Brillouin radius.

Brewer compared some of his experimental beam profiles with the predictions of Pierce and Walker [4]. The temperature used was the cathode temperature multiplied by the area compression ratio of the gun. Brewer demonstrated that this was reasonable by examining the radial spread of electrons passing through an aperture located on the axis of the beam. The observed temperature was within 20% of the calculated temperature. Brewer's measurement of the current profile for the beam edge compared favorably with the Pierce and Walker theory near the nominal beam radius. At larger radii, there was a low current-density "tail" on the current profile that was far from the prediction of the Pierce and Walker theory. This tail was thought to have resulted from the translaminar streams of electrons produced by gun aberrations.

The reports of Gilmour *et al.* [2, 3] and Brewer [4] show that beams focused for near Brillouin flow conditions do, in fact, behave very nearly as predicted by laminar flow theory. On the other hand, Ashkin [6] and Harker [7] have reported that electron flow in some beams, which had been thought to be laminar, was actually very nonlaminar. Following the works of Ashkin and Harker, Herrmann presented an optical theory for thermal velocity effects that discarded the assumption of laminarity [8].

Herrmann's theory treats beams as perfect optical systems. The thermal velocities of the electrons emitted in all directions by the cathode are assumed to have a Maxwellian distribution. Electron trajectories are assumed to obey the laws of geometrical optics of ideal optical systems. A comparison of electron trajectories in Herrmann's optical model with those for a laminar flow model is shown in Figure 7-27. For laminar flow, as has been discussed earlier in this chapter, the electrons move side by side without ever crossing each other's paths. For imbalanced space charge and magnetic forces, the beam radius oscillates with a scallop wavelength, λ_s.

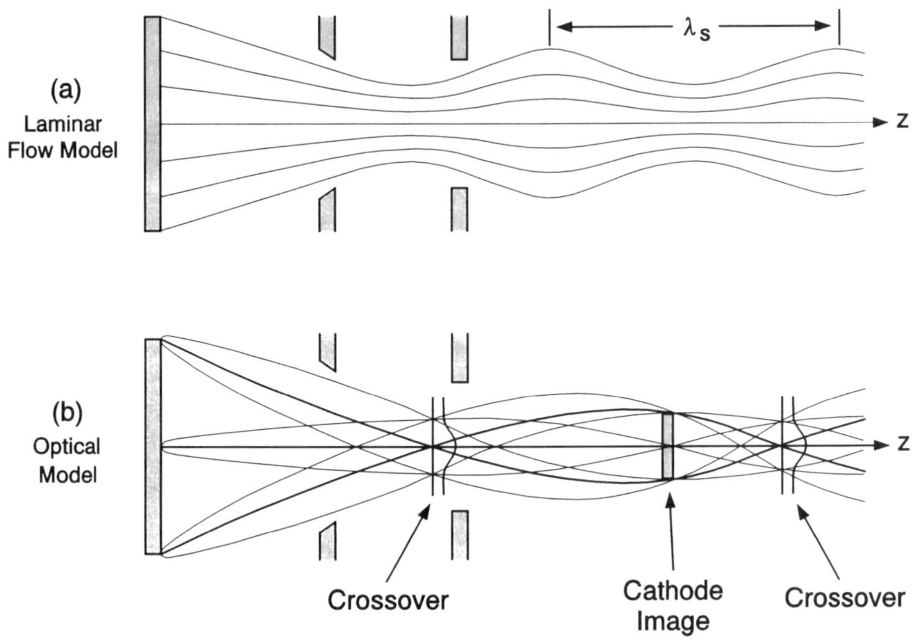

Figure 7-27 Comparison of laminar and optical models of electron flow. Adapted from Gabriel Herrmann, *Jour. Appl. Phys.*, February 1958.

With the optical theory, electrons emitted from a point on the cathode focus periodically to form cathode images. As explained by Herrmann: "One finds along the axis a succession of cathode images alternating with crossovers which impart to the beam a distinct structure which depends primarily on the internal motion of the electrons rather than on the shape of a boundary electron, as is the case in laminar theory."

Amboss compared the predictions of Herrmann's theory with measurements on a highly convergent, low-perveance (0.015×10^{-6}) beam focused by a flux density over an order of magnitude above the Brillouin value [9]. As shown in Figure 7-28, he found excellent agreement with the predictions of the optical theory. In this figure, λ_I is the distance between cathode images (the first and second images are shown). Note that the current profile changes from rectangular to Gaussian in the crossover region between images.

The Amboss results show the validity of the optical theory under conditions far removed from normal beam focusing conditions. For focusing fields closer to the Brillouin level, Brewer found that the behavior of the translaminar streams of electrons, which were produced by gun aberrations, was similar to that predicted by the optical theory. In Gilmour's beam studies, the only evidence of optical-theory behavior was that shown in Figure 5-49. In that figure, an image of the spiral-wound heater used for heating the cathode was shown. It was the magnetic flux lines generated by the heater current that perturbed the flow of electrons from the cathode and, apparently, led to the optical behavior of the subsequent electron trajectories.

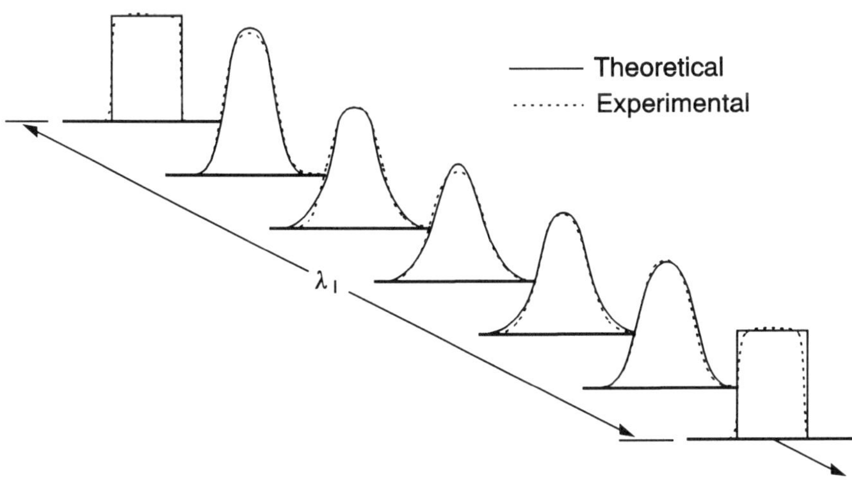

Figure 7-28 Comparison of optical theory predictions with experiment. Adapted from K. Amboss, *IEEE Trans. on Electron Devices*, October 1964. © 1964 IEEE.

Other studies of nonlaminar beams include those of Kirstein [10, 11] and Hechtel [12]. Kirstein used essentially the same approach as Herrmann. Axially symmetric and two-dimensional strip beams were treated. Unfortunately, Kirstein's results are not in a form that can readily be used. Hechtel treated beams having a Gaussian current distribution in the radial direction at the exit from the gun. The reason for considering this type of beam is the, at millimeter-wave frequencies where beams are very small, thermal velocity effects in the electron gun result in a beam having a current distribution that is nearly Gaussian. Analytical as well as numerical methods were used. Expressions are given for the magnetic field required to achieve various percentages of beam transmission through a tunnel.

7.4 PERIODIC PERMANENT MAGNET (PPM) FOCUSING

7.4.1 Overview

Periodic permanent magnets are frequently used when it is necessary to minimize the weight and size of a TWT. The weight reduction of the focusing system, when compared with a solenoid, can be one to two orders of magnitude. Appendix D contains a discussion of periodic permanent magnet focusing systems and the reasons for the large weight reductions that result from their use.

An electron beam focused by a periodic permanent magnet is illustrated in Figure 7-29. The basic principle of PPM focusing is straightforward. As an electron beam enters the field of one of the magnet sections, the forces on the electrons act in exactly the same way as was described in the section on Brillouin focusing. That is, the beam starts to rotate and the interaction of the resulting rotational motion with the axial field produces a radial force that compresses the beam.

As the beam leaves the magnet section, rotation stops and the focusing forces vanish. As a result, the beam expands under the influence of space charge forces. The beam then enters another magnet section with the field in the opposite direction from the previous one. As a result, the beam rotates in the opposite direction, but is focused just as it was in the previous section.

The overall result, as the beam traverses the alternating field of the periodic magnet structure, is that its rotational motion oscillates back and forth, producing alternating periods of magnetic focusing and beam expansion. The appearance of the beam, as shown in Figure 7-29, is like that of a scalloping Brillouin beam.

7.4.2 Laminar Flow

While the qualitative explanation of PPM focusing is straightforward, the analysis is complex and not so straightforward. Still, much can be learned from an examination of the beam equation when PPM fields are used. To proceed with the analysis, it is assumed, first, that the magnetic field is a function of axial position, z, only and that no radial variations occur. It can be shown with Laplace's equation that this assumption is realized along the axis when the magnetic field varies sinusoidally with distance. For the purposes of this analysis, it will be assumed that

$$B_z = B_p \cos \frac{2\pi z}{L} \tag{7-79}$$

where B_p is the peak field and L is the magnet period, or twice the distance between adjacent pole pieces as shown in Figure 7-29.

Next, assuming that no magnetic flux links the cathode, the beam equation may be written as

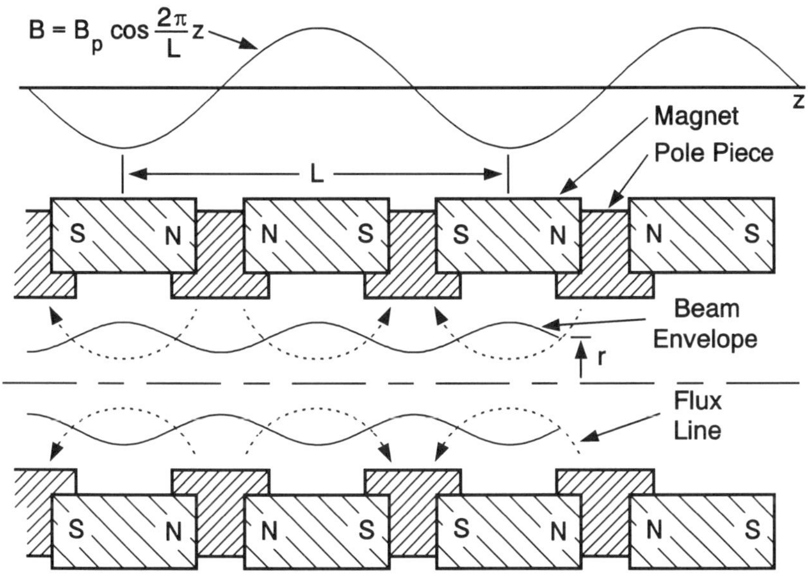

Figure 7-29 Periodic focusing of an electron beam with permanent magnets.

$$\ddot{b} + b\left[\frac{\eta}{2}B_z\right]^2 - \frac{\eta I}{2\pi\epsilon_o u_o b} = 0 \tag{7-80}$$

or as

$$\ddot{b} + b\left[\frac{\eta}{2}B_p \cos\frac{2\pi z}{L}\right]^2 - \frac{\eta I}{2\pi\epsilon_o u_o b} = 0 \tag{7-81}$$

To obtain the equation of motion as a function of distance for an electron at the surface of the beam, we let $z = u_0 t$. Then,

$$\frac{d^2 b}{dz^2} + b\left[\frac{\eta B_p}{2u_o}\cos\frac{2\pi z}{L}\right]^2 - \frac{\eta I}{2\pi\epsilon_o u_o^3 b} = 0 \tag{7-82}$$

Now, following the procedure used in analyzing Brillouin beams, it will be assumed that there is the equivalent of a Brillouin flux density, B_B, that produces an equilibrium radius, a, for the beam. The relation between a and B_B is the same as for Brillouin flow:

$$a = \frac{1}{\frac{\eta}{2} B_B} \left[\frac{\eta I}{2 \pi \epsilon_o u_o} \right]^{1/2} \tag{7-83}$$

Using this equilibrium radius, the equation of motion may be written as

$$\frac{d^2 b}{dz^2} + b \left[\frac{\eta}{2u_o} \right]^2 \left[B_p^2 \cos^2 \frac{2 \pi z}{L} - B_B^2 \frac{a^2}{b^2} \right] = 0 \tag{7-84}$$

The relation between B_p and B_B may be found by equating the magnitudes of the restoring forces at the maximum and minimum excursions of the beam radius. At the maximum radius, let $b = b_M$. There, the flux density is at a maximum so $\cos 2\pi z/L = 1$ and so

$$\frac{d^2 b_M}{dz^2} = - b_M \left[\frac{\eta}{2u_o} \right]^2 \left[B_p^2 - B_B^2 \frac{a^2}{b_M^2} \right] \tag{7-85}$$

At the minimum beam radius, let $b = b_m$. Here, the flux density passes through zero so $\cos 2\pi z/L = 0$ and so

$$\frac{d^2 b_m}{dz^2} = b_m \left[\frac{\eta}{2u_o} \right]^2 B_B^2 \frac{a^2}{b_m^2} \tag{7-86}$$

Now, equating the magnitudes of the restoring forces,

$$b_M \left[B_p^2 - B_B^2 \frac{a^2}{b_M^2} \right] = b_m B_B^2 \frac{a^2}{b_m^2} \tag{7-87}$$

When the amplitude of the ripple on the beam is small, $b_M \approx b_m \approx a$ so

$$B_p^2 - B_B^2 = B_B^2 \tag{7-88}$$

or

$$B_p = \sqrt{2}\, B_B \tag{7-89}$$

Thus, the RMS value of the periodic field is the Brillouin field. As a result, the PPM focusing discussed here is similar to Brillouin focusing in that the space charge forces are balanced by the magnetic forces and no field is present at the cathode.

Next, the stability of PPM focusing will be examined. The stability question is important because there are ranges of values of flux density, magnet period, L, beam current and beam voltage for which the beam diverges so that no beam current is transmitted through the PPM structure. Following the works of Pierce [13] and Mendel *et al.* [14], with the substitutions

$$T = \frac{2\pi z}{L} \tag{7-90}$$

and

$$\omega = \frac{2\pi u_o}{L} \tag{7-91}$$

which is the frequency at which the field appears to the moving electrons to alternate, (7-82) may be rewritten as

$$\frac{d^2 b}{dT^2} + b\left[\frac{\eta B_p}{2\omega}\right]^2 \cos^2 T - \frac{\eta I}{\omega^2 2\pi\epsilon_o u_o b} = 0 \tag{7-92}$$

The quantity, $\eta B_p/2$, is the peak value of the Larmor frequency, ω_L, and

$$\frac{\eta I}{2\pi\epsilon_o u_o} = \frac{\eta\rho a^2}{2\epsilon_o} = \frac{\omega_p^2 a^2}{2} \tag{7-93}$$

where ρ is an average charge density for the beam and ω_p is an average plasma frequency.

Now,

$$\frac{d^2 b}{dT^2} + b\left[\frac{\omega_L}{\omega}\right]^2 \cos^2 T - \frac{\omega_p^2 a^2}{2\omega^2 b} = 0 \tag{7-94}$$

Letting $b/a = \sigma$ be normalized beam radius and, since $\cos^2\theta = 1/2(1 + \cos 2\theta)$, then

$$\frac{d^2\sigma}{dT^2} + \frac{1}{2}\left[\frac{\omega_L}{\omega}\right]^2 (1 + \cos 2T)\sigma - \frac{1}{2}\left[\frac{\omega_p}{\omega}\right]^2 \frac{1}{\sigma} = 0 \tag{7-95}$$

Finally, letting α be the magnetic field coefficient given by

$$\alpha = \frac{1}{2}\left[\frac{\omega_L}{\omega}\right]^2 \tag{7-96}$$

and β be the space charge coefficient given by

$$\beta = \frac{1}{2}\left[\frac{\omega_p}{\omega}\right]^2 \tag{7-97}$$

the equation for the normalized beam radius becomes

$$\frac{d^2\sigma}{dT^2} + \alpha(1 + \cos 2T)\sigma - \frac{\beta}{\sigma} = 0 \tag{7-98}$$

Solutions to this equation have been obtained with the aid of an analog computer [13, 14]. It was assumed that the electrons were injected into the magnetic field with no radial velocity at $z = 0$, which was where the magnetic field was a maximum. Typical beam envelope curves for three values of the magnetic field are shown in Figure 7-30. Similar curves were obtained for a wide range of values of β and from these curves the universal plot for optimum focusing given in Figure 7-31 was constructed. Note that optimum focusing occurs when

$$\alpha = \beta \tag{7-99}$$

or when

$$\omega_L = \omega_p \tag{7-100}$$

When substitutions are made for flux density, charge density and velocity, this relation is seen to be that for Brillouin flow given in (7-83). This result is not surprising since, on average, the space charge forces are the same as for the Brillouin case and the RMS field is the Brillouin field and so the centrifugal and focusing forces are the same as for the Brillouin case.

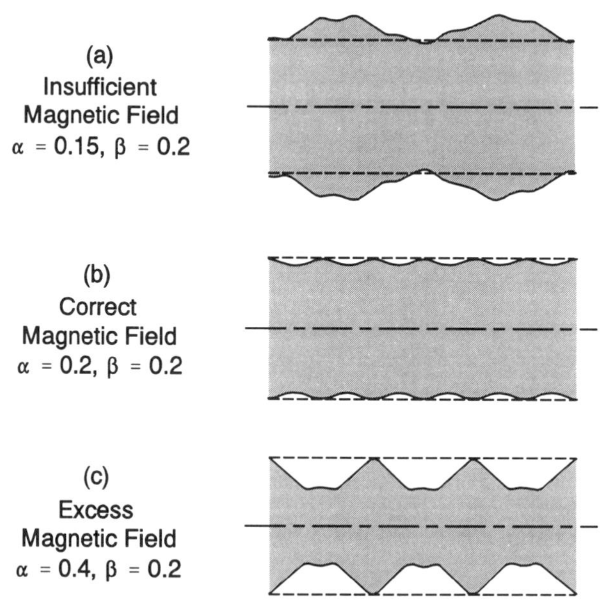

Figure 7-30 Typical analog computer plots of beam contours. From J.T. Mendel *et al.*, *Proc. IRE*, May 1954. © 1954 IRE (now IEEE).

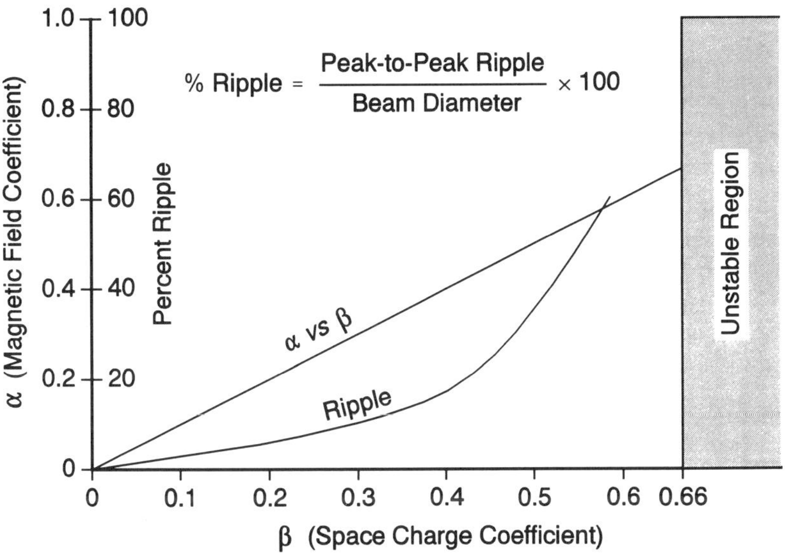

Figure 7-31 Optimum focusing conditions and minimum ripple amplitude. Adapted from J.T. Mendel *et al.*, *Proc. IRE*, May 1954. © 1954 IRE (now IEEE).

The minimum ripple for optimum focusing is also shown in Figure 7-31. For values of β below about 0.35, the ripple amplitude is reasonably small. As β increases above 0.35, the ripple amplitude becomes very large and, in fact, for $\beta > 0.66$, the beam diverges.

We can learn more about the onset of the unstable region in Figure 7-31 by further examining the equation for the normalized beam radius (7-98). The term containing β is the space charge term and represents a diverging force. If this term is neglected and the remainder of the equation diverges, then the solutions including space charge forces must diverge. Now, ignoring the space charge term, (7-98) is

$$\frac{d^2\sigma}{dT^2} + \alpha\left(1 + \cos 2T\right)\sigma = 0 \tag{7-101}$$

This equation is a form of the Mathieu equation, which is characterized by solutions for α that are periodic in T for some ranges of α and unstable for other ranges of α. Figure 7-32 shows the ranges of α for which the solutions are unstable. Note that, as α increases from zero, the first unstable region starts at

$$\alpha = 0.66 \tag{7-102}$$

For $\alpha = \beta$ then, the unstable region for β would be expected for $\beta > 0.66$ as shown in Figure 7-31.

From (7-96) and (7-97), the relations between ω_L, ω_p and ω at which the unstable region starts are

$$\omega_L = 1.15\,\omega \tag{7-103}$$

and

$$\omega_p = 1.15\,\omega \tag{7-104}$$

Thus, if the rotation frequency of the beam, ω_L, is close to the frequency at which the field appears (to the electrons) to alternate, ω, the beam diverges. Similarly, if the natural frequency of oscillation of the electrons, ω_p, is close to ω, the beam diverges. In each case, the periodic impulses provided by the field are in near synchronism with a natural frequency of the beam and this results in oscillations that grow with time.

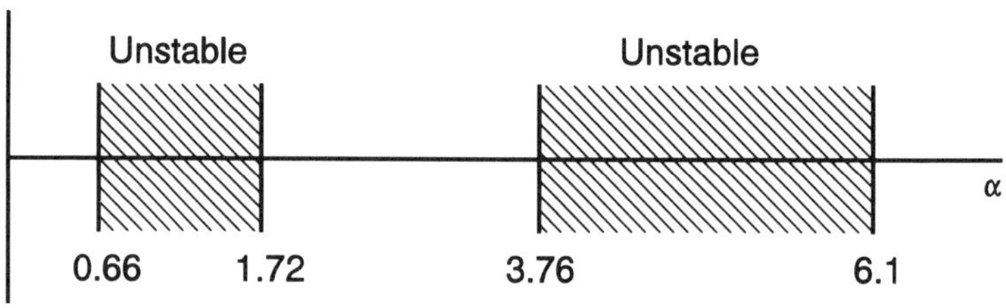

Figure 7-32 Regions of α for which (7-101) is unstable.

In practice, PPM structures are designed to operate well below the unstable region. As a result, when the parameters contained in ω_L and ω of the magnetic field coefficient (7-96) are used, it is found that

$$\frac{L^2 B_p^2}{V_o} < 2.37 \times 10^{-9} \tag{7-105}$$

where V_0 is the dc beam voltage. Also, from the space charge coefficient (7-97)

$$\frac{\lambda_p}{L} > 1.15 \tag{7-106}$$

This quantity is known as the *beam stiffness factor*.

In this section we have examined PPM focusing of beams from electron guns with cathodes that are shielded from magnetic flux. It is also possible to achieve the equivalent of confined flow focusing with PPM fields and with flux through the cathode. Several researchers including Chang [15], Mendel [16] and Harker [17] have treated periodic focusing from partially shielded cathodes.

Confined flow PPM focusing achieves many of the advantages of confined flow focusing with uniform fields. For example, resistance to ion and RF defocusing is increased and the control of beams from gridded guns is improved. A deficiency in the

PPM case that is not present in uniform-field focusing is that the peak rotational energy of the beam increases with magnetic field level.

The reason for the increase of rotational energy with the magnetic field can be understood by reconsidering the discussion of uniform-field confined flow focusing given earlier in this chapter. As shown in Figure 7-10, magnetic flux links the cathode, and as the beam leaves the gun and enters the main focusing field, the *increase* in flux density must be sufficient to produce the magnetic focusing force that counterbalances the space charge and centrifugal forces. Because the axial field is larger than for Brillouin flow, proper focusing occurs when the beam rotates more slowly than for the Brillouin case.

For confined flow PPM focusing, the entrance condition of the beam into the first magnet of the PPM stack is the same as for uniform-field focusing. As the beam leaves the first magnet and enters the second, the electrons encounter a change in flux density that is twice the peak magnet flux density, and this peak flux density must be greater than the peak density for the Brillouin case for confined flow to occur. The large change in flux density causes a reversal and large change in the rate at which the beam rotates. Thus, the rotation changes from a rate well below that for the Brillouin case to a rate in the opposite direction that is far above that for Brillouin flow.

Figure 7-33 shows the variation of peak rotational beam energy with magnetic flux density and with microperveance (perveance $\times 10^6$) [18]. Note that, for reasonable values of confinement, a large fraction of the beam energy is in rotation for microperveances above ~ 0.5. As a result, efficiency is degraded. Thus, confined flow PPM focusing is limited to low-perveance designs. Hence, it is particularly suitable for millimeter-wave tubes for which perveance is usually limited by cathode loading considerations and in which beam interception problems are particularly critical.

A primary reason for using confined flow focusing is to limit the expansion of the beam during RF operation. In [18], Ayers *et al.* present computer calculations of beam expansion factors (the ratio of maximum beam radius with RF drive to the dc radius) for a 0.45 microperveance TWT and for a 0.036 microperveance TWT. The results for the 0.036 microperveance case are shown in Figure 7-34. Some of the more important aspects of this figurc arc

1. Beam expansion is dramatically decreased as magnetic field is increased.
2. There is little change in beam expansion as λ_p/L is increased.
3. The ripple, which is defined as

$$\%\text{ripple} = \frac{r_{\max} - r_{\min}}{r_{\max} + r_{\min}} \times 100 \tag{7-107}$$

   decreases rapidly as λ_p/L is increased.
4. The maximum flux density and, to some extent, the energy product of the magnets, limits operation at high magnetic fields and high values of λ_p/L.

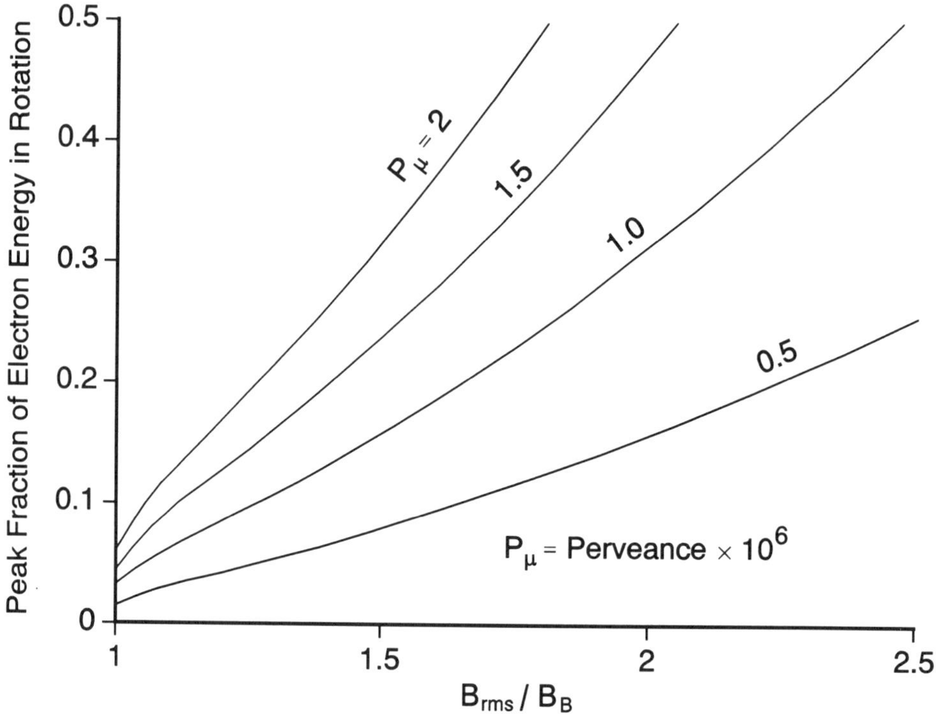

Figure 7-33 Dependence of peak transverse energy on magnetic field for various microperveances. From W.R. Ayers *et al.*, *Technical Digest*, 1985 IEDM. © 1985 IEEE.

7.4.3 Nonlaminar Flow

In the chapters on cathodes and guns, many sources of beam nonlaminarities were discussed. The most important were the grids in gridded electron guns. Because a large fraction of PPM-focused TWTs use gridded electron guns and because the nonlaminarities produced by these guns are usually far more significant than other nonlaminarities, we will deal primarily with beams from gridded guns in this chapter.

To gain an appreciation for the nonlaminar nature of the beam from a gridded gun, consider Figure 7-35. This figure contains computer-generated electron trajectories for a PPM focused beam from a shadow gridded gun [19]. The radial scale has been expanded to ten times the axial scale to help illustrate the apparently random nature of the electron trajectories.

The most definitive procedure that has been reported for dealing with PPM focused beams and gridded guns is that by True [19–21]. In this section, we will describe the basic elements of True's theory. There are, basically, three steps to True's procedure:

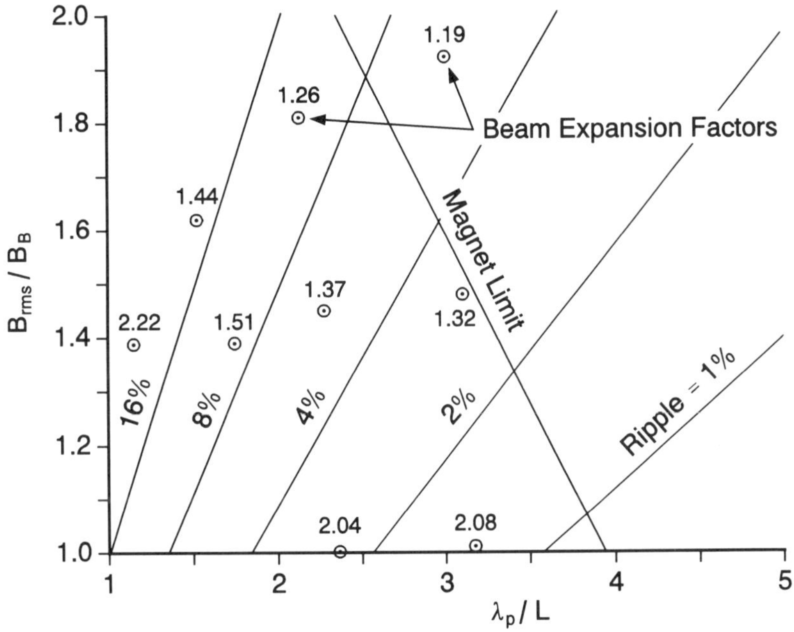

Figure 7-34 Beam expansion factors for 0.036 microperveance TWT. From W.R. Ayers *et al.*, *Technical Digest*, 1985 IEDM. © 1985 IEEE.

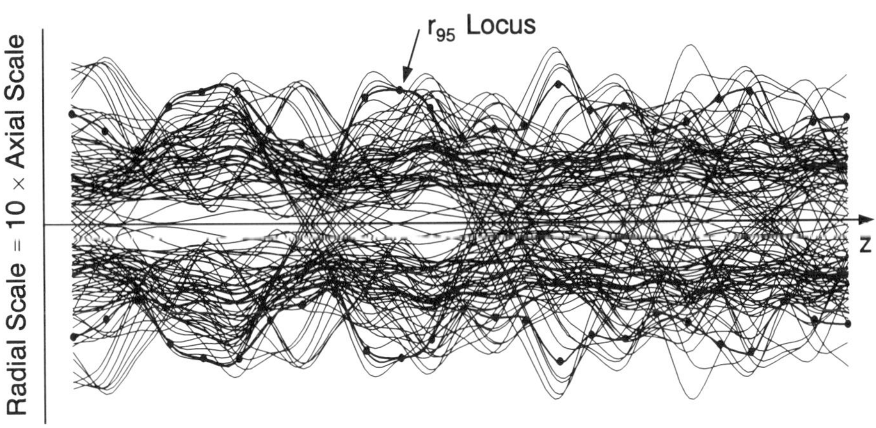

Figure 7-35 PPM focused beam from shadow gridded gun. Adapted from Richard True, *IEEE Trans. on Electron Devices*, February 1987. © 1987 IEEE.

1. Most important, the transverse velocity content of the beam at its entrance into the magnetic focusing field is found. This is a measure of the transverse velocities of the electrons and is expressed in normalized form as

$$\sigma = \frac{\overline{v_r}}{u_o} \tag{7-108}$$

where u_o is the dc velocity of the beam.
2. The transverse velocities along with space charge forces must be controlled by the magnetic focusing structure. In effect, True treats the transverse velocity content as an additional force (in addition to the space charge force) that must be focused. This is done by finding an effective microperveance, P_μ^+, for the beam (microperveance equals perveance multiplied by 10^6), which is the sum of the regular beam microperveance, P_μ, and a scattering microperveance, $P_{\mu s}$, that results from the transverse velocity content. That is,

$$P_\mu^+ = P_\mu + P_{\mu s} \tag{7-109}$$

3. In the absence of magnetic flux through the cathode, the field level required to focus an ideal beam is the Brillouin field, and this is proportional to the square root of perveance. True assumes that the increased focusing field, B^+, made necessary by transverse electron velocities should be at least

$$\frac{B^+}{B_B} = \left[\frac{P_\mu^+}{P_\mu}\right]^{1/2} \tag{7-110}$$

The ratio, B^+/B_B is referred to as the *magnetic field intensification ratio*. True finds that the perveance ratio, P_μ^+/P_μ, must be multiplied by a factor of 1.65 to minimize beam ripple in actual magnet designs.

The normalized transverse velocity content of the beam, σ, is composed of two parts. One, σ_T, results from the thermal velocity content of the beam, given by

$$\sigma_T = \frac{r_c}{r_{95}}\left[\frac{kT}{2eV_o}\right]^{1/2} \tag{7-111}$$

where r_c is the disc radius of the cathode and r_{95} is the beam radius containing 95% of the beam current (see, for example, Figure 7-35). The other part of σ is the nonthermal part, σ_{NT}, which results from gun aberrations and scattering due to grid wires. This nonthermal part can be found only by a computer simulation that includes grid wire details. True finds that a good approximation for the way in which the thermal and nonthermal values of σ combine is

$$\sigma = \left(\sigma_{NT}^2 + \sigma_T^2\right)^{1/2} \tag{7-112}$$

For gridded guns, $\sigma_{NT} >> \sigma_T$ and so $\sigma \approx \sigma_{NT}$. In the remainder of this section, we will treat only gridded guns and so we will ignore thermal effects. We will also substitute σ for σ_{NT}.

The value for σ is found at or near the beam minimum from the weighted standard deviation of the slopes of the electron trajectories. Although σ is defined in terms of slopes, it is essentially equal to $\bar{v}_r/u_0$. True has found, as shown in Figure 7-36, that σ can be approximated by

$$\sigma = 0.0015\,\theta \tag{7-113}$$

where θ is the half angle in degrees of the electron gun. As a side note, from Figure 7-36, it is observed that guns which have small half angles also have low values of σ. The reason for this was given in Chapter 6 when a transverse electron crossing a convergent electron stream was considered. The transverse velocity increase was found to be inversely proportional to the radius compression ratio.

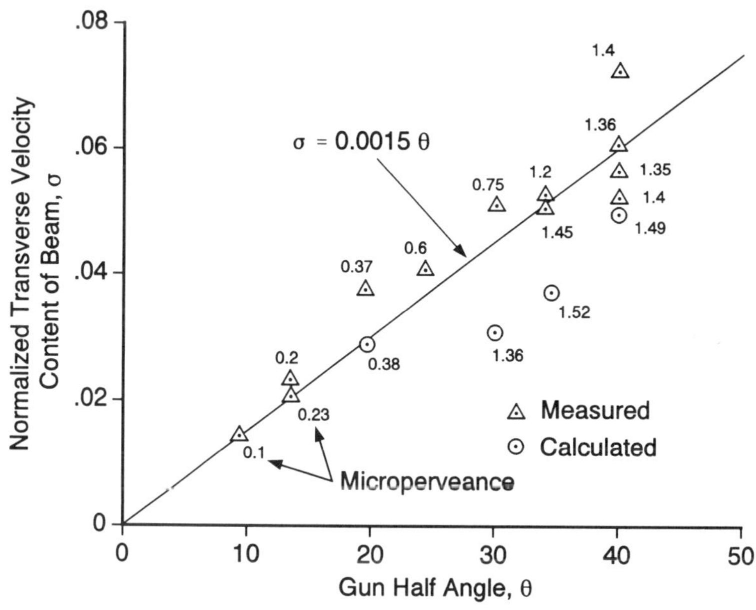

Figure 7-36 Justification for $\sigma = 0.0015\,\theta$. From Richard True, *IEEE Trans. on Electron Devices*, March 1984. © 1984 IEEE.

The next step in the analysis is to incorporate the effect of transverse velocities on the spreading of the electron beam. As pointed out earlier in this section, True treats the transverse velocity content as a force that, along with the space charge force, must be focused by the magnetic field. This is done by finding an effective microperveance, which is $P_\mu^+ = P_\mu + P_{\mu s}$. The scattering microperveance, $P_{\mu s}$, is found to be related to σ by

$$P_{\mu s} = 197.6\,\sigma^2 \tag{7-114}$$

This relation is derived by considering electron behavior and the rate of beam spread near the beam minimum. Thus, the effective microperveance is

$$P_{\mu}^{+} = P_{\mu} + 197.6\,\sigma^2 \tag{7-115}$$

The effective microperveance is next used to find the magnetic field intensification factor, B^+/B_B. The magnetic field is proportional to the square root of perveance. Also, allowance must be made for effects such as spherical aberration, nonideal entrance conditions and gun-design imperfections, and so True assumes that the intensification factor should take the form

$$\frac{B^+}{B_B} = \left(F\,\frac{P_{\mu}^{+}}{P_{\mu}} \right)^{1/2} \tag{7-116}$$

where a value for F of 1.65 is chosen on the basis of numerical and experimental data. Now, substituting the relation for effective microperveance, the intensification factor becomes

$$\frac{B^+}{B_B} = \left[F \left(1 + \frac{197.6\,\sigma^2}{P_{\mu}} \right) \right]^{1/2} \tag{7-117}$$

To make use of this equation, the relation between σ and P_{μ} is needed. Equation 7-113 was an empirical relation between σ and the gun half angle, θ. The exact relation between perveance and θ was discussed in Chapter 6. For our purposes here, a much simpler equation developed by True can be used. That is,

$$P_{\mu} = 0.006 \left(\theta\,\frac{r_c}{r_{95}} \right)^2 \tag{7-118}$$

By substituting this equation and that for σ into the intensification factor,

$$\frac{B^+}{B_B} = \left[1.65 \left(1 + 0.0741\,\frac{r_c}{r_{95}} \right) \right]^{1/2} \tag{7-119}$$

This result is plotted in Figure 7-37. It is interesting that the intensification factor is a function of only the radius compression ratio. We now see why beams from high convergence guns require field levels that are significantly higher above the Brillouin level than do beams from low convergence guns.

The magnetic field intensification factor may be combined with r_{95}/a, where a is the tunnel radius (so that r_{95}/a is the fraction of the tunnel filled by the beam) to obtain

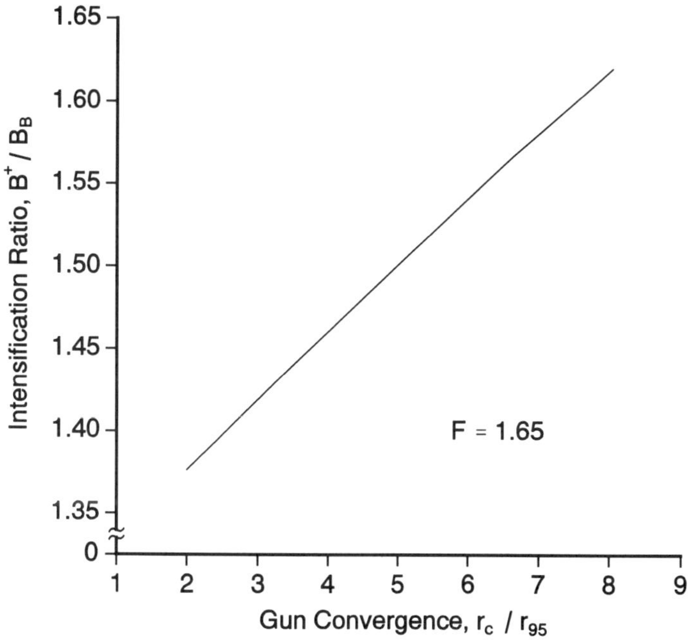

Figure 7-37 PPM field intensification ratio *versus* gun convergence. From Richard True, *IEEE Trans. on Electron Devices*, March 1984. © 1984 IEEE.

a normalized focusing factor, denoted by True as FOC F_1 and given by

$$\text{FOC } F_1 = \frac{B^+/B_B}{r_{95}/a} \tag{7-120}$$

Acceptable limits for this parameter for good tube design are shown in Figure 7-38.

In the investigation of the stability of PPM focused beams, we noted that the beam stiffness factor, λ_p/L, should be greater than 1.15 but we gave no indication of how much greater it should be. True states that for pulse tubes, λ_p/L should exceed 2.7 and for CW tubes, it should be greater than 3.2. These values are large enough so that the effects of the individual lenses of the PPM field blend together so that the beam approximates a beam focused by a uniform magnetic field.

Before leaving the subject of nonlaminar PPM focused beams, it is important to point out that factors other than RF drive level may cause a beam to increase in diameter as it drifts through a tube. This, in turn, would lead to higher interception than is expected in the absence of increased beam diameter. These factors may also affect the rate at which a beam expands as it enters the collector and this, in turn, affects the collector design. To deal with these phenomena, True [20, 21] has defined the term *tunnel emittance*, which is given by

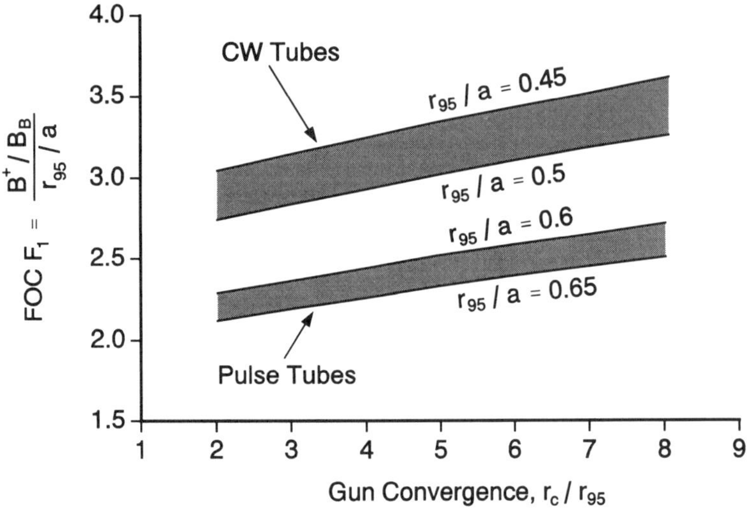

Figure 7-38 Relation of field intensification and beam filling factors to gun convergence. From Richard True, *IEEE Trans. on Electron Devices*, March 1984. © 1984 IEEE.

$$\text{Tunnel emittance} = \sigma \frac{r_{95}}{a} \tag{7-121}$$

Figure 7-39 shows the growth of tunnel emittance with distance. This growth is due partly to an increase in normalized transverse velocity (σ) and partly to a small increase in beam size.

One important factor that affects tunnel emittance growth is the current profile of the beam leaving the electron gun. The ideal beam shape is shown in Figure 7-40(a). This profile is shown slightly dished to match the radial variation in the magnetic field. The profiles shown in Figure 7-40(b) and (c) are less desirable shapes because of the concentration of charge in the center of the Gaussian distribution or in the ears of the "rabbit ears" of the hollow distribution. These charge concentrations contain potential energy that is converted to increased transverse electron velocities as the beam flows through the tube.

Another factor that affects tunnel emittance growth is the entrance point of the beam into the magnetic field. True considers the beam minimum position (the beam waist) to be the point where tunnel emittance is at a minimum. The axial position of the waist should be at the point where the first magnetic field peak has risen to $1/\sqrt{2}$ of its maximum value.

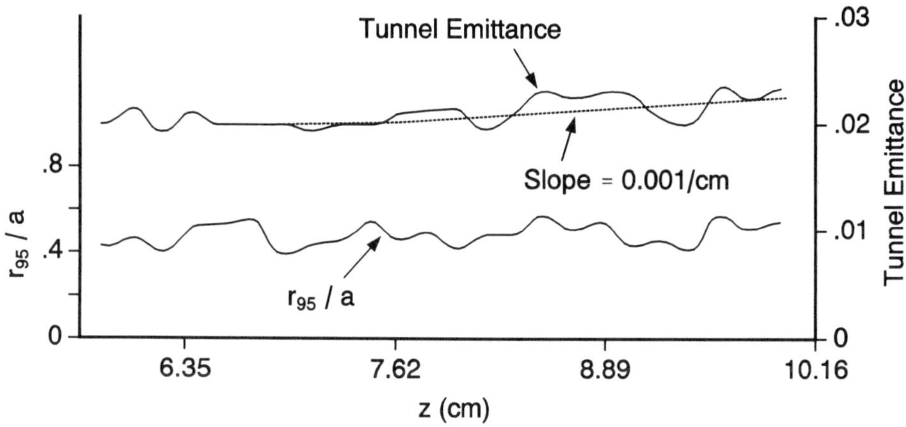

Figure 7-39 Illustration of growth of tunnel emittance with distance. From Richard True, *Technical Digest*, 1985 IEDM. © 1985 IEEE.

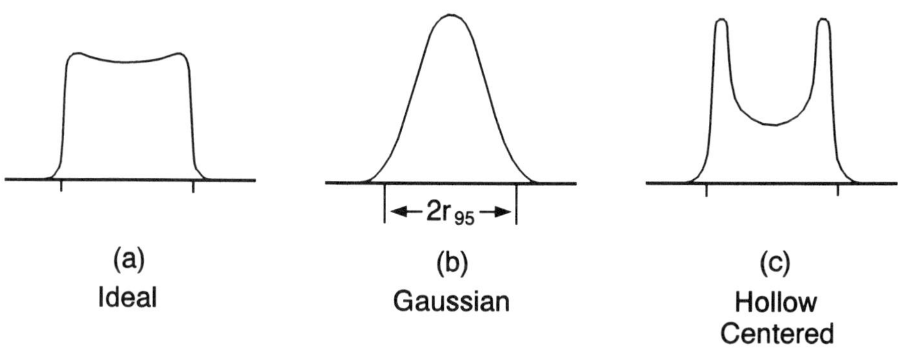

Figure 7-40 Current density *versus* radius at beam minimum. From Richard True, *IEEE Trans. on Electron Devices*, February 1987. © 1987 IEEE.

7.5 ION EFFECTS IN ELECTRON BEAMS [22]

Gas in a microwave tube is ionized by the electron beam and may degrade performance if the ion density becomes excessive. Linear beam tubes (klystrons and TWTs) are far more susceptible to ion effects than crossed field tubes because potential wells exist in which ions are trapped. Ion density increases with time so that ion effects are more severe during operation under long pulse or CW conditions. The use of confined flow focusing is effective in reducing or eliminating instabilities that result from ion formation. Ions may affect tube performance in at least the following three ways:

1. Ions neutralize electron space charge potentials within a beam. Beam diameter

becomes smaller (*ion focusing* is said to occur), and so power output is reduced. Also, interception current decreases with time. In extreme cases, the electron beam remains focused in the collector and melts a hole through the collector wall and destroys the tube.

2. Because the electron space charge potential is reduced by ions, the electron beam velocity is increased. This causes a reduction in the phase shift through a tube. The change in beam potential caused by ions may be on the order of 1%. Since beam velocity is proportional to the square root of potential, the change in velocity may be on the order of 1/2%. For a tube phase length of 10,000°, the decrease in phase would be about 50° and would vary with tube operating conditions.
3. Ions may produce electron beam oscillations or noise, which appears as modulation on the power output of a tube (*ion sidebands* are said to occur). The frequency of the oscillations is dependent on electron beam density and ion mass. Frequencies of MHz to tens of MHz are common.

Figures 7-41 and 7-42 contain graphic illustrations of some of the effects of ions in an electron beam. In Figure 7-41, an electron beam cross section is shown with a significant peak in electron density in the center of the beam. The peak in density results from positive ions in the center of the beam near the electron gun being accelerated toward the cathode. In the region near the center of the cathode the ions cause sufficient neutralization of the electron space charge cloud so that additional electrons leave the cathode.

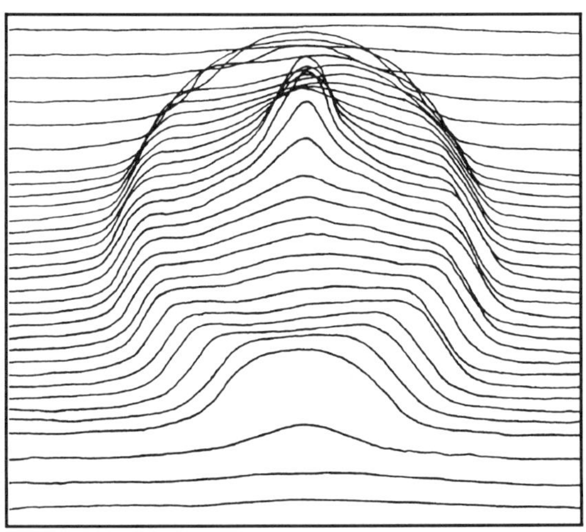

Figure 7-41 Electron beam cross section showing effect of ions.

When the ion concentration becomes sufficient, oscillations may occur as indicated by Figure 7-42(a). This figure shows the amplitude of the current to a probe in a pulsed Brillouin-focused electron beam in the presence of gas (hydrogen at a pressure of about 10^{-6} torr). In the absence of gas, the probe current was as shown in Figure 7-42(b).

At the beginning of the pulse with gas present, several cycles of a relatively high frequency of oscillation occur (~8 MHz). After about 2 μs, a lower frequency of oscillation (~4.5 MHz) occurs. Then, the higher oscillation frequency returns. Finally, at the end of the pulse, the oscillations become very noisy. Detailed experimental analyses showed that two distinct modes of ion induced beam oscillations occur. Measurements were made of the position and shape of the electron beam at various times during a pulse of the type shown in Figure 7-42.

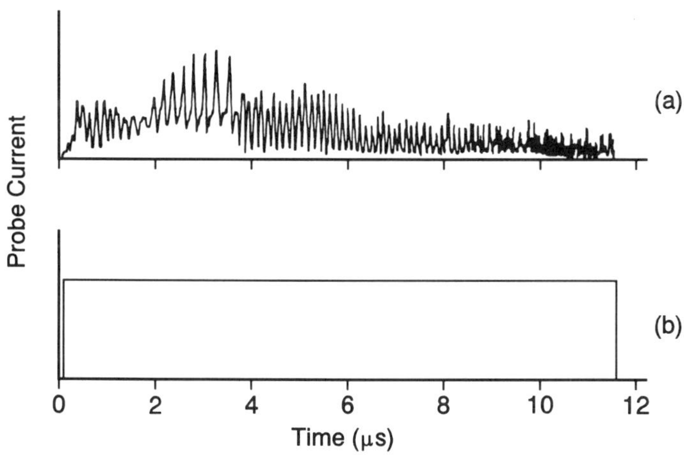

Figure 7-42 Probe current in electron beam analyzer with (a) and without (b) ion oscillations. From A.S. Gilmour, Jr., RADC-TDR-63-465, vol.1, December 1963.

7.5.1 Transverse Oscillations

During the low-frequency oscillation portions of the pulse, a transverse oscillation of the electron beam was found. That is, the peaks and valleys in current in the 2–3 μs time frame of Figure 7-42 were found to correspond to alternating transverse positions of the electron beam shown in Figure 7-43.

Ion oscillations have been analyzed by Peterson and Puthoff [23]. The transverse mode of oscillation may be explained with the aid of Figure 7-44, which shows the potential distribution in a drift tube containing an electron beam. Using Gauss's law, the electric field within the beam is found to be

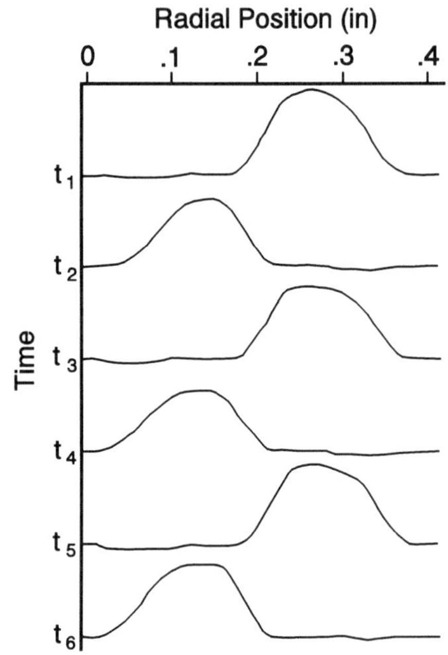

Figure 7-43 Probe current as a function of radial position for transverse mode of oscillation. From A.S. Gilmour, Jr., RADC-TDR-63-465, vol. 1, December 1963.

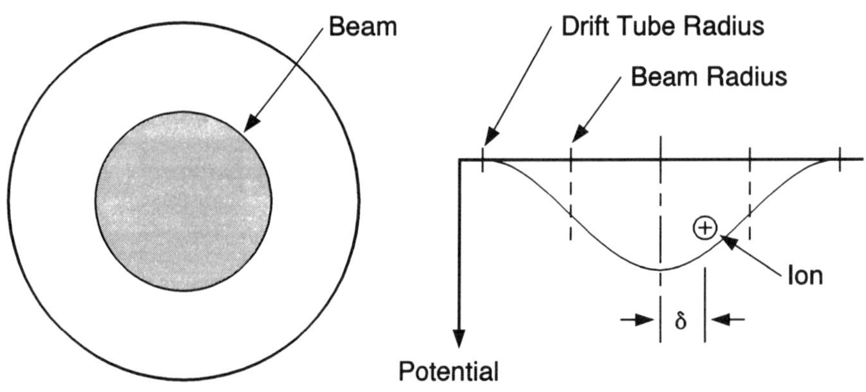

Figure 7-44 Potential distribution in a drift tube containing an electron beam.

$$E_r = -\frac{r\rho_e}{2\epsilon_o} \tag{7-122}$$

where ρ_e is the electron charge density in the beam.

Assume that a positive ion is located distance, δ, from the center of the beam as is indicated in Figure 7-44. If that ion has a single electronic charge, e, and mass, m_i then the force equation for the ion may be written as

$$m_i \frac{d^2\delta}{dt^2} = eE_r = -\frac{e\delta\rho_e}{2\epsilon_o} \tag{7-123}$$

or

$$\frac{d^2\delta}{dt^2} + \frac{\omega_{ip}^2}{2}\delta = 0 \tag{7-124}$$

where

$$\omega_{ip} = \left[\frac{e\rho_e}{m_i\epsilon_o}\right]^{1/2} = \left[\frac{e}{m_i\epsilon_o}\frac{I}{\pi b^2 u_o}\right]^{1/2} \tag{7-125}$$

is the ion plasma frequency, that is, the natural frequency of oscillation of the ion. Of course, the beam current is I and the beam velocity is u_o. Equation 7-124 describes simple harmonic motion and has a solution of the form

$$\delta = \delta_o \sin \frac{\omega_{ip}}{\sqrt{2}} t \tag{7-126}$$

When an ion beam of radius c (as indicated in Figure 7-45) with charge density equal to the electron density is considered, then the analysis is much more complex than that given previously, but the equation of motion for small displacements, δ, of the center of the ion beam from the center of the electron beam is still straightforward and is

$$\delta = \delta_o \sin \frac{\omega_{ip}}{\sqrt{2}}\left[1 - \frac{c^2}{d^2}\right]^{1/2} t \tag{7-127}$$

where d is the radius of the drift tube. Thus, the frequency of ion oscillation is reduced. If the diameter of the ion beam increases with time as ions are generated by the electron beam, it would be expected that the ion oscillation frequency would decrease. In fact, this effect has been observed as is indicated by Figure 7-46.

7.5.2 Radial Oscillations

During high-frequency portions of the pulse shown in Figure 7-42, a radial oscillation of the electron beam was found. That is, the beam diameter changed periodically from a relatively large value to a smaller value, as shown in Figure 7-47. This figure shows

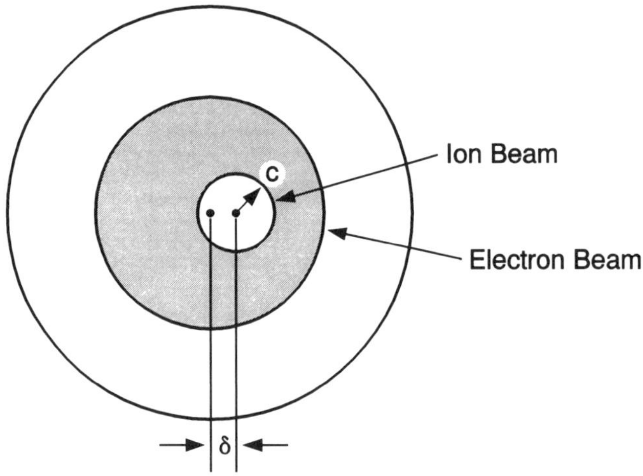

Figure 7-45 Ion beam contained within electron beam.

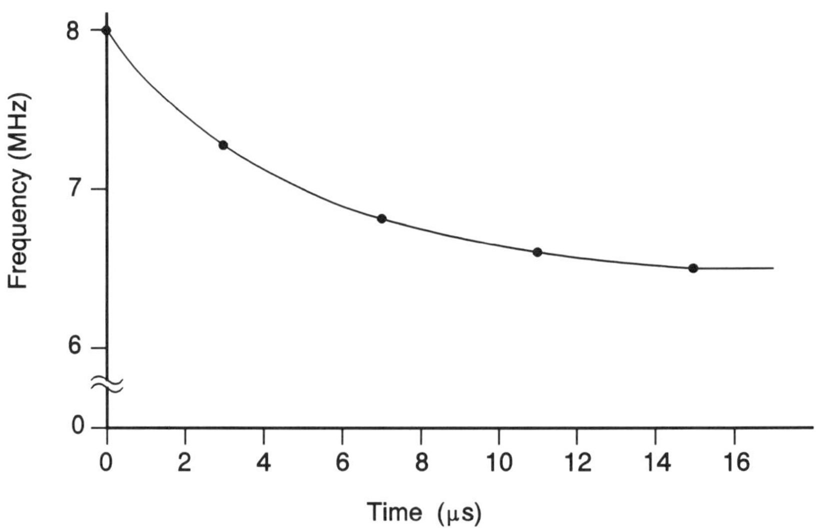

Figure 7-46 Frequency of transverse oscillations as a function of time after the beginning of oscillations. From A.S. Gilmour, Jr., *Proc.* 5th International Congress on Microwave Tubes.

probe current as a function of radial probe position in the beam at times of maximum and minimum beam diameter. The time increments were equally spaced at the half period of the oscillation frequency.

The radial mode of oscillation may be explained with the aid of Figure 7-48. At

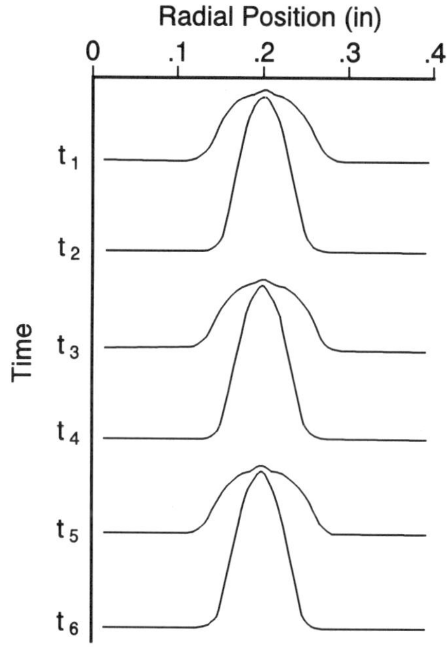

Figure 7-47 Probe current as a function of radial position for radial mode of oscillation. From A.S. Gilmour, Jr., RADC-TDR-63-465, vol. 1, December 1963.

some point in time, the positive ions are nearly uniformly distributed throughout the potential well of the electron beam. The ions are attracted and flow to the center of the beam (bottom of the well). The resulting concentration of ions produces a repulsive force (associated with a positive peak in the potential well not shown in Figure 7-48), which causes the ions to reverse direction and flow radially outward. As the ions flow outward, the potential well deepens, causing a strong enough electrostatic force to pull them inward again. The radial electrostatic forces in the electron beam depend on the ion concentration and so the electron beam oscillates radially in synchronism with the ions.

The frequency at which the radial oscillations are expected to occur may be determined with the aid of Figure 7-49. It is assumed that an ion beam is concentric with the electron beam and that, initially, the ion density, ρ_i, equals the electron density, ρ_e. For a given length, ℓ, of the beams and within the radius, r_o, of the ion beam, the total ion charge, Q_i, and the total electron charge, Q_e, are

$$Q_i = \pi r_o^2 \ell \rho_i \tag{7-128}$$

and

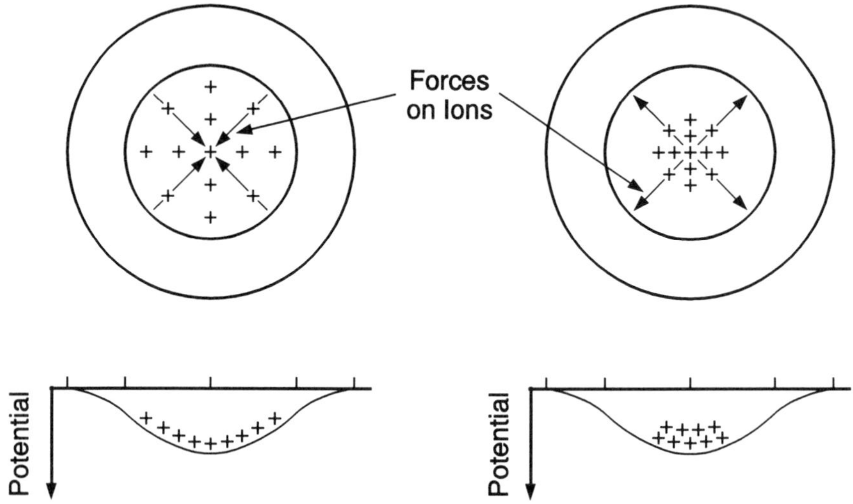

Figure 7-48 Ion distribution in radial mode of oscillation.

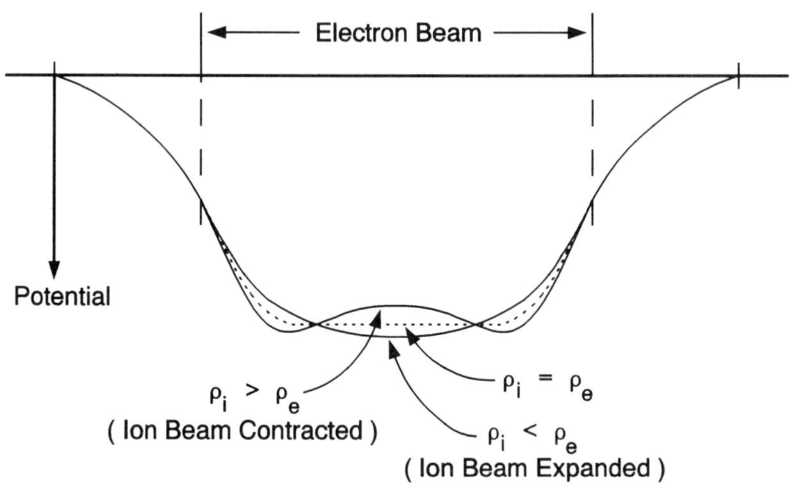

Figure 7-49 Potential profiles for radial ion oscillations.

$$Q_e = -\pi r_o^2 \ell \rho_e \tag{7-129}$$

Next, assume that the ion beam expands to r_i with no increase in the total number of ions. Thus, the resulting ion density is less than the electron density, and so there is a net electric field tending to cause the ion beam to contract. That field can be found from Gauss's law as follows:

$$\oint_s E(r_i)\, dS = 2\pi r_i \ell E(r_i) = \frac{\pi r_o^2 \ell \rho_i - \pi r_i^2 \ell \rho_e}{\epsilon_o} \tag{7-130}$$

but $\rho_e = \rho_i$ so

$$E(r_i) = \frac{\rho_e}{2\epsilon_o}\left[\frac{r_o^2 - r_i^2}{r_i}\right] \tag{7-131}$$

This electric field may be used in the force equation as follows to find the motion of the ion beam:

$$m_i \frac{d^2 r_i}{dt^2} = eE(r_i) = \frac{e\rho_e}{2\epsilon_o}\left[\frac{r_o^2 - r_i^2}{r_i}\right] \tag{7-132}$$

or

$$\frac{d^2 r_i}{dt^2} + \frac{e\rho_e}{m_i \epsilon_o}\left[\frac{r_i^2 - r_o^2}{2r_i}\right] = 0 \tag{7-133}$$

Next, letting $r_i/r_o = 1 + \delta$, then

$$\frac{r_i^2 - r_o^2}{2r_i} = \frac{1}{r_o}\frac{(1 + \delta)^2 - 1}{2(1 + \delta)} \approx \frac{\delta}{r_o} \tag{7-134}$$

and so the force equation becomes

$$\frac{d^2 \delta}{dt^2} + \omega_{ip}^2 \delta = 0 \tag{7-135}$$

The only difference between this equation and the equation for the transverse oscillation for the single ion is that the $2^{1/2}$ is missing so that the frequency of radial oscillations is expected to be $2^{1/2}$ greater than the frequency of transverse oscillations (neglecting ion beam diameter effects).

The radial oscillation frequency, like the transverse oscillation frequency, is a function of ion mass as well as electron beam density and density profile. Oscillation frequencies from the megahertz range to several tens of megahertz are common.

The presence of ion oscillations is indicated by ion sidebands on the output frequency of a microwave tube. Also, the oscillation frequency is a function of the square root of beam voltage (see Figure 7-50). This provides a means for checking to determine if gas is the cause of spurious tube output signals.

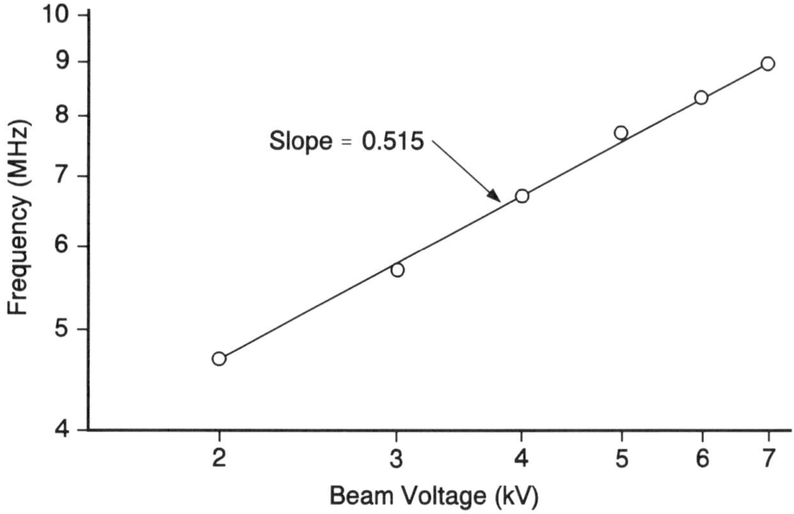

Figure 7-50 Ion induced beam-oscillation frequency as a function of beam voltage. From A.S. Gilmour, Jr., RADC-TDR-63-465, vol. 1, December 1963.

7.5.3 Low Frequency Instabilities

At low pressures ($< 1\times10^{-8}$ torr) and long-pulse or CW operation, a low frequency (hundreds of hertz to kilohertz) instability is observed in some linear beam tubes. An example is shown in Figure 7-51 [24]. These results were obtained with a Brillouin-focused klystron designed for CW operation. The tube was equipped with a shadow-gridded electron gun. The modulator used provided pulse lengths from 1 μs to 100 ms.

At the beginning of each pulse, after an initial turn-on transient, the tube operated stably for about 15 ms. Then the tube exhibited an ion induced instability with a period of about 5 ms. As shown in Figure 7-51, transients in the output power were the evidence of the instability. It was found that, if the beam was pulsed off for a period as short as 1 μs, the instability would not recur until about 15 ms after the beam had been pulsed off. Apparently, the ions in the beam dissipated during the 1 μs off period.

The tube was not instrumented to permit an investigation of the beam behavior so the details of tube behavior during the transients is not known. It appears, however, that during the 15 ms period after beam turn-on, the ion density in the beam increased to the point where the electron space charge potential was neutralized. Then, as ion formation continued, the net charge density in the beam became positive so that some of the ions could dissipate to the tube walls, causing the net charge density to become negative again. Then, in the 5 ms period between transients, as ionization continued, the net charge density became positive again so that ions could again escape to the walls of the tube.

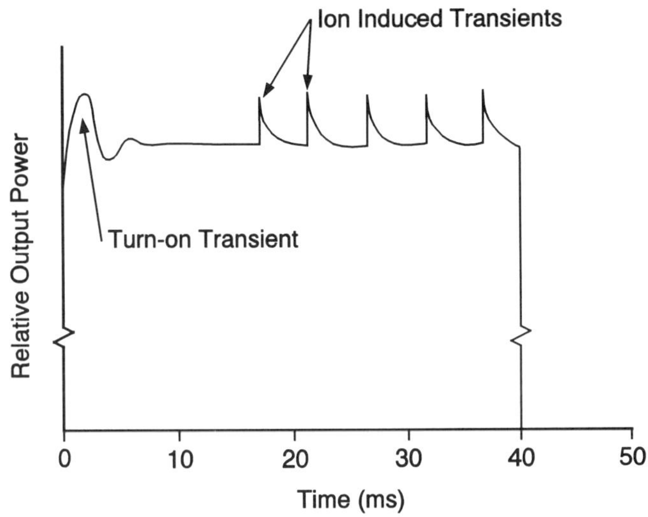

Figure 7-51 RF output with 40 ms beam pulse width. From E.W. McCune, *Technical Digest*, 1983 IEDM. © 1983 IEEE.

An important conclusion of McCune's work is that beam pulsed linear beam amplifiers can be pulsed off periodically for a very short period (as short as 1 μs) and be inherently free of ion oscillation problems assuming that the tube has been adequately evacuated.

REFERENCES

[1] Recommended for further reading: M. Chodorow and C. Susskind, *Fundamentals of Microwave Electronics*, New York: McGraw-Hill, 1964; J.F. Gittins, *Power Travelling Wave Tubes*, New York: American Elsevier, 1965; W.J. Kleen, *Electronics of Microwave Tubes*, New York: Academic Press, 1958.

[2] A.S. Gilmour, Jr., G.C. Dalman, and L.F. Eastman, "The current and velocity distributions in a velocity-modulated Brillouin-focused electron beam," *Proc. 4th International Congress on Microwave Tubes*, Eindhoven, The Netherlands: Centrex, 1963, pp. 675–688.

[3] A.S. Gilmour, Jr., and D. Hallock, "A demountable beam analyzer for studying magnetically confined electron beams," *Advances in Electron Tube Techniques*, vol. 2, New York: Pergammon Press, 1963, pp. 153–157.

[4] George R. Brewer, "Some characteristics of a magnetically focused electron beam," *Jour. Appl. Phys.*, vol. 30, no. 7, July 1959, pp. 1022–1038.

[5] J.R. Pierce and L.R. Walker, ""Brillouin flow" with thermal velocities," *Jour. Appl. Phys.*, vol. 24, no. 10, October 1953, pp. 1328–1330.

[6] A. Ashkin, "Electron beam analyzer," *Jour. Appl. Phys.*, vol. 28, no. 5, May 1957, pp. 564–569.

[7] Kenneth J. Harker, "Nonlaminar flow in cylindrical electron beams," *Jour. Appl. Phys.*, vol. 28, no. 6, June 1957, pp. 645–650.

[8] Gabriel Herrmann, "Optical theory of thermal velocity effects in cylindrical electron beams," *Jour. Appl. Phys.*, vol. 29., no. 2, February 1958, pp. 127–136.

[9] K. Amboss, "Verification and use of Herrmann's optical theory of thermal velocity effects in electron beams in the low perveance regime," *IEEE Trans. on Electron Devices*, October 1964, pp. 479–485.

[10] Peter T. Kirstein, "On the effects of thermal velocities in two-dimensional and axially symmetric beams," *IEEE Trans. on Electron Devices*, March 1963, pp. 69–80.

[11] P.T. Kirstein, "Thermal velocity effects in axially symmetric solid beams," *Jour. Appl. Phys.*, vol. 34, no. 12, December 1963, pp. 3479–3490.

[12] J. Richard Hechtel, "Magnetic focusing of electron beams in the presence of transverse velocity components," *IEEE Trans. on Electron Devices*, vol. ED-28, no. 5, May 1981, pp. 473–482.

[13] J.R. Pierce, "U.S. Patent 2,847,607," filed April 29, 1953, issued August 12, 1958.

[14] J.T. Mendel, C.F. Quate, and W.H. Yocum, "Electron beam focusing with periodic permanent magnet fields," *Proc. IRE*, May 1954, pp. 800–810.

[15] Kern K.N. Chang, "Beam focusing by periodic and complementary fields," *Proc. IRE*, January 1955, pp. 62–71.

[16] J.T. Mendel, "Magnetic focusing of electron beams," *Proc. IRE*, March 1955, pp. 327–331.

[17] K.J. Harker, "Periodic focusing of beams from partially shielded cathodes," *IEEE Trans. on Electron Devices*, October 1955, pp. 13–19.

[18] W.R. Ayers, G.A. Kalkanis, A.L. Nordquist and R.W. Vranas, "RF beam spread in PPM confined-flow TWTs," *Technical Digest*, 1985 International Electron Devices Meeting, pp. 357–360.

[19] Richard True, "A theory for coupling gridded gun design with PPM focusing," *IEEE Trans. on Electron Devices,* vol. ED-31, no. 3, March 1984, pp. 353–362.

[20] Richard True, "Emittance and the design of beam formation, transport, and collection systems in periodically focused TWTs," *IEEE Trans. on Electron Devices*, vol ED-34, no. 2, February 1987, pp. 473–485.

[21] Richard True, "Tunnel emittance growth in PPM focused TWTs," *Technical Digest*, 1985 International Electron Devices Meeting, pp. 192–195.

[22] A.S. Gilmour, Jr., "Transverse and radial oscillations induced in electron beams by ions," *Proc. 5th International Congress on Microwave Tubes*. Paris: Dunod, 1965, pp. 453–456.

[23] W.W. Peterson and H.E. Puthoff, "A theoretical study of ion plasma oscillations," *IEEE Trans. on Electron Devices*, October 1959, pp. 372–377.

[24] E.W. McCune, "Ion oscillations for pulsed klystron amplifiers," *Technical Digest*, 1983 International Electron Devices Meeting, pp. 148–150.

Chapter 8

Beam-Gap Interactions

8.1 GRIDDED (PLANAR) GAPS

8.1.1 Beam Modulation

Consider the interaction of an electron beam with the electric field in a gridded (planar) gap as shown in Figure 8-1. The velocity of the beam entering the gap is u_o and is modified by an amount Δu as the beam passes through the gap. The time and position at the entrance to the gap are, respectively, t_o and $z = 0$. The width of the gap is ℓ and the voltage applied to the gap is

$$V = V_1 \sin \omega t \tag{8-1}$$

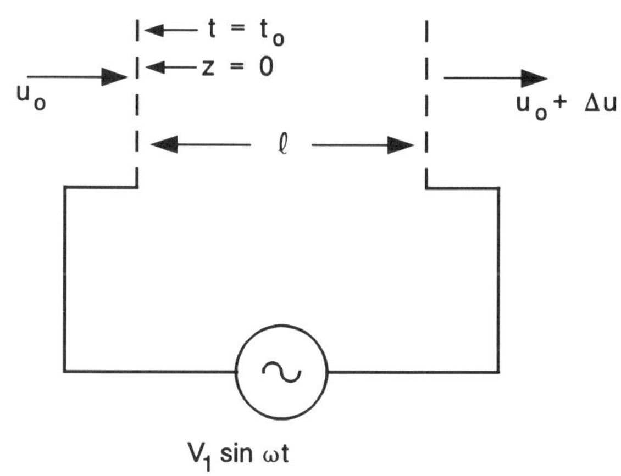

Figure 8-1 Velocity modulation of an electron beam.

The equation of motion for electrons in the gap is

$$\ddot{z} = - \eta E = \eta \frac{V_1}{\ell} \sin \omega t \tag{8-2}$$

When this is integrated, the equation for electron velocity is found to be

$$\dot{z} = - \frac{\eta V_1}{\omega \ell} \cos \omega t + C_1 \tag{8-3}$$

The constant can be evaluated using the initial conditions at the entrance of the electron into the gap, that is, $t = t_o$ and $\dot{z} = u_o$, and so

$$C_1 = u_o + \frac{\eta V_1}{\omega \ell} \cos \omega t_o \tag{8-4}$$

Thus, the velocity may be written as

$$\dot{z} = u_o + \frac{\eta V_1}{\omega \ell} \left(\cos \omega t_o - \cos \omega t \right) \tag{8-5}$$

Next, it is useful to introduce the following notation. The depth of modulation, α, is defined as follows:

$$\alpha = \frac{V_1}{V_o} \tag{8-6}$$

where V_o is the dc acceleration voltage for the electrons in the electron gun. The dc transit time through the gap is ℓ / u_o and the dc transit angle, ϕ_o, is defined as follows:

$$\phi_o = \omega \frac{\ell}{u_o} \tag{8-7}$$

Since $u_o^2 = 2 \eta V_o$,

$$\frac{\eta V_1}{\omega \ell} = \frac{\eta V_1 V_o}{\omega \ell V_o} = \frac{\alpha \eta u_o^2}{\omega \ell 2 \eta} = \frac{\alpha u_o}{2 \phi_o} \tag{8-8}$$

and so

$$u = u_o + \frac{\alpha u_o}{2 \phi_o} \left(\cos \omega t_o - \cos \omega t \right) \tag{8-9}$$

For small signal levels, the electrons do not move far from their dc positions as they pass through the gap. Thus, the time $t = t_1$ at which electrons exit from the gap is approximately t_o plus the dc transit time, ℓ / u_o. As a result,

$$\omega t_o + \phi_o = \omega t_1 \tag{8-10}$$

and so

$$u = u_o + \frac{\alpha u_o}{2\phi_o}\left[\cos(\omega t_1 - \phi_o) - \cos \omega t_1\right] \tag{8-11}$$

Using the trigonometric identity

$$\cos(\omega t_1 - \phi_o) - \cos \omega t_1 = 2 \sin \frac{\phi_o}{2} \sin\left[\omega t_1 - \frac{\phi_o}{2}\right] \tag{8-12}$$

the expression for velocity may be written as

$$u = u_o + \frac{\alpha u_o}{2} M \sin\left[\omega t_1 - \frac{\phi_o}{2}\right] \tag{8-13}$$

where

$$M = \frac{\sin \frac{\phi_o}{2}}{\frac{\phi_o}{2}} \tag{8-14}$$

is called the *modulation coefficient* or the *gap coupling coefficient*. This function is shown in Figure 8-2. The significance of M can be explained by first writing the velocity u as

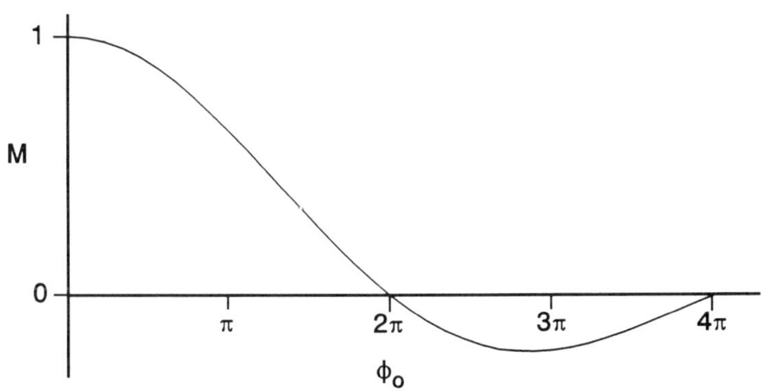

Figure 8-2 Gap coupling coefficient as a function of gap transit angle.

$$u = u_o + u_1 \tag{8-15}$$

where u_1 is the ac component of velocity produced by the gap. This velocity, normalized with respect to u_o is

$$\frac{u_1}{u_o} = \frac{\alpha}{2} M \sin\left[\omega t_1 - \frac{\phi_o}{2}\right] \tag{8-16}$$

Now, as is indicated in Figure 8-3, for a very narrow gap, ϕ_o is small and so the voltage across the gap is essentially constant as an electron traverses the gap. For this case, the gap coupling coefficient is maximum and is nearly unity. As the gap is widened, the voltage varies as an electron traverses the gap and so the effective gap voltage (V_{eff} in Figure 8-3) is less than the maximum gap voltage (V_{max}). The effectiveness of the gap in accelerating the electron decreases (i.e., the gap coupling coefficient goes down). If the gap width is increased to the point where $\phi_o = 2\pi$, the applied voltage varies through one complete cycle as an electron traverses the gap and the net acceleration of the electron is zero. In this case, $M = 0$.

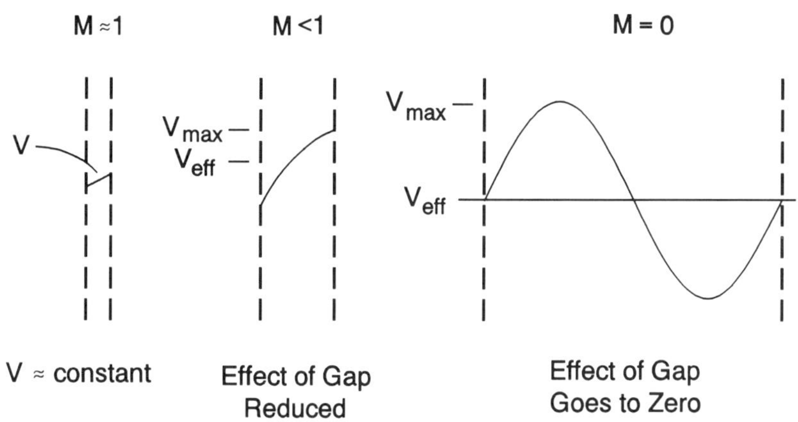

Figure 8-3 Effect of gap width on gap coupling coefficient.

The usefulness of the gap coupling coefficient is that, in the analysis, the gap of finite width ℓ can be replaced with an infinitesimal gap with voltage reduced by M without affecting the calculated velocity. In addition, the reference plane where $z = 0$ can be shifted from the first grid of the gap to the center of the gap where $z = \ell/2$. There,

$$\omega t = \omega t_1 - \frac{\phi_o}{2} \tag{8-17}$$

and so

$$u = u_o\left[1 + \frac{\alpha M}{2}\sin \omega t\right] \tag{8-18}$$

The modulation process described here is referred to as *velocity modulation*. The reason, which is clear from (8-18), is that the velocities of electrons are "modulated" as they traverse a gap to which an alternating voltage is applied. This is in contrast to the situation in a triode, tetrode, klystrode, *et cetera*, where the *density* of the electrons is modulated. With velocity modulation, the electrons are first accelerated to high velocity and then subjected to the RF modulating field rather than being modulated at low (nearly zero) velocity and then accelerated. The result is that the transit time limitations of the triode, *et cetera*, are overcome with the velocity modulation technique and operation at much higher frequencies and power levels is made possible.

Before proceeding with an analysis of electron motion after velocity modulation takes place, it is important to examine the nonlinear nature of the basic velocity modulation process. Consider a narrow gap for which $M \approx 1$. Then, the velocity of an electron after modulation is

$$u = \left[2\eta V_o(1 + \alpha \sin \omega t)\right]^{1/2}$$

$$= \sqrt{2\eta V_o}\left[1 + \frac{\alpha}{2}\sin \omega t - \frac{\alpha^2}{8}\sin^2\omega t + \cdots\right] \tag{8-19}$$

$$= \sqrt{2\eta V_o}\left[1 + \frac{\alpha}{2}\sin \omega t - \frac{\alpha^2}{4}(1 - \cos\ 2\ \omega t) + \cdots\right]$$

Thus, for large modulation signals, a second harmonic term appears in the resulting velocity modulation. This is not surprising when an example is considered. In Figure 8-4, a normalized (with respect to V_o) voltage of $1 + 0.5\sin\omega t$ is plotted along with the square root of that voltage (which is proportional to velocity). The distortion in the velocity term is evident. The positive peak is 1 + .22 while the negative peak is 1 - .29. Thus, the velocity wave is not symmetrical. A properly phased second harmonic component is required to describe most of the distortion.

8.1.2 Current Induction

Consider, first, a sheet of electron charge passing between two planar parallel grids as shown in Figure 8-5. The surface charge density on the charge sheet is assumed to be σ_e coulombs per unit area. Electric fields E_L and E_R are produced by σ_e. Positive charge densities σ_L and σ_R are induced on the left and right grids, and as the charge sheet moves, electron current I_i flows in the circuit connecting the grids.

Neglecting fringing fields at the edges of the grids and charge sheet and using Gauss's law, the charge density induced in area A of the left grid by field E_L may be found from

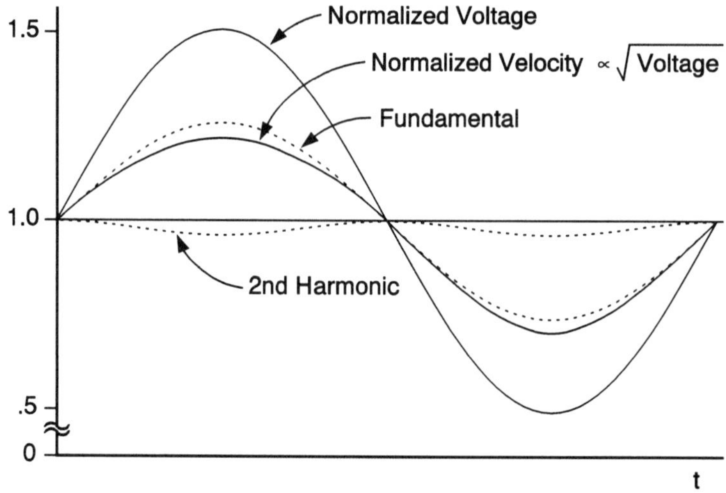

Figure 8-4 Illustration of nonlinear behavior of velocity modulation.

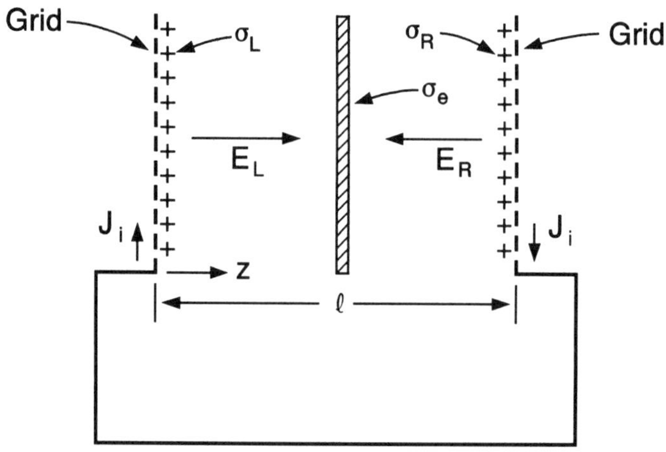

Figure 8-5 Electrode and charge-sheet configuration for current induction analysis.

$$\oint_S \boldsymbol{E_L} \cdot \boldsymbol{dS} = \frac{\sigma_L A}{\epsilon_o} \tag{8-20}$$

and is

$$\sigma_L = \epsilon_o E_L \tag{8-21}$$

Similarly, the charge induced in the right grid by field E_R is

$$\sigma_R = \epsilon_o E_R \tag{8-22}$$

The electric fields E_L and E_R produced by the charge sheet may also by found using Gauss's law (and neglecting fringing fields) as follows:

$$\oint_S \boldsymbol{E} \cdot \boldsymbol{dS} = -\left(E_L + E_R\right) A = -\frac{\sigma_e A}{\epsilon_o} \tag{8-23}$$

or

$$E_L + E_R = \frac{\sigma_e}{\epsilon_o} \tag{8-24}$$

The total voltage across the gap is zero and so the voltages to the left and to the right of the charge sheet must add to zero. Thus, the electric fields may be related by

$$-E_L z + E_R\left(\ell - z\right) = 0 \tag{8-25}$$

Combining this result with that from (8-5), the fields are found to be

$$E_L = \frac{\sigma_e}{\epsilon_o}\left[1 - \frac{z}{\ell}\right] \tag{8-26}$$

and

$$E_R = \frac{\sigma_e}{\epsilon_o}\frac{z}{\ell} \tag{8-27}$$

With these fields, the charge densities on the grids are found to be

$$\sigma_L = \sigma_e\left[1 - \frac{z}{\ell}\right] \tag{8-28}$$

and

$$\sigma_R = \sigma_e \frac{z}{\ell} \tag{8-29}$$

At all positions,

$$\sigma_L + \sigma_R = \sigma_e \tag{8-30}$$

As the position of the charge sheet changes, σ_L and σ_R vary as shown in Figure 8-6.

As the charge sheet moves incremental distance δz, the changes in σ_L and σ_R are

$$\delta\sigma_L = -\sigma_e \frac{\delta z}{\ell} \tag{8-31}$$

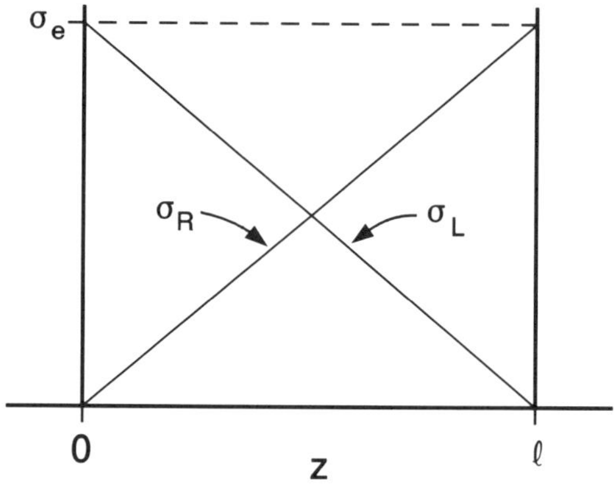

Figure 8-6 Variation of charge densities induced on left and right grids by charge sheet at position z.

and

$$\delta \sigma_R = \sigma_e \frac{\delta z}{\ell} \tag{8-32}$$

If the movement occurs in time δt then, in the limit for very small δt, the electron current density, J_i, induced in the grids is

$$J_i = -\frac{\delta \sigma_L}{\delta t} = \frac{\delta \sigma_R}{\delta t} = \frac{\sigma_e}{\ell}\frac{\delta z}{\delta t} \tag{8-33}$$

or

$$J_i = \frac{\sigma_e}{\ell} u \tag{8-34}$$

where u is the velocity of the charge sheet. The circuit current is, of course, $I_i = J_i A$.

The preceding analysis is a form of Ramo's theorem. The importance of this analysis is that it shows that, immediately upon the entrance of a charge into a region, current starts to flow in nearby electrodes. The charge can be emitted from one of the electrodes and collected by the other or can enter and leave the region through grids. This analysis can be applied to RF gaps by assuming that RF magnetic fields are negligible (which is the case in conventional microwave tubes) and that field propagation effects are negligible (also valid because the regions in which current induction occurs are small compared with the wavelength of the RF field).

There is much to be learned from this simple relationship (8-34). First, and most important, the current waveform induced in the grids and circuit may be very different than the current waveform of the charge sheet. Since the thickness of the charge sheet has been assumed to be infinitesimal, the volumetric charge density, ρ_e, must be infinite. The current density of the sheet, $J_e = \rho_e u$, is therefore infinite. On the other hand, for constant velocity, the current density in the grids is constant as the sheet of charge traverses the gap. Thus, J_i has a rectangular waveform with a duration $\tau = d/u$, which is the transit time of the charge sheet across the gap. The waveforms for J_i and J_e are illustrated in Figure 8-7 for the charge sheet located at position z in the gap.

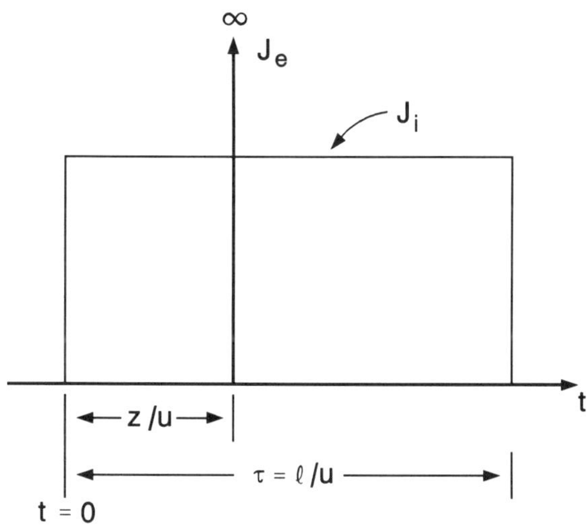

Figure 8-7 Current waveform induced in circuit (J_i) and current waveform of charge sheet (J_e).

If the gap width, ℓ, is made smaller and smaller then, in the limit, as $\ell \to 0$, J_i becomes the same as J_e (they are both delta functions having zero width and infinite amplitudes).

Next, assume that the circuit connecting the plates has impedance as indicated in Figure 8-8. If the electron velocity is high enough that the voltage produced by the induced current has negligible affect on velocity, then the waveform of the induced current is the same as it is with no impedance. This is the assumption that is made in small signal theories of interactions between beams and circuits.

If the charge density in the sheet is high enough to produce significant electric fields then the potential between the plates will vary with sheet position and the velocity of the sheet will vary. As a result, induced current will vary as the charge sheet traverses the gap. To illustrate this, assume, as is indicated in Figure 8-9, that the electrons have been accelerated to velocity u_o by voltage V_o before passing the left grid and entering the gap. Assume, also, that the right grid is also at V_o.

The potential at position z is

$$V(z) = V_o - E_L z - V_o - \frac{\sigma_e}{\epsilon_o}\left[1 - \frac{z}{\ell}\right] z \tag{8-35}$$

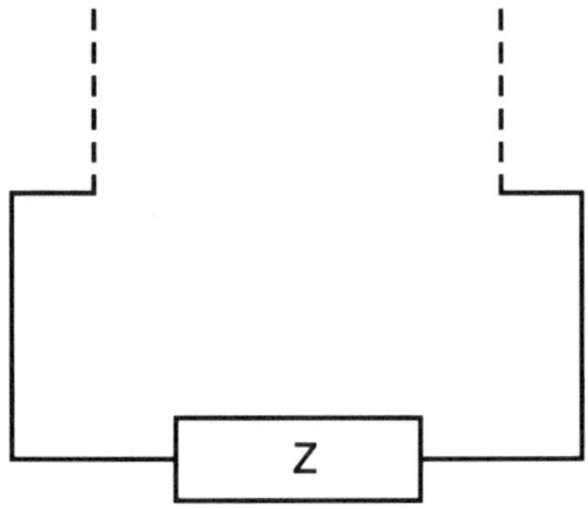

Figure 8-8 Circuit with impedance connecting grids.

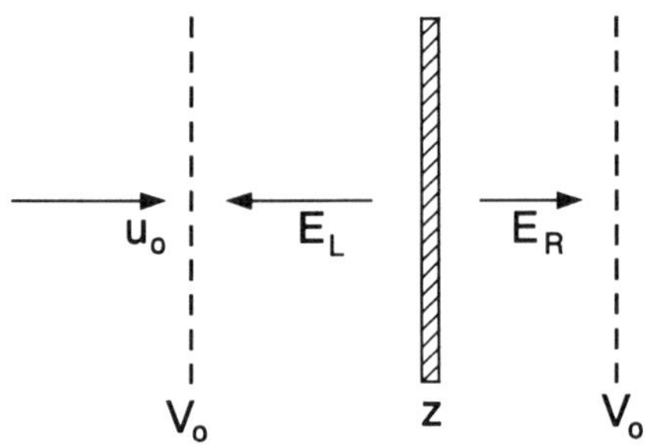

Figure 8-9 Situation assumed for calculating affect of charge density on potential distribution and on current waveform.

or

$$\frac{V(z)}{V_o} = 1 - \frac{\sigma_e \ell}{\epsilon_o V_o} \left(1 - \frac{z}{\ell}\right) \frac{z}{\ell} \tag{8-36}$$

From energy considerations, the velocity of the sheet of charge is $u(z) = (2\eta V(z))^{1/2}$ and so

$$\frac{u(z)}{u_o} = \left[1 - \frac{\sigma_e \ell}{\epsilon_o V_o} \left(1 - \frac{z}{\ell}\right) \frac{z}{\ell}\right]^{1/2} \tag{8-37}$$

As an example, assume that the charge sheet contains 10^{10} electrons/cm^2. This could arise from a volumetric density of 10^{11}, which is typical for an electron beam, over a thickness of 0.1 cm compressed into a single sheet. Assume a gap width of 1 cm and a voltage, V_o, of 10 kV. Then the potential and velocity variations with position of the charge sheet in the gap are as shown in Figure 8-10.

If, in computing the time at which the electron sheet arrives at a position z, it is

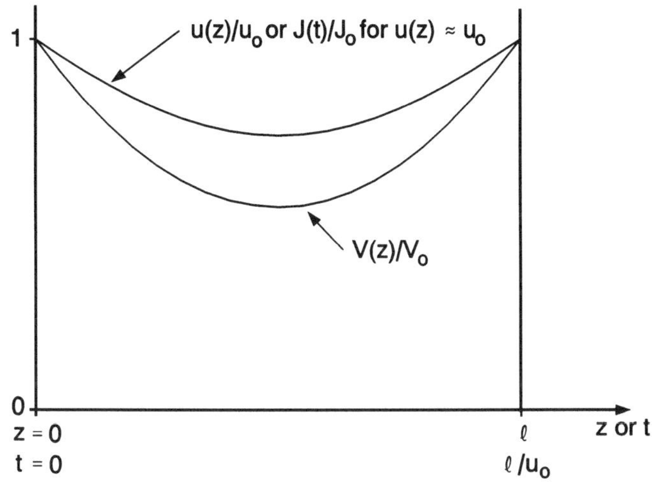

Figure 8-10 Effect of charge of electron sheet on potential and on velocity of electron sheet.

assumed that the velocity is u_o, then the current induced in the grids has the same variation with time as the velocity has with distance. This exercise illustrates the effect of a potential depression caused by electrons on induced currents in an RF gap in a microwave tube.

Next, consider the situation when the voltage between the grids is large enough to change the electron velocity. Then, the induced current waveform is changed. For example, assume, as is indicated in Figure 8-11, that the left grid is the cathode of a diode that emits a sheet of charge at zero velocity and that the right grid is an anode at potential V. When the charge in the sheet is small enough that the electric field is approximately V/ℓ, the velocity of the electrons can be found from the force equation

$$\frac{d^2z}{dt^2} = \eta\frac{V}{\ell} \tag{8-38}$$

and is

$$u = \frac{dz}{dt} = \eta\frac{V}{\ell}t \tag{8-39}$$

Thus, the velocity of the charge sheet increases linearly with time until the sheet arrives at the anode. The induced current density is

$$J_i = \frac{\sigma_e}{\ell}\eta\frac{V}{\ell}t \tag{8-40}$$

and has a triangular waveform, as shown in Figure 8-12.

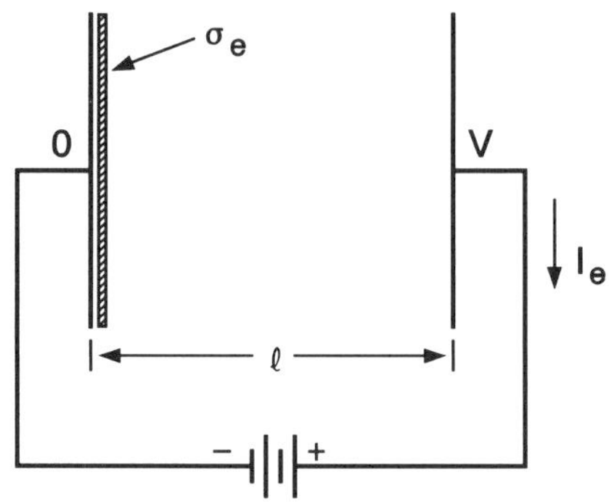

Figure 8-11 Charge sheet emitted by cathode with accelerating voltage on anode.

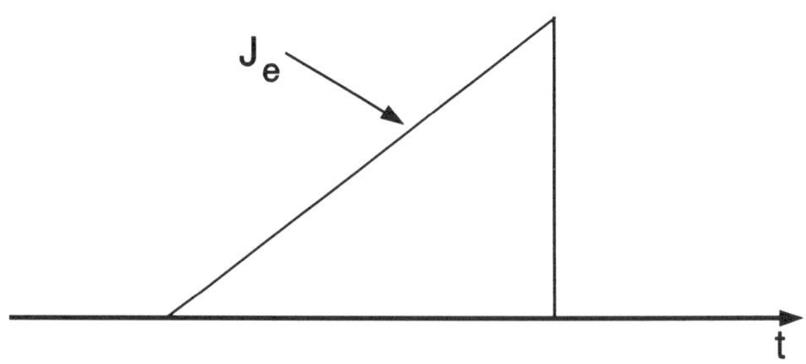

Figure 8-12 Current waveform induced by charge sheet accelerated from zero velocity to anode potential in a diode.

If several sheets of charge are considered, then, as long as interactions between them are ignored, the current waveforms produced are superimposed. For the case of several sheets of electrons emitted at random times and leaving the cathode with zero velocity and being accelerated across the gap to an anode, the current pulses might appear as shown in Figure 8-13. Here, it has been assumed that the charge sheets have different charge densities and so the current pulses have different amplitudes. This situation is fundamental to the analysis of shot noise in the current emitted by a cathode.

To determine the current induced by a continuous electron beam having volumetric charge density $\rho(z)$ as shown in Figure 8-14, the beam will be divided into slices with thickness Δz having density $\sigma(z) = \rho(z)\Delta z$. The increment of current density, ΔJ_i, induced in the grids by each slice of charge is

$$\Delta J_i = \frac{\rho(z)\,\Delta z}{\ell}\,u \tag{8-41}$$

But the current density in the beam, $J_e = \rho(z)u$ and so

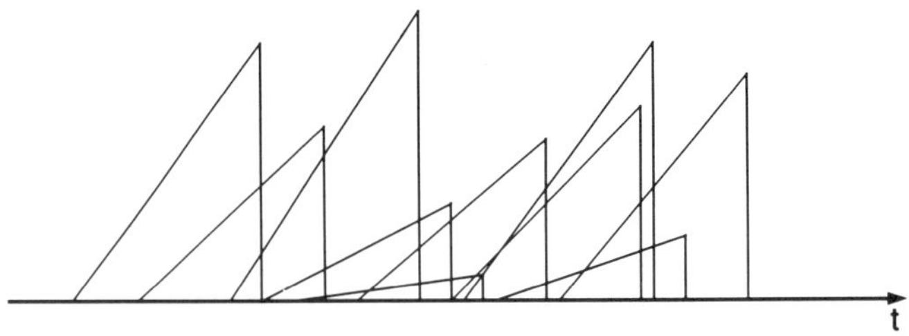

Figure 8-13 Superimposed current pulses induced by several charge sheets having different charge densities.

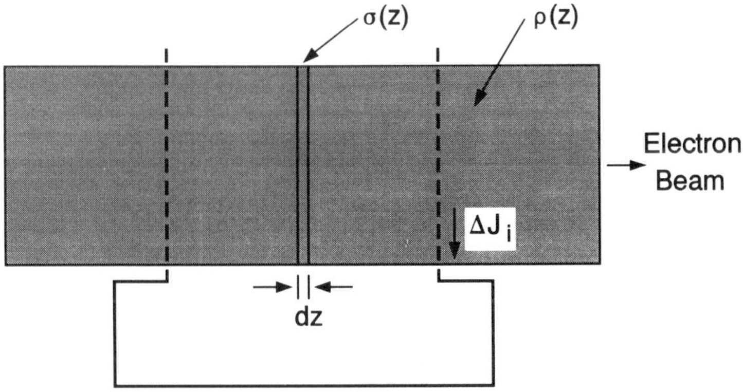

Figure 8-14 Situation assumed for determining current induced by electron beam.

$$\Delta J_i = \frac{1}{\ell} J_e \Delta z \tag{8-42}$$

In the limit, as $\Delta z \to dz$, the induced current density becomes

$$J_i = \frac{1}{\ell} \int_0^{\ell} J_e \, dz \tag{8-43}$$

As an example of the application of this relation, assume (as is the case in the small signal analysis of the modulation of electron beams) that the beam current contains an RF current component that varies sinusoidally with time and distance so that the total beam current is

$$J_e = J_o + J_1 \sin\left[\omega t - \frac{\omega}{u_o} z\right] = J_o + J_1 \sin\left(\omega t - \beta_e z\right) \tag{8-44}$$

where $\beta_e = \omega/u_o$ is referred to as the propagation constant of the electron beam. Thus, the current induced in the circuit is

$$J_i = \frac{1}{\ell} \int_0^{\ell} \left[J_o + J_1 \sin\left(\omega t - \beta_e z \right) \right] dz \tag{8-45}$$

or

$$J_i = J_o + \frac{1}{\beta_e \ell} J_1 \left[\cos\left(\omega t - \beta_e \ell \right) - \cos \omega t \right] \tag{8-46}$$

But

$$\beta_e \ell = \frac{\omega \ell}{u_o} = \phi_o \tag{8-47}$$

and so, by combining the two cosine terms, the induced current density becomes

$$J_i = J_o + J_1 M \sin \left(\omega t - \frac{\phi_o}{2} \right) \tag{8-48}$$

where M is modulation coefficient or the gap coupling coefficient given by

$$M = \frac{\sin \dfrac{\phi_o}{2}}{\dfrac{\phi_o}{2}} \tag{8-49}$$

This is the same coefficient that is in the expression for the velocity modulation produced by a gap, discussed at the beginning of the chapter.

In the following chapter, it will be shown that

$$J_1 \propto u_1 \tag{8-50}$$

where u_1 is the RF component of electron velocity. Since $u_1 \propto M$ (8-18), then

$$J_1 \propto M \tag{8-51}$$

This means that the induced current must contain M^2. (This assumes that the modulating gap and the current induction gap are the same. If the two gaps are different, then the induced current contains the product of the modulation coefficients.) Thus,

$$J_i = J_o + C M^2 \sin \left(\omega t - \frac{\phi_o}{2} \right) \tag{8-52}$$

where C is an amplitude factor. For wide gaps, therefore, the induced current is much smaller than for narrow gaps. This effect is even more pronounced in gridless gaps (to be treated in the following section), where M is a function of the radial as well as the axial dimensions of the gaps.

8.2 GRIDLESS (NONPLANAR) GAPS [1]

8.2.1 Beam Modulation

Gridless gaps are used for velocity modulation in virtually all coupled cavity TWTs and klystrons. These gaps are usually formed by two facing cylinders and the fields produced are somewhat as shown in Figure 8-15.

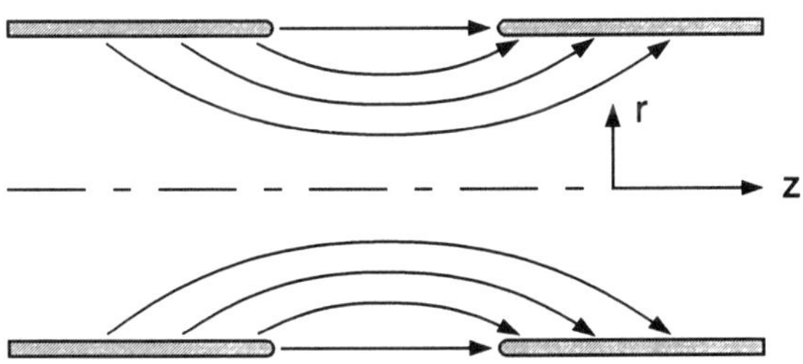

Figure 8-15 Electric field lines in gridless gap.

The electric field extends over a large region inside the cylinders and has variations in both the axial and radial directions. When an electron beam passes through the gap, electrons near the axis must travel further to "see" the full gap voltage than electrons near the gap. The field on axis may go through several cycles while the field near the gap varies through part of a cycle as the electrons pass through. As a result, electron on axis may receive a small net velocity change compared with those near the gap. Thus, the modulation coefficient varies with radius.

The analysis of the interaction of the beam in the gridless gap is significantly more complicated than in the gridded gap. In carrying out the analysis, it will be of value to rewrite the force equation in the following manner:

$$\frac{du}{dt} = \frac{\partial u}{\partial t} + u\frac{\partial u}{\partial z} \approx \frac{\partial u}{\partial t} + u_o\frac{\partial u}{\partial z} = -\eta E \tag{8-53}$$

The velocity u_o is, as usual, the dc velocity of the electrons. In replacing u by u_o, the assumption is made that the electron velocity does not vary a great deal from the dc velocity.

The electric field, E, will be assumed to vary with z and with t as follows:

$$E = E(z)\, e^{j\omega t} \tag{8-54}$$

where, as is standard practice, it is understood that either the real or the imaginary part of $E(z)e^{j\omega t}$ will be used throughout the analysis. In our case, for continuity with the analysis in the first part of this chapter, we will use the imaginary part, that is, $\sin\omega t$, so that

$$E = \operatorname{Im} E(z)\, e^{j\omega t} = E(z) \sin\omega t \tag{8-55}$$

Because E varies as $e^{j\omega t}$, then u also does. In addition, u starts at u_o as the electron enters the gap and acquires an ac component that varies with z and with t. Thus,

$$u = u_o + u(z)\, e^{j\omega t} \tag{8-56}$$

Now, the force equation may be rewritten as

$$j\omega u(z)\, e^{j\omega t} + u_o e^{j\omega t} \frac{\partial u(z)}{\partial z} = -\eta E(z)\, e^{j\omega t} \tag{8-57}$$

or

$$\frac{du(z)}{dz} + j\beta_e u(z) = -\frac{\eta}{u_o} E(z) \tag{8-58}$$

where

$$\beta_e = \frac{\omega}{u_o} \tag{8-59}$$

and is referred to as the *propagation constant* for the electron beam.

Next, the force equation may be rewritten as

$$\frac{d}{dz} e^{j\beta_e z} u(z) = -\frac{\eta}{u_o} e^{j\beta_e z} E(z) \tag{8-60}$$

and integrated and rearranged to obtain

$$u(z) = -\frac{\eta}{u_o} e^{-j\beta_e z} \int_{z_o}^{z} e^{j\beta_e z} E(z)\, dz \tag{8-61}$$

The lower limit of integration, z_o, is the position at which the electron enters the field. To find the total effect of the gap, the upper limit is the position at which the electron leaves the field.

As an example of the application of (8-61), consider the gridded gap case, which was treated at the beginning of this chapter. Assuming that $z = 0$ at the center of the gap, then the electron enters the gap at $z_o = -\ell/2$ and leaves the gap at $z = \ell/2$. The electric field, $E(z)$, was constant and was $-V_1/\ell$. Thus, at the exit from the gap, the velocity is

$$u(\ell/2) = \frac{\eta V_1}{u_o \ell} e^{-j\beta_e \ell/2} \int_{-\ell/2}^{\ell/2} e^{j\beta_e z} dz \tag{8-62}$$

or

$$u(\ell/2) = \frac{2\eta V_1}{u_o \beta_e \ell} e^{-j\beta_e \ell/2} \left[\frac{e^{j\beta_e \ell/2} - e^{-j\beta_e \ell/2}}{2j} \right] \tag{8-63}$$

But,

$$\frac{\beta_e \ell}{2} = \frac{\omega \ell}{2u_o} = \frac{\phi_o}{2} \tag{8-64}$$

and

$$\frac{\eta V_1}{u_o} = \frac{u_o \eta V_1}{2\eta V_o} = \frac{\alpha u_o}{2} \tag{8-65}$$

so

$$u(\ell/2) = \frac{\alpha u_o}{2} \frac{\sin \frac{\phi_o}{2}}{\frac{\phi_o}{2}} e^{-j\phi_o/2} = \frac{\alpha u_o}{2} M e^{-j\phi_o/2} \tag{8-66}$$

where, of course, M is the modulation coefficient. Now, the total electron velocity is

$$u = u_o + \frac{\alpha u_o}{2} M e^{j(\omega t_1 - \phi_o/2)} \tag{8-67}$$

The time, t_1, at the exit from the gap is

$$t_1 = t + \frac{\ell}{2u_o} \tag{8-68}$$

where t is the time at the center of the gap, so

$$\omega t_1 = \omega t + \frac{\phi_o}{2} \tag{8-69}$$

Finally,

$$u = u_o \left[1 + \frac{\alpha M}{2} e^{j\omega t} \right] \tag{8-70}$$

and the imaginary part of this is

$$u = u_o \left[1 + \frac{\alpha M}{2} \sin \omega t \right] \tag{8-71}$$

which is the result obtained at the beginning of the chapter.

Now, returning to the analysis of the gridless gap, the problem is to find an expression for the electric field within the gap. As was the case for the gridded gap, space charge forces will be ignored. Also, because gap dimensions are small compared with a wavelength, it is possible to assume that the fields are static; that is, the fields do not propagate. Under these conditions, from Gauss's law,

$$\nabla \cdot \boldsymbol{E} = \frac{\rho}{\epsilon_o} = 0 \tag{8-72}$$

Taking the gradient of this expression, using the vector identity

$$\nabla \left(\nabla \cdot \boldsymbol{E} \right) = \nabla^2 \boldsymbol{E} + \nabla \times \nabla \times \boldsymbol{E} \tag{8-73}$$

and realizing that $\nabla \times \nabla \times \boldsymbol{E} = 0$ because the fields do not propagate, then

$$\nabla^2 \boldsymbol{E} = 0 \tag{8-74}$$

Only the axial forces on the electrons are of interest at this point and so only the z component, E_z, of the electric field is of interest. In cylindrical coordinates with axial symmetry (no theta variations),

$$\frac{\partial^2}{\partial r^2} E_z + \frac{1}{r} \frac{\partial}{\partial r} E_z + \frac{\partial^2}{\partial z^2} E_z = 0 \tag{8-75}$$

The electric field is an even function (symmetrical about $z = 0$ so $E(z) = E(-z)$) and so solutions to this equation are of the form

$$\mathrm{I}_0 \left(pr \right) \cos pz \tag{8-76}$$

where I_0 is a modified Bessel function of the first kind. If E_z is periodic (it is not periodic for the single gap case under consideration) with period L, then the complete solution can be written as a Fourier series as follows:

$$E_z = \sum_n A_n \mathrm{I}_0 \left[\frac{2\pi n}{L} r \right] \cos \frac{2\pi n}{L} z \tag{8-77}$$

where the values of A_n are determined by the boundary conditions.

For the single gap case, the period goes to ∞ and the Fourier series becomes the Fourier integral:

$$E_z = \int_{-\infty}^{\infty} A(p) \mathrm{I}_0 \left(pr \right) \cos pz \, dp \tag{8-78}$$

where $A(p)$ is evaluated at the boundary (the walls of the cylinders and the gap between cylinders where the electric field is known).

In the equation for velocity (8-61) the limits of integration were z_o and z. To find the velocity of an electron moving all the way through a single gridless gap, the limits go to $-\infty$ and ∞ and so the velocity equation becomes

$$u(z) = -\frac{\eta}{u_o} e^{-j\beta_e z} \int_{-\infty}^{\infty} e^{j\beta_e z} E(z)\, dz \tag{8-79}$$

It is interesting to note that this integral expression is a Fourier transform of the field. Thus, the electron is, in effect, doing a Fourier analysis of the field.

When (8-78) for the electric field is used with (8-79), the velocity becomes

$$u(z) = -\frac{\eta}{u_o} e^{-j\beta_e z} \int_{-\infty}^{\infty} e^{j\beta_e z} \int_{-\infty}^{\infty} A(p)\, \mathrm{I}_0(pr) \cos pz\, dp\, dz \tag{8-80}$$

Now, there are two considerations that make the solution to this double integral straightforward:

1. For $p \neq \beta_e$, the solution is zero. Only for $p = \beta_e$ is there a nonzero solution. (The electron is selecting that portion of the electric field with the same periodicity as the electron wave.)
2. Only the $\cos \beta_e z$ portion of $e^{j\beta ez}$ is needed because the $\sin \beta_e z$ portion integrates to zero.

Thus,

$$u(z) = -\frac{\eta}{u_o} e^{-j\beta_e z} \int_{-\infty}^{\infty} \cos \beta_e z \int_{-\infty}^{\infty} A(\beta_e)\, \mathrm{I}_0(\beta_e r) \cos \beta_e z\, d\beta_e\, dz \tag{8-81}$$

The first integration is a Fourier cosine transform of $A(\beta_e)\mathrm{I}_0(\beta_e r)$, and the second integration recovers this quantity from the transform and so the result is

$$u(z) = -\frac{\eta}{u_o} e^{-j\beta_e z} 2\pi A(\beta_e)\, \mathrm{I}_0(\beta_e r) \tag{8-82}$$

To determine the value of $A(\beta_e)$, we will assume that an electron passes very close to the inner wall of the cylinders forming the modulating gap as shown in Figure 8-16(a). Here, the electric field is zero until the electron reaches the gap, and then the field can be approximated as being constant in the gap as shown in Figure 8-16(b). (The field actually varies somewhat in the gap, but this variation has little effect on the field pattern within the cylinders where the electron beam will pass.)

The field experienced by the electron traveling close to the walls of the cylinders is the same as the field in the gridded gap, and so the electron velocity leaving the gap (at $z = \ell/2$) is given by (8-66). The equation for the velocity throughout the entire gap (8-82) must reduce to (8-66) at $z = \ell/2$ and $r = a$. As a result,

$$A(\beta_e) = -\frac{\frac{\alpha u_o}{2} M}{\frac{\eta}{u_o} 2\pi \mathrm{I}_0(\beta_e a)} \tag{8-83}$$

and so

$$u(z) = \frac{\alpha u_o}{2} M \frac{\mathrm{I}_0(\beta_e r)}{\mathrm{I}_0(\beta_e a)} e^{-j\beta_e z} \tag{8-84}$$

(a)

(b)

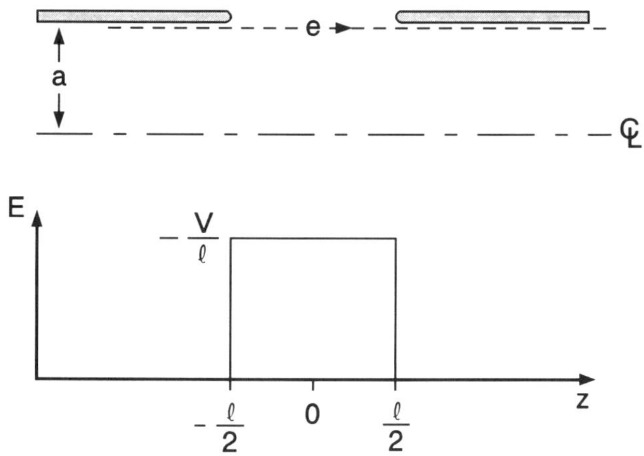

Figure 8-16 (a) Electron path and (b) electric field used for determining the value of $A(\beta_e)$.

Thus, the modulation coefficient for the gridless gap is M modified by the I_0 Bessel functions and is as shown in Figure 8-17. This shows what is expected for a gridless gap, that the modulation is weaker on axis than at the wall. The variation with radius can be determined by examining the series expansion for I_0, which is

$$\mathrm{I}_0(\beta_e a) = 1 + \frac{(\beta_e a)^2}{2^2(1!)^2} + \frac{(\beta_e a)^4}{2^4(2!)^2} + \cdots \tag{8-85}$$

For $\beta_e a \approx 1$, $\mathrm{I}_0(\beta_e a) \approx 1.25$ and so the variation in modulation with radius is not too large. For larger values of $\beta_e a$, say 2 or more, $\mathrm{I}_0(\beta_e a) > 2$ and so the variation in modulation with radius becomes very large. For this reason, it is usual practice to limit $\beta_e a$ to $\approx$ 1 to 1.5.

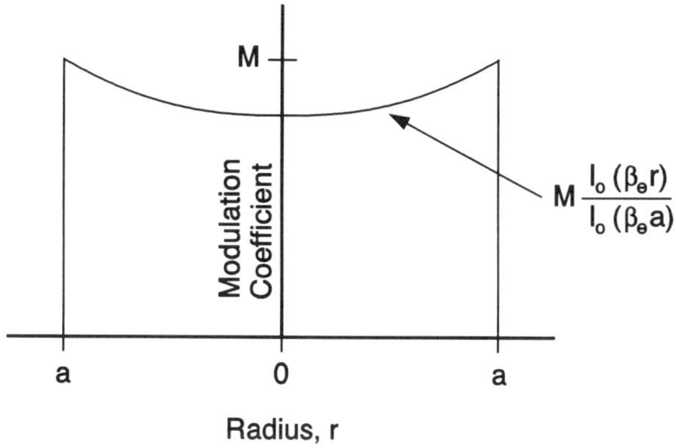

Figure 8-17 Modulation coefficient for gridless gap.

8.2.2 Current Induction

Because the modulation coefficient of the modulating gap varies with radius, the velocity and current generated on the axis are smaller than at the edge of the beam. As the modulated beam passes through a second gridless gap, the current on axis is not only smaller than that at the beam edge but is also farther from the walls of the gap and is, for these two reasons, less effective in inducing current in the second gap.

When the analysis of current induction for gridless gaps is carried out, the currents induced by filaments of beam current at various radial positions are added. The result is that the modulation coefficient enters the calculation twice (once for current generation and once for current induction) just as it did for the gridded gap case (see (8-52)), and the total induced current is obtained by integrating the current filaments over the cross sectional area of the beam.

REFERENCE

[1] M. Chodorow and C. Susskind, *Fundamentals of Microwave Electronics*, New York: McGraw-Hill, 1964.

Chapter 9

Electron Bunching

To properly deal with RF interactions of waves on circuits and electron beams, it is necessary to include the effects of space charge. Before doing that, however, it is instructive to analyze electron behavior in the presence of RF fields while ignoring the repulsive forces between electrons. This will provide background for understanding the fundamental relations, necessary approximations, *et cetera* that will be encountered when space charge effects are taken into consideration.

9.1 BALLISTIC BUNCHING

The term "ballistic bunching" is used to describe the bunching of electrons when space charge forces are neglected. Consider, again, the equation for the velocity of electrons leaving a modulating gap. This equation was developed in Chapter 8 (8-18) and is

$$u = u_o \left[1 + \frac{\alpha M}{2} \sin \omega t \right] \tag{9-1}$$

This relation shows, as is expected, that when an electron beam is passed through a modulating gap, some electrons have their velocities increased, some have their velocities decreased and some have their velocities unchanged. After the beam leaves the modulating gap, those electrons that have had their velocity increased gradually overtake the slower electrons. This process is illustrated in Figure 9-1, which is known as an *Applegate diagram.* Axial positions of electrons are plotted as a function of time, and so the slopes of the various lines represent electron velocities.

The trajectories of electrons leaving the modulating gap when the voltage is negative (indicated at the bottom of the figure) are represented by lines of reduced slope. Similarly, the trajectories of accelerated electrons are represented by lines of increased slope.

In Figure 9-1, space charge forces between electrons are ignored, and so the electron trajectories are shown to cross. In practice, at small and medium drive signal levels, space charge forces prevent electron trajectories from crossing. That situation is discussed later in this chapter. At large signal levels, electron trajectories do cross,

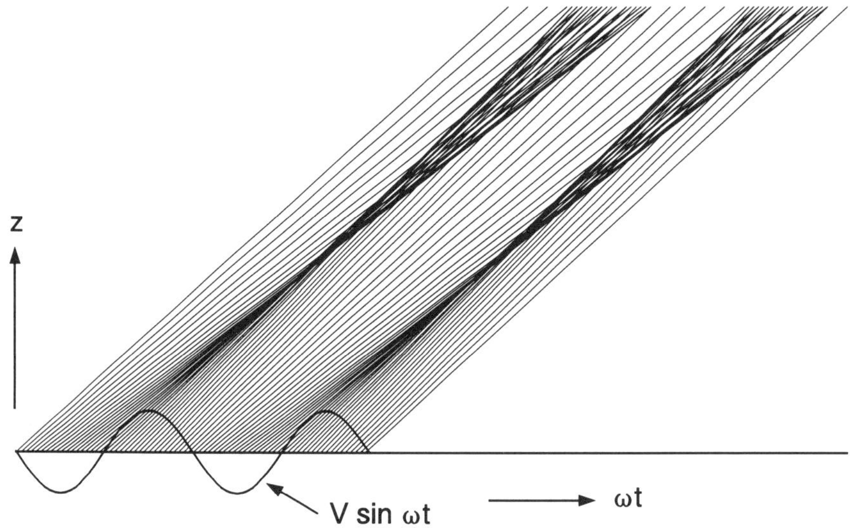

Figure 9-1 Distance-time curves for the electrons in a velocity-modulated beam in a field-free drift space (Applegate diagram).

and so the following discussion is relevant to that situation.

As a result of the convergence of electron trajectories illustrated by the Applegate diagram, electron "bunching" is said to occur. The traveling electron bunches constitute an RF current in the electron beam. In the discussion that follows, that RF current is computed.

Assume that the time of departure of an electron from the modulating gap is t_1. The time of arrival, $t(z)$, at any position, z, is t_1 plus the transit time to position z, which is z/u. As a result,

$$t(z) = t_1 + \frac{z}{u} = t_1 + \frac{z}{u_o\left[1 + \dfrac{\alpha M}{2} \sin \omega t_1\right]} \tag{9-2}$$

or

$$t(z) \approx t_1 + \frac{z}{u_o} - \frac{z}{u_o}\frac{\alpha M}{2} \sin \omega t_1 \tag{9-3}$$

when $\alpha M/2 << 1$.

Thus, the time of arrival of electrons at $t(z)$ as a function of the time of departure from the modulating gap, t_1, may be plotted as shown in Figure 9-2. Note in particular, that some electrons leaving the modulating gap later than others may arrive earlier at some distance z from the modulating gap. This was also illustrated in Figure 9-1.

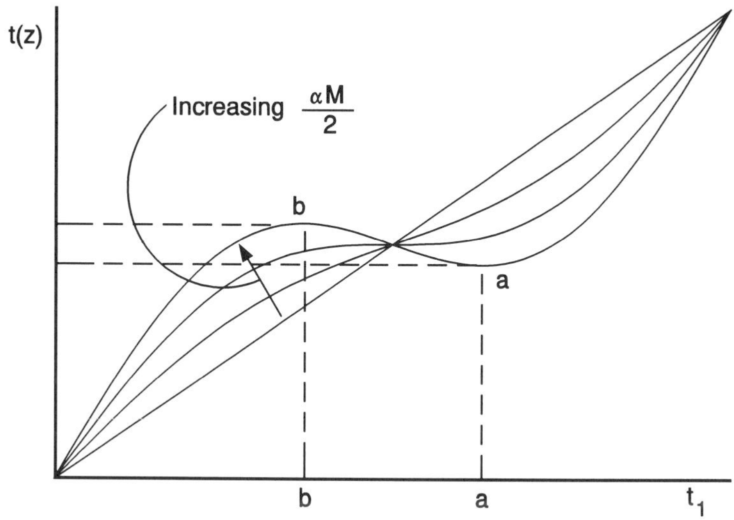

Figure 9-2 Time of arrival of electrons at position z as a function of $\alpha M/2$.

Next, with the help of Figure 9-3, consider an incremental quantity of charge dq at times t_1 and $t(z)$. Since dq remains constant, $i_1 dt_1 = i(z)dt(z)$. This means that

$$i(z) = i_1 \left| \frac{dt_1}{dt(z)} \right| \tag{9-4}$$

but

$$\left| \frac{dt(z)}{dt_1} \right| = \left| 1 - \frac{\omega z}{z_o} \frac{\alpha M}{2} \cos \omega t_1 \right| \tag{9-5}$$

so

$$i(z) = \frac{i_1}{\left| 1 - \frac{\omega z}{u_o} \frac{\alpha M}{2} \cos \omega t_1 \right|} = \frac{i_1}{\left| 1 - X \cos \omega t_1 \right|} \tag{9-6}$$

where

$$X = \frac{\omega z}{u_o} \frac{\alpha M}{2} \tag{9-7}$$

is called the bunching parameter. If α is small, then $i_1 \approx I_o$, that is, i_1 is approximately the dc beam current.

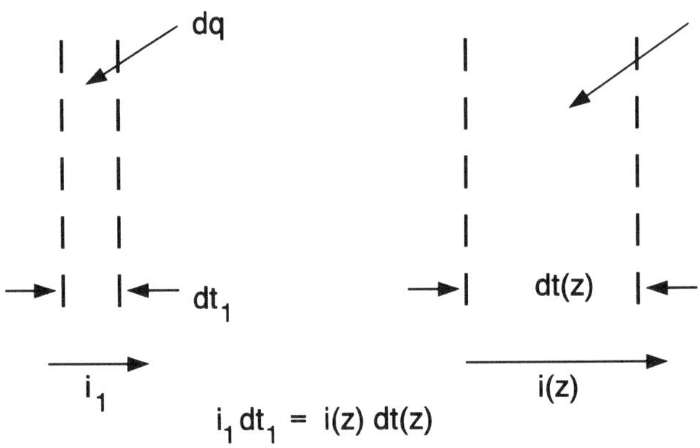

Figure 9-3 Relation of charge and current at t_1 and at $t(z)$.

The current waveforms as a function of axial position are sketched in Figure 9-4 (see also Reference 1). Note that, because of the cosine function in the denominator with an arbitrarily large multiplier, infinite current peaks are possible with this analysis. In practice, very large current peaks may occur but, as will be shown later, space charge forces prevent these peaks from becoming infinite.

The current waveforms shown in Figure 9-4 are rich in harmonics but the only component of interest is the fundamental component. This may be obtained [2] by doing a Fourier analysis of the waveforms.

To carry out the analysis, it is assumed, first, that the current can be expressed as follows:

$$i(z) = \frac{a_o}{2} + a_1 \cos\left[\omega t(z) - \frac{\omega}{u_o}z\right] + a_2 \cos 2\left[\omega t(z) - \frac{\omega}{u_o}z\right] + \cdots \tag{9-8}$$

$$+ b_1 \sin\left[\omega t(z) - \frac{\omega}{u_o}z\right] + b_2 \sin 2\left[\omega t(z) - \frac{\omega}{u_o}z\right] + \cdots$$

The technique for evaluating the coefficients, a_n and b_n, is to multiply both sides of this equation by the appropriate cos or sin term and then to integrate over the period. Thus, for example, a_n can be found as follows:

$$a_n = \frac{1}{\pi}\int_{-\pi}^{\pi} i(z) \cos n\left[\omega t(z) - \frac{\omega}{u_o}z\right] d\omega t(z) \tag{9-9}$$

But $i(z)d\,\omega t(z) \approx I_o d\,\omega t_1$ and, from (9-3) and (9-7)

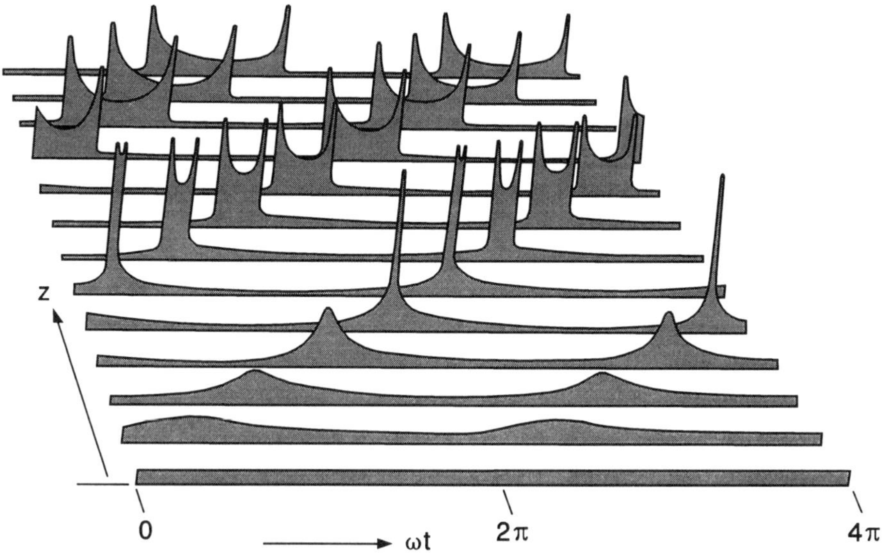

Figure 9-4 Current waveforms as a function of distance. Adapted from *Vacuum Tubes* by Karl R. Spangenberg, copyright 1948 by McGraw-Hill Book Company, Inc.

$$\omega t(z) - \frac{\omega z}{u_o} = \omega t_1 - X \sin \omega t_1 \tag{9-10}$$

so

$$a_n = \frac{I_o}{\pi} \int_{-\pi}^{\pi} \cos n \left(\omega t_1 - X \sin \omega t_1 \right) d\omega t_1 = 2 I_o \mathrm{J_n}\, nX$$

$$a_o = 2 I_o \tag{9-11}$$

$$b_n = \frac{I_o}{\pi} \int_{-\pi}^{\pi} \sin n \left(\omega t_1 - X \sin \omega t_1 \right) d\omega t_1 = 0$$

Finally,

$$i(z) = I_o \left[1 + 2 \sum_{n=1}^{\infty} \mathrm{J_n}\, nX \cos n \left(\omega t(z) - \frac{\omega z}{u_o} \right) \right] \tag{9-12}$$

In these expressions, the $\mathrm{J_n}$ functions are Bessel functions of the first kind.

Now, the fundamental component of RF current, i_f, is

$$i_f = 2\,\mathrm{J}_1 X \tag{9-13}$$

and is as shown in Figure 9-5. Note that the peak value of i_f is larger than the dc current. The value of X for which i_f is maximum is 1.84, and so the axial position, z, at which the peak value occurs is

$$z = 1.84 \frac{2u_o}{\omega \alpha M} \tag{9-14}$$

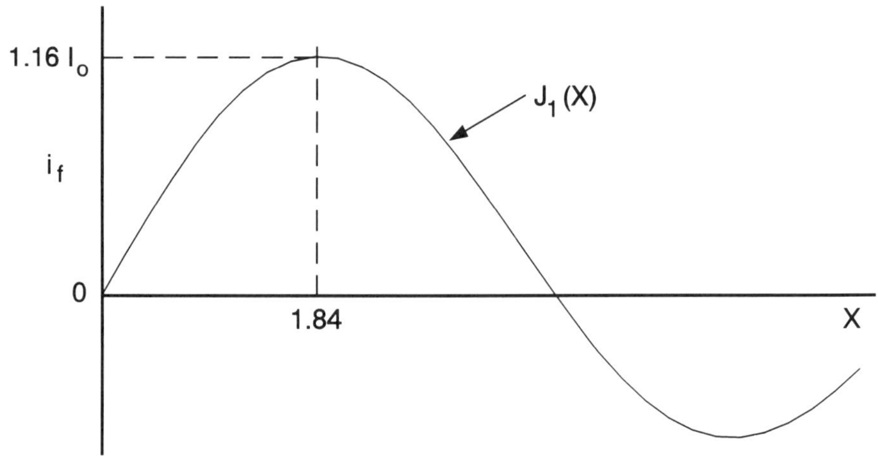

Figure 9-5 Amplitude of fundamental component of RF current.

At this point it should be noted that, in a device like a two-cavity klystron where it is desirable to locate a second gap at the point of maximum fundamental current, a problem arises because of (9-14). The problem is that the position of maximum current is inversely proportional to the depth of modulation, α, and therefore to the RF drive level. Once a position has been chosen, then for larger or smaller drive levels, the peak current in the beam remains the same but the axial position of the peak changes. Also, it should be noted that, once a drive level and position have been chosen, the relationship between drive level and RF current is not linear because of the J_1 function. Fortunately, this problem is alleviated when space charge effects are taken into consideration.

9.2 BUNCHING WITH SPACE CHARGE FORCES

The Applegate diagram in Figure 9-1 illustrates electron bunching after a beam is passed through a modulating gap when space charge forces are ignored. For small signal levels, when space charge forces are considered, the process of velocity modulation within the modulating gap is altered very little. This is because the electrons barely move from their dc positions as the beam passes through the gap.

As electrons converge in the bunching process after the beam leaves the modulating gap, space charge forces build up, impeding the rate at which fast electrons overtake slow electrons. In addition, slow electrons have their velocities increased because of the bunch that forms behind them (toward the modulating cavity). The resulting electron trajectories are illustrated in Figure 9-6, which is a portion of an Applegate diagram modified to show the effects of space charge forces. Trajectories in the absence of space charge forces are also shown.

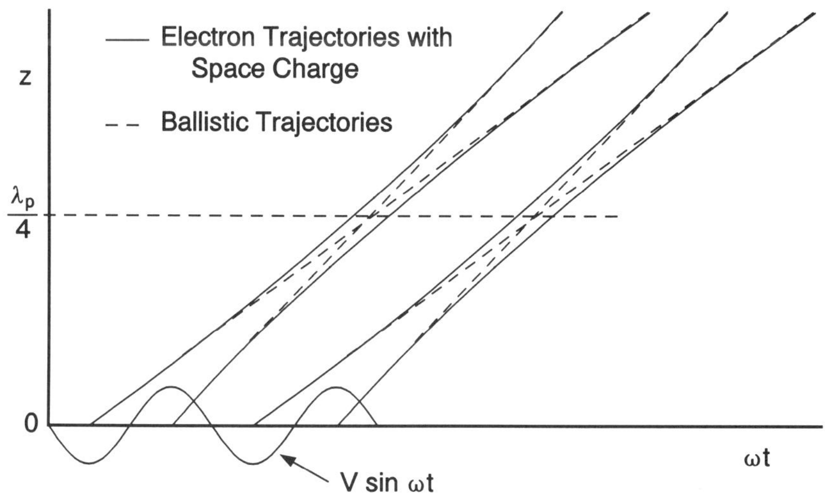

Figure 9-6 Modified Applegate diagram indicating space charge effects.

At the axial position denoted by $\lambda_p/4$, the fast electrons have been slowed to the dc velocity of the beam and the slow electrons have been accelerated to the dc velocity of the beam. Thus, at that point, all electrons have the same velocity. Also, at $\lambda_p/4$, the RF electron density and the RF current reach maximum values. For small to medium RF signal levels, the RF current is essentially sinusoidal.

A very important characteristic of the bunching process with space charge forces is that *all* electrons are either speeded up or slowed down to the same velocity (the dc beam velocity) at the same axial position ($\lambda_p/4$). In addition, even if the amplitude of the modulating field is changed so that initial electron velocities are changed, the axial position of the bunch remains the same. This result is extremely important to the klystron engineer because, unlike the situation when space charge forces are ignored, the cavity location for maximum RF beam current is not a function of signal level, of gap width, or of frequency of operation. The distance, λ_p, is the electron plasma wavelength. The origin of this wavelength is the natural frequency of oscillation (the electron plasma frequency) of a cloud of electrons and the velocity with which the cloud of electrons is moving.

9.2.1 Electron Plasma Oscillations

To understand electron plasma oscillations, it is useful to start by considering an electron cloud that, as a whole, is not moving. The transverse dimensions of the cloud are assumed to extend to infinity and stationary ions are assumed to be uniformly distributed throughout the cloud to neutralize the average charge of the electrons. (These ions will not affect the electron plasma oscillations.)

Initially, assume that the electrons are uniformly distributed as shown in Figure 9-7(a). Assume further, that some electrons have been accelerated to the left and some have been accelerated to the right as indicated by the arrows. (This could be accomplished by moving a pair of parallel grids, between which an RF voltage is applied, through the cloud. This is similar to what happens in a modulating gap in a klystron or TWT except that, in that case, the electrons move and the gap is stationary.)

After acceleration, electrons move to the right and to the left as indicated in Figure 9-7(b). Regions of high density (bunches) form and the resulting space charge forces in these bunches cause the electron movement to stop and then reverse directions. Next, as indicated in Figure 9-7(c), the electrons become uniformly distributed again but are moving in directions opposite to those shown in Figure 9-7(a). As the electrons continue to move, new bunches form as indicated in Figure 9-7(d). Again the space charge forces that build up in the bunches cause the electron movement to stop and reverse directions. After that, the distribution shown in Figure 9-7(a) returns and the process starts over again. The electrons continue to move back and forth at a frequency known as the *electron plasma frequency*.

The calculation of the electron plasma frequency is straightforward and may be carried out with the aid of Figure 9-8. A slab of charge of thickness dz will be considered. Electron positions for uniform charge distribution (as in Figure 9-7(a)) are represented by the equally spaced dots at the top of the figure. The positions of these electrons at a later time in a region of compression (as in Figure 9-7(b)) are shown at the bottom of the figure.

The distance moved by the electrons is s, which is a function of z. As the electrons are displaced from their dc positions, an electric restoring field, E, develops, which tends to force the electrons back to their dc positions. From the force equation, the relation between the electron acceleration and electric field is

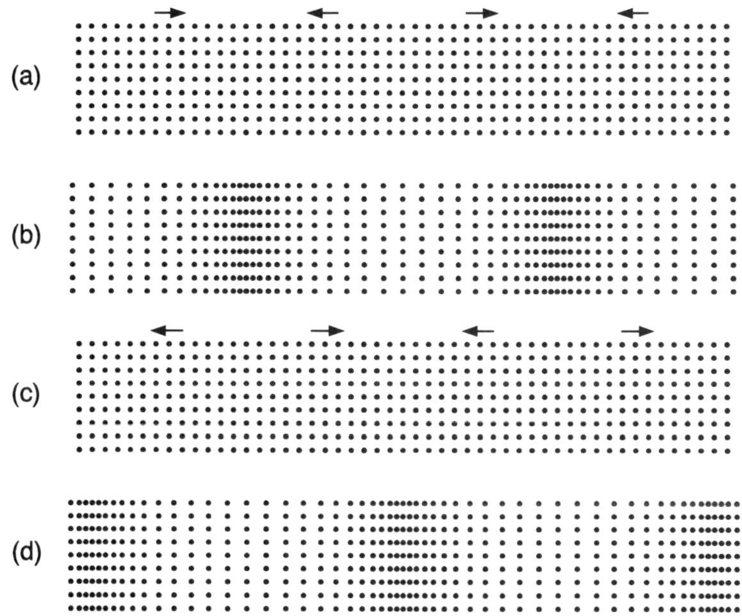

Figure 9-7 Electron distributions during plasma oscillations.

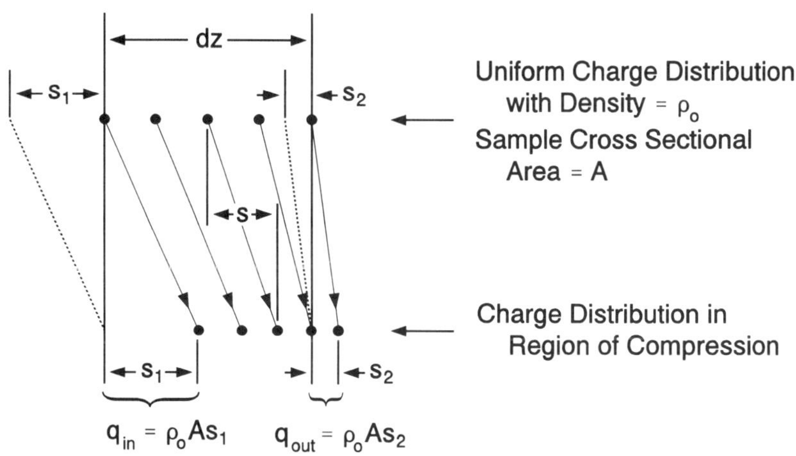

Figure 9-8 Displacement of electrons during plasma oscillation.

$$\ddot{s} = -\eta E \tag{9-15}$$

The electric field can be found from Gauss's law, which, because there are variations only in the z direction, is

$$\nabla \cdot \boldsymbol{E} = \frac{\partial}{\partial z} E = -\frac{\rho}{\epsilon_o} \tag{9-16}$$

where ρ is the net charge density resulting from the movement of electrons. To find ρ, we may examine the electron charge, q_{in}, moving into the slab of thickness dz and the electron charge, q_{out}, moving out of the slab. The resulting net charge, q, is related to q_{in}, q_{out} and ρ by

$$q = q_{in} - q_{out} = \rho A\, dz \tag{9-17}$$

The charge entering z from the left consists of those electrons in distance s before being displaced where the charge density is ρ_o. Thus,

$$q_{in} = \rho_o A s_1 \tag{9-18}$$

where A is an arbitrary cross sectional area of the sample of electrons being considered. Similarly, the charge leaving dz from the right is

$$q_{out} = \rho_o A s_2 \tag{9-19}$$

But the distance s_2 is s_1 plus the amount by which s has changed in distance dz; that is,

$$s_2 = s_1 + \frac{\partial s}{\partial z} dz \tag{9-20}$$

and so

$$q_{out} = \rho_o A \left[s_1 + \frac{\partial s}{\partial z} dz \right] \tag{9-21}$$

Now, by subtracting q_{out} from q_{in}, the net charge density is found to be

$$\rho = -\rho_o \frac{\partial s}{\partial z} \tag{9-22}$$

Substituting this result into Gauss's law, we obtain

$$\frac{\partial E}{\partial z} = \frac{\rho_o}{\epsilon_o} \frac{\partial s}{\partial z} \tag{9-23}$$

After integration, the field is

$$E = \frac{\rho_o}{\epsilon_o} s \tag{9-24}$$

where the constant of integration is zero because $E = 0$ when $s = 0$. With this field, the force equation becomes

$$\ddot{s} + \eta \frac{\rho_o}{\epsilon_o} s = 0 \tag{9-25}$$

which is the equation for simple harmonic motion. Solutions are of the form

$$s = s_o \sin \omega_p t \tag{9-26}$$

where

$$\omega_p = \left[\eta \frac{\rho_o}{\epsilon_o} \right]^{1/2} \tag{9-27}$$

is the frequency of oscillation of the electron cloud (the electron plasma frequency). It should be noted that the plasma frequency is dependent *only* on the electron density. We will see later that ω_p is an extremely important quantity in the analysis and understanding of the wave motion in a TWT.

To obtain an appreciation for the magnitude of the plasma frequency in a TWT, consider a 10,000 V, 1 A electron beam with a diameter of 2 mm. The electron plasma frequency, f_p, is

$$f_p = 0.82 \quad \text{GHz} \tag{9-28}$$

which is well below the operating frequency of the TWT containing that beam.

The preceding analysis of electron plasma oscillations was carried out for a stationary cloud of electrons. When the cloud is moving so that an electron beam is formed, the plasma oscillations are not affected (unless the beam velocity is high enough that relativistic effects are important). Then, referring back to Figure 9-6, the situation at $z = 0$ is like that in Figure 9-7(a). At $\lambda_p/4$, the electrons are compressed as shown in Figure 9-7(b). This is one quarter of a cycle of oscillation after $z = 0$ and so

$$\frac{\lambda_p}{4} = \frac{u_o}{f_p/4} \tag{9-29}$$

or

$$\lambda_p = \frac{u_o}{f_p} = \frac{2\pi u_o}{\omega_p} \tag{9-30}$$

For the 10,000 V, 1 A beam considered previously

$$\lambda_p = 4.5 \quad \text{mm} \tag{9-31}$$

This length is a about twice the beam diameter. Because λ_p and beam diameter are comparable in typical beams, boundary effects (the relatively small diameter of the beam and the presence of a nearby metallic structure) have a considerable impact on the magnitude of the electric fields in the bunches and on the values of ω_p and λ_p.

9.2.2 Qualitative Discussion of Space Charge Waves

In the case of TWTs, the velocity of propagation of electron bunches in a beam is of significant concern. The fundamental aspects of bunch propagation may be understood by considering the analogy with acoustic-wave propagation shown in Figure 9-9. In Figure 9-9(a), back-to-back acoustic transducers launch acoustic pressure (and density) waves in the air. The waves result from motion of the diaphragms in the transducers, which periodically compress the air. The waves propagate to the right and to the left at a velocity that is dependent on air pressure. With increased pressure, wave velocity increases.

In Figure 9-9(b), a grid is shown inserted in an electron cloud that is not moving. We will assume that the grid does not collect electrons. As the voltage on the grid oscillates from positive to negative and back to positive again, electrons are attracted, then repelled and then attracted again. As a result, the electron density near the grid is alternately reduced and increased and electron density waves are launched to the right and to the left. The velocities of these waves are dependent on the electron density. With increased density, wave velocities increases. These electron density waves are called *space charge waves*.

Next, let the electron column move to the right at a velocity that is much higher than the velocities of the space charge waves. The moving electron column is, of course, an electron beam. Both space charge waves are moving to the right with the beam. One is moving faster than the beam and is called a *fast space charge wave*. The other is moving slower than the beam and is called a *slow space charge wave*. This slow space charge wave is used in traveling wave tubes to produce amplification. The dependence of the velocity of space charge waves on electron density and on beam and circuit geometries are analyzed in the following section. The interaction of the slow space charge wave with an RF circuit is examined in the following chapter on traveling wave interaction.

9.2.3 Space Charge Waves in Infinite Beams

In the ballistic analysis there is no need to consider propagating electric and magnetic fields. However, when space charge effects are considered these fields must be included in the analysis. These fields are described by Maxwell's equations, which, when electron charge and current are present, may be written as follows:

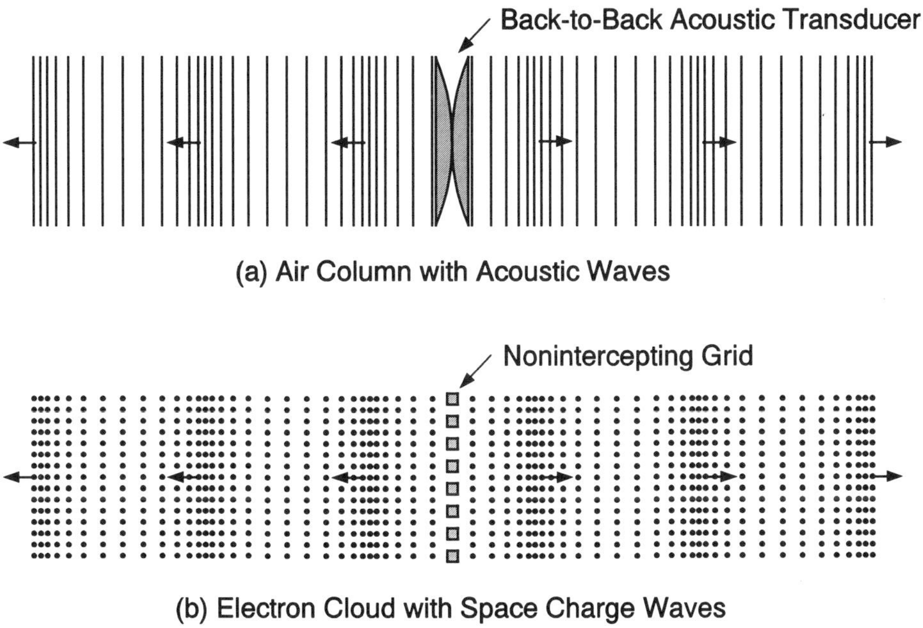

Figure 9-9 Arrangements for launching acoustic waves and electron density (space charge) waves.

$$
\begin{aligned}
\nabla \cdot \boldsymbol{E} &= -\frac{\rho}{\epsilon_o} \\
\nabla \times \boldsymbol{E} &= -\frac{\partial \boldsymbol{B}}{\partial t} = -\mu \frac{\partial \boldsymbol{H}}{\partial t} \\
\nabla \times \boldsymbol{H} &= -\boldsymbol{J} + \frac{\partial \boldsymbol{D}}{\partial t} = -\boldsymbol{J} + \epsilon \frac{\partial \boldsymbol{E}}{\partial t} \\
\nabla \cdot \boldsymbol{B} &= 0
\end{aligned}
\tag{9-32}
$$

In addition,

$$\nabla \cdot \nabla \times \boldsymbol{H} = -\nabla \cdot \boldsymbol{J} + \epsilon \frac{\partial}{\partial t} \nabla \cdot \boldsymbol{E} = 0 \tag{9-33}$$

so

$$\nabla \cdot \boldsymbol{J} = -\frac{\partial \rho}{\partial t} \tag{9-34}$$

which is the equation of continuity.

In general, the notation that will be used is the following:

$$F(z,t) = F_o(z) + F_1(z,t) \tag{9-35}$$

where the subscript, o, indicates a dc quantity and the subscript, 1, indicates an ac quantity. In most of the analyses, it will be assumed that signal levels are small and that all ac quantities vary sinusoidally with time.

Now, to examine the effect of space charge on bunching, an infinitely wide electron beam will be considered first. It will be assumed that the dc space charge in the beam is neutralized by stationary ions so that potential depression problems do not exist. This simplifies the analysis a great deal because no transverse variations of quantities need to be considered. From Maxwell's equations, the following may be written for the ac quantities in the beam.

$$\begin{aligned} \nabla \times \boldsymbol{E}_1 &= -\mu_o \frac{\partial \boldsymbol{H}_1}{\partial t} \\ \nabla \times \boldsymbol{H}_1 &= -\boldsymbol{J}_1 + \epsilon_o \frac{\partial \boldsymbol{E}_1}{\partial t} \end{aligned} \tag{9-36}$$

and so

$$\begin{aligned} \nabla \times \nabla \times \boldsymbol{E}_1 &= -\mu_o \nabla \times \frac{\partial \boldsymbol{H}_1}{\partial t} \\ &= \mu_o \frac{\partial}{\partial t} \boldsymbol{J}_1 - \mu_o \epsilon_o \frac{\partial^2}{\partial t^2} \boldsymbol{E}_1 \end{aligned} \tag{9-37}$$

With the vector identity

$$\nabla \times \nabla \times \boldsymbol{E} = \nabla(\nabla \cdot \boldsymbol{E}) - \nabla^2 \boldsymbol{E} \tag{9-38}$$

and with Gauss's law

$$\nabla \cdot \boldsymbol{E}_1 = -\frac{\rho_1}{\epsilon_o} \tag{9-39}$$

9-37 becomes the wave equation, which is

$$\nabla^2 \boldsymbol{E}_1 - \mu_o \epsilon_o \frac{\partial^2}{\partial t^2} \boldsymbol{E}_1 = -\mu_o \frac{\partial}{\partial t} \boldsymbol{J}_1 - \frac{1}{\epsilon_o} \nabla \rho_1 \tag{9-40}$$

This equation is essential to the analysis of the behavior of electron beams when space charge effects are taken into consideration.

When it is assumed that all quantities vary sinusoidally with time (a small-signal assumption in the case of the TWT), that is, all quantities vary as $e^{j\omega t}$, then the wave equation becomes

$$\nabla^2 \boldsymbol{E_1} + k^2 \boldsymbol{E_1} = -j\omega\mu_o \boldsymbol{J_1} - \frac{1}{\epsilon_o}\boldsymbol{\nabla}\rho_1 \tag{9-41}$$

where

$$\mu_o \epsilon_o = \frac{1}{c^2} \quad \text{and} \quad \frac{\omega^2}{c^2} = k^2 \tag{9-42}$$

Here, c is the speed of light and k is the propagation constant of a wave traveling at the speed of light.

For the case under consideration, there are no variations of quantities in the x and y directions and so the vector quantities $\boldsymbol{E_1}$ and $\boldsymbol{J_1}$ may be written as

$$\boldsymbol{E_1} = E_z \quad \text{and} \quad \boldsymbol{J_1} = J_z \tag{9-43}$$

Also,

$$\frac{\partial^2}{\partial x^2} E_z = \frac{\partial^2}{\partial y^2} E_z = 0 \tag{9-44}$$

and so the wave equation for this case becomes

$$\frac{\partial^2}{\partial z^2} E_z + k^2 E_z = -j\omega\mu_o J_z - \frac{1}{\epsilon_o}\frac{\partial}{\partial z}\rho_1 \tag{9-45}$$

Next, it will be assumed that all quantities vary sinusoidally with distance (another small-signal assumption). This may be done by assuming that all quantities vary as $e^{-j\beta z}$ where β is the propagation constant. Values for β depend on beam and circuit parameters, *et cetera*. With this assumption, the wave equation becomes

$$\left(\beta^2 - k^2\right) E_z = j\omega\mu_o J_z - \frac{j\beta}{\epsilon_o}\rho_1 \tag{9-46}$$

The current density, J_z, may be found with the aid of the definition of current, $J = \rho u$, which can be expanded as follows:

$$\begin{aligned} J = J_o + J_1 &= \left(\rho_o + \rho_1\right)\left(u_o + u_1\right) \\ &= \rho_o u_o + \rho_1 u_o + \rho_o u_1 + \rho_1 u_1 \end{aligned} \tag{9-47}$$

But $J_o = \rho_o u_o$ and $\rho_1 u_1$ is the product of small quantities and, for small signal levels, may be neglected. Also, $J_1 = J_z$ and $u_1 = u_z$ so

$$J_z = \rho_o u_z + \rho_1 u_o \tag{9-48}$$

Also, from the continuity equation

$$\boldsymbol{\nabla} \cdot \boldsymbol{J} = -\frac{\partial \rho}{\partial t} \quad \text{or} \quad -j\beta J_z = -j\omega\rho_1 \tag{9-49}$$

so

$$\rho_1 = \frac{\beta}{\omega} J_z \tag{9-50}$$

and so charge density can be eliminated from the current equation (9-48) to obtain

$$J_z = \frac{\rho_o}{1 - \dfrac{\beta u_o}{\omega}} u_z \tag{9-51}$$

The velocity, u_z, may be eliminated by use of the force equation. Thus,

$$\frac{d}{dt} u_z = \frac{\partial}{\partial t} u_z + u_o \frac{\partial}{\partial z} u_z = j \omega u_z - u_o j \beta u_z = -\eta E_z \tag{9-52}$$

so

$$u_z = -\frac{\eta}{j(\omega - \beta u_o)} E_z \tag{9-53}$$

As a result, the current density becomes

$$J_z = -\frac{\rho_o \eta}{j(\omega - \beta u_o)\left[1 - \dfrac{\beta u_o}{\omega}\right]} E_z$$

$$= j \frac{\omega \rho_o \eta}{(\omega - \beta u_o)^2} E_z \tag{9-54}$$

When the plasma frequency, ω_p, of the electrons (9-27) is substituted into this expression, the current density becomes

$$J_z = j \omega \epsilon_o \frac{\omega_p^2}{(\omega - \beta u_o)^2} E_z \tag{9-55}$$

This relation between beam current density and electric field is often referred to as the "electronic equation."

This electronic equation and the relation for charge density (9-50) can now be combined with the wave equation (9-46) to obtain

$$
\left(\beta^2 - k^2\right) E_z = \left[j\omega\mu_o - \frac{j\beta^2}{\omega\epsilon_o} \right] j\omega\epsilon_o \frac{\omega_p^2}{\left(\omega - \beta u_o\right)^2} E_z
$$

$$
= \left(\beta^2 - k^2\right) \frac{\beta_p^2}{\left(\beta_e - \beta\right)^2} E_z \tag{9-56}
$$

or

$$
\left(\beta^2 - k^2\right) \left[1 - \frac{\beta_p^2}{\left(\beta_e - \beta\right)^2} \right] E_z = 0 \tag{9-57}
$$

where $\beta_e = \omega/u_o$ and $\beta_p = \omega_p/u_o$. This equation, which is often called the *determinantal equation*, can be solved for the propagation constant, β. The solutions are

$$
\beta = \pm k \qquad \text{or} \qquad \beta = \beta_e \pm \beta_p \tag{9-58}
$$

Now that the values for the propagation constant, β, have been determined, the question is: What do they mean? When $\beta = \pm k$, then, the electric field varies as

$$
E_z \propto e^{j(\omega t - kz)} \tag{9-59}
$$

and this represents waves that propagate in free space. The remaining two values of β show that

$$
E_z \propto e^{j[\omega t - (\beta_e \pm \beta_p)z]} \tag{9-60}
$$

The significance of this expression may be realized by letting $\omega t = (\beta_e \pm \beta_p)z$. Then the velocity, z/t, of the waves is

$$
\frac{z}{t} = \frac{\omega}{\beta_e \pm \beta_p} = \frac{\omega u_o}{\omega \pm \omega_p} = \frac{u_o}{1 \pm \dfrac{\omega_p}{\omega}} \tag{9-61}
$$

Thus, these are waves that propagate just below (+) and just above (-) the velocity of the electron beam. These are the space charge waves discussed qualitatively in the previous section.

In general, the electric field contains both fast and slow space charge waves and may be written as

$$
E_z = A_E e^{j[\omega t - (\beta_e + \beta_p)z]} + B_E e^{j[\omega t - (\beta_e - \beta_p)z]} \tag{9-62}
$$

where A_E and B_E are amplitudes to be determined from initial conditions.

The velocity is proportional to the electric field so

$$u_z = A_u e^{j[\omega t - (\beta_e + \beta_p)z]} + B_u e^{j[\omega t - (\beta_e - \beta_p)z]} \tag{9-63}$$

For the case of velocity modulation in a gridded gap, the ac velocity component at $z = 0$ is

$$u = u_o \frac{\alpha M}{2} e^{j\omega t} \tag{9-64}$$

But, at $z = 0$,

$$u_z = A_u e^{j\omega t} + B_u e^{j\omega t} \tag{9-65}$$

and so

$$A_u + B_u = u_o \frac{\alpha M}{2} \tag{9-66}$$

Another initial condition is needed to determine A_u and B_u. At $z = 0$, even though the velocities of the electrons have been modulated, the electrons have not yet moved from their dc positions. This means that the charge density is still the dc charge density and that $\rho_1 = 0$. Since, in general, $J_z = \rho_o u_z + \rho_1 u_o$, then, at $z = 0$,

$$J_z = \rho_o u_z \tag{9-67}$$

so

$$J_z = \rho_o \left(A_u + B_u\right) e^{j\omega t} \tag{9-68}$$

Also, from (9-51),

$$J_z = \frac{\rho_o}{1 - \dfrac{\beta u_o}{\omega}} u_z = \frac{\beta_e \rho_o}{\beta_e - \beta} u_z = \frac{\beta_e \rho_o}{\beta_e - (\beta_e \pm \beta_p)} u_z = \frac{\beta_e \rho_o}{\mp \beta_p} u_z \tag{9-69}$$

With the substitution of velocity from (9-63),

$$J_z = -\frac{\beta_e \rho_o}{\beta_p} A_u e^{j[\omega t - (\beta_e + \beta_p)z]} + \frac{\beta_e \rho_o}{\beta_p} B_u e^{j[\omega t - (\beta_e - \beta_p)z]} \tag{9-70}$$

which, at $z = 0$, is

$$J_z = \frac{\beta_e \rho_o}{\beta_p} \left(-A_u + B_u\right) e^{j\omega t} \tag{9-71}$$

The combination of this with (9-68) yields

$$A_u + B_u = \frac{\beta_e}{\beta_p}\left(-A_u + B_u\right) \tag{9-72}$$

or

$$B_u = \frac{\beta_e + \beta_p}{\beta_e - \beta_p} A_u \tag{9-73}$$

Now it is noted that, since $\omega_p << \omega$, then $\beta_p << \beta_e$ and so

$$B_u \approx A_u \tag{9-74}$$

This approximation is equivalent (see (9-71)) to assuming that $J_z = 0$ at $z = 0$.

With the approximation in (9-74), the value of A_u is found to be

$$A_u = u_o \frac{\alpha M}{4} \tag{9-75}$$

so

$$u_z = u_o \frac{\alpha M}{4}\left\{e^{j[\omega t - (\beta_e + \beta_p)z]} + e^{j[\omega t - (\beta_e - \beta_p)z]}\right\} \tag{9-76}$$

This may be written as

$$u_z = u_o \frac{\alpha M}{2}\left[\frac{e^{-j\beta_p z} + e^{j\beta_p z}}{2}\right] e^{j[\omega t - \beta_e z]} = u_o \frac{\alpha M}{2} \cos\beta_p z \, e^{j[\omega t - \beta_e z]} \tag{9-77}$$

The exponential term remaining in the velocity is cos + j sin and either the real or the imaginary part can be chosen as long as we are consistent in using this choice throughout the analysis. We will choose to use the real part and so

$$u_z = u_o \frac{\alpha M}{2} \cos\beta_p z \cos\left(\omega t - \beta_e z\right) \tag{9-78}$$

If we move along the beam at a velocity such that $\omega t = \beta_e z$ or at $z/t = u_o$ (the dc beam velocity), then the ac velocity, u_z, of the electrons appears as shown by curve 1 of Figure 9-10. If we choose a time such that $\omega t = \beta_e z - \pi$, then u_z appears as shown by curve 2. At other times, u_z lies between curves 1 and 2 as indicated by the dashed lines. Note that, regardless of ωt, u_z is zero for all electrons at $\beta_p z = \pi/2$ or at

$$z = \frac{\pi}{2\beta_p} = \frac{\pi}{2}\frac{u_o}{\omega_p} = \frac{u_o}{4f_p} = \frac{\lambda_p}{4} \tag{9-79}$$

where λ_p is referred to as the plasma wavelength. Thus, all electrons have the same velocity (u_o) at $\lambda_p/4$. The reason for this is that, as the fast electrons approach the slow ones at $z = \lambda_p/4$, a bunch is formed. Furthermore, the bunching action causes the fast electrons to slow to the dc velocity, and the slow electrons to speed up to the dc velocity. As the electrons drift beyond $\lambda_p/4$, the space charge forces within the bunch cause the electrons in the front of the bunch to speed up and those in the rear of the bunch to slow down.

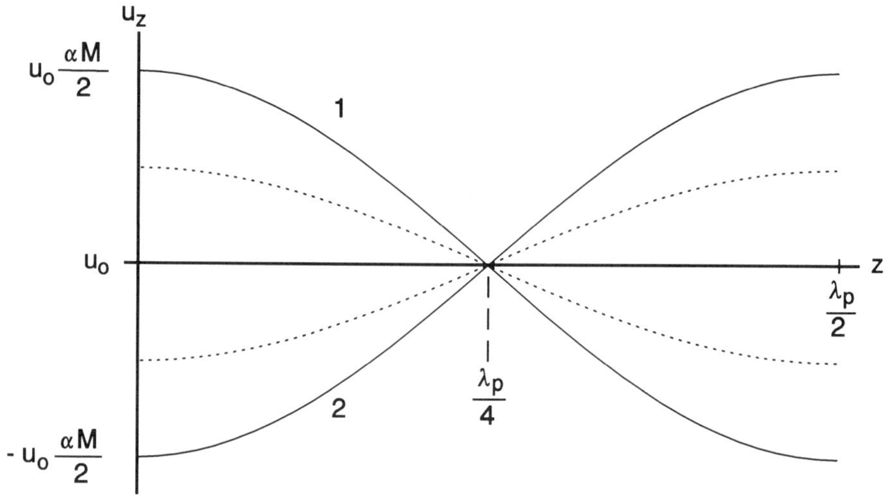

Figure 9-10 Velocities of electrons as a function of position.

With the substitution of A_u and B_u into (9-70), the current density is

$$J_z = \frac{\beta_e \rho_o}{\beta_p} \frac{u_o \alpha M}{2} \left[\frac{-e^{-j\beta_p z} + e^{j\beta_p z}}{2} \right] e^{j[\omega t - \beta_e z]}$$

$$= \frac{\beta_e \rho_o}{\beta_p} \frac{u_o \alpha M}{2} \sin \beta_p z \; j e^{j[\omega t - \beta_e z]} \tag{9-80}$$

$$= \frac{\beta_e \rho_o}{\beta_p} \frac{u_o \alpha M}{2} \sin \beta_p z \, \sin(\omega t - \beta_e z)$$

Now, $J_z = 0$ at $z = 0$ and varies sinusoidally with distance as shown in Figure 9-11. It should be noted that the maximum current occurs when $z = \lambda_p/4$, which corresponds with the position where the ac velocity goes to zero. If the electrons are permitted to drift beyond $\lambda_p/4$, then the bunch disperses and the current density decreases. The bunch and current density disappear at $\lambda_p/2$ and then reappear as the beam approaches $3\lambda_p/4$.

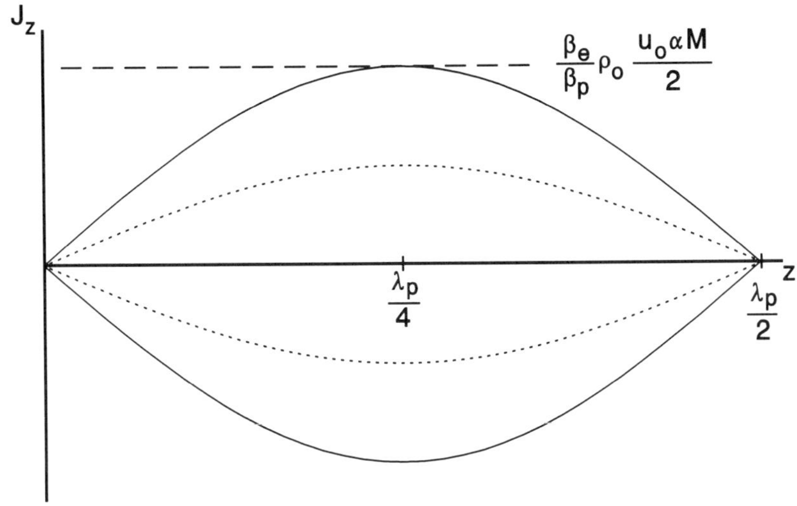

Figure 9-11 Current density as a function of position (as seen by an observer moving at the beam velocity).

The variations of current density shown in Figure 9-11 are those that an observer moving at the beam velocity would see. If, instead, at an instant in time, a "snapshot" of current density as a function of position is considered, it would appear as shown in Figure 9-12. ($\omega t = 0, 2\pi, 4\pi$, *et cetera* for this figure.) The current peaks labeled 1, 2 and 3 result from bunches of electrons that are in the process of forming. Beyond $\lambda_p/4$, the bunches disperse and so the current peaks (4 and 5) decrease.

For the plot in Figure 9-12, $\omega = 10\omega_p$ and so within a distance λ_p there would be ten cycles of RF current and ten bunches of electrons in various stages of formation (five bunches are indicated in $\lambda_p/2$ in Figure 9-12).

It is instructive to carry the analysis of the current waveform shown in Figure 9-12 one step further as is indicated in Figure 9-13. The trigonometric identity for the product of sine waves in the current waveform ((9-80) at $\omega t = 0$) is

$$\sin \beta_p z \sin \beta_e z = \frac{1}{2} \cos(\beta_e z - \beta_p z) - \frac{1}{2} \cos(\beta_e z + \beta_p z) \tag{9-81}$$

Thus, the current waveform can be represented as the beat pattern of two waves with propagation constants $\beta_e \pm \beta_p$. The wave propagating above the beam velocity ($\beta_e - \beta_p$) is the fast space charge wave, and the one propagating below the beam velocity ($\beta_e + \beta_p$) is the slow space charge wave. For the case considered here (velocity modulation by a gap and current density approximately equal to zero at $z = 0$) these waves have equal amplitudes. With other modulation techniques, these waves can be independently excited. As has been pointed out previously, in a TWT, the slow space charge wave interacts with the wave on the circuit to produce amplification.

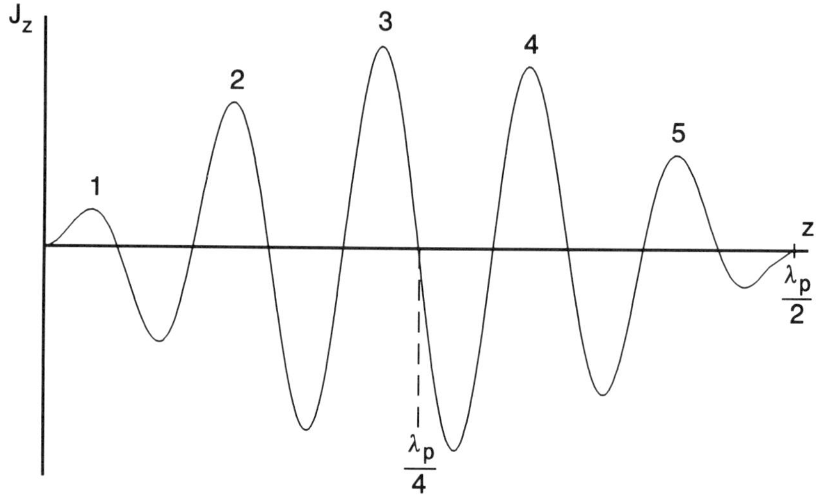

Figure 9-12 Instantaneous current density as a function of position.

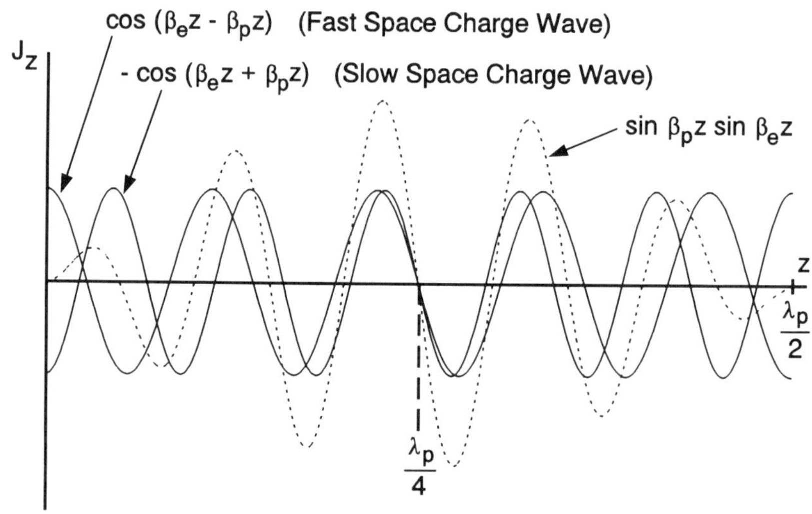

Figure 9-13 Fast and slow space charge waves.

9.2.4 Space Charge Waves in Confined Flow

In the previous section, the transverse dimensions of the electron beam were infinite. Next, the usual case of a cylindrical beam in a cylindrical tunnel as shown in Figure 9-14 will be considered. A very large axial magnetic field is assumed to ensure that electrons move only in the z direction.

We now seek a solution of the wave equation that makes E_z finite on the axis of the system and zero at the metallic wall and that obeys the standard boundary conditions at $r = b$ (the boundary of the electron beam). From the symmetry of the problem, the fields have no θ variation. The z component of the wave equation

$$\nabla^2 \boldsymbol{E}_1 - \mu_o \epsilon_o \frac{\partial^2}{\partial t^2} \boldsymbol{E}_1 = -\mu_o \frac{\partial}{\partial t} \boldsymbol{J}_1 - \frac{1}{\epsilon_o} \nabla \rho_1 \tag{9-82}$$

is therefore

$$\frac{1}{r}\frac{\partial}{\partial r}\left[r \frac{\partial E_z}{\partial r} \right] + \frac{\partial^2}{\partial z^2} E_z - \mu_o \epsilon_o \frac{\partial^2}{\partial t^2} E_z = -\mu_o \frac{\partial}{\partial t} J_z - \frac{1}{\epsilon_o}\frac{\partial}{\partial z}\rho_1 \tag{9-83}$$

Again, it will be assumed that all functions vary as $e^{j(\omega t - \beta z)}$ so

$$\frac{\partial^2}{\partial z^2} E_z - \mu_o \epsilon_o \frac{\partial^2}{\partial t^2} E_z = \left(\beta^2 - k^2\right) E_z \tag{9-84}$$

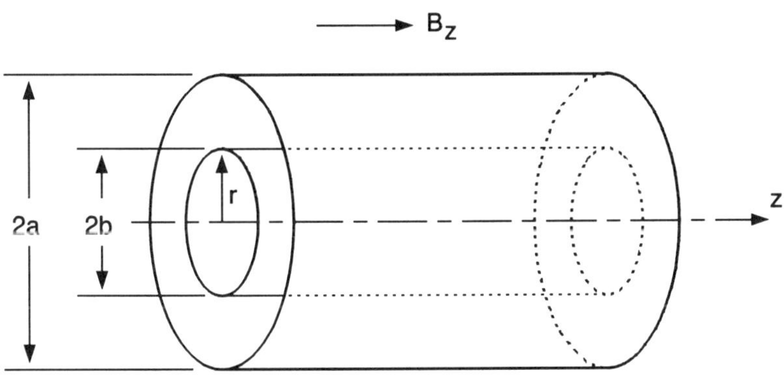

Figure 9-14 Cylindrical beam in a cylindrical tunnel.

Since there are no r or θ components

$$J_z = \frac{\omega}{\beta}\rho_1 \tag{9-85}$$

Also,

$$u_z = -\frac{\eta}{j(\omega - \beta u_o)} E_z \tag{9-86}$$

These relations are the same as for the infinite beam case, and so the righthand side of the wave equation (9-83) is still

$$-\mu_o \frac{\partial}{\partial t} J_z - \frac{1}{\epsilon_o} \frac{\partial}{\partial z} \rho_1 = \left(\beta^2 - k^2\right) \frac{\beta_p^2}{\left(\beta_e - \beta\right)^2} E_z \tag{9-87}$$

Thus, the wave equation becomes

$$\frac{1}{r} \frac{\partial}{\partial r} \left\{ r \frac{\partial E_z}{\partial r} \right\} - \left(\beta^2 - k^2\right) \left[1 - \frac{\beta_p^2}{\left(\beta_e - \beta\right)^2} \right] E_z = 0 \tag{9-88}$$

or

$$\frac{1}{r} \frac{\partial}{\partial r} \left\{ r \frac{\partial E_z}{\partial r} \right\} + \gamma_r^2 E_z = 0 \tag{9-89}$$

where

$$\gamma_r^2 = -\left(\beta^2 - k^2\right) \left[1 - \frac{\beta_p^2}{\left(\beta_e - \beta\right)^2} \right] \tag{9-90}$$

At this point, it should be noted that, for the infinite beam, $\gamma_r = 0$.

Next, the region inside the beam will be denoted by I and the region outside the beam will be denoted by II. Because of the plus sign in (9-89), the solution inside the beam (region I) is

$$E_{zI} = A\mathrm{J}_0(\gamma_r r) + B\mathrm{Y}_0(\gamma_r r) \tag{9-91}$$

where J_0 is a Bessel function of the first kind and Y_0 is a Bessel function of the second kind. At $r = 0$, $\mathrm{Y}_0(0) = \infty$, which means that $B = 0$. Thus,

$$E_{zI} = A\mathrm{J}_0(\gamma_r r) \tag{9-92}$$

Outside the beam (region II), there is no current, so $\omega_p^2 = 0$ and

$$\frac{1}{r} \frac{\partial}{\partial r} \left\{ r \frac{\partial E_{zII}}{\partial r} \right\} - \left(\beta^2 - k^2\right) E_{zII} = 0 \tag{9-93}$$

The general solution to this equation is

$$E_{zII} = C\mathrm{I}_0(\gamma_o r) + D\mathrm{K}_0(\gamma_o r) \tag{9-94}$$

where $\gamma_o^2 = (\beta^2 - k^2)$, I_0 is a modified Bessel function of the first kind, and K_0 is a modified Bessel function of the second kind.

Next, the boundary conditions may be used to determine the relations between A, C and D. At $r = a$, $E_{zII} = 0$ and so

$$\frac{D}{C} = -\frac{I_0(\gamma_o a)}{K_0(\gamma_o a)} \tag{9-95}$$

At $r = b$, $E_{zI} = E_{zII}$ and so

$$\begin{aligned} A J_0(\gamma_r b) &= C I_0(\gamma_o b) + D K_0(\gamma_o b) \\ &= C I_0(\gamma_o b) - C K_0(\gamma_o b) \frac{I_0(\gamma_o a)}{K_0(\gamma_o a)} \end{aligned} \tag{9-96}$$

or

$$\frac{A}{C} = \frac{1}{J_0(\gamma_r b)} \left[I_0(\gamma_o b) - K_0(\gamma_o b) \frac{I_0(\gamma_o a)}{K_0(\gamma_o a)} \right] \tag{9-97}$$

To find A in terms of C, another boundary condition is needed. At the edge of the beam, E_r is continuous. An expression for E_r may be obtained by going back to the wave equation and solving for E_r. From this and the boundary condition at the edge of the beam, the following relation between A and C may be obtained.

$$\frac{A}{C} = \frac{-\gamma_o b}{\gamma_r b\, J_1(\gamma_r b)} \left[I_1(\gamma_o b) + \frac{I_0(\gamma_o a)}{K_0(\gamma_o a)} K_1(\gamma_o b) \right] \tag{9-98}$$

Therefore,

$$-\gamma_r b \frac{J_1(\gamma_r b)}{J_0(\gamma_r b)} = \gamma_o b \left[\frac{I_1(\gamma_o b) K_0(\gamma_o a) + I_0(\gamma_o a) K_1(\gamma_o b)}{I_0(\gamma_o b) K_0(\gamma_o a) - I_0(\gamma_o a) K_0(\gamma_o b)} \right] \tag{9-99}$$

This is a transcendental equation that, for the general case, must be solved numerically.

For the special case when $a = b$, that is, the beam just fills the tunnel, we find that (9-99) reduces to

$$J_0(\gamma_r b) = 0 \tag{9-100}$$

The zeros of the J_0 Bessel function occur at

$$\gamma_r b = 2.405, \quad 5.520, \quad 8.654, \quad \cdots \tag{9-101}$$

Thus, an infinite set of waves is available for matching initial conditions. As an example of an application, assume that a gridded gap as indicated in Figure 9-15(a) is used to modulate the beam. Then, the ac velocity (Figure 9-15(b)) is constant across the beam. In order to match this initial condition, the ac velocity must be written as

$$u_z = \sum_{n=0}^{\infty} u_{zn} \mathrm{J}_o(\gamma_{rn} r) \tag{9-102}$$

to obtain the various Fourier components to make up the rectangular velocity distribution as shown in Figure 9-15(c).

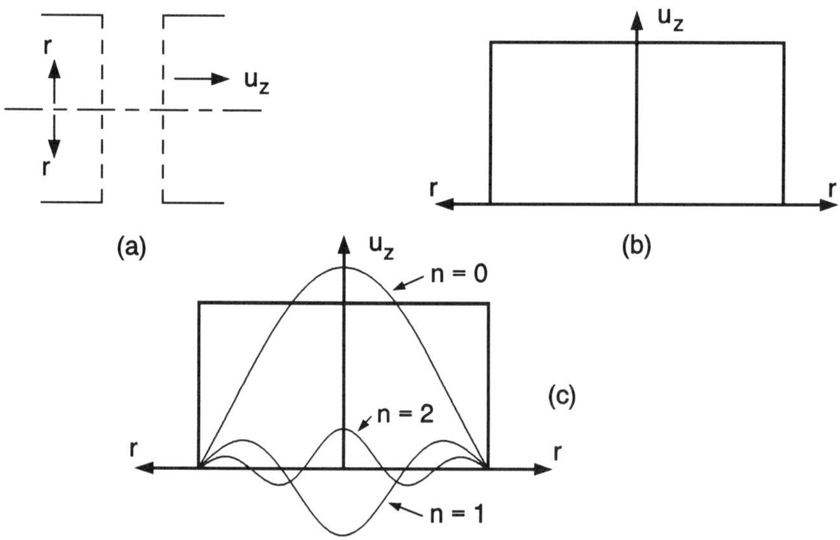

Figure 9-15 Velocity modulation by gridded gap.

For the general case, the numerical solutions of (9-99) may be found by assuming values of $\gamma_o b$ and a/b and then finding γ_r. There is an infinite set of solutions but the $n = 0$ solution is the one of primary interest because it corresponds to the dominant wave. Now, assuming that γ_r has been found as a function of γ_o, we proceed by assuming that the solutions for the propagation constant are similar to those found with the infinite beam. Thus,

$$\beta = \frac{\omega \pm F\omega_p}{u_o} \tag{9-103}$$

That is, it is assumed that the major change in propagation constant that occurs results from a change in plasma frequency by some factor F. Then, because $k^2 << \beta^2$, from (9-90),

$$\gamma_r^2 = \beta^2 \left[\frac{\beta_p^2}{(\beta_e - \beta)^2} - 1 \right] = (\beta_e \pm F\beta_p)^2 \left[\frac{1}{F^2} - 1 \right] \tag{9-104}$$

Since $F\beta_p << \beta_e$, F can be written as

$$F = \left[\frac{1}{1 + \gamma_r^2 / \beta_e^2} \right]^{1/2} \tag{9-105}$$

But $\beta_e \approx \gamma_o$ and γ_r has been found as a function of γ_o, so F can be plotted as a function of γ_o (or β_e) and a/b. The result is shown in Figure 9-16. In this figure, $n = a/b$, that is, the ratio of the tunnel radius to the beam radius. The reduced plasma frequency, $F\omega_p$, is usually denoted by ω_q. Note that, if $\gamma_o b = 1$ and $n = 1.5$, the value of F is about 0.5 and so the plasma frequency is one half of the value calculated for an infinite beam. Correspondingly, the plasma wavelength for the confined beam, λ_q, is twice the value calculated for an infinite beam. Thus, the maximum RF current in the beam occurs at a distance from the modulating cavity that is twice that for the infinite beam.

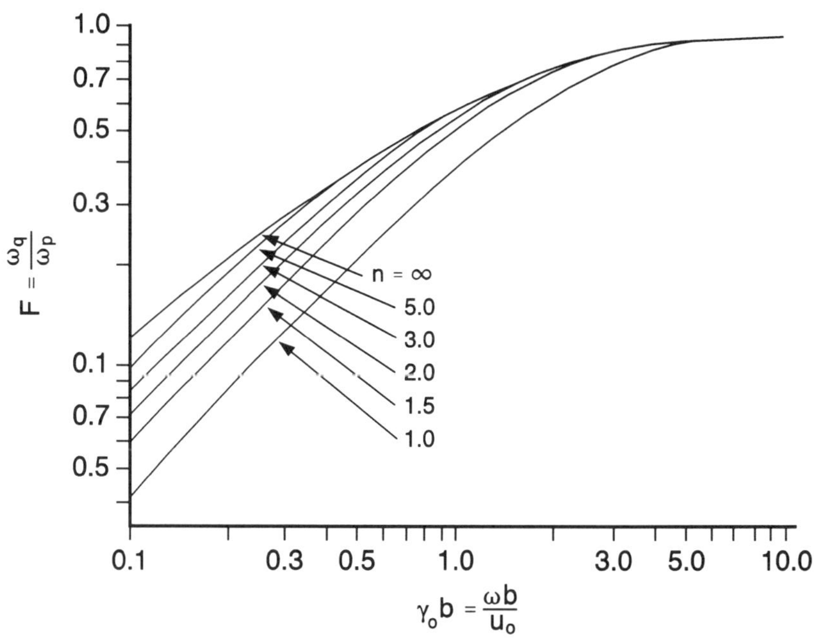

Figure 9-16 Plasma frequency reduction factor for a confined beam.

9.2.5 Space Charge Waves in Brillouin Flow

When the dc focusing of Brillouin beams was examined, it was found that a change in current in the beam caused a significant perturbation of the beam radius (as current increased, radius increased and vice versa). It would be expected, therefore, that the behavior of a modulated Brillouin beam would be quite different from that of a confined beam and, in fact, that is the case as the following analysis shows. A nonscalloping Brillouin beam will be assumed.

The components of the Lorentz force equation are

$$\text{(radial)} \qquad \ddot{r} - r\dot{\theta}^2 = -\eta\left(E_r + B_z r\dot{\theta}\right)$$

$$\text{(theta)} \qquad \frac{1}{r}\frac{d}{dt}\left(r^2\dot{\theta}\right) = \eta B_z \dot{r} \tag{9-106}$$

$$\text{(axial)} \qquad \ddot{z} = -\eta E_z$$

The dc solutions to these equations for the nonscalloping case are as follows:

$$\begin{aligned}
\dot{r} &= 0 \\
\dot{\theta} &= \omega_L = \frac{\eta}{2} B_B \\
\dot{z} &= u_o \\
\eta E_r &= -r\,\omega_L^2 \\
E_z &= 0 \\
\rho_o &= \frac{2\epsilon_o \omega_L^2}{\eta}
\end{aligned} \tag{9-107}$$

If the initial positions of an electron in the beam are r_o, θ_o and z_o, then at time t, the coordinates are

$$r = r_o$$

$$\theta = \theta_o + \omega_L t \tag{9-108}$$

$$z = z_o + u_o t$$

When the beam is modulated, these coordinates are perturbed. The perturbed positions can be found as follows: The theta equation of motion can be integrated to obtain

$$\frac{d}{dt}\left(r^2\dot{\theta}\right) = \frac{\eta}{2}B_B\frac{d}{dt}r^2 \tag{9-109}$$

or

$$r^2\dot{\theta} = \frac{\eta}{2}B_B r^2 + C, \quad \text{where} \quad C = 0 \tag{9-110}$$

so

$$\dot{\theta} = \omega_L \tag{9-111}$$

and so the angular velocity of the electrons remains constant, even when the beam is modulated. Thus, the radial force equation may be written as

$$\ddot{r} - r\omega_L^2 = -\eta\left(E_r + B_z r\,\omega_L\right) \tag{9-112}$$

or

$$\ddot{r} + r\omega_L^2 = -\eta E_r \tag{9-113}$$

Assuming that the perturbed values of r, z and E_r are

$$r = r_o + r_1$$

$$z = z_o + z_1 \tag{9-114}$$

$$E_r = E_{r_o} + E_{r_1}$$

then the force equations become

$$\ddot{r}_1 + \left(r_o + r_1\right)\omega_L^2 = -\eta\left(E_{r_o} + E_{r_1}\right) \tag{9-115}$$

and

$$\ddot{z} = -\eta E_{z1} \tag{9-116}$$

It must be remembered that, if the radial position of an electron varies, then the electron experiences a dc electric field which is

$$-\eta E_{r_o} = \omega_L^2 (r_o + r_1) \tag{9-117}$$

The reason for this is clear from the plot of E_r *versus* r given in Figure 9-17. It should be noted that a small error is introduced when electrons move outside the outer diameter of the beam. This will be neglected. Now the radial force equation reduces to

$$\ddot{r}_1 = -\eta E_{r_1} \tag{9-118}$$

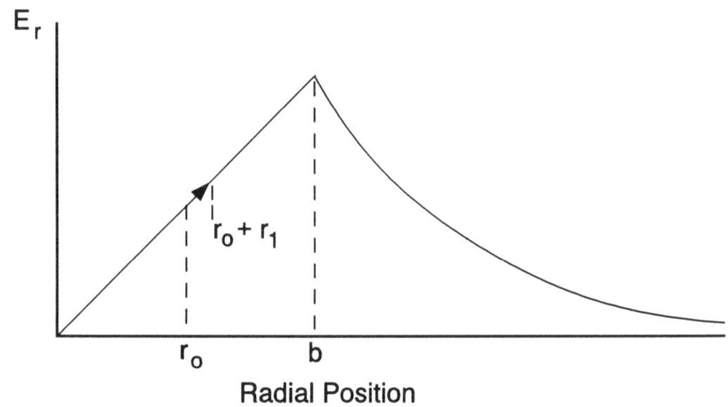

Figure 9-17 E_r inside and outside the beam.

Assuming that all RF quantities are small and that they vary as $e^{j(\omega t - \beta z)}$, then

$$\frac{d}{dt} = j u_o (\beta_e - \beta) \tag{9-119}$$

and so

$$\dot{r}_1 = j \frac{\eta}{u_o(\beta_e - \beta)} E_{r_1} \tag{9-120}$$

In addition, from the axial force equation,

$$\dot{z}_1 = j \frac{\eta}{u_o(\beta_e - \beta)} E_{z_1} \tag{9-121}$$

which is the same as it was for the infinite beam case and for the confined flow case.

The continuity equation may be written as

$$\nabla \cdot \boldsymbol{J} = \nabla \cdot (\rho \boldsymbol{u}) = -\frac{\partial \rho}{\partial t} \tag{9-122}$$

or as

$$\boldsymbol{u}_o \cdot \nabla \rho_1 + \rho_o \nabla \cdot \boldsymbol{u}_1 = -\frac{\partial \rho_1}{\partial t} \tag{9-123}$$

where, as usual, products of ac quantities have been assumed to be small enough that they can be neglected. This may be rewritten as

$$\rho_o \nabla \cdot \boldsymbol{u}_1 = -\frac{\partial \rho_1}{\partial t} - u_o \frac{\partial \rho_1}{\partial z} = -\frac{d}{dt}\rho_1 \tag{9-124}$$

so

$$\rho_1 = j \frac{\rho_o}{u_o(\beta_e - \beta)} \nabla \cdot \boldsymbol{u}_1 \tag{9-125}$$

But

$$\boldsymbol{u}_1 = \dot{r}_1 + \dot{z}_1 = j \frac{\eta}{u_o(\beta_e - \beta)} \left(\boldsymbol{E}_{r_1} + \boldsymbol{E}_{z_1}\right) = j \frac{\eta}{u_o(\beta_e - \beta)} \boldsymbol{E}_1 \tag{9-126}$$

so

$$\nabla \cdot \boldsymbol{u}_1 = j \frac{\eta}{u_o(\beta_e - \beta)} \nabla \cdot \boldsymbol{E}_1 \tag{9-127}$$

and so

$$\rho_1 = -\frac{\rho_o \eta}{u_o^2 (\beta_e - \beta)^2} \nabla \cdot \boldsymbol{E}_1 \tag{9-128}$$

or

$$\rho_1 = \frac{\beta_p^2}{(\beta_e - \beta)^2} \rho_1 \tag{9-129}$$

This equation has two possible solutions:

$$\beta = \beta_e \pm \beta_p \quad \text{and} \quad \rho_1 = 0 \tag{9-130}$$

The first solution, $\beta = \beta_e \pm \beta_p$, is the result found for infinite planar flow. This result is difficult to accept because it is expected that the overall beam behavior should not differ appreciably from that for confined flow. That is, we expect

$$\beta = \beta_e \pm F\beta_p \tag{9-131}$$

where F is the plasma frequency reduction factor, and we expect $F \approx F_c$ where F_c is the reduction factor for a beam with confined flow.

The second solution, $\rho_1 = 0$, states that the charge density within the modulated beam is ρ_o. This means, of course, that the RF current must result from modulation of the beam radius as shown in Figure 9-18. This result is extremely important because it shows that the RF current in a Brillouin beam is on the surface of the beam. There is no RF current in the center of the beam. As the depth of modulation increases, the amplitude of the ripple on the surface of the beam increases and so the RF current increases. As the amplitude of the ripple increases, the probability that interception will occur also increases and so beam focusing becomes extremely critical.

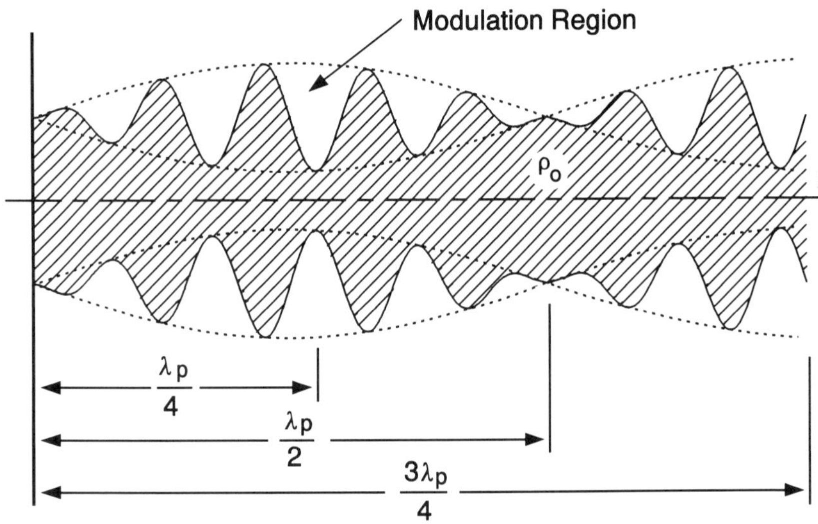

Figure 9-18 RF modulation of beam radius for a Brillouin beam.

With the result that $\rho_1 = 0$, it is possible to set up and solve the wave equation in a manner similar to that carried out for the confined beam. When this is done, the plasma frequency reduction factor may be found, and it is shown in Figure 9-19. It should be noted that the values of F for the Brillouin beam are somewhat smaller than for confined flow. For example, for $n = 1.5$ and $\gamma_o b = 1$, the value of F is about 0.4 whereas, for the confined beam, the value was 0.5. The reason for the smaller values of F can be explained with the aid of Figure 9-20, where the electric field pattern for a Brillouin beam is compared to that for the confined beam. Note that the field lines can bow inward as well as outward from the surface bunch on the Brillouin beam. As a result, the axial component of the field and the space charge forces are smaller than for the confined beam. This results in a lower plasma frequency and larger plasma wavelength for the Brillouin beam than for the confined beam.

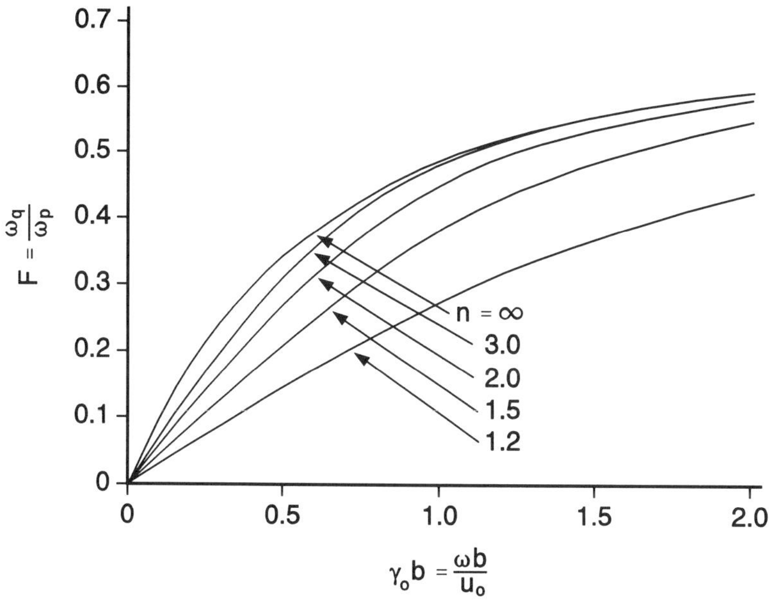

Figure 9-19 Plasma frequency reduction factor for a Brillouin beam.

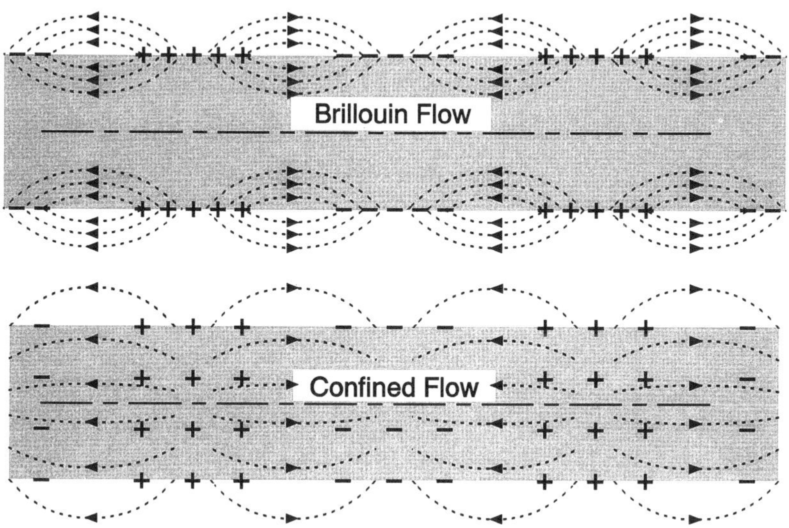

Figure 9-20 Comparison of RF electric field patterns near bunches in a Brillouin beam and a confined beam.

9.3 EXPERIMENTAL VERIFICATION

Experiments have been performed to verify that the RF current is on the surface of a Brillouin beam [3]. The RF beam analyzer is shown in Figure 9-21. This was the analyzer used for the dc beam measurements described in the chapter on linear beams (Chapter 7). The analyzer was modified for RF current measurements. The beam from a shielded Pierce gun was focused with a solenoid. A vacuum ball valve was used to permit changes to be made in the measurement apparatus without exposing the gun and cathode to air (which could have changed the beam characteristics).

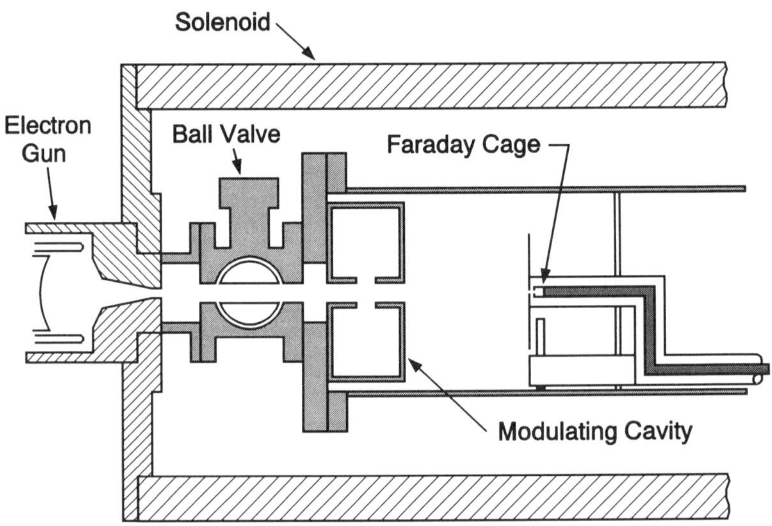

Figure 9-21 RF beam analyzer for measuring surface currents in a Brillouin beam. Adapted from A.S. Gilmour, Jr., *IEEE Trans. on Electron Devices,* July 1972. © 1972 IEEE.

The beam was velocity-modulated at approximately 2000 MHz. A beam-scanning mechanism with a .010 in aperture was used for sampling currents at various transverse and axial positions. The circuit used for detecting average and RF currents is given in Figure 9-22. Either the fundamental, second harmonic, or third harmonic current could be selected for measurement with a bandpass filter.

Plots of the average and fundamental components of current as functions of the transverse and axial position in the beam are shown in Figure 9-23. The depth of modulation, α, was about .05. Because of the small RF drive signal, the average current distribution was very similar to that observed in a dc beam.

The RF current was entirely on the surface of the beam. The predicted space-charge wave behavior of the electrons is clearly illustrated by four half plasma wavelengths of variation of the beam current.

The fundamental current distributions at higher RF drive levels are shown in Figure

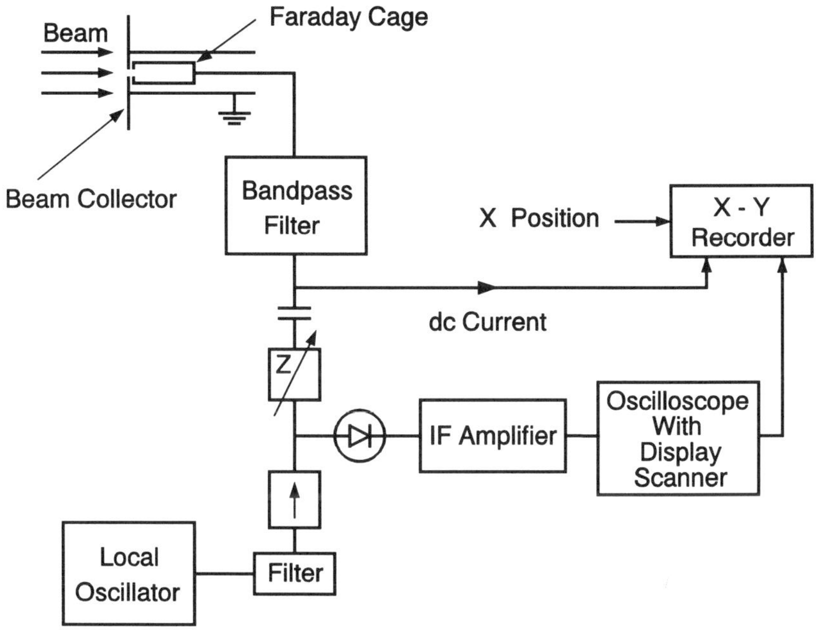

Figure 9-22 Circuit for detecting average and RF currents in a velocity modulated Brillouin beam. Adapted from A.S. Gilmour, Jr., *IEEE Trans. on Electron Devices,* July 1972. © 1972 IEEE.

9-24 [4]. The RF drive powers of 10 W, 50 W and 200 W corresponded to approximate values of α of 0.1, 0.25 and 0.5, respectively. Note that at the medium drive level the space-charge wave behavior became severely distorted. At the highest drive level, the space-charge wave behavior disappeared completely.

Electron energy measurements were made to help explain the beam behavior shown in Figure 9 24. First, the peak values of the electron energies were determined at relatively low values of α using the circuit shown in Figure 9-25. The Faraday cage used for collecting the electrons passing through the .010-in aperture in the beam-scanning mechanism was biased to a negative voltage (4800 V) comparable to the dc beam voltage. In addition, a 10,000 MΩ resistor was placed in series with the bias batteries. With this large value of resistance, only a fraction of a microampere of current to the Faraday cage caused an additional bias of several hundred to several thousands of volts. A fraction of a microampere was less than 0.1% of the normal peak Faraday cage current and consisted, therefore, of only those electrons having the maximum energy.

Shown in Figure 9-26 are the peak energies (expressed in terms of the bias voltage on the Faraday cage) up to an α of approximately 0.1. At small values of α, the variation of energy was a series of half sinusoids, as expected. When α exceeded 0.05,

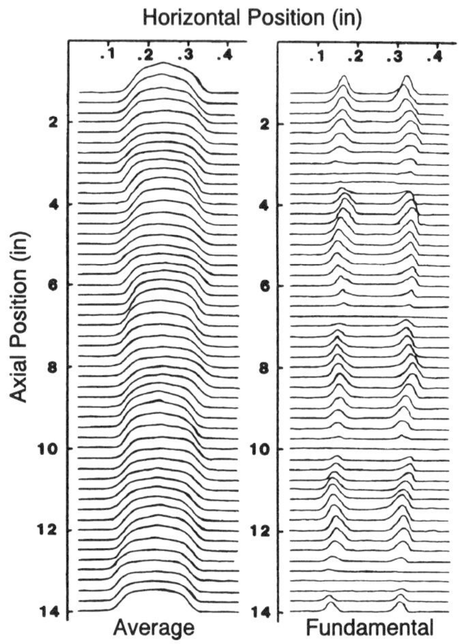

Figure 9-23 Average and fundamental current distributions for $\alpha \approx .05$. From A.S. Gilmour, Jr. *Proc. High Power Microwave Tube Symposium,* The Hexagon, 1962.

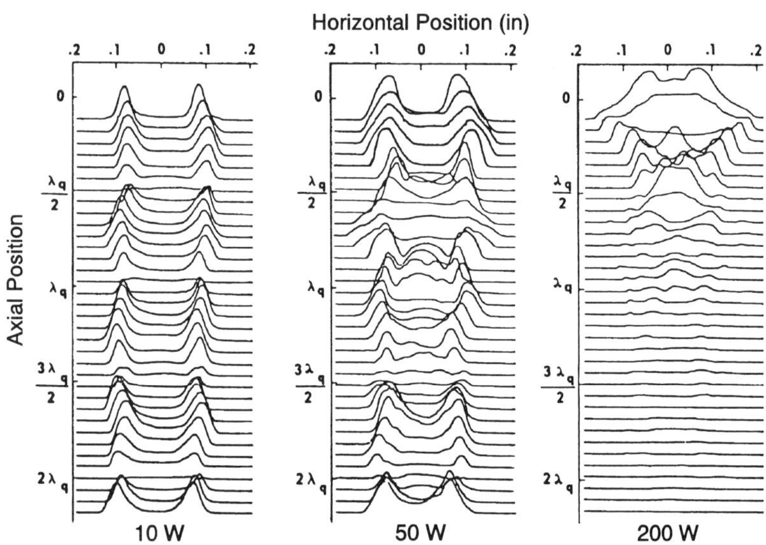

Figure 9-24 Distributions of fundamental RF current for various modulation powers. From A.S. Gilmour, Jr., *IEEE Trans. on Electron Devices,* July 1972. © 1972 IEEE.

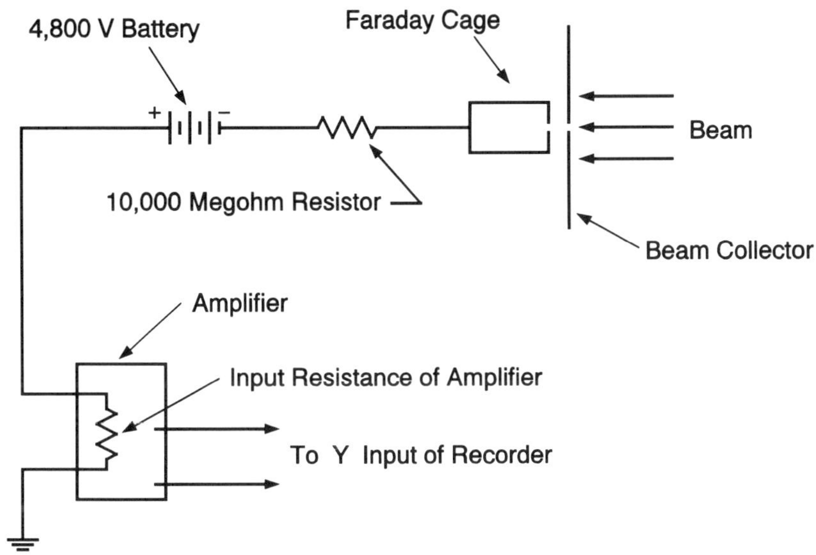

Figure 9-25 Circuit used to provide peak beam energy indication to X-Y recorder. Adapted from A.S. Gilmour, Jr., *IEEE Trans. on Electron Devices,* July 1972. © 1972 IEEE.

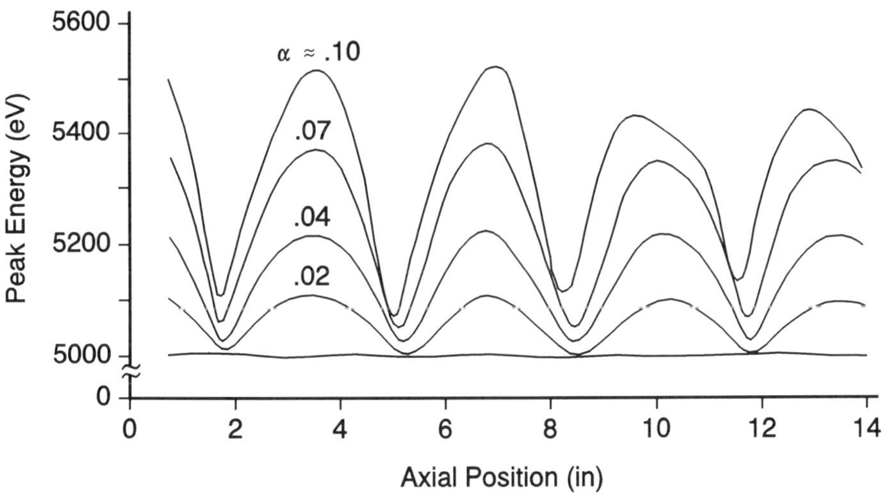

Figure 9-26 Peak beam energy as a function of axial position and RF drive level for Brillouin flow conditions. From A.S. Gilmour, Jr., *Proc. High Power Microwave Tube Symposium*, The Hexagon, 1962.

the energy profiles became distorted. At values of α much above 0.1, the profiles became so severely distorted as to be essentially meaningless. This is better illustrated by Figure 9-27, which shows the energy distributions of the electrons at various axial position and for three drive power levels. As a reminder, the RF drive powers of 10 W, 50 W and 200 W corresponded to approximate values of α of 0.1, 0.25 and 0.5, respectively.

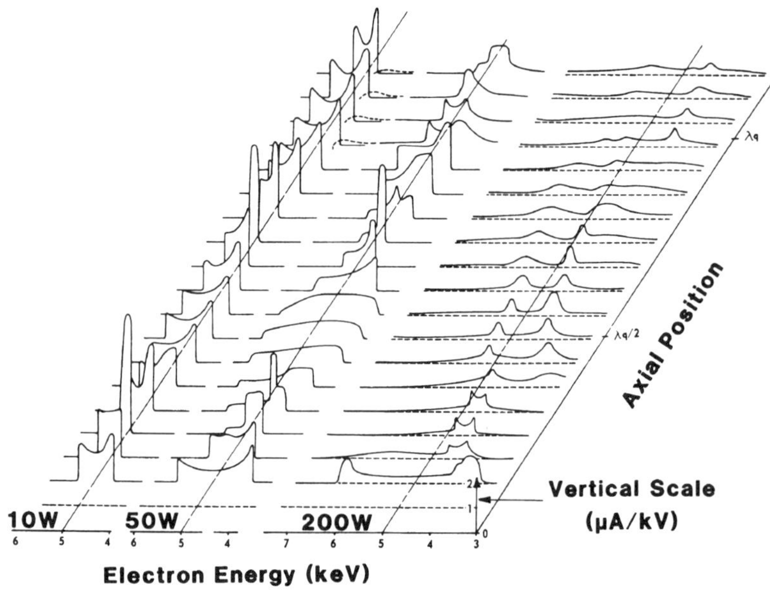

Figure 9-27 Energy distribution of current as a function of axial position and drive power. From A.S. Gilmour, Jr., *IEEE Trans. on Electron Devices*, July 1972. © 1972 IEEE.

Data for Figure 9-27 were obtained by sweeping the Faraday cage voltage and measuring cage current as a function of voltage. The resulting curve was then differentiated to determine the energy distribution of the current. In Figure 9-27, energy, expressed in terms on the Faraday cage potential in kilovolts, is the horizontal axis. The vertical scale is in microamperes of current to the cage per kilovolt of potential applied to the cage. The third coordinate is axial position in the beam.

The 10 W ($\alpha \approx 0.1$) curves show that the electrons left the modulating cavity (at axial positions equal to zero) with an estimated energy distribution from 4.5 to 5.5 keV. At $\lambda_q/4$, where the first bunch occurred, the horizontal spread of the curve narrowed to $\approx$ 5 keV $\pm$ 100 eV, so that all of the electrons essentially had the dc velocity of the beam as is indicated by the peak in current. As the beam drifted beyond $\lambda_q/4$, the energy distribution broadened to a maximum at $\lambda_q/2$ and then returned to a minimum value at $3\lambda_q/4$, where another bunch was formed. The distributions shown in Figure 9-27(a) are very nearly as would be expected for space charge wave behavior of a beam.

In Figure 9-27(b) (50 W and $\alpha \approx 0.25$), the electrons left the modulating cavity with an energy distribution of over 2 keV. At $\lambda_q/4$, the energy distribution does not become narrow as it did at 10 W. Instead, it appears that high energy electrons move through the bunch that was being formed. This is even more evident in Figure 9-27(c), where there are no major peaks in current, indicating that the electrons have a broad energy distribution at all axial positions. Thus, electron velocities were apparently large enough so that they were not significantly affected by space charge forces.

REFERENCES

[1] W.J. Kleen, *Electronics of Microwave Tubes,* New York: Academic Press, 1958.

[2] D.L. Webster, "Cathode-ray bunching," *J. Appl. Phys.,* 10, 1939, pp. 501–508.

[3] A.S. Gilmour, Jr., G.C. Dalman and L.F. Eastman, "The current and velocity distributions in a velocity-modulated Brillouin-focused electron beam, "*Proc. 4th International Congress on Microwave Tubes,* Eindhoven: Centrex, 1963, pp. 675–688.

[4] A.S. Gilmour, Jr., "Radial and axial current and velocity distributions in large-signal velocity-modulated electron beams," *IEEE Trans. on Electron Devices,* vol. ED-19, no. 7, July 1972, pp. 886–890.

Chapter 10

Traveling Wave Interaction

10.1 PIERCE THEORY

There are many approaches to the analysis of the interaction of an electron beam with a traveling wave on a circuit. The classical small-signal theory given by J.R. Pierce [1] is a logical extension of the qualitative description of traveling wave interaction given in the Introduction. Also, many of the parameters defined by Pierce are now part of the accepted vocabulary of the TWT industry. For these reasons and because the Pierce theory provides considerable physical insight into the mechanisms of TWT operation, the Pierce theory will be given here.

The analysis is divided into several distinct parts. First the RF current in the beam is found. (The relation used is the electronic equation, which was derived in connection with the analysis of space charge effects on bunching in the previous chapter.) Then the slow wave circuit is analyzed. The circuit equation is then combined with the electronic equation to determine the propagation constants. Operation under synchronous (beam velocity equal to cold circuit phase velocity) and nonsynchronous (beam velocity not equal to cold circuit phase velocity) conditions are examined. Finally, the effects of circuit loss and of space charge are considered.

The Pierce analysis is a small signal analysis, and it is assumed that quantities vary as $e^{j(\omega t - \beta z)}$. (It should be noted that Pierce used $(j\omega t - \Gamma z)$ and so his Γ is the same as the $j\beta$ used here.)

10.1.1 RF Current in a Beam

The current in the beam as a function of the electric field was found in the analysis of the effect of space charge on electron bunching. The approach was to combine the equation for the definition of current ($J = \rho u$), the continuity equation and the force equation. The result was the electronic equation:

$$J_z = j\omega\epsilon_o \frac{\omega_p^2}{(\omega - \beta u_o)^2} E_z \tag{10-1}$$

For the following analysis, it is convenient to rewrite this equation by replacing ω_p^2 by $\eta\rho_o/u_o$, ρ_o by J_o/u_o, and u_o^2 by $2\eta V_o$. Also, the current densities, J_z and J_o, may be replaced by the corresponding currents, i_z and I_o. The resulting equation is

$$i_z = \frac{j\beta_e I_o}{2V_o(\beta_e - \beta)^2} E_z \tag{10-2}$$

The form of this equation given by Pierce is

$$i = \frac{j\beta_e I_o \Gamma V}{2V_o(j\beta_e - \Gamma)^2} \tag{10-3}$$

where V is the RF voltage in the beam. Equation 10-2 may be converted to (10-3) with the substitution of Γ for $j\beta$ and the realization that Pierce uses conventional current for i and electron current for I_o. He also uses the value for the electronic charge ($-e$) rather than the magnitude (e) so that the force equation contains $+eE$ rather than $-eE$.

10.1.2 Circuit Equation

In the Pierce theory, the RF circuit of the TWT is modeled as a transmission line as shown in Figure 10-1. This model serves as a starting point, but then the analysis is generalized so that the exact form of the model is not important. The line has distributed inductance and capacitance per unit length of L_ℓ and C_ℓ respectively. The beam with current i_z is assumed to pass very close to the circuit and so, from Ramo's theorem, the current induced in the circuit is $I = i_z$. In an incremental length, Δz, the current induced is $\Delta I = \Delta i_z$.

The current induced, ΔI, contributes to the current in the inductance, I_L, and to the current in the capacitance, I_C, as indicated in Figure 10-2.

Using the convention that all currents are electron currents, then the currents at A can be summed as follows:

$$\Delta i_z + I_L = I_L + \frac{\partial I_L}{\partial z}\Delta z + I_C \tag{10-4}$$

but

$$I_C = -\frac{V}{X_C} = -j\omega C_\ell \Delta z V \tag{10-5}$$

so

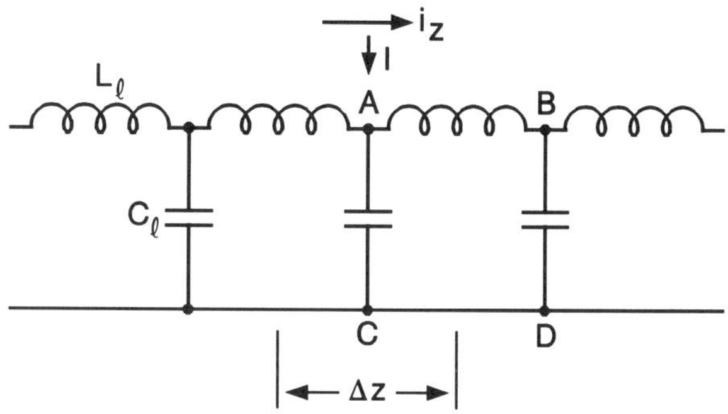

Figure 10-1 Transmission line model for a slow wave circuit.

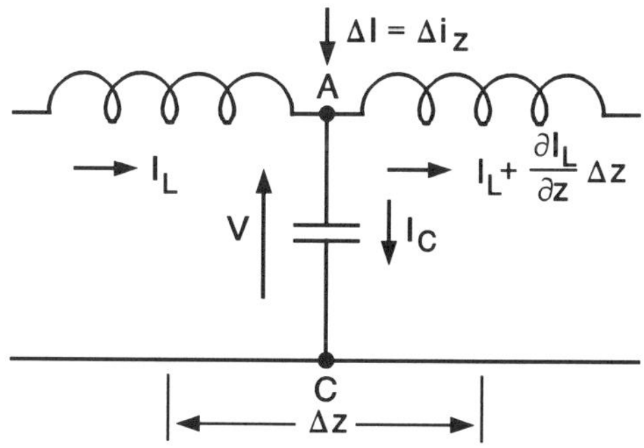

Figure 10-2 Currents at point A.

$$\Delta i_z = \frac{\partial I_L}{\partial z}\Delta z - j\omega C_\ell V\Delta z \tag{10-6}$$

In the limit, as $\Delta z \to 0$,

$$\frac{\partial I_L}{\partial z} = \frac{\partial i_z}{\partial z} + j\omega C_\ell V \tag{10-7}$$

Next, the voltages around loop ABCD, as shown in Figure 10-3, may be added. The result is

$$V = -j\omega L_\ell \Delta z I_L + V + \frac{\partial V}{\partial z}\Delta z \tag{10-8}$$

or

$$\frac{\partial V}{\partial z} = j\omega L_\ell I_L \tag{10-9}$$

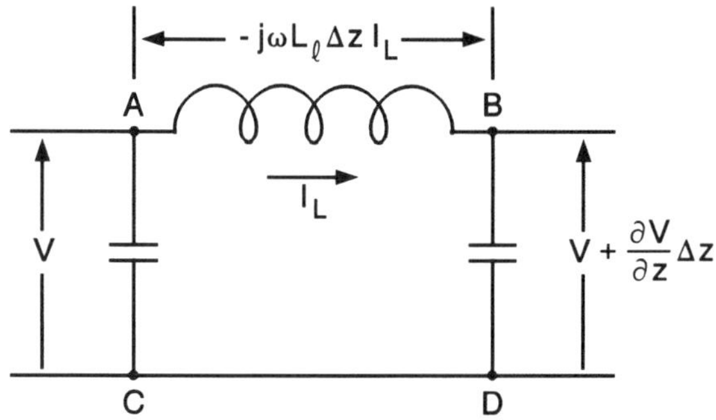

Figure 10-3 Voltages around ABCD.

Since all quantities are assumed to vary as $e^{j(\omega t - \beta z)}$, the current and voltage equations may be written as

$$-j\beta I_L = -j\beta i_z + j\omega C_\ell V \tag{10-10}$$

and

$$-j\beta V = j\omega L_\ell I_L \tag{10-11}$$

Eliminating I_L,

$$V = \frac{\beta \omega L_\ell i_z}{\left(\beta^2 - \omega^2 L_\ell C_\ell\right)} \tag{10-12}$$

Now, assuming that the electrons pass very close to the circuit, then the electric field, E_z, acting on them is

$$E_z = -\frac{\partial V}{\partial z} = j\beta V = j\frac{\beta^2 \omega L_\ell i_z}{\left(\beta^2 - \omega^2 L_\ell C_\ell\right)} \tag{10-13}$$

The quantities L_ℓ and C_ℓ can be eliminated by considering the properties of the circuit. In the absence of an electron beam (so $i_z = 0$), the phase velocity and characteristic impedance of the circuit are

$$v_p = \left[\frac{1}{L_\ell C_\ell}\right]^{1/2} \quad \text{and} \quad Z_o = \left[\frac{L_\ell}{C_\ell}\right]^{1/2} \tag{10-14}$$

and so

$$\beta^2 = \beta_c^2 = \omega^2 L_\ell C_\ell \tag{10-15}$$

and

$$\omega L_\ell = \beta_c Z_o \qquad (10\text{-}16)$$

With these circuit quantities (Z_o and β_c) the equation for the electric field of the circuit that is acting on the electrons (10-13) becomes

$$E_z = j \frac{\beta^2 \beta_c Z_o}{\left(\beta^2 - \beta_c^2\right)} i_z \qquad (10\text{-}17)$$

The only difference between this equation and Pierce's circuit equation, other than the difference in propagation constants, is his use of K rather than Z_o for circuit impedance.

10.1.3 The Determinantal Equation

The electric field produced by space charge has been neglected in deriving (10-3). The electronic equation (10-2) and the circuit equation (10-17) provide two relationships between i_z and E_z. These may be combined in the absence of space charge forces to obtain the following determinantal equation from which the propagation constants can be obtained:

$$1 = \frac{\beta_e I_o}{2V_o(\beta_e - \beta)^2} \frac{\beta^2 \beta_c Z_o}{\left(\beta_c^2 - \beta^2\right)} \qquad (10\text{-}18)$$

For reasons that will become clear later, Pierce defines a quantity C, called the *gain parameter*, by

$$C^3 = \frac{Z_o}{4V_o/I_o} \qquad (10\text{-}19)$$

Thus, if V_o/I_o is called the *beam impedance*, then C^3 is one quarter of the ratio of the circuit impedance to the beam impedance. Because Z_o is usually on the order of tens of ohms and V_o/I_o is on the order of thousands to tens of thousands of ohms, C^3 is a very small quantity and so C is usually in the 0.01 to 0.1 range.

Using C^3, the determinantal equation may be written as

$$\frac{\beta_e}{(\beta - \beta_e)^2} \frac{\beta^2 \beta_c}{\left(\beta^2 - \beta_c^2\right)} 2C^3 + 1 = 0 \qquad (10\text{-}20)$$

This equation is valid for any electron velocity, given by β_e, and any wave velocity and attenuation, specified by the real and imaginary parts of the circuit propagation constant β_c.

The determinantal equation is a fourth degree equation, which means that there will be four possible solutions for β, and these will represent four natural modes of propagation along the electron beam and the circuit. As Pierce points out, the circuit alone would have two modes of propagation, and this is consistent with the fact that the voltages at the two ends can be specified independently, and so two boundary conditions must be satisfied. Four boundary conditions must be satisfied with the combination of circuit and electron stream. These may be taken as the voltages at the two ends of the circuit and the RF velocity and current of the electron beam at the point where the beam enters the circuit. The four waves resulting from the determinantal equation make it possible to satisfy these boundary conditions.

10.1.4 Synchronous Operation

The primary interest in solving the determinantal equation is in a wave that propagates in the direction of and with the approximate speed of the electron beam. This is the wave that interacts with the electron beam to produce gain. To find this wave, we will first assume that the electron speed is made equal to the speed of the wave in the absence of electrons, so

$$\beta_e = \beta_c \tag{10-21}$$

This is the condition for synchronous operation.

Next, it is assumed that β differs from β_e by only a small amount, ξ, so

$$\beta = \beta_e + \xi = \beta_c + \xi \tag{10-22}$$

With these assumptions, the determinantal equation becomes

$$\frac{\beta_e^2\left(\beta_e^2 + 2\beta_e\xi + \xi^2\right)}{\xi^2\left(2\beta_e\xi + \xi^2\right)} 2C^3 + 1 = 0 \tag{10-23}$$

Because β is expected to be close to β_e, $\xi << \beta_e$. As a result, $\beta_e\xi$ and ξ^2 can be neglected in comparison to β_e^2 in the numerator and ξ^2 can be neglected in comparison to $\beta_e\xi$ in the denominator. With these approximations

$$\xi^3 = -\beta_e^3 C^3 \tag{10-24}$$

or

$$\xi = (-1)^{1/3}\beta_e C \tag{10-25}$$

The quantity $(-1)^{1/3}$ has three roots, which represent three forward waves. To find these three roots, we may write

$$(-1)^{1/3} = \left[e^{j(2n-1)\pi}\right]^{1/3} = \cos(2n-1)\frac{\pi}{3} + j\sin(2n-1)\frac{\pi}{3} \tag{10-26}$$

where n = 0, 1 and 2. The three roots are shown graphically in Figure 10-4.

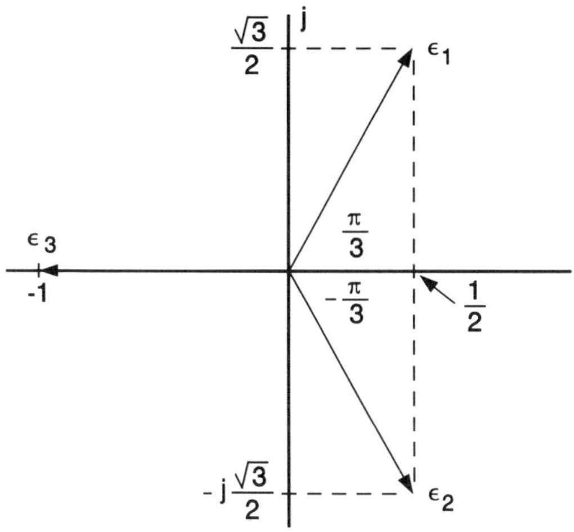

Figure 10-4 The three roots of -1.

The three roots, ϵ_1, ϵ_2 and ϵ_3, are

$$\begin{aligned} \epsilon_1 &= \frac{1}{2} + j\frac{\sqrt{3}}{2} \\ \epsilon_2 &= \frac{1}{2} - j\frac{\sqrt{3}}{2} \\ \epsilon_3 &= -1 \end{aligned} \tag{10-27}$$

and the corresponding propagation constants, β_1, β_2 and β_3, are

$$\begin{aligned} \beta_1 &= \beta_e + \epsilon_1 \beta_e C = \beta_e + \frac{\beta_e C}{2} + j\frac{\sqrt{3}}{2}\beta_e C \\ \beta_2 &= \beta_e + \epsilon_2 \beta_e C = \beta_e + \frac{\beta_e C}{2} - j\frac{\sqrt{3}}{2}\beta_e C \\ \beta_3 &= \beta_e + \epsilon_3 \beta_e C = \beta_e - \beta_e C \end{aligned} \tag{10-28}$$

It should be noted that the values of ϵ used here are related to the incremental propagation constants (δ) used by Pierce by

$$\epsilon = j\delta \tag{10-29}$$

The propagation constant for the fourth wave, β_4, can be found by assuming that the wave travels in the backwards direction (just as one of the waves does in the absence of the electron beam) with a velocity very close to the velocity in the absence of the electron beam. The result is

$$\beta_4 = -\beta_e + \beta_e \frac{C^3}{4} \tag{10-30}$$

The meaning of the propagation constants can easily be examined by remembering the assumption that all waves vary as $e^{(j\omega t - \beta z)}$. Thus, for example, for β_1,

$$e^{j(\omega t - \beta_1 z)} = e^{j\left[\omega t - \left(\beta_e + \frac{\beta_e C}{2}\right) z\right]} e^{\frac{\sqrt{3}}{2}\beta_e C z} \tag{10-31}$$

Now, β_1, as well as β_2 and β_3, contain the propagation constant of the electron beam, β_e, along with a small quantity (small because it is multiplied by C). As a result, each wave is a forward wave (traveling in the same direction as the electron beam) propagating at a velocity close to the beam velocity.

In β_1 and β_2, the quantity $+\beta_e C/2$ indicates that the waves travel slower than the electron beam. In β_3, the quantity $-\beta_e C$ indicates that the wave travels faster than the electron beam. The dominant quantity in the fourth wave is $-\beta_e$, which shows that the wave travels in the direction opposite to that of the electron beam. In β_4, the quantity C^3 is so small that the speed of the fourth wave is only slightly greater than the beam speed. In β_1 and β_2, the imaginary part of the propagation constant indicates that the wave is either a growing wave (+ in β_1) or a decaying wave (- in β_2). Finally, the propagation constants β_1 through β_4 represent the four waves illustrated in Figure 10-5.

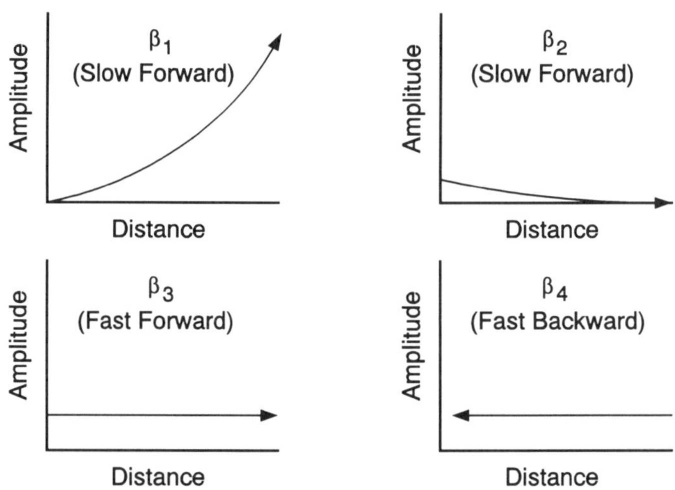

Figure 10-5 Waves represented by β_1 through β_4.

It is of particular interest to determine the rate at which the increasing wave increases. This may be determined from the $e^{\sqrt{3}/2\,\beta e C z}$ term in (10-30). In a number of wavelengths, N,

$$\beta_e z = \frac{\omega}{u_o} z = \frac{2\pi f}{u_o} z = 2\pi \frac{z}{\lambda} = 2\pi N \tag{10-32}$$

Power is proportional to the square of the electric field so the power gain is

$$10 \log_{10} \left(e^{\frac{\sqrt{3}}{2} C 2\pi N} \right)^2 = 47.3\, CN \quad \text{dB} \tag{10-33}$$

The reason that C is called the *gain parameter* is now clear. Note that gain is directly proportional to C.

Equation 10-33 gives the gain for the lossless case. In addition to being the lossless case, this does not include losses that occur when a signal is launched on an electron beam. Since three forward waves are launched on the beam, a signal applied to the input of a TWT must divide in some way among these three forward waves and so only a portion of the input signal is used to launch the growing wave.

To determine the relative amplitudes of the launched waves, the boundary conditions at the input must be matched. Here, the three waves must add to give the correct RF electric field, velocity and current. If E_{in}, u_{in} and i_{in} are the RF field, velocity and current at the beginning of a section of the slow wave circuit, then

$$\begin{aligned} E_1 + E_2 + E_3 &= E_{in} \\ u_1 + u_2 + u_3 &= u_{in} \\ i_1 + i_2 + i_3 &= i_{in} \end{aligned} \tag{10-34}$$

From the force equation (9-41)

$$u_1 = -\frac{\eta}{j(\omega \quad \beta_1 u_o)} E_1 = j \frac{\eta}{u_o(\beta_e \quad \beta_1)} E_1 \tag{10-35}$$

and similar relations can be written for u_2 and u_3. Now, substituting for β_1, (i.e., $\beta_1 = \beta_e + \epsilon_1 \beta_e C$),

$$u_1 = j \frac{\eta}{u_o(\beta_e - \beta_e - \epsilon_1 \beta_e C)} E_1 = -j \frac{\eta}{u_o \epsilon_1 \beta_e C} E_1 \tag{10-36}$$

and so

$$u_1 + u_2 + u_3 = -j \frac{\eta}{u_o \beta_e C} \left[\frac{E_1}{\epsilon_1} + \frac{E_2}{\epsilon_2} + \frac{E_3}{\epsilon_3} \right] = u_{in} \tag{10-37}$$

From the electronic equation (10-2)

$$i_1 = j \frac{\beta_e I_o}{2V_o(\beta_e - \beta_1)^2} E_1 = j \frac{\beta_e I_o}{2V_o(\epsilon_1 \beta_e C)^2} E_1 \tag{10-38}$$

and similar relations can be written for i_2 and i_3. As a result,

$$i_1 + i_2 + i_3 = j \frac{I_o}{2V_o \beta_e C^2} \left[\frac{E_1}{\epsilon_i^2} + \frac{E_2}{\epsilon_2^2} + \frac{E_3}{\epsilon_3^2} \right] = i_{in} \tag{10-39}$$

Now, (10-34), (10-37) and (10-39) can be solved for E_1, E_2, and E_3 in terms of the initial values E_{in}, u_{in} and i_{in}. In general (for example, after a sever or attenuator in the slow wave circuit) u_{in} and i_{in} are not zero, but at the input to the TWT, the values for u_{in} and i_{in} are zero. As a result, (10-34), (10-37) and (10-39) become

$$E_1 + E_2 + E_3 = E_{in}$$

$$\frac{E_1}{\epsilon_1} + \frac{E_2}{\epsilon_2} + \frac{E_3}{\epsilon_3} = 0 \tag{10-40}$$

$$\frac{E_1}{\epsilon_1^2} + \frac{E_2}{\epsilon_2^2} + \frac{E_3}{\epsilon_3^2} = 0$$

When these are combined, E_1 is found to be

$$E_1 = \frac{E_{in}}{\left[1 - \dfrac{\epsilon_2}{\epsilon_1}\right]\left[1 - \dfrac{\epsilon_3}{\epsilon_1}\right]} \tag{10-41}$$

The corresponding equations for E_2 and E_3 can be obtained by transposing suffixes. For example, to obtain E_2, write 2 for 1 and 1 for 2 in (10-41).

When the values for ϵ_1, ϵ_2 and ϵ_3 are used, it is found that

$$E_1 = E_2 = E_3 = \frac{1}{3} E_{in} \tag{10-42}$$

That is, the input signal divides evenly among the three waves.

Taking this into account, the power gain, G, becomes

$$G = 10 \log_{10} \left[\frac{1}{3} e^{\frac{\sqrt{3}}{2} C 2\pi N} \right]^2 \tag{10-43}$$

or

$$G = -9.54 + 47.3CN \ \text{dB} \tag{10-44}$$

Taking all three waves into account, the circuit field grows as a function of distance as

$$E_z = \frac{1}{3} E_{in} e^{-j\beta_e z}\left(e^{-j\frac{1}{2}\beta_e Cz} e^{\frac{\sqrt{3}}{2}\beta_e Cz} + e^{-j\frac{1}{2}\beta_e Cz} e^{-\frac{\sqrt{3}}{2}\beta_e Cz} + e^{j\beta_e Cz}\right)$$

$$= \frac{1}{3} E_{in} e^{-j\beta_e(1 - C)z}\left[1 + e^{-j\frac{3}{2}\beta_e Cz}\left(e^{\frac{\sqrt{3}}{2}\beta_e Cz} + e^{-\frac{\sqrt{3}}{2}\beta_e Cz}\right)\right] \tag{10-45}$$

$$= \frac{1}{3} E_{in} e^{-j\beta_e(1 - C)z}\left\{ 1 + 2 e^{-j\frac{3}{2}\beta_e Cz} \cosh \frac{\sqrt{3}}{2}\beta_e Cz \right\}$$

The power gain, $10\log_{10} | E_z/E_{in} |^2$, can now be calculated and is shown in Figure 10-6.

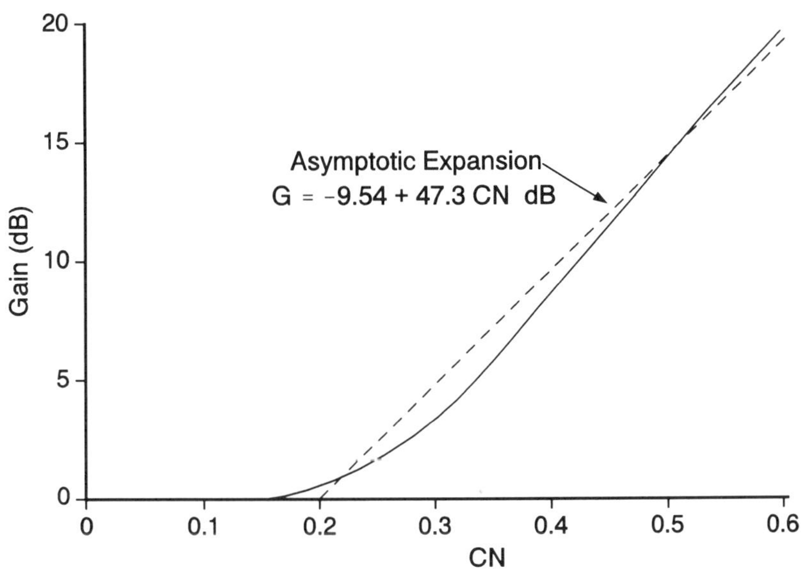

Figure 10-6 Power gain in a TWT for the synchronous case with zero loss and no space charge. Adapted from *Traveling Wave Tubes*, by J.R. Pierce, copyright 1950 by D. Van Nostrand Company.

Initially, the voltage does not change with distance and there is no gain. This is because the electron beam has no RF current and cannot induce a signal into the circuit until it becomes bunched. Eventually, the increasing wave predominates over the other two and the slope of the line, B, must correspond to that for the growing wave, that is, 47.3. The dashed line represents the increasing wave, which starts at - 9.54 dB and has the slope 47.3. Pierce writes an asymptotic expression for the gain as

$$G = A + BCN \tag{10-46}$$

where, for this synchronous case with no loss, no space charge and no initial RF velocity or current on the beam,

$$\begin{aligned} A &= -9.54 \text{ dB} \\ B &= 47.3 \end{aligned} \tag{10-47}$$

This result is reasonably good for $CN > 0.2$.

10.1.5 Nonsynchronous Operation

When the electron beam travels at a velocity other than the velocity of the wave on the circuit in the absence of the beam, solutions to the quartic equation for the propagation constants are obtained numerically.

At this point, it is helpful to introduce Pierce's velocity parameter, b. This parameter is used to relate the dc electron velocity, u_o, to the phase velocity of the circuit in the absence of electrons, v_p, as follows:

$$u_o = (1 + Cb)v_p \qquad \text{or} \qquad b = \frac{u_o - v_p}{C v_p} \tag{10-48}$$

and

$$\beta_c = \beta_e(1 + Cb) \tag{10-49}$$

For $b > 0$, the electrons go faster than the undisturbed wave and for $b < 0$, the electrons go slower than the undisturbed wave. For $b = 0$, the synchronous case considered in the previous section occurs.

When (10-49) and $\beta = \beta_e + \beta_e C\epsilon$ (from (10-28)) are used in the determinantal equation (10-20) and the quantities $C\epsilon$ and Cb are neglected in comparison with unity, it is found that

$$\epsilon^2(\epsilon - b) = -1 \tag{10-50}$$

In terms of Pierce's δ, this equation is

$$\delta^2(\delta + jb) = -j \tag{10-51}$$

In dealing with this equation, it is noted that δ is, in general, a complex number and so Pierce writes

$$\delta = x + jy \tag{10-52}$$

In terms of Pierce's x and y, the ϵ used here is

$$\epsilon = jx - y \tag{10-53}$$

Because $\beta = \beta_e + \beta_e C\epsilon$, the complete propagation constant is

$$\beta = \beta_e(1 - Cy) + j\beta_e Cx \tag{10-54}$$

and so the waves propagate as

$$e^{j[\omega t - \beta_e(1 - Cy)z]} \, e^{\beta_e Cxz} \tag{10-55}$$

The amplitude of the waves varies as $e^{\beta_e Cxz}$ and so, if x is positive, the waves are growing. Their velocity is $u_o/(1 - Cy)$ and so, if y is positive, they are traveling faster than the electron beam.

Equations (10-51) and (10-52) have been solved numerically, and the result is given in Figure 10-7. The values of x (*i.e.*, x_1) and y (*i.e.*, y_1) for the growing wave only are shown in Figure 10-8. The important points that Pierce notes are

1. For $b < (3/2)(2)^{1/3}$, there are two waves for which $x \neq 0$ and one unattenuated wave. The increasing and decreasing waves have equal and opposite values of x and, since $y < 0$, they travel more slowly than the electrons, even when the electrons travel more slowly than the undisturbed wave. In fact, the electrons must travel faster than the increasing wave to give energy to it.
2. For $b > (3/2)(2)^{1/3}$, there are three unattenuated waves, two traveling faster than the electrons and one more slowly.
3. For large positive or negative values of b, two waves have nearly the electron speed and one travels with the speed of the undisturbed wave.

Pierce also notes that, in a tube in which the total gain is large, a change in b of ± 1 about $b = 0$ can make a change of several dB in gain. This change produces a difference between the phase velocity of the undisturbed wave, v_p, and the electron velocity, u_o, by $\sim \pm C$. Thus, the allowable difference between v_p, which may be a function of frequency, and u_o, which, of course, is not, is of the order of C.

10.1.6 Effect of Circuit Loss

In addition to the losses that occur naturally in RF circuits, additional losses may be introduced to attenuate the backward wave. The backward wave normally originates from reflections either inside or outside the tube and can serve as a feedback mechanism to cause oscillations or large variations in gain (because of beating effects with the input signal).

The effect of loss may be taken into account by introducing Pierce's circuit attenuation (loss) parameter, d, in the circuit propagation constant as follows:

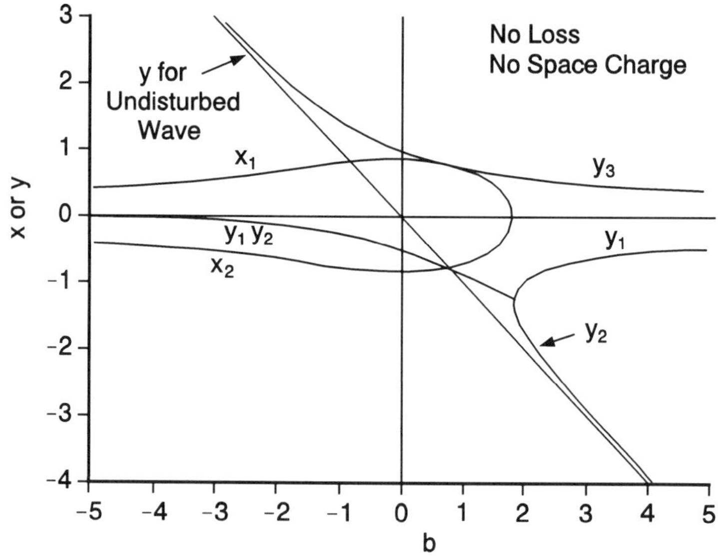

Figure 10-7 Components of incremental propagation constant as a function of velocity parameter. Adapted from *Traveling Wave Tubes*, by J.R. Pierce, copyright 1950 by D. Van Nostrand Company, Inc.

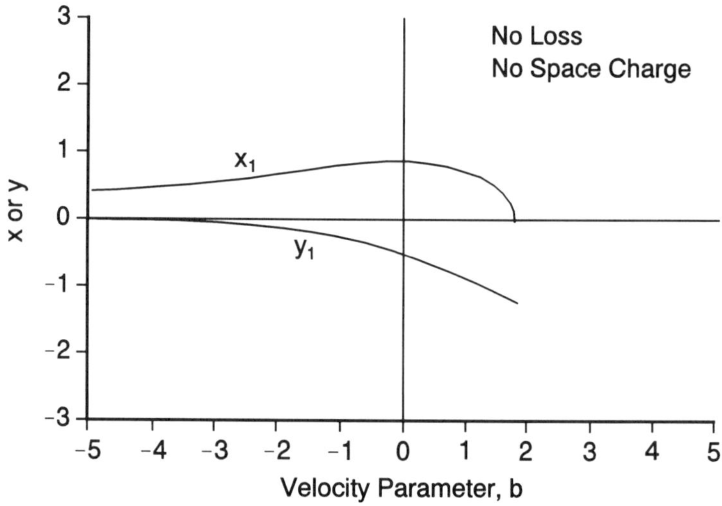

Figure 10-8 Components of incremental propagation constant for growing wave.

$$\beta_c = \beta_e(1 + Cb) - j\beta_e Cd \tag{10-56}$$

The circuit wave then decays as $e^{-\beta_e C d z}$. Letting $\beta_e z = 2\pi N$, then the loss is

$$\text{Loss} = 20 \log_{10} e^{-2\pi N C d} \tag{10-57}$$

and so the loss per wavelength, L, is

$$\begin{aligned} L &= 20 (\log_{10} e) 2\pi C d \\ &= 54.5 C d \ \text{dB/wavelength} \end{aligned} \tag{10-58}$$

from which

$$d = 0.0184 \frac{L}{C} \tag{10-59}$$

As an example, when $C = .025$ and $d = 1$, the loss is 1.36 dB/wavelength.

When β_c from (10-56) and $\beta = \beta_e + \beta_e C\epsilon$ (from (10-28)) are used in the determinantal equation (10-20) and the quantities $C\epsilon$, Cb and Cd are neglected in comparison with unity, it is found that

$$\epsilon^2(\epsilon - b + jd) = -1 \tag{10-60}$$

In terms of Pierce's δ, the equation is

$$\delta^2(\delta + jb + d) = -j \tag{10-61}$$

This equation has been solved numerically for the values of x and y and a typical case is shown in Figure 10-9. The most significant effect of attenuation is that, as expected, the gain of the growing wave is decreased. This may be shown clearly by examining the effect of the reduction in x_1 (proportional to the gain of the growing wave) on the value of B in the gain equation. Figure 10-10 illustrates the effect of d on B. Note that the maximum value of B occurs very close to $b = 0$.

Pierce shows that, for small losses, the reduction in gain for the growing wave is about one third of the circuit loss. The effect on the net gain of a TWT can be calculated only by taking into account the initial conditions in the presence of loss.

10.1.7 Effect of Space Charge

At the beginning of this chapter the electronic equation for the beam current as a function of electric field was combined with the circuit equation, which gave the circuit field acting on the electron beam. There was no consideration of space charge forces and their contribution to the electric field.

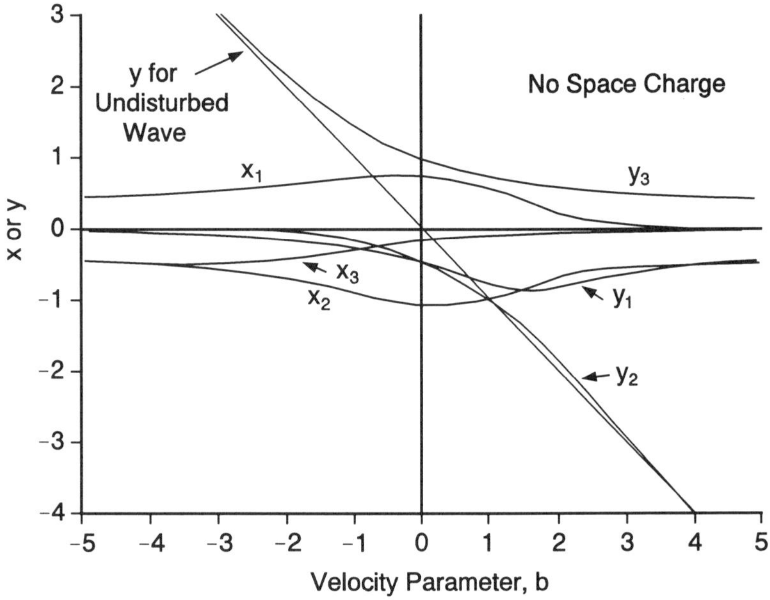

Figure 10-9 The values of x and y for a circuit with loss ($d = 0.5$). Adapted from *Traveling Wave Tubes*, by J.R. Pierce, copyright 1950 by D. Van Nostrand Company, Inc.

The electronic equation, (10-2), is

$$i_z = j \frac{\beta_e I_o}{2 V_o (\beta_e - \beta)^2} E_z \tag{10-62}$$

but now, to include the effects of space charge, E_z will be written as

$$E_z = E_{zc} + E_{zs} \tag{10-63}$$

where

$$E_{zc} = j \frac{\beta^2 \beta_c Z_o}{\left(\beta^2 - \beta_c^2\right)} i_z \tag{10-64}$$

which was (10-17). The subscript c has been added to E_{zc} to denote circuit field.

The space charge field, E_{zs}, is given by Poisson's equation; that is,

$$\frac{\partial E_{zs}}{\partial z} = -\frac{\rho_1}{\epsilon_o} \tag{10-65}$$

or, because all quantities are assumed to vary as $e^{j(\omega t - \beta z)}$,

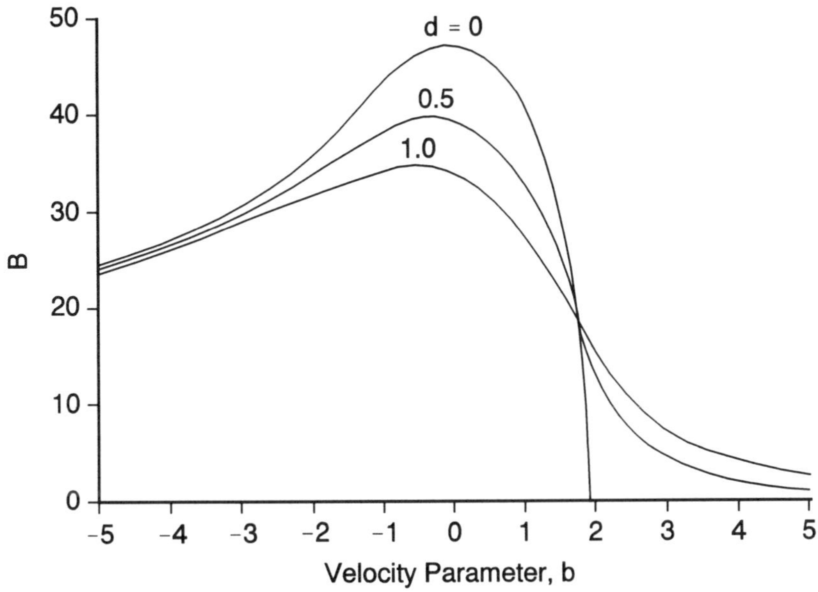

Figure 10-10 Effect of loss on B factor in gain of growing wave. Adapted from *Traveling Wave Tubes*, by J.R. Pierce, copyright 1950 by D. Van Nostrand Company, Inc.

$$E_{zs} = \frac{\rho_1}{j\beta\epsilon_o} \tag{10-66}$$

Also,

$$\nabla \cdot \boldsymbol{J} = \frac{\partial J_z}{\partial z} = -\frac{\partial \rho_1}{\partial t} \tag{10-67}$$

and so

$$\rho_1 = \frac{\beta}{\omega} J_z \tag{10-68}$$

Thus, the space charge field may be written as

$$E_{zs} = \frac{J_z}{j\omega\epsilon_o} = -j\frac{i_z}{\omega\epsilon_o A} \tag{10-69}$$

where A is the cross sectional area of the beam.

Now, the total electric field in the beam is

$$E_z = \left[\frac{j\beta^2 \beta_c Z_o}{\left(\beta^2 - \beta_c^2\right)} - j \frac{1}{\omega \epsilon_o A} \right] i_z \tag{10-70}$$

At this point, we should examine Pierce's method of accounting for space charge and compare his total electric field with that found here. In his analysis, Pierce writes the space charge field as

$$E_{zs} = -j \frac{\Gamma^2}{\omega C_1} i_z \tag{10-71}$$

where C_1 is a new parameter to be defined. Here, it must be remembered that i_z is conventional current, not electron current and that Pierce's $\Gamma = j\beta$.

The quantity C_1 has the dimensions of capacitance per unit length, and this suggested to Pierce that the effect of space charge could be represented by capacitances between the beam and the circuit, as shown in Figure 10-11. The current per unit length that is coupled into the circuit is the current per unit length in the beam; that is, $-\partial i_z/\partial z$ or Γi_z. The voltage developed across the capacitance is $\Gamma i_z / j\omega C_1$. The equivalent voltage acting on the beam is the circuit voltage, V_c, plus this space charge voltage or

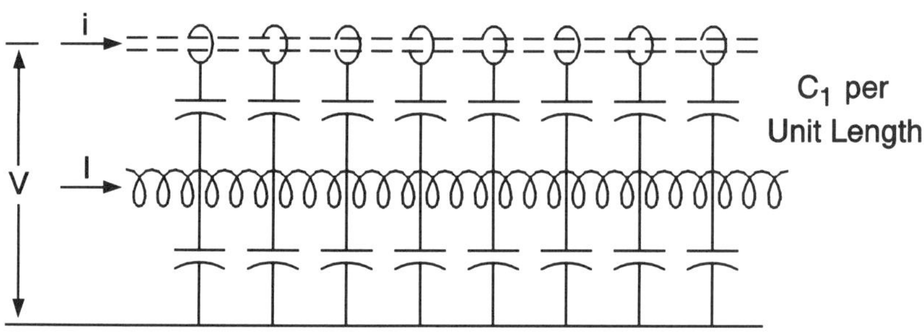

Figure 10-11 Pierce's method of representing the effect of space charge.

$$V = V_c - j \frac{\Gamma^2}{\omega C_1} i_z \tag{10-72}$$

The electric field acting on the electrons is

$$E_z = -\frac{\partial V}{\partial z} = \Gamma V_c - j \frac{\Gamma^2}{\omega C_1} i_z \tag{10-73}$$

By comparing (10-70) with (10-73) and remembering that the currents in the two equations have opposite signs,

$$\frac{\Gamma^2}{C_1} = -\frac{1}{\epsilon_o A} \tag{10-74}$$

For now, it is just noted that this relationship exists. More will be said about it later.

Next, combining the new equation for the total field (10-70) with the electronic equation (10-62), the following is obtained:

$$1 = j\frac{\beta_e I_o}{2V_o(\beta_e - \beta)^2}\left[j\frac{\beta^2\beta_c Z_o}{\left(\beta^2 - \beta_c^2\right)} - j\frac{1}{\omega\epsilon_o A}\right] \tag{10-75}$$

With the substitution of Pierce's gain parameter, this equation may be written as

$$1 = \frac{\beta_e}{(\beta_e - \beta)^2}\left[\frac{-\beta^2\beta_c}{\left(\beta^2 - \beta_c^2\right)}2C^3 + \frac{I_o}{2V_o\omega\epsilon_o A}\right] \tag{10-76}$$

But

$$\frac{I_o}{2V_o\omega\epsilon_o A} = \frac{J_o}{2V_o\omega\epsilon_o} = \frac{\rho_o}{2V_o\beta_e\epsilon_o} = \frac{\eta\rho_o}{u_o^2\beta_e\epsilon_o} = \frac{\omega_p^2}{u_o^2\beta_e} = \frac{\beta_p^2}{\beta_e} \tag{10-77}$$

and so (10-76), which is the determinantal equation, may be written as

$$\frac{\beta_e}{(\beta_e - \beta)^2}\left[\frac{\beta^2\beta_c}{\left(\beta^2 - \beta_c^2\right)}2C^3 - \frac{\beta_p^2}{\beta_e}\right] + 1 = 0 \tag{10-78}$$

Now, it should be noted that the only difference between this equation and the determinantal equation for the case when space charge is ignored, (10-20), is the quantity β_p^2 over β_e and this is a measure of the significance of the space charge forces.

Letting $\beta = \beta_e(1 + C\epsilon)$ and $\beta_c = \beta_e(1 + Cb) - j\,\beta_e Cd$, and ignoring $C\epsilon$, Cb and Cd compared with unity, then

$$\epsilon^2 = -\frac{1}{\epsilon - b + jd} + \left[\frac{\beta_p}{C\beta_e}\right]^2 \tag{10-79}$$

In terms of Pierce's δ, this equation is

$$\delta^2 = \frac{1}{j\delta - b + jd} - \left[\frac{\beta_p}{C\beta_e}\right]^2 \tag{10-80}$$

Pierce writes this equation as

$$\delta = \frac{1}{j\delta - b + jd} - 4QC \tag{10-81}$$

and so

$$4QC = \left[\frac{\beta_p}{C\beta_e} \right]^2 \tag{10-82}$$

Pierce calls Q the *space charge parameter*. Using his capacitance approach, its value is

$$Q = \frac{\beta_e}{2\omega C_1 Z_o} \tag{10-83}$$

From (10-74), the value of C_1 is

$$C_1 = -\Gamma^2 \epsilon_o A \tag{10-84}$$

and so Pierce's relation for $4QC$ is

$$4QC = -\frac{\beta_e^2}{\Gamma^2} \left[\frac{\beta_p}{C\beta_e} \right]^2 = \frac{\beta_e^2}{\beta^2} \left[\frac{\beta_p}{C\beta_e} \right]^2 \tag{10-85}$$

Note that the difference between Pierce's $4QC$ (derived on the basis of the capacitor analogy for space charge) and the $4QC$ derived using Poisson's equation is $\beta_e{}^2/\beta^2$. This difference is very small (usually on the order of a few percent).

The meanings of Q and of QC are not straightforward. However, by rewriting $4QC$ as follows, the physical significance can be inferred.

$$4QC = \frac{(\beta_p/\beta_e)^2}{C^2} = \frac{(\omega_p/\omega)^2}{C^2} \tag{10-86}$$

Now, the ratio of plasma frequency to RF frequency is a measure of space charge debunching forces. Pierce's gain parameter, C, is a measure of the intensity of the interaction between the slow wave circuit and the electron beam. As a result, QC is a measure of the relative strength of the debunching forces to the bunching forces. Thus, QC has been adopted to describe the seriousness of space charge effects in traveling wave tubes.

It should be noted that, for finite beams, the plasma frequency, ω_q, is less than it is for the infinite beam (ω_p) and that ω_q should be used in the expression for $4QC$. Then

$$4QC = \left[\frac{\omega_q}{C\omega} \right]^2 = \left[\frac{\beta_q}{C\beta_e} \right]^2 \tag{10-87}$$

The incremental propagation constants can be found by solving (10-80) numerically. Figure 10-12 shows the values of x and y as a function of b for $QC = 0.25$ and for no loss ($d = 0$). Figure 10-13 shows the values of x and y for the growing wave only for several values of QC. In comparing Figures 10-12 and 10-13 with Figures 10-7 and 10-8 several effects of space charge are apparent. The most important is that the range of beam velocities (indicated by b) over which gain occurs (x_1 positive) is severely restricted. Furthermore, the velocity at which there is maximum gain in the increasing wave is now higher than in the absence of space charge. Also, the maximum gain is reduced.

Slow, $\beta_e(1 + \omega_p/\omega)$, space charge waves are shown by the dashed lines in Figure 10-13. Since waves propagate as $\beta_e(1 - Cy)$ then

$$y = -\frac{\omega_p}{\omega C} = -\sqrt{4QC} \tag{10-88}$$

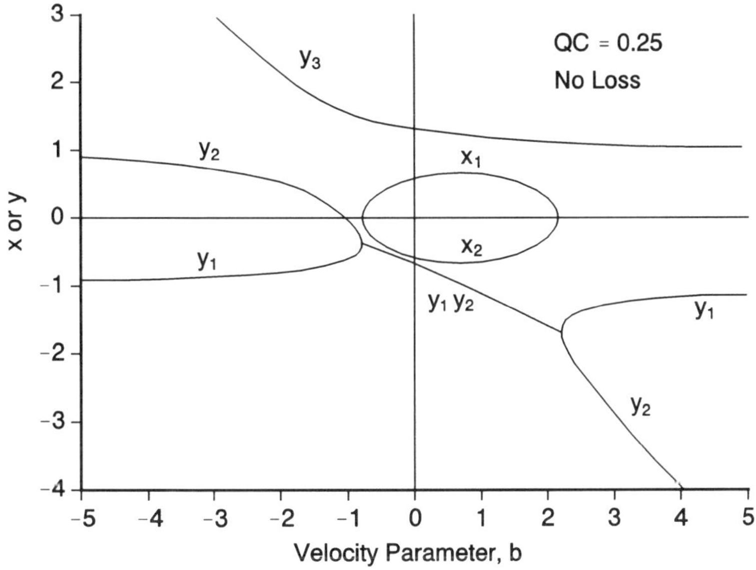

Figure 10-12 Incremental propagation constants when space charge is taken into account. Adapted from *Traveling Wave Tubes*, by J.R. Pierce, copyright 1950 by D. Van Nostrand Company, Inc.

The physical explanation for the restricted ranges of y_1 and b over which gain occurs for the growing wave may be found by examining the space charge wave behavior of the beam. In a slow space charge wave, the electron bunches move slower than the electron beam. Electrons behind a bunch catch up with the bunch and are

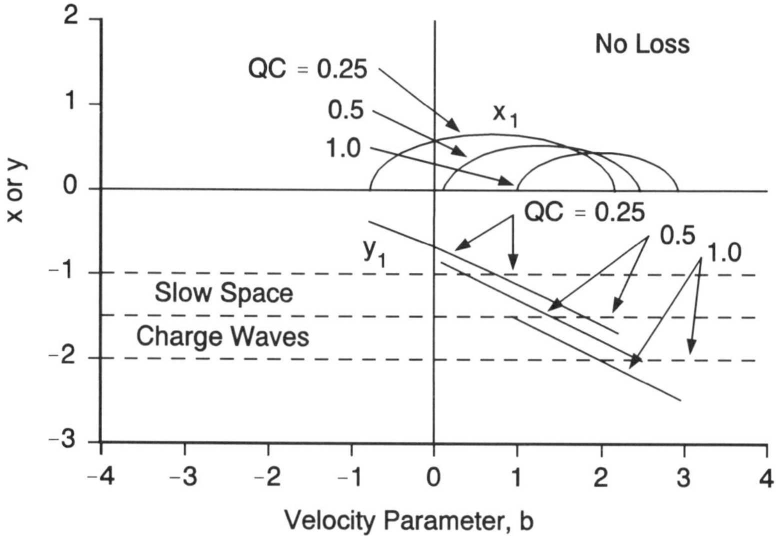

Figure 10-13 Effect of space charge on incremental propagation constants for growing wave.

slowed by space charge forces. Electrons in the front of the bunch are expelled by space charge forces.

Figure 10-13 shows that it is the bunching associated with the slow space charge wave that produces gain in a TWT. The phase velocity of the signal on the circuit must be close to the velocity of the bunches in the slow space charge wave for gain to occur. Note, for example, that, for $QC = 0.25$, if y_1 becomes very much different than -1 the gain (proportional to x_1) goes to zero.

In Figure 10-13, it is also shown that the maximum gain decreases as the beam density (and QC) increases. This is to be expected because, as the beam density increases, space charge forces increase and the intensity of the bunching process decreases, which, in turn, means that the RF current in the beam decreases.

10.2 HIGH LEVEL INTERACTION

10.2.1 Discussion of Interactions

Before considering high level (nonlinear) interactions in a TWT, it is of value to briefly review the bunching action that occurs at small signal levels. Figure 1-8 from the Introduction is repeated here as Figure 10-14 for the sake of convenience. As the beam enters the RF circuit of a TWT, the axial component of the electric field from the circuit accelerates some electrons and decelerates others to start the formation of bunches as indicated in Figure 10-14(a). The resulting RF current in the beam induces a current and voltage on that circuit the lags behind the initial voltage by 90°.Thus, the induced wave travels more slowly than the initial wave. As the bunches and the induced wave grow, decelerating field regions develop where the bunches are located as shown in Figure 10-14(b). The energy extracted from the decelerating bunches is transferred to the circuit field, thereby producing amplification of that field. As long as small signal conditions exist, Pierce's theory applies and the various amplified waves on the beam and circuit grow exponentially with distance.

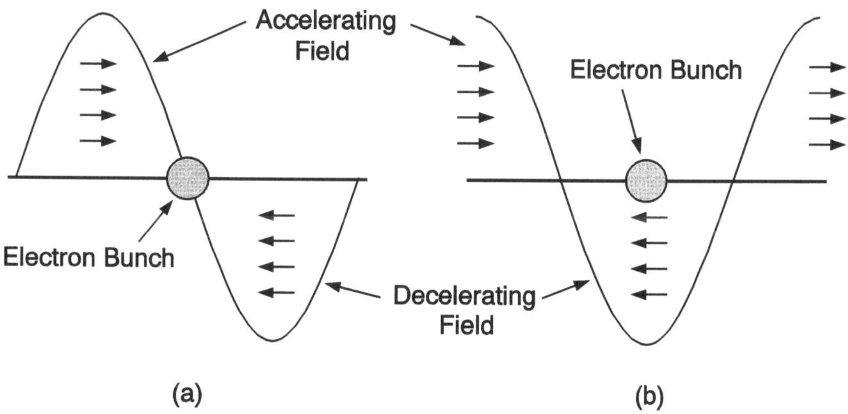

Figure 10-14 Axial field the bunches and extracts energy from beam (a) as beam enters circuit and (b) after interaction has occurred.

As the waves grow, the interaction between the beam and circuit becomes nonlinear. Perhaps the clearest paper ever published describing nonlinear phenomena in a TWT was that by Cutler [2]. A large scale model of a TWT was constructed. Using an electron deflection system, the electron velocities and currents in the beam leaving the TWT interaction region were measured as a function of phase position relative to the circuit wave. Some of Cutler's results showing the events leading up to saturation are shown in Figure 10-15. The RF drive levels are in dB below the level that produced saturation. The circuit voltage is shown at the top of each of the frames of Figure 10-15. RF current is indicated by the width of the shaded region and RF velocity is indicated by the deflection of the shaded region, from the center horizontal line in each frame. Thus, in each frame there is a complete story of fundamental circuit voltage, electron velocity and current as a function of phase, for a particular drive level.

In Figure 10-15(a) the drive level was well below saturation and the current and velocity patterns were representative of small-signal conditions. Cutler notes that the velocity and current modulations at small signal levels check calculated values well. The phase of the voltage relative to the RF electron velocity was computed from

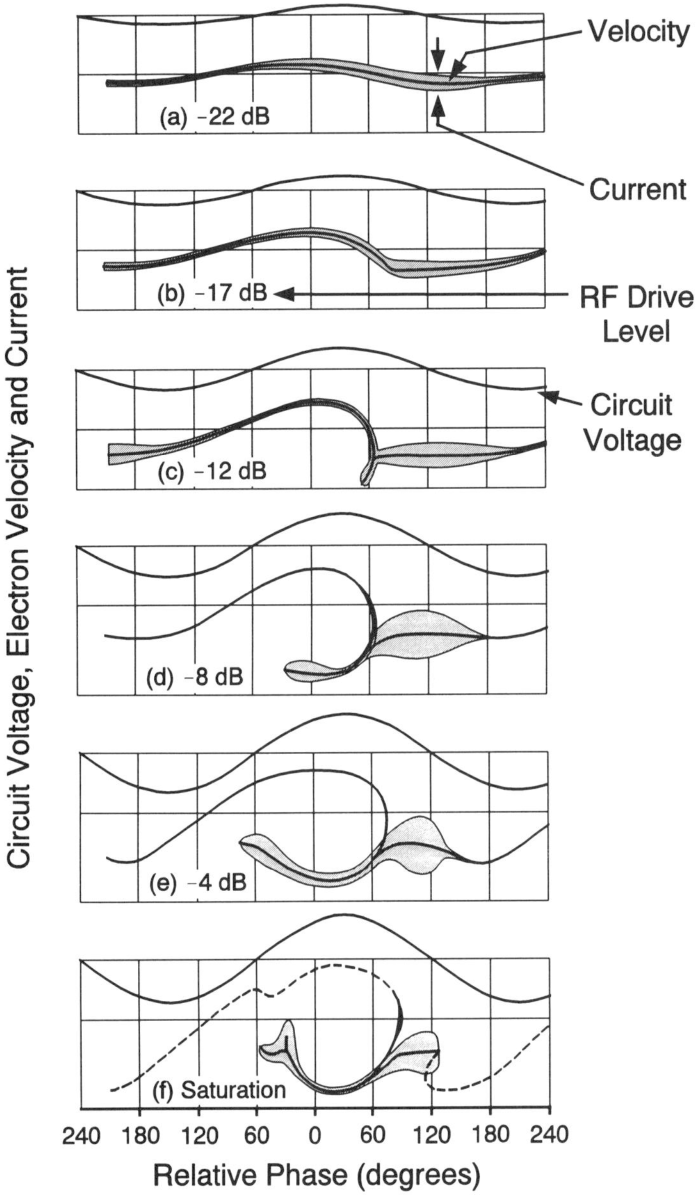

Figure 10-15 Electron bunch behavior as a function of drive level relative to saturation drive. Adapted from C.C. Cutler, *Bell Sys. Tech. Jour.*, July 1956.

small-signal theory, but everything else in Figure 10-15, including subsequent variations in phase, were measured.

The first evidence of nonlinear behavior occurred at a drive level 17 dB below the saturation drive level (Figure 10-15(b)). The velocity and current are no longer sinusoidal, but (in Cutler's words) "show the beginnings of a cusp in the velocity curve and a definite non-sinusoidal bunching of the electrons in the retarding field region (between +30° and 210°)."

The next pattern at 12 dB below the saturation drive level, Figure 10-15(c), shows the development of a spur on the velocity curve and further development of the electron bunch between +60° and + 180°. Cutler observes that the velocity within the bunch has not changed significantly from lower drive levels. For this to be true, since the bunch is in a decelerating portion of the circuit field, the space charge field must just compensate for the circuit field. Near 60°, the decelerating field must increase very rapidly to produce the sharp deceleration of electrons and the development of the velocity spur. Furthermore, Cutler observes that the decelerating field must be much larger than the circuit field because electrons just behind the cusp are much more sharply decelerated than those preceding the cusp. He concludes that there are very sharply defined space charge fields much stronger than the circuit field.

The next two patterns (Figures 10-15(d) and (e)) show continuing development of the spur, continued movement of electrons from the high velocity regions and continued consolidation of the main bunch just ahead of the spur. Apparently, because the electron velocities within the bunch are still nearly the same as at lower drive levels, the space charge and its associated field in the bunch has increased enough to compensate for the circuit field so that the net field remains near zero.

At 4 dB below the saturation drive level, the velocities of a large fraction of the electrons in the spur have increased because the spur has moved well into the accelerating region. The main bunch of electrons to the right of the spur has been concentrated into the 60°–120° phase interval. Apparently, the space charge field in advance of this region no longer compensates for the decelerating circuit field and so electron velocity is decreasing.

At saturation, the tip of the spur has formed into a bunch that is nearly equal to the main bunch and is in a circuit field that is nearly equal and opposite to that in the main bunch. The few electrons (indicated by the dashed line) ahead of the main bunch are being decelerated into a second low-velocity loop. The reason for saturation is that the power being extracted from the circuit by the spur bunch is nearly the same as the power being added by the main bunch.

10.2.2 Estimates of Maximum Efficiency

Early estimates of the maximum efficiency that might be achieved in a TWT were made without the aid of Cutler's experimental results and without the aid of computer models. Slater [3] suggested that energy extraction from the beam in a TWT could be analyzed by observing the motions of electrons while traveling along at the speed of the circuit wave. We will use one of the frames from Cutler's results (Figure 10-15(d)) to illustrate Slater's suggestion. That frame is shown in Figure 10-16. The initial beam velocity is u_o. The velocity of the circuit wave is Δu_o below the beam velocity. Slater suggested that electrons eventually become trapped and oscillate in the region of the maximum decelerating circuit field. This trapping is clearly evident in Figure 10-16, where a bunch is shown to have formed ahead of the velocity cusp in the decelerating field.

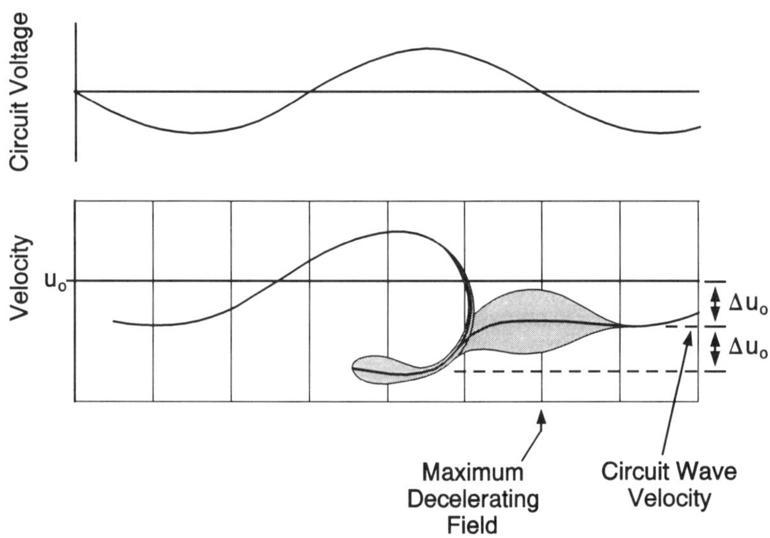

Figure 10-16 Considerations in Slater's estimate of maximum TWT efficiency.

If the electrons initially have an average velocity, u_o, greater than that of the circuit wave by Δu_o, and they are reflected by the decelerating field, they cannot acquire a velocity backwards relative to the wave greater than Δu_o. Those electrons in the small bunch to the left of the velocity cusp in Figure 10-16 fall into this category. These have been decelerated below the initial beam velocity by $2\Delta u_o$. The maximum efficiency would occur if all of the electrons (rather than the small fraction shown in Figure 10-16) were reflected and decelerated by the velocity increment $2\Delta u_o$. Then the maximum efficiency would be

$$\eta = \frac{\frac{1}{2} m u_o^2 - \frac{1}{2} m (u_o - 2\Delta u_o)^2}{\frac{1}{2} m u_o^2} \tag{10-89}$$

Pierce [1] followed Slater's line of reasoning and used the Pierce small-signal expression for the velocity of the growing circuit wave to determine Δu_o. Pierce notes that "such considerations are complicated by the fact that the phase velocity of the wave in the large-signal region will not be the same as its phase velocity in the small-signal region. It is interesting, however, to see what limiting efficiencies this leads to."

From 10-55 (and the paragraphs following 10-55) the velocity of the growing wave is

$$v = \frac{u_o}{1 - Cy_1} \tag{10-90}$$

where C is Pierce's gain parameter and y_1 is the part of the incremental propagation constant that relates the speed of the circuit wave to the initial speed of the beam. For the maximum gain, as shown in Figures 10-7, 10-9, 10-12 and 10-13, y_1 is a negative quantity that varies from about -1/2 to -2. The value of Δu_o is

$$\Delta u_o = u_o - v = -\frac{u_o C y_1}{1 - Cy_1} \tag{10-91}$$

and so the maximum efficiency is predicted to be

$$\eta = -\frac{4\, C y_1}{\left(1 - C y_1\right)^2} \tag{10-92}$$

For $Cy_1 << 1$

$$\eta \approx -4\, C y_1 \tag{10-93}$$

As an example, for a value of y_1 of -1 and C of .05, the maximum efficiency is predicted to be about 20%, which is high but not totally unrealistic.

It should be noted in Cutler's work (Figure 10-15(f)) that the maximum efficiency (saturation) actually occurred when the electron bunch to the left of the velocity cusp was nearly equal to the main bunch. The velocities of the two bunches were nearly the same and were about the velocity of the circuit wave (Δu_o below u_o). As a result, the efficiency in Cutler's experiment is predicted to be about half that in (10-93) or

$$\eta = -2\, C y_1 \tag{10-94}$$

10.2.3 Comment on Computer Modeling

Cutler's results give a clear picture of events leading up to saturation. An important impact on theoretical considerations is that the velocity becomes multivalued, and it does so at a relatively low drive level (see Figure 10-15(c) for example). Thus, the theoretical techniques (usually referred to as the *Eulerian approach*) that treat the electron beam as a fluid cannot be used. For example, our technique for analyzing space charge wave behavior in Chapter 9 and the Pierce theory of traveling wave interaction in the chapter both treat the beam as a fluid. These approaches are not valid when velocity becomes multivalued.

Instead of the fluid approach, the theory must be able to follow individual particles or groups of particles. This is often referred to as the Lagrangian approach to the analysis. The ballistic theory of velocity modulation in Chapter 8 and of ballistic bunching in Chapter 9 used this approach. Of course, space charge forces, which were ignored in those calculations, must be included.

No algebraic theory is available for use with the large-signal particle approach. Instead, numerical techniques must be used. It is beyond the scope of this book to examine the computer modeling of interactions in microwave tubes. This is not to degrade the importance of computer modeling. All modern TWTs are designed with the aid of computer codes. Most of the very important TWT characteristics such as power, electronic and overall efficiencies and bandwidth can be accurately predicted with these codes. There is even a code available (the IBC code [4]) that runs on a personal computer and provides a great deal of insight into the performance of a TWT as various operating and design parameters are varied).

From the early numerical computations of Nordsieck [5], who ignored space charge effects, there was a steady development of computational techniques in the late 1950s and 1960s. Some of the more important of these works are listed as [6–15]. Some results from Hess's analysis [13] are given in Figure 10-17. Various quantities are plotted as a function of distance along the circuit in half wavelengths. The shaded areas of the distance scale indicate regions of decelerating circuit field. Note the similarity of the overall appearance of the electron velocity plots with those given by Cutler (Figure 10-15).

The earliest indication of nonlinear interaction in Hess's results appears in the electron density plot, when density becomes nonsinusoidal. A few half wavelengths later, a spur, like the Cutler spur, develops on the velocity plot and so the velocity becomes multivalued. At the same distance, the electron density bunch splits and, in subsequent half wavelengths, a portion falls back into an accelerating field. The resulting acceleration of electrons is shown by the increase in velocity of electrons in the spur. Finally, saturation occurs (the efficiency and circuit voltage reach maximum values) when a sufficient number of electrons have drifted back into the accelerating field.

10.2.4 Velocity Tapering

A major reason for saturation of the power output in TWTs is the reduction in average beam velocity caused by the transfer of electron energy to the circuit fields. An example of this reduction in average beam velocity is shown in Hess's results in Figure 10-17. The dashed line in the electron velocity plot shows average beam velocity. As

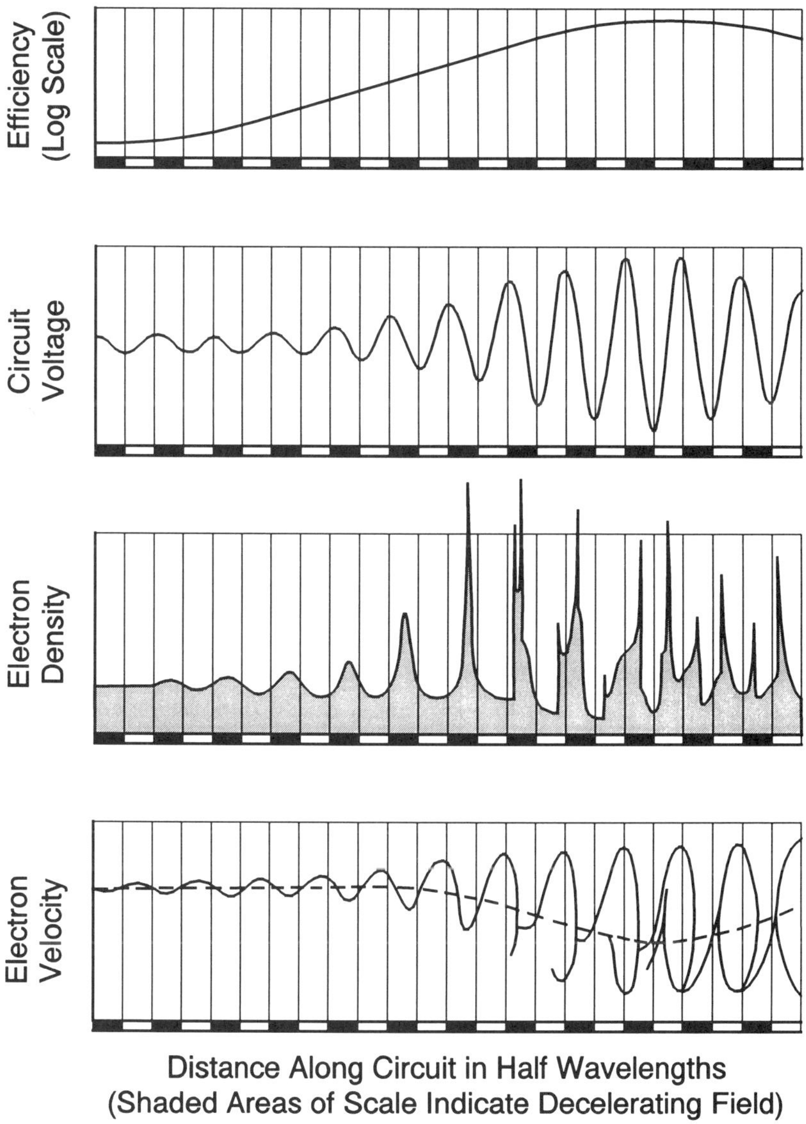

Figure 10-17 Characteristics of interactions in a TWT determined by computer analysis. Adapted from P. Hess, Ph.D. dissertation, University of California at Berkeley, 1960.

we have discussed in a previous section, partly as a result of the reduced velocity near saturation, electrons fall back out of phase with the circuit field, move into accelerating regions and extract energy from the circuit.

It has been recognized since the original invention of the TWT (see the varying pitch of the helix in Figure 1-1 from Lindenblad's patent) that the efficiency might be improved if the circuit is modified so that the velocity of the circuit wave decreases along with the beam velocity. Then, even though the bunches are slowed down as they lose energy to the circuit, they will remain in phase with the circuit wave and continue to deliver energy.

This technique of modifying the circuit to enhance efficiency (and often other TWT characteristics) is known as *velocity tapering*. Early attempts to use velocity tapering were not successful and produced little or no improvement in efficiency over untapered circuits. The advent of computer analyses of traveling wave interaction (see [14] for example) resulted in the design of tapers that produced appreciable improvements in efficiency. In [14], Meeker and Rowe report an efficiency improvement from 18 to 35% through the use of a variable-pitch helix.

A recent example of a computer designed taper is reported in [16]. In a NASA development of a 20 GHz coupled-cavity TWT, a phase-adjusted taper was used to achieve an RF efficiency of 22.6% and a peak power of 1000 W. Previously, the interaction structure for this particular tube had a piecewise linear taper that consisted of sequences of cavities of equal length and cavity sequences of linearly tapering length. The RF efficiency of that tube was 9.6% and the peak power output was 420 W. In the phase-adjusted taper, the phase of the electron bunches with respect to the phase of the circuit field changes gradually from a value (about 0°) conducive to the formation of strong bunches at the input end of the taper to a value (about 90°) conducive to strong transfer of beam power to the circuit at the output end of the taper. A comparison of the efficiencies of the TWTs with the piecewise linear taper and with the phase-adjusted taper is shown in Figure 10-18.

In addition to enhanced efficiency, a properly designed velocity taper can be used to enhance the linearity of the power output *versus* power input characteristic of a TWT. This is particularly important in communications applications of TWTs, where low distortion is required. An example of a linearizing taper is the dynamic velocity taper (DVT) developed by Kosmahl and reported in [17]. The basic philosophy of the DVT may be understood by examining the relation for Pierce's velocity parameter (10-48), which is

$$b = \frac{u_o - v_p}{C v_p} \tag{10-95}$$

where v_p is the phase velocity of the undisturbed (no electron beam) wave on the circuit.

In the small signal region, the gain of a TWT is constant and corresponds directly to the value of b. As the signal level grows and the TWT approaches nonlinear

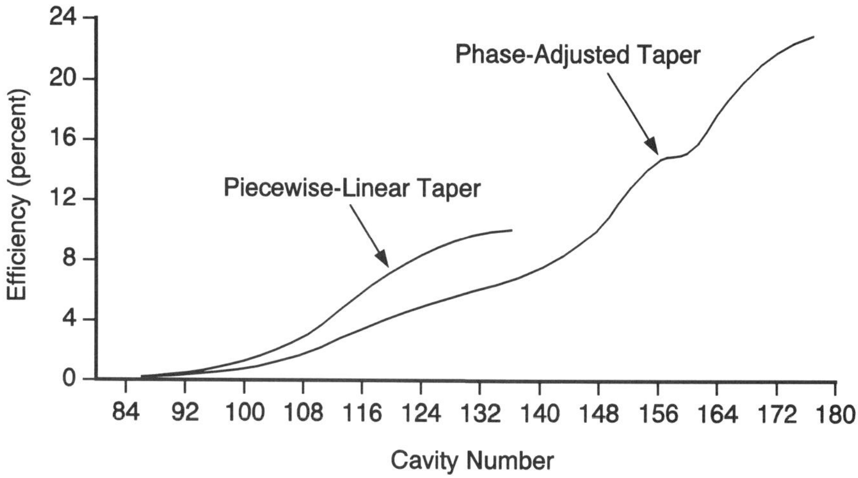

Figure 10-18 Computed RF efficiencies for a 20 GHz TWT with a phase-adjusted taper and one with a piecewise linear taper. From *NASA Tech Briefs*, September 1992.

operation the beam gives up energy to the circuit. The average velocity of the beam decreases and so the velocity parameter decreases. To prevent the gain from decreasing, it is necessary to modify the circuit and v_p so that the velocity parameter remains constant.

Results are presented in the referenced report [17] for TWTs with and without DVTs. For one of the TWTs with the DVT (the 12 GHz, 50 W Hughes 8802 experimental TWT), the pitch variation is shown in Figure 10-19 and the resulting improvement in linearity and power output are shown in Figure 10-20.

Note in Figure 10-19 that the pitch decreases linearly (this decrease is actually the first part of an exponential decrease) by about 7% over the final 18% of the length of the helix. The results of this change in pitch are impressive as shown in Figure 10-20. The linearity near saturation is significantly improved, and the power characteristic approaches that of a hard limiter at saturation. (The discrepancy between computed and measured powers resulted from the use of cold rather than hot helix loss values in the computation.) The power output is increased about 1.5 dB and the RF efficiency (efficiency of conversion of beam power to RF output power) is about 16%, which is 5% better than without the DVT. The relative phase delay near saturation is about 30° and the AM to PM conversion does not exceed 3°/dB, which is nearly the same as that in an unmodified TWT.

Finally, Kosmahl and Peterson note that, with an RF efficiency of 16%, a well-designed multistage depressed collector should raise the overall efficiency above 50%. By replacing the beryllium oxide (BeO) support rods for the helix with diamond, losses could be reduced, the interaction impedance would be improved and the overall efficiency would increase to about 65%.

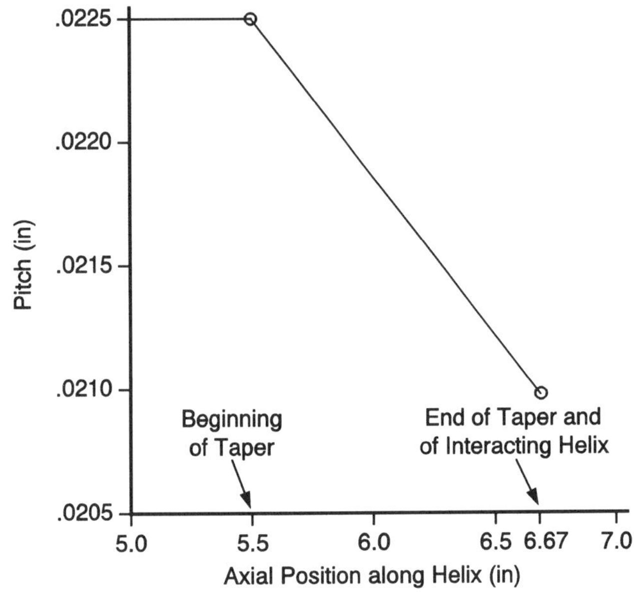

Figure 10-19 Pitch variation in the dynamic velocity taper (DVT) for the Hughes 8802 TWT. Adapted from H.G. Kosmahl and J.C. Peterson, NASA Tech. Memo. 83590, March 1984.

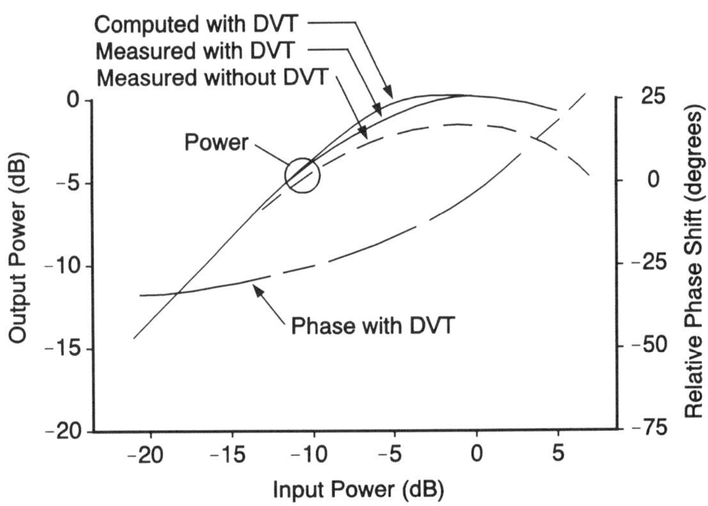

Figure 10-20 Effect of DVT on power and phase for the Hughes 8802. From H.G. Kosmahl and J.C. Peterson, NASA Tech. Memo. 83590, March 1984.

REFERENCES

[1] J.R. Pierce, *Traveling Wave Tubes,* Princeton, N.J.: D. Van Nostrand Company, Inc., 1950.

[2] C.C. Cutler, "The nature of power saturation in traveling wave tubes," *Bell Sys. Tech. Jour.*, vol. 35, July, 1956, pp. 841–876.

[3] J.C. Slater, *Microwave Electronics*, Princeton, N.J.: D. Van Nostrand Company, Inc., 1950.

[4] I.J. Morey and C.K. Birdsall, "Traveling-wave-tube simulation; the IBC code," Memorandum No. UCB/ERL M89/116, Electronics Research Laboratory, University of California at Berkeley, September 26, 1989.

[5] A. Nordsieck, "Theory of the large signal behavior of traveling wave amplifier," *Proc. IRE*, vol. 41, May 1953, pp. 630–647.

[6] H.C. Poulter, "Large signal theory of the traveling wave tube," Tech. Report No. 73, Electronics Research Laboratory, Stanford University, January 1954.

[7] P.K. Tien, L.R. Walker and V.M. Wolontis, "A large signal theory of traveling wave amplifiers," *Proc. IRE*, 43, March 1955, pp. 260–277.

[8] J.E. Rowe, "A large signal analysis of the traveling wave amplifier," Tech. Report No. 19, Electron Tube Laboratory, University of Michigan.

[9] J.E. Rowe, "Design information on large-signal traveling-wave amplifiers," *Proc. IRE*, February 1956, pp. 200–210 and correction in June 1956, pp. 818–819.

[10] P.K. Tien, "A large signal theory of traveling wave amplifiers including the effects of space charge and finite C," *Bell Sys. Tech. Jour.*, vol. 34, March 1956, pp. 349–374.

[11] J.E. Rowe, "One-dimensional traveling-wave tube analyses and the effect of radial electric field variations," *IRE Trans. on Electron Devices*, January 1960, pp. 16–21.

[12] H. Sobol and J.E. Rowe, "Theoretical power output and bandwidth of traveling-wave amplifiers," *IRE Trans. on Electron Devices*, April 1960, pp. 84–94.

[13] P. Hess, "Traveling wave tube large signal theory with application to amplifiers having dc voltage tapered with distance," Ph.D. dissertation, University of California at Berkeley, 1960.

[14] J.G. Meeker and J.E. Rowe, "Phase focusing in linear-beam devices," *IRE Trans. on Electron Devices*, May 1962, pp. 257–266.

[15] H.K. Detweiler, "Characteristics of magnetically focused large signal traveling-wave amplifiers," Ph.D. dissertation, University of Michigan, Ann Arbor, October 1968, also RADC Tech. Rep. TR-68-433; ASTIA Doc. AD-842733.

[16] NASA Lewis Research Center, "Coupled-cavity traveling-wave tube has phase-adjusted taper," *NASA Tech Briefs*, September 1992, pp. 26–30.

[17] H.G. Kosmahl and J.C. Peterson, "A TWT amplifier with a linear power transfer characteristic and improved efficiency," NASA Technical Memorandum 83590, prepared for the Tenth Communications Satellite Systems Conference, Orlando, Florida, March 18–22, 1984.

Chapter 11

Wave Velocities and Dispersion

For amplification to occur in a traveling wave tube, the axial component of the velocity of the wave on the RF circuit must be close to the velocity of the bunches of electrons in the beam. If, as frequency is varied, the velocity of the wave on the circuit moves away from the electron bunch velocity, then gain will decrease. Thus, the variation of circuit wave velocity with frequency is an extremely important consideration in the design of a TWT.

A circuit in which the wave velocity varies with frequency is said to have *dispersion* or to be *dispersive*. Before discussing dispersion in more detail, it is important to examine wave velocities and, in particular, the concepts of the group and phase velocities of a wave.

11.1 GROUP AND PHASE VELOCITY

The simple example illustrated in Figure 11-1 may serve to clarify the concepts of group and phase velocity. Waves are shown breaking at an angle θ on a beach. The velocity, v, with which the wave (and the energy of the wave) is moving is indicated by a vector perpendicular to the crest of the wave. The velocity with which the wave crest sweeps along the beach is indicated by a vector parallel to the beach. This velocity is the velocity at which any point on the wave (or any particular phase of the wave) sweeps along the beach. This velocity is referred to as the *phase velocity* and is usually denoted by v_p. A person walking along the beach and desiring to keep up with the wave crest (or with any other point on the wave) would have to walk at the phase velocity. Similarly, a person on a surfboard desiring to ride the crest of the wave in a path parallel to the beach would have to move at this same phase velocity.

It is important to realize that the magnitude of the phase velocity depends on the direction chosen for phase velocity relative to the direction of motion of the wave. In Figure 11-1(b) the angle between the crest of the wave and the beach (and the direction chosen for the phase velocity) is smaller than in Figure 11-1(a). As a result, the wave crest in (b) sweeps along the beach more rapidly than in (a). Thus, the phase velocity has increased.

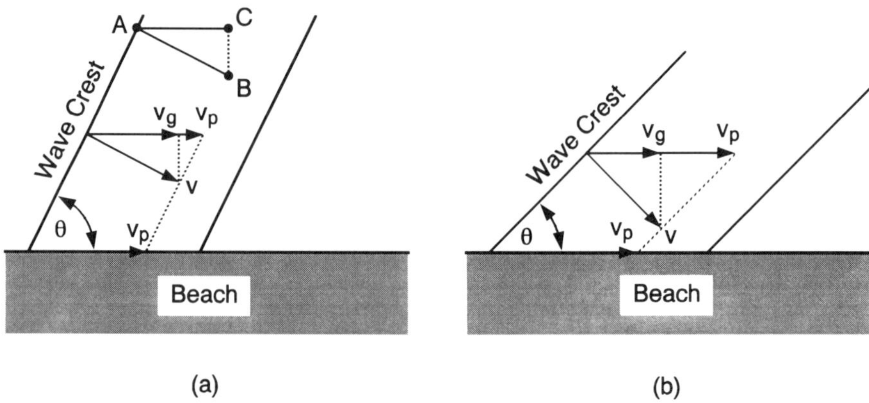

Figure 11-1 Illustration of the difference between phase velocity and group velocity.

The group velocity, v_g, of the wave is the velocity of the energy of the wave in the direction of interest. For example, in Figure 11-1(a), as the energy in the wave at point A moves to point B, the distance by which that energy moves in a direction parallel to the beach (and to the phase velocity) is from A to C. Thus, in this case, the group velocity is the component of the wave velocity in the direction of the phase velocity. Like the phase velocity, the magnitude of the group velocity depends on direction. This can be seen by comparing Figures 11-1(a) and (b). The group velocity in (b) is smaller than that in (a).

Group velocity and phase velocity can have opposite directions. Consider, for example, the situation shown in Figure 11-2. The waves are moving to the left and the group velocity is to the left. For this example, it will be assumed that there is a seawall with equally spaced openings along the shoreline. It is assumed that a person walking along the beach would be able to see the waves only when passing one of the openings in the seawall. The person could walk from right to left and pass successive openings at the same time that one of the wave crests did. Assuming that the angle θ is the same as in Figure 11-1(a) then the magnitudes of velocities would all be the same as in Figure 11-1(a) but the directions would be reversed.

Alternatively, the person could walk from left to right (from 1 to 2, for example) and pass the openings in synchronism with successive wave crests moving from right to left (from 3 to 2, for example). In this case, the phase velocity is to the right while the group velocity is to the left. For a person on a surfboard to move to the right and continue to be accelerated by the wave crests (which are moving to the left), that person would have to jump from wave crest to wave crest!

When the group velocity and the phase velocity are in opposite directions, interaction with the wave is referred to as a *backward wave interaction*. An important aspect of this interaction, which was evident in discussion in the preceding paragraph, is that the interaction is periodic rather than continuous as it is when the group velocity and the phase velocity are in the same directions.

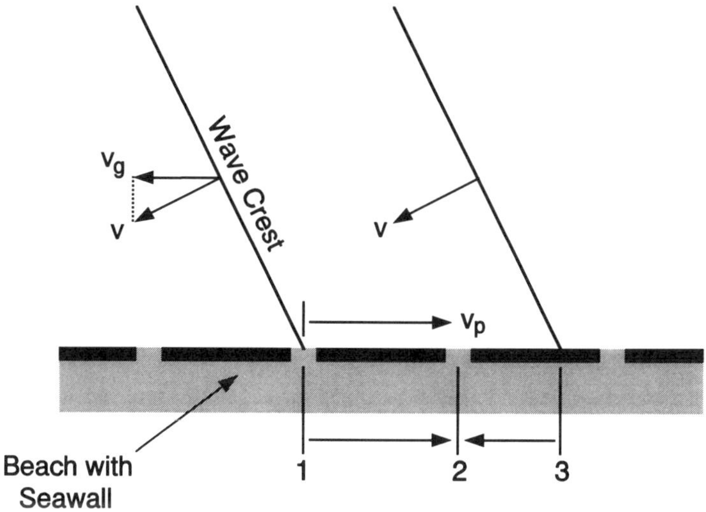

Figure 11-2 Illustration of backward wave interaction.

11.2 DISPERSION

As noted at the beginning of this chapter, a circuit in which the wave velocity varies with frequency is said to have dispersion or to be dispersive. Two ways are commonly used to illustrate the dispersion characteristics of a circuit. One, shown in Figure 11-3, is straightforward. The vertical axis represents phase velocity and the horizontal axis represents frequency. The plot for a circuit with no dispersion is simply a horizontal line. For a circuit with dispersion, the line is not horizontal.

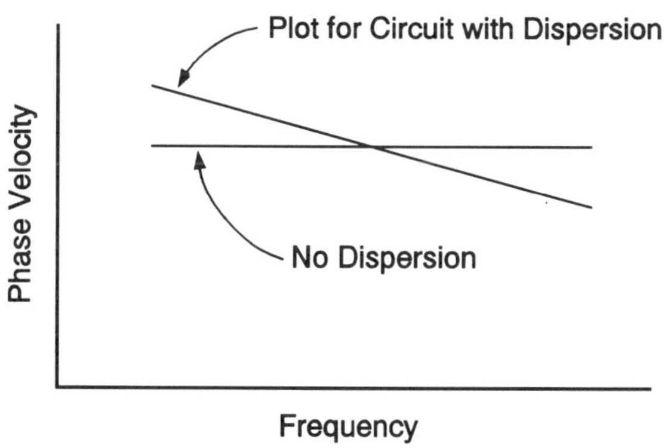

Figure 11-3 Straightforward representation of dispersion.

The second way to show dispersion characteristics is with a Brillouin diagram, which is often referred to as an ω-β diagram. Here, ω is angular frequency and β is the wave propagation constant. This diagram is not as straightforward as the first (phase velocity *versus* frequency). However, its use permits the display of considerably more information than the first. Because of the importance and widespread use of ω-β diagrams in the microwave component industry, their development is considered here.

A convenient way to develop an understanding of ω-β diagrams is to consider, first, simple microwave circuits and how their characteristics are represented. For this purpose, we will use a coaxial transmission line, which is not dispersive, and a waveguide, which is dispersive. Following these two examples, the more complicated example of a periodically loaded waveguide is examined.

11.2.1 Coaxial Transmission Line

In the fundamental TEM mode in a simple vacuum insulated coaxial transmission line, shown in Figure 11-4, the wave velocity, the electric and magnetic field lines are in planes perpendicular to the axis of the transmission line. The group velocity and the phase velocity are the same and are the velocity of light, c, for all frequencies. Thus, the coaxial line has no dispersion.

The propagation constant, β, is

$$\beta = k = \frac{\omega}{c} \tag{11-1}$$

The plot of ω *versus* β for the coaxial line is given in Figure 11-5. The plot is a straight line with slope c. In general, those portions of an ω-β diagram with constant slope represent frequency regions of operation of a circuit in which phase velocity is constant and so there is no dispersion.

11.2.2 Rectangular Waveguide

The phase velocity of the wave in a rectangular waveguide varies with frequency and so the waveguide is dispersive. To understand why phase velocity varies with frequency and how phase velocity along with group velocity are shown on an ω-β diagram, it is useful to examine the configuration of waves propagating in a rectangular waveguide. A convenient place to start is with the consideration of two waves of equal amplitude traveling in free space, as indicated in Figure 11-6. The direction of propagation of one is at an angle $+\alpha$ with respect to the z axis and the other is at an angle $-\alpha$. The solid lines represent position of wave crests (maximum value of E field out of the figure) and the dashed lines represent positions of wave troughs (minimum value of E field into the figure).

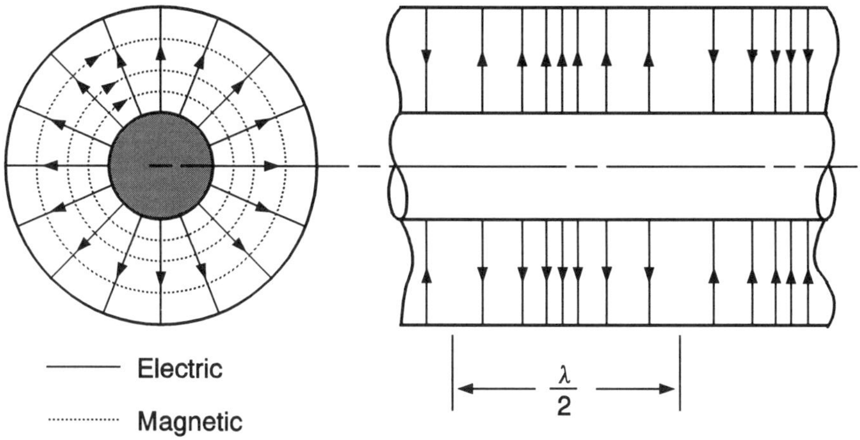

Figure 11-4 The TEM mode in a coaxial transmission line.

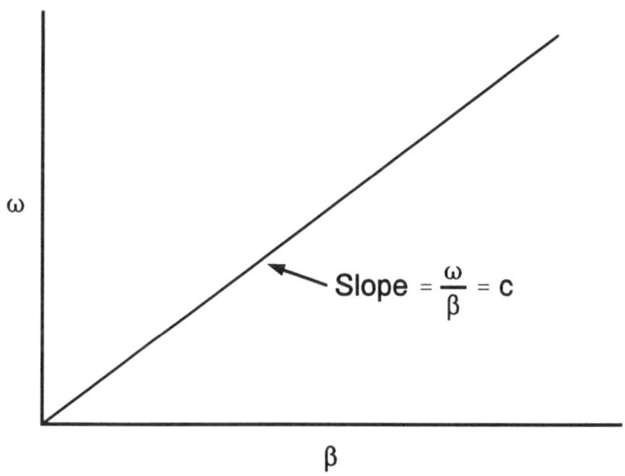

Figure 11-5 The ω-β diagram for the TEM mode in a coaxial transmission line.

Wherever wave crests and troughs (solid and dashed lines) intersect, the electric field is zero and only a magnetic field exists. In fact, all along a plane through zero-field points and parallel to the y-z plane, the electric field is zero. Not only is the E field zero for the instant of time implied in Figure 11-6, but also at other times as long as the waves propagate at the angles shown. Thus, conductors can be inserted along these planes without disturbing the E fields as long as the currents in the conductors correspond to those that would be induced by the magnetic fields, which, it should be noted, add so that the net fields, H_z, are parallel to the conductors.

The magnetic field lines are also parallel to any x-z plane, and the electric field lines are perpendicular to these planes. As a result, conductors containing currents and charges corresponding to those that would be induced by the fields can be placed in the

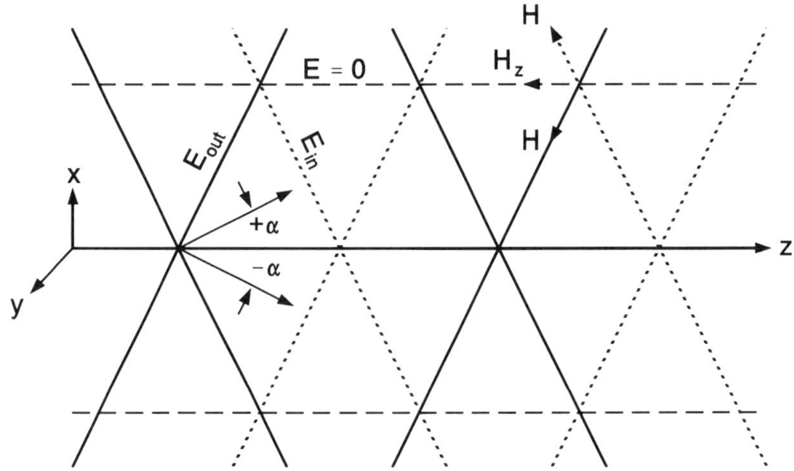

Figure 11-6 Plane waves of equal amplitude propagating at angles of $\pm\alpha$ with respect to the z direction.

x-z planes.

Finally, two conductors in the y-z plane can be joined to the conductors in the x-z plane to form a rectangular pipe (waveguide) containing portions of the two waves. These two waves can propagate as though they were in free space, except that, in reality, they are continually reflected back and forth from one side of the guide to the other. In fact, there is only one wave because each wave is a reflection of the other.

The velocity with which the wave propagates along the waveguide (the group velocity) can easily be determined with the aid of Figure 11-7. Each component of the wave travels at the speed of light, c, at an angle α with respect to the axis of the guide. As a result, the group velocity is smaller than c and, in fact, is

$$v_g = c \cos \alpha \tag{11-2}$$

The phase velocity of the wave (the velocity of a particular phase or amplitude of the wave) is indicated by v_p. Thus, for example, as the crest of the wave propagates at angle α a distance proportional to the velocity of light, c, the crest sweeps along the wall of the waveguide a distance proportional to higher velocity, v_p, which is

$$v_p = \frac{c}{\cos \alpha} \tag{11-3}$$

If an electron traveling parallel to the wall of the waveguide were to "ride" the crest of the wave and continually be deflected into the figure by the electric field, it would have to move at velocity v_p. Of course, this is not possible because, in this example, $v_p > c$. TWT circuits are designed to be "slow wave" circuits so that $v_p < c$ and so electrons can move at a velocity equal to the phase velocity.

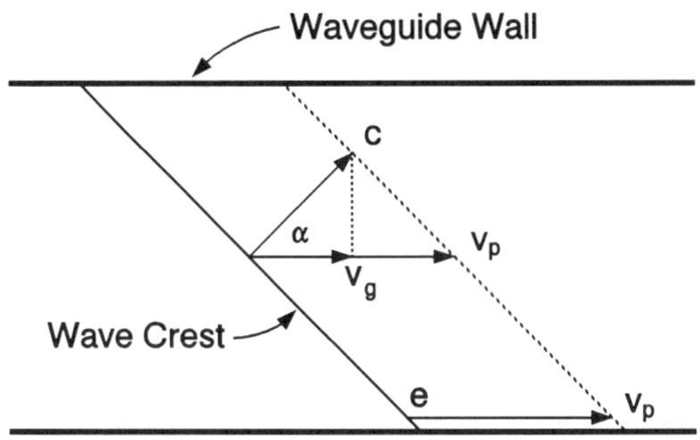

Figure 11-7 Wave velocities in a waveguide.

To examine the dispersion of waveguide, it is necessary to consider the change in wave configuration that takes place in a waveguide when frequency is varied. Shown in Figure 11-8 are wave configurations for three different frequencies $f_1 > f_2 > f_3$. Group velocities and phase velocities are also indicated. Note that, as frequency decreases, wavelength increases and so the angle α must increase for the wave to fit in the waveguide. Note also that, as $\lambda/2$ approaches the width of the guide, a, α approaches 90° and most of the wave motion is back and forth across the waveguide rather than along the waveguide. When $\alpha = 90°$, wave propagation down the guide stops and the condition known as *cutoff* occurs. The wavelength and frequency at which cutoff occurs are often denoted by λ_c and f_c. As frequency decreases from f_1 through f_2 to f_3 and then to f_c, the phase velocity steadily increases and, in fact, becomes infinite at cutoff. At the same time, the group velocity steadily decreases and becomes zero at cutoff.

The ω-β relation for the rectangular waveguide may be derived by examining the wave velocities as a function of frequency (or wavelength) with the aid of Figure 11-9. Thus,

$$\left[\frac{\lambda_c}{2}\right]^2 = \left[\frac{\lambda}{2}\right]^2 + \left[\frac{\lambda_c}{2}\cos\alpha\right]^2 \tag{11-4}$$

or

$$1 = \left[\frac{\lambda}{\lambda_c}\right]^2 + \cos^2\alpha \tag{11-5}$$

Also,

$$\cos\alpha = \frac{c}{v_p} \tag{11-6}$$

but

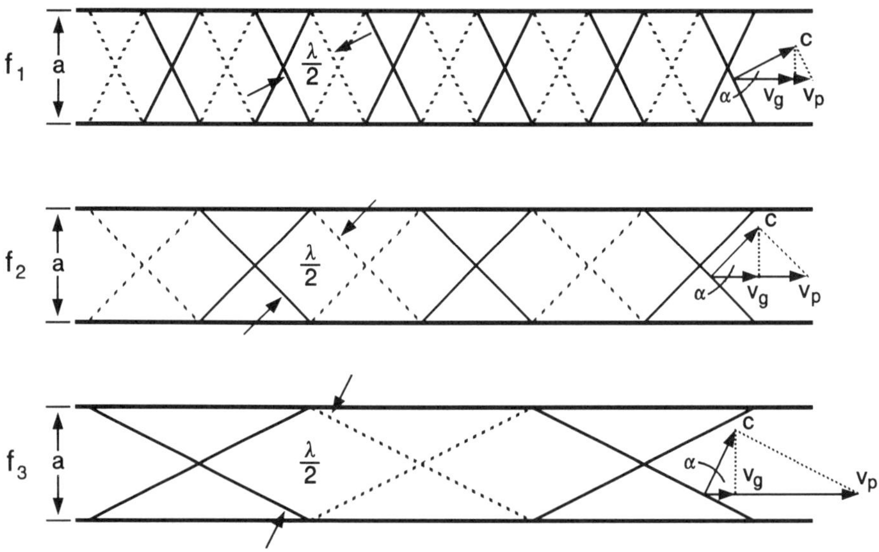

Figure 11-8 Wave configurations for frequencies $f_1 > f_2 > f_3$.

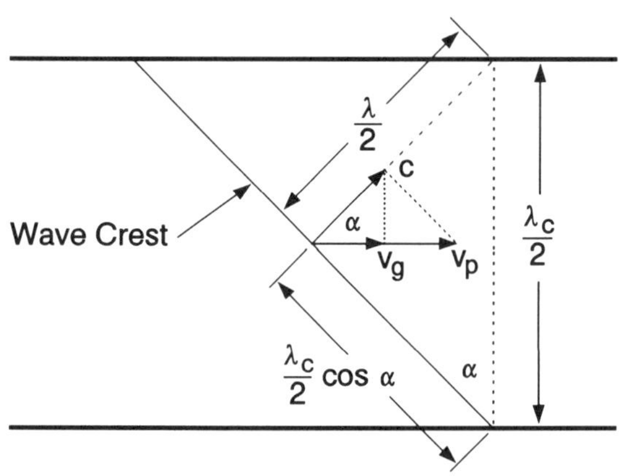

Figure 11-9 Quantities used in derivation of an ω-β relation.

$$\frac{\lambda}{\lambda_c} = \frac{c}{f}\frac{f_c}{c} = \frac{\omega_c}{\omega} \tag{11-7}$$

and

$$\frac{c}{v_p} = \frac{\omega}{v_p}\frac{c}{\omega} = \beta\frac{c}{\omega} \tag{11-8}$$

so

$$1 = \left[\frac{\omega_c}{\omega} \right]^2 + \beta \left[\frac{c}{\omega} \right]^2 \tag{11-9}$$

which may be rewritten to obtain the ω-β relation

$$\omega = \sqrt{\omega_c^2 + (\beta c)^2} \tag{11-10}$$

and which is plotted in Figure 11-10. Since $\beta = \omega/v_p$, v_p at any ω and β is simply the slope of a line drawn from the corresponding point on the ω-β curve through the origin. Note that the phase velocities for the various frequencies used in Figure 11-8 are indicated in Figure 11-10.

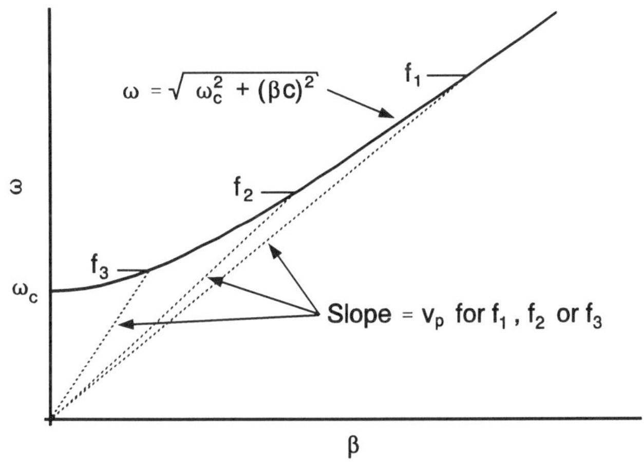

Figure 11-10 The ω-β diagram for rectangular waveguide.

The group velocity of the wave may also be displayed on an ω-β diagram. To determine how this is done, consider Figure 11-11, which shows the propagation constants k and β in vector form and the changes that occur (Δk and $\Delta\beta$) when the frequency is varied by $\Delta\omega$. First of all, from Figure 11-9,

$$\cos\alpha = \frac{c}{v_p} = \frac{c/\omega}{v_p/\omega} = \frac{1/k}{1/\beta} = \frac{\beta}{k} \tag{11-11}$$

which is why the relations of β, k and α are as indicated in Figure 11-11. When ω is reduced by $\Delta\omega$, α increases by $\Delta\alpha$, k decreases by Δk and β decreases by $\Delta\beta$. But, for a small $\Delta\omega$ and a small $\Delta\alpha$,

$$\frac{\Delta k}{\Delta \beta} \approx \cos \alpha \tag{11-12}$$

and, from Figure 11-10,

$$v_g = c \cos \alpha \tag{11-13}$$

so

$$v_g = \frac{\Delta \omega}{\Delta \beta} \tag{11-14}$$

Thus, group velocity can be shown on the ω-β diagram as indicated in Figure 11-12. The group velocity is the slope of the ω-β curve. Group velocities for the various wave configurations in Figure 11-8 along with the cutoff condition are also indicated in Figure 11-12.

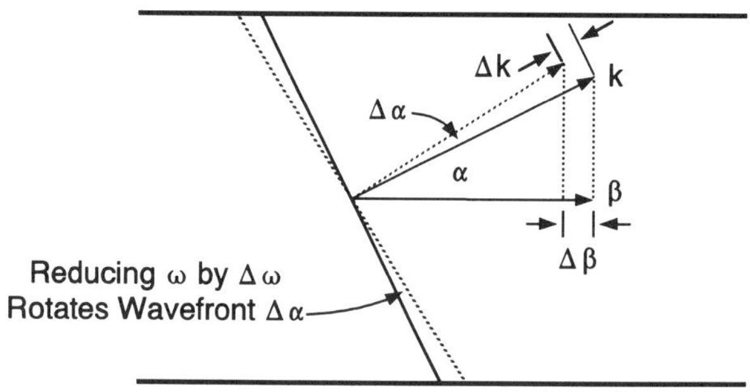

Figure 11-11 Changes in the propagation constant as frequency is varied.

In summary,

1. Phase velocity is the ratio of ω to β;
2. Group velocity is the slope of the ω-β curve;
3. For a waveguide, as $\omega \rightarrow \infty$, $\beta \rightarrow k$, $v_p \rightarrow c$ and $v_g \rightarrow c$.

In this section, we have treated the wave in a rectangular waveguide as though it were composed of two waves. As noted previously, there is only one wave, because each is a reflection of the other. The amplitude of the resulting electric field as functions of axial and transverse positions in the waveguide is shown in Figure 11-13. It should be noted that this is the fundamental TE_{01} mode of propagation. TE refers to the configuration of the electric field (E), which is transverse (T) to the axis of the guide. There are no variations of field across the narrow dimension of the guide (thus, the subscript 0) and one-half sinusoidal variation in the broad dimension (subscript 1). At higher frequencies, other modes not of interest at this point in the discussion can propagate.

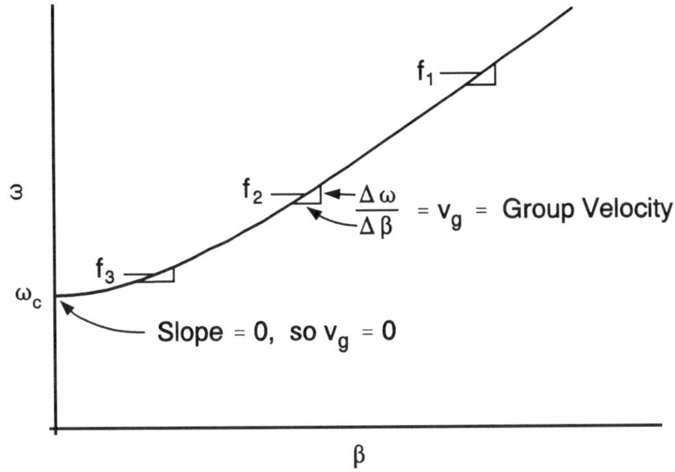

Figure 11-12 Group velocity is determined by the slope of the ω-β curve.

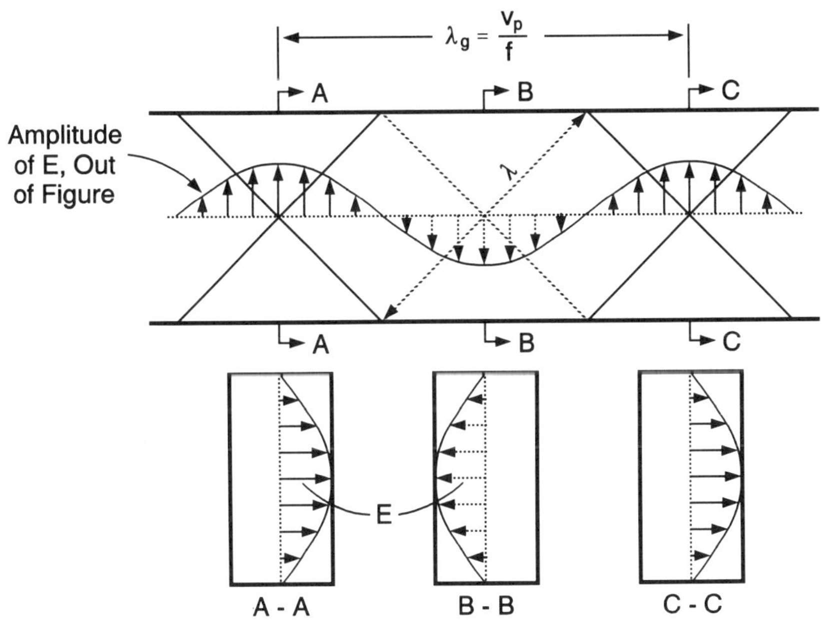

Figure 11-13 Electric field distributions in fundamental TE_{01} mode in rectangular waveguide.

The wavelength of the total wave is the guide wavelength, λ_g, which is

$$\lambda_g = \frac{v_p}{f} \tag{11-15}$$

The guide wavelength is longer than the free-space wavelength, λ, at all normal operating frequencies. In applications of waveguides, only the guide wavelength is of value. The free-space wavelength was an artifice of use in explaining waveguide behavior but plays no role in the use of waveguide. Thus, in following discussions where wavelength in the waveguide is referred to, it is to be assumed that guide wavelength is the quantity is question even though the term *wavelength* is used.

Finally, in considering the ω-β diagram for waveguide, it is important to note that waves can propagate with identical characteristics in either direction and so the ω-β diagram is symmetrical about the ω axis as shown in Figure 11-14. At higher frequencies, in either direction, the group and phase velocities approach the velocity of light.

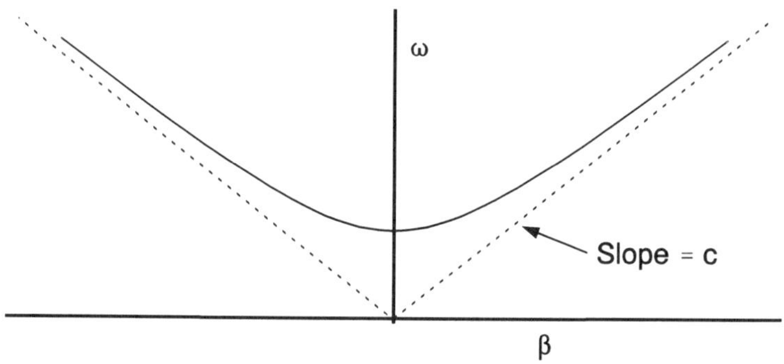

Figure 11-14 The ω-β diagram for waves propagating in either direction in a waveguide.

11.2.3 Periodically Loaded Waveguide

An obstacle, such as a post or a metal vane in a waveguide, produces a change in impedance and is said to load the waveguide. This loading causes part of a propagating wave to be reflected. If equally spaced obstacles are used, the waveguide is said to be periodically loaded. In an infinitely long waveguide, if the reflections from the obstacles add in phase, complete reflection of a wave will eventually occur. Thus, the wave is "stopped" and, in a frequency range where complete reflection occurs, a "stop band" is produced.

An example of a periodically loaded waveguide is shown in Figure 11-15. The obstacles shown are metal vanes attached to the top and bottom of the waveguide. The impedance change produced by these vanes is capacitive. The phase shift that occurs in the electric field reflected from each of the vanes is 180°. At frequencies near ω_c, the phase velocity is high and the wavelength is much longer than the spacing between the vanes, L. Wave reflections from the vanes do not add in phase and so there is no significant affect on wave propagation. As frequency is increased and the wavelength decreases and approaches $2L$ so that the phase shift between vanes is 180°, reflections from the vanes add in phase, as is illustrated in Figure 11-16.

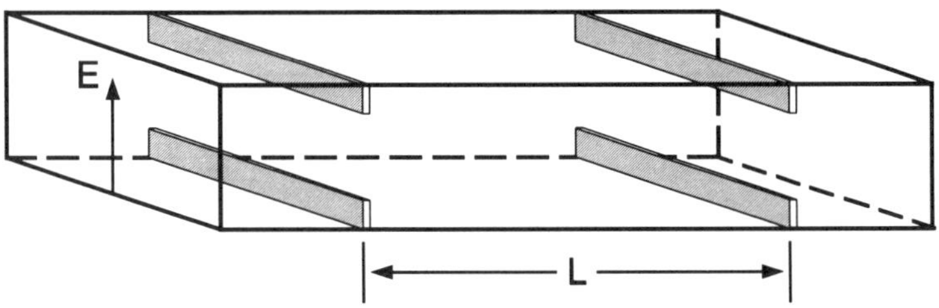

Figure 11-15 Use of vanes to periodically load waveguide.

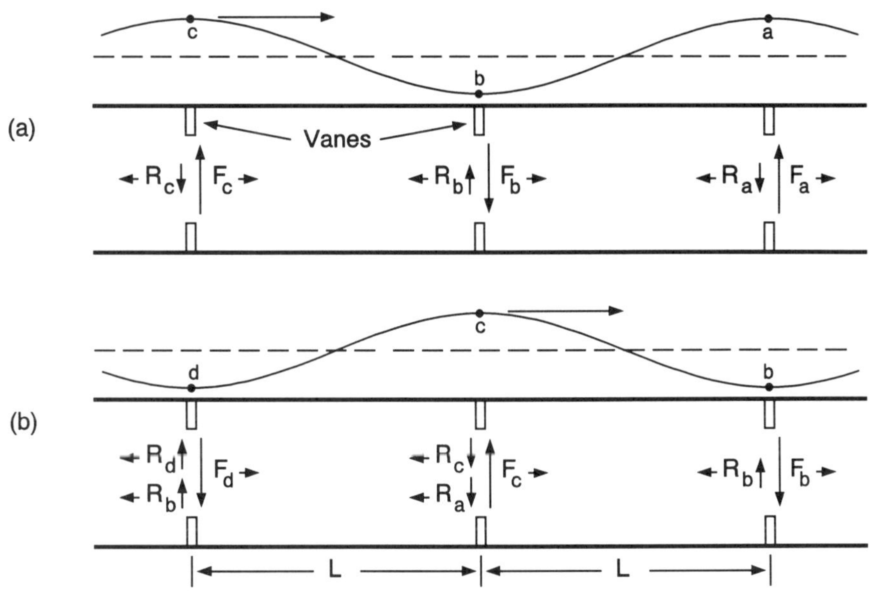

Figure 11-16 Forward (F) and reflected (R) waves for $\beta L = \pi$.

In Figure 11-16(a), the sinusoid indicates the amplitude of the electric field in the waveguide. The forward and reflected waves at each of the vanes are indicated by vertical arrows. F_a, F_b and F_c indicate the forward waves and R_a, R_b and R_c indicate the reflected waves at vanes 1, 2 and 3.

For Figure 11-16(b), it is assumed that the forward wave has propagated a distance λ/2 = L (π rad) to the right. At the same time reflections from the vanes in Figure 11-16(a) have propagated π rad to the left. Thus, for example, R_a arrives at vane 2 at the same time that F_c does and adds in phase with the reflection, R_c, from vane 2. Eventually, in an infinitely long waveguide, complete reflection of the forward wave occurs. There is no propagation so the group velocity goes to zero. The resulting effect on the ω-β diagram is shown in Figure 11-17. Here, instead of plotting β in radians/meter, the horizontal axis is βL, which is simply the phase shift in radians from vane to vane.

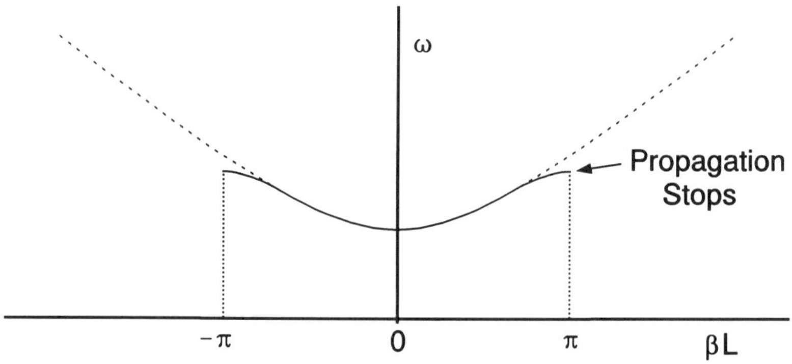

Figure 11-17 Effect of periodically loaded waveguide on an ω-β diagram.

The next problem to be considered is the shape of the ω-β characteristic for values of $\beta L > \pi$. Certainly, the curve does not stop at $\beta L = \pi$. Consider, first, a value of βL somewhat less than π. For this case, the wave pattern is as indicated in Figure 11-18(a). The phase velocity is higher than at $\beta L = \pi$ and the wavelength is longer.

It is useful, at this point, to consider an electron moving along the axis of the guide. If the electron moves at the phase velocity of the wave then, of course, the electron will move from vane to vane in the same time that it takes the wave crest to move from vane to vane. The electron is said to maintain synchronism with the wave.

To consider a value of $\beta L > \pi$, assume, as indicated in Figure 11-18(b), that a wave with the same frequency as that assumed for Figure 11-18(a) is traveling in the backward direction. The wavelength and phase velocity (in the backward direction) are the same as those for the forward wave in (a). An electron moving to the right with the same velocity as in (a) will arrive at the next vane before the next backward-moving wave crest does. The velocity of the electron must be reduced for the electron to arrive at the next vane when the next backward-moving wave crest does (for the electron to be synchronous with the backward wave). Thus, the phase velocity of the wave as seen by the electron in (b) is lower than the phase velocity of the forward wave in (a).

As a result of these considerations, it is seen that as βL increases above π, a backward wave exists and so the ω-β curve must turn downward as shown in Figure 11-19. In fact, for values of βL in the range $\pi < \beta L < 2\pi$, the wave is a backward wave and the ω-β characteristic is the same as for $-\pi < \beta L < 0$.

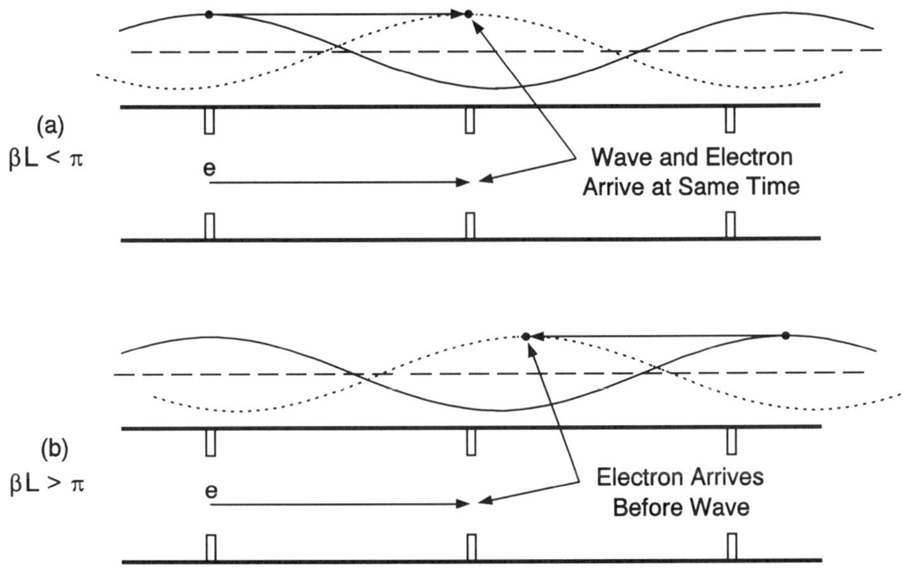

Figure 11-18 Forward (a) and backward (b) waves for βL near π.

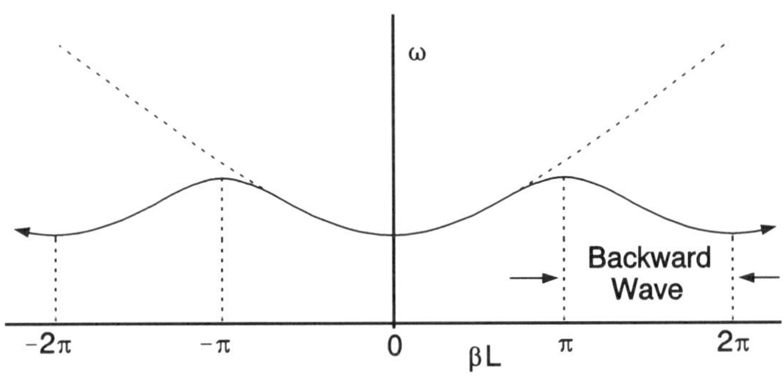

Figure 11-19 Extension of ω-β curve to $\pi < \beta L < 2\pi$.

As βL approaches 2π, the backward wave approaches the cutoff condition with wavelength and phase velocity (in the reverse direction) approaching infinity. For an electron velocity (and phase velocity in the forward direction) such that $\beta L = 2\pi$, the electron moves from vane to vane in synchronism with wavecrests of the backward wave that are an infinite distance apart but moving with infinite velocity.

Finally, as βL increases to values greater than 2π, a forward wave again exists and the ω-β diagram turns upward again. The shape of the curve from $\beta L = 2\pi$ to $\beta L = 3\pi$ is the same as for $0 \leq \beta L \leq \pi$. The part of the ω-β diagram shown in

Figure 11-19 is referred to as a *passband* because waves can propagate. This band will be of primary interest when coupled cavity TWTs are considered.

Before coupled cavity structures can be considered (in Chapter 13), it should be noted that there are other passbands for periodically loaded waveguides. There are bands of frequencies in which reflections from the vanes do not add in phase and so wave propagation takes place. However, whenever the wavelength approaches a condition where the distance between vanes is equal to several half wavelengths, a cutoff condition occurs and this places a limit on the upper or lower frequency of the band in which propagation occurs. As a result, the ω-β diagram contains an infinite set of passbands, four of which are shown in Figure 11-20.

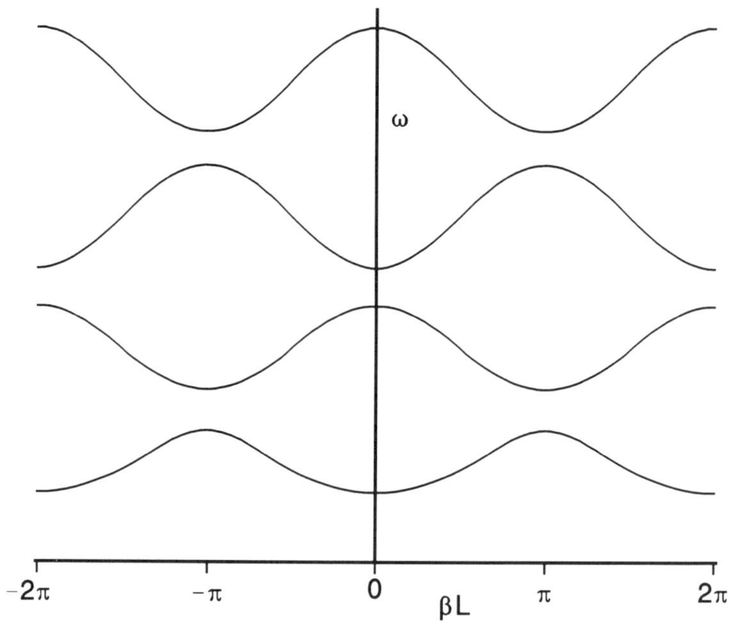

Figure 11-20 The ω-β diagram for a vane loaded waveguide.

Chapter 12

Helix TWTs

As was indicated in Chapter 1, the TWT invented by Lindenblad used a helix as the slowwave circuit. During the formative years of TWT development, hundreds of other propagation structures were investigated, but the helix remains as the circuit of choice when a broad bandwidth is required. The purpose of this chapter is to describe the characteristics of helix TWTs that have made them the mainstay of the TWT industry.

Major applications for helix TWTs include communications and electronic countermeasures (ECM). Shown in Figure 12-1 is a cross section of a helix TWT like that used for many ECM applications. Many ECM TWTs contain the most advanced characteristics of all helix TWTs. Those shown in Figure 12-1 are the electron gun with a shadow grid, PPM focusing, multistage collector and integral equalizer and phase conditioner. Other important characteristics not indicated in Figure 12-1 include techniques for suppressing backward wave oscillations (BWO), velocity tapers and dispersion control structures.

12.1 HELIX BANDWIDTH

The extremely broad bandwidth of the helix is the single most important property of this circuit. No other circuit has ever rivaled the helix in bandwidth and most do not even come close. Helix TWTs can now be made with a bandwidth of over two octaves, as is shown in Figure 12-2.

One of the problems with the very wide bandwidth shown in Figure 12-2 is the high harmonic power that is generated at low frequencies. This results from the wide bandwidth of the tube and the fact that the velocity modulation and bunching processes are highly nonlinear. At the low end of the operating band of a broadband helix TWT, the second and even the third harmonic signals generated in the beam are within the bandwidth of the tube and are thus amplified. At the lower end of the band, harmonic power can rival fundamental power.

One way to reduce harmonic output power is with a technique known as *harmonic injection*. An indication of the effectiveness of this technique is illustrated in Figure 12-3. When a second harmonic signal of the proper phase and amplitude is injected,

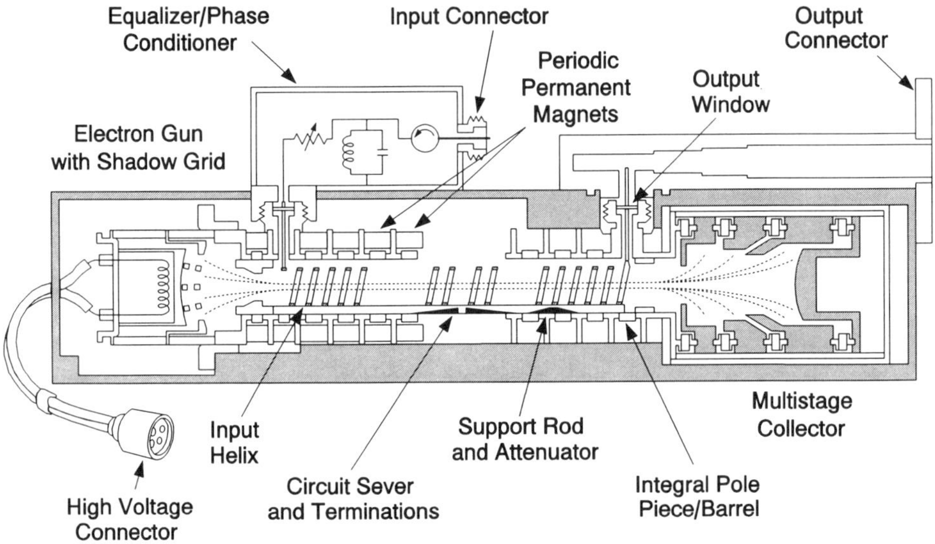

Figure 12-1 Cross section of helix TWT for electronic countermeasures use. Courtesy of Varian Associates.

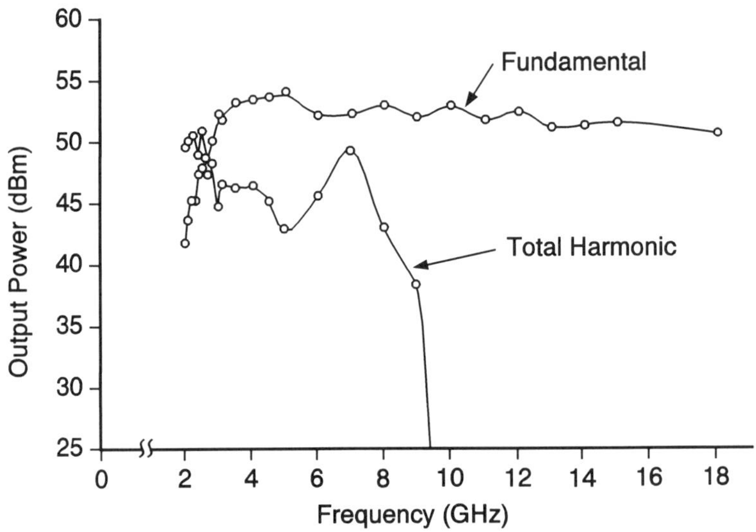

Figure 12-2 Saturated output power *versus* frequency for the Varian VTU-6292A1 helix TWT. Courtesy of Varian Associates.

along with the fundamental signal, the harmonic output power drops and the fundamental power increases. This occurs when the injected second harmonic signal is 180° out of phase with the second harmonic signal generated by nonlinear processes inherent in the beam-circuit interaction process. With proper design of the input circuit, the cancellation process can be made reasonably noncritical.

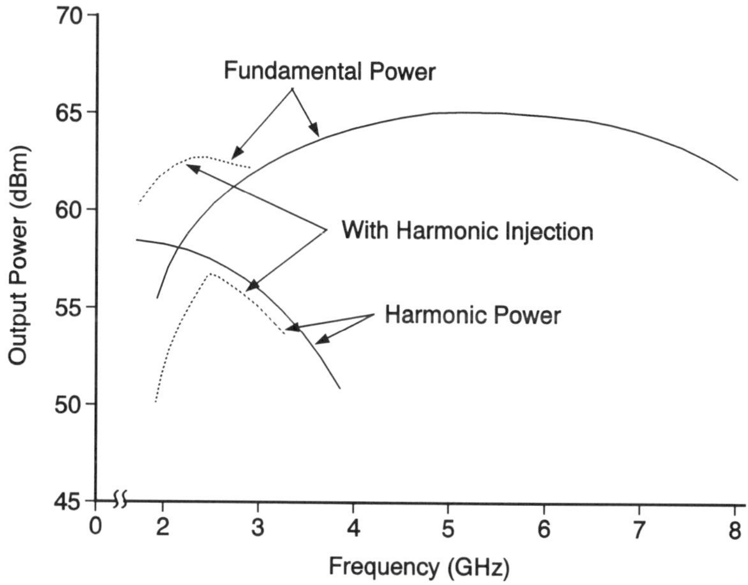

Figure 12-3 Affect of harmonic injection on output power for a wideband helix TWT. From Hughes Aircraft Co., *TWT and TWTA Handbook*.

The broad bandwidth of the helix results from its low dispersion. The helical circuit can be thought of as being derived from a single-wire transmission line for which the dispersion is zero. Shown in Figure 12-4(a) is a helix and in (b) a view of the helix when it has been cut at the points marked by x and unrolled. A wave propagating along a wire with the velocity of light, c, advances in the axial direction with phase velocity

$$v_p \approx c \sin \psi \tag{12-1}$$

The angle ψ is the helix pitch angle given by

$$\cot \psi = \frac{2\pi a}{p} \tag{12-2}$$

where a is the radius and p is the pitch of the helix.

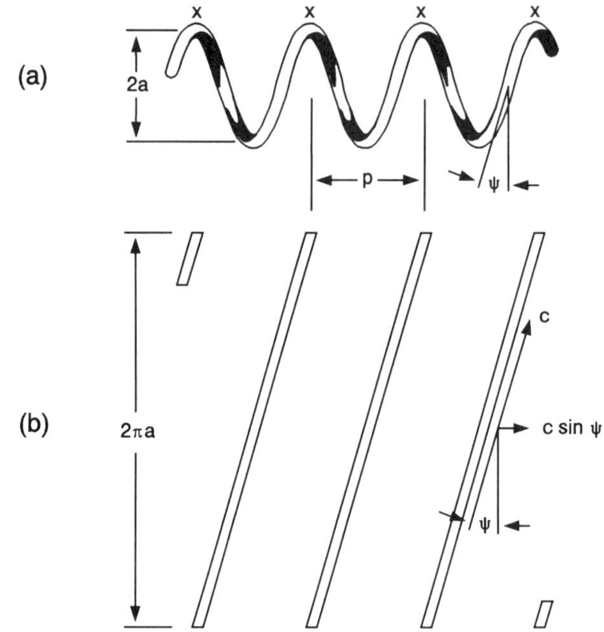

Figure 12-4 Helix (a) and view of helix cut at each x and unrolled (b).

12.1.1 Dispersion

Unless corrective steps are taken, phase velocity varies somewhat with frequency for two primary reasons:

1. As frequency is reduced, the number of turns of the helix per wavelength is increased. As a result, the coupling of electric and magnetic fields between turns changes. For example, as shown in Figure 12-5, when current flows in the same direction in adjacent turns, there is cancellation of the magnetic flux between turns and so the flux generated per unit current (the inductance) decreases as the wavelength (and number of turns per wavelength) increases. Thus, as frequency decreases, inductance decreases and wave velocity increases. As the wave velocity around the helix increases, then, of course, the phase velocity increases. This is evident in Figure 12-6, which is an ω-β diagram for a helix with a 10° pitch angle. The phase velocity at high frequencies is shown to be essentially constant, but increases at low frequencies. The helix in a TWT is supported by an insulating structure that often takes the form of three equally spaced dielectric rods as indicated in Figure 12-6. These rods and the ways in which they are incorporated in a TWT are discussed in a

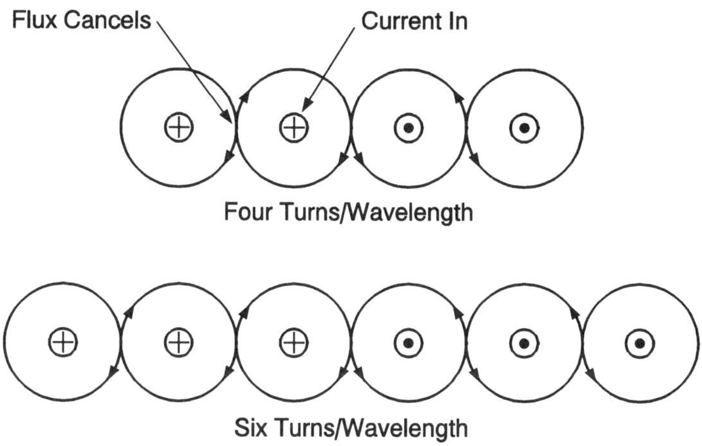

Figure 12-5 Illustration of cancellation of magnetic flux between turns.

following section. The effect of the dielectric material of the rods is also evident in Figure 12-6. Wave velocity is inversely proportional to the square root of dielectric constant so, when dielectric material is added in the region surrounding the helix, wave velocity decreases.

2. At high frequencies (and a relatively small number of turns per wavelength), the electric field is concentrated between turns of the helix. As the frequency is reduced and the number of turns per wavelength is increased, the electric field extends farther from the helix, as is indicated in Figure 12-7. Therefore, a metal shell (barrel) surrounding the helix will intercept a greater portion of the electric field (and also the magnetic field, which is not shown in Figure 12-7) at low frequencies than at high frequencies. (Note in Figure 12-7 that only the position of the barrel is shown. The distortion of the electric field pattern that the barrel produces is not shown.) As the barrel is brought closer to the helix, the effect increases, with the result that the wave velocity on the helix decreases. The decrease in velocity is greater at low frequencies than at high frequencies.

The increase in phase velocity resulting from the reduction in inductance at low frequencies can be partially offset by bringing the barrel closer to the helix. Unfortunately, while the resulting decrease in dispersion is desirable, there is an accompanying decrease in circuit interaction impedance that is undesirable because gain and efficiency are reduced. The decrease in circuit impedance results from inductive coupling of the helix currents to the barrel and the flow of current in the circumferential direction in the barrel.

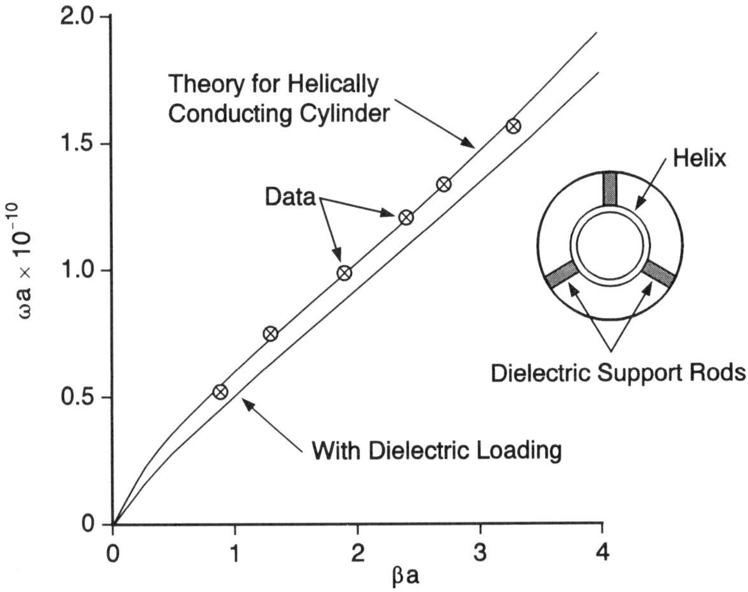

Figure 12-6 The ω-β diagram for a wire helix with a 10° pitch angle. From *Power Travelling Wave Tubes* by J.F. Gittins, published 1965 by American Elsevier, Inc.

12.1.2 Dispersion Control

Ideally, if a metallic shell that conducted only in the axial direction could be placed around the helix, there would be no inductive coupling and no circumferential currents could flow in the shell. Still, the shell would have a greater effect on the electric field at low frequencies than at high frequencies. As a result, control of dispersion would be possible.

In practice, an axially conducting shell is approximated by one of the techniques illustrated in Figure 12-8 [1]. One of the four configurations shown (labeled *unloaded*) is used for comparison purposes and has only the outer metal shell, which supports the helix through dielectric support rods. In the other three configurations, either solid, vane-shaped, or T-shaped metallic elements, which are frequently referred to as *anisotropic loading elements*, are used to provide the desired loading.

Experiments have been performed to provide a basis for the choice of the best loading technique to use for controlling dispersion. Shown in Figure 12-9 are phase velocity and Pierce's velocity parameter as functions of frequency for three of the configurations in Figure 12-8. Tests were performed with two T-shaped elements having different spacings from the helix. The results were obtained in tubes containing these configurations, which were nearly identical in all other respects.

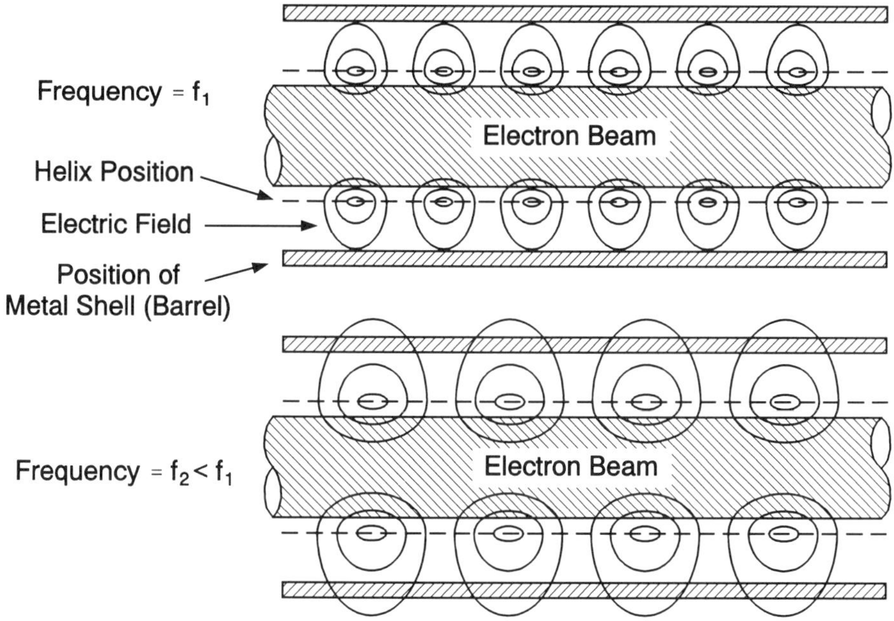

Figure 12-7 Variation of the electric field pattern with frequency.

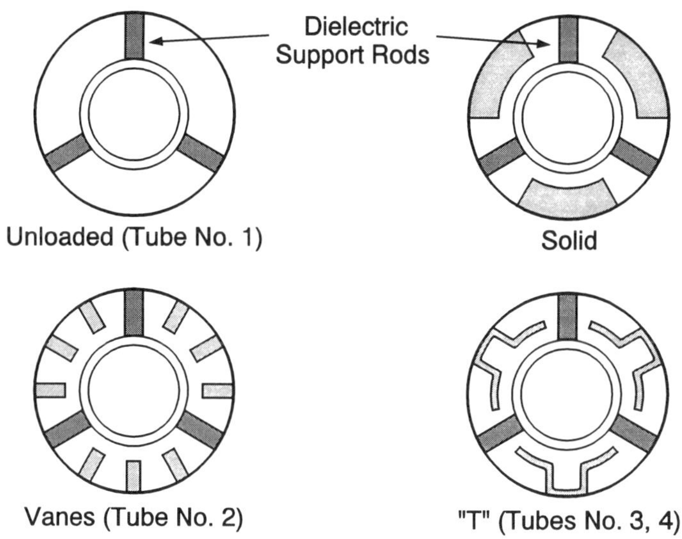

Figure 12-8 Techniques for dispersion control by approximating an axially conducting shell. From J.L. Putz and M.J. Casconc, *Technical Digest*, 1979 IEDM. © 1979 IEEE.

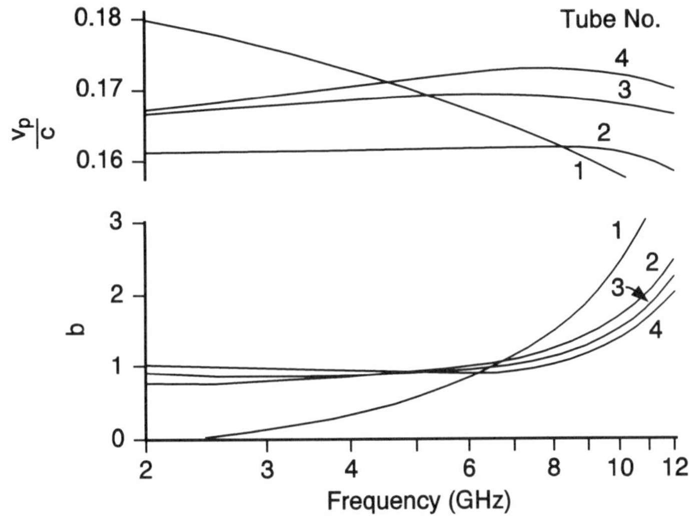

Figure 12-9 Normalized phase velocity and Pierce's velocity parameter as functions of frequency. From J.L. Putz and M.J. Cascone, *Technical Digest*, 1979 IEDM. © 1979 IEEE.

With tube no. 1, phase velocity increases and b decreases as frequency is reduced. In tube no. 2, the vanes produced a virtually flat phase-velocity characteristic over nearly the entire frequency range. The T-shaped segments produced negative dispersion characteristics. That is, the phase velocity decreased at low frequencies.

The small-signal gain characteristics of the four tubes tested are shown in Figure 12-10. The maximum gain for tube no. 1 occurred in the 6 to 7 GHz range. This frequency corresponds to a wavelength such that there are about four helix turns per wavelength. This is in agreement with the prediction of the small-signal theory of helix TWTs. At higher frequencies, gain decreases because the field lines are concentrated closer to the helix (this was shown in Figure 12-7) and the component available for interaction with the beam decreases. At lower frequencies, gain decreases because the length of the helix, in terms of the number of wavelengths, decreases. Of the four tubes tested, tube no. 4, which had a closer helix-to-loading-element spacing than tube no. 3, had the best small-signal bandwidth (in excess of two octaves).

Shown in Figure 12-11 is a cutaway view of a typical PPM-focused helix TWT containing T-shaped loading elements. In this example, the tungsten tape helix is supported by anisotropic pyrolytic boron nitride (APBN) rods. The reason for using APBN and the techniques for assembling the helix and supported-rod structure are discussed in the section on helix support techniques.

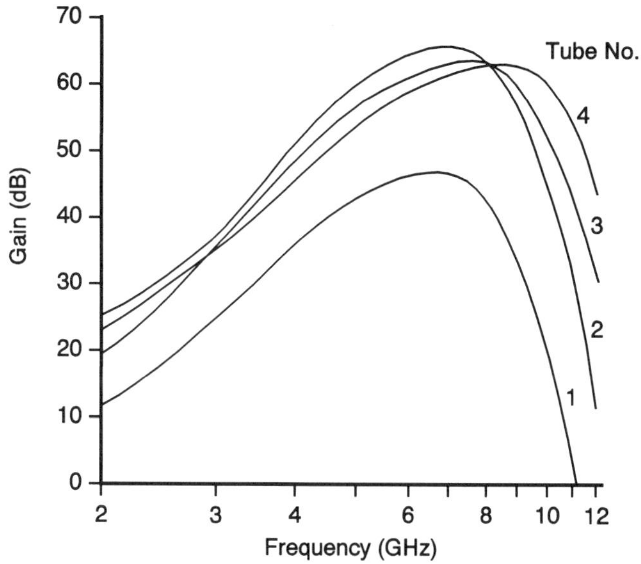

Figure 12-10 Small-signal gain for various loading configurations. From J.L. Putz and M.J. Cascone, *Technical Digest*, 1979 IEDM. © 1979 IEEE.

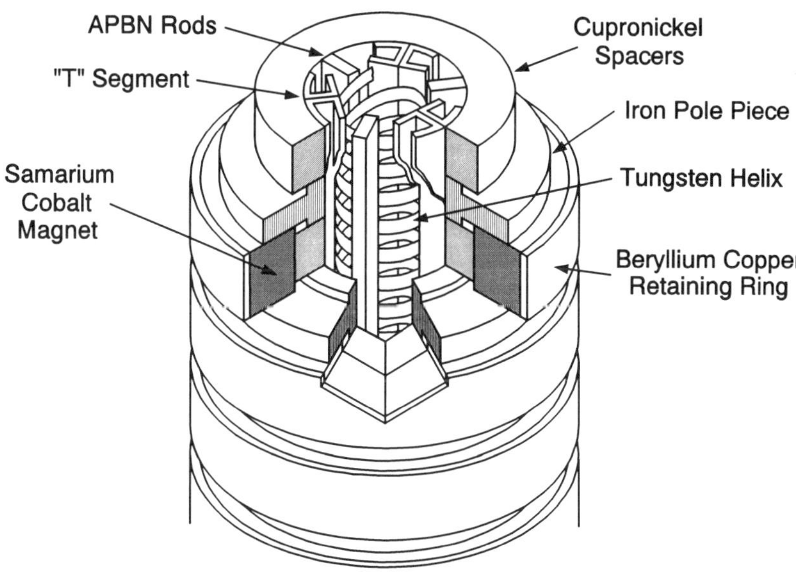

Figure 12-11 PPM focused helix TWT containing T-shaped loading elements. Courtesy of Varian Associates.

12.1.3 Backward Wave Oscillations

A major problem that limits the upper frequency of operation and the peak power capability of helix TWTs is backward wave oscillations (BWO). At a frequency such that the electron transit angle from turn to turn is 180° there are two turns per wavelength or 180° per turn and a backward wave interaction can occur. This can be understood with the aid of Figure 12-12. Consider electrons near the helix moving to the right (designated by *e*) and a backward wave moving to the left. Electrons are initially accelerated as is indicated by the arrow representing force (F) in Figure 12-12(a). These electrons experience essentially no force 180° later (Figure 12-12(b)) because the wave has moved to the left 180° and the nulls in the electric field are on the same side of the helix as the electrons. When the electrons have moved another 180° to the right while the wave has moved another 180° to the left (Figure 12-12(c)) acceleration again occurs.

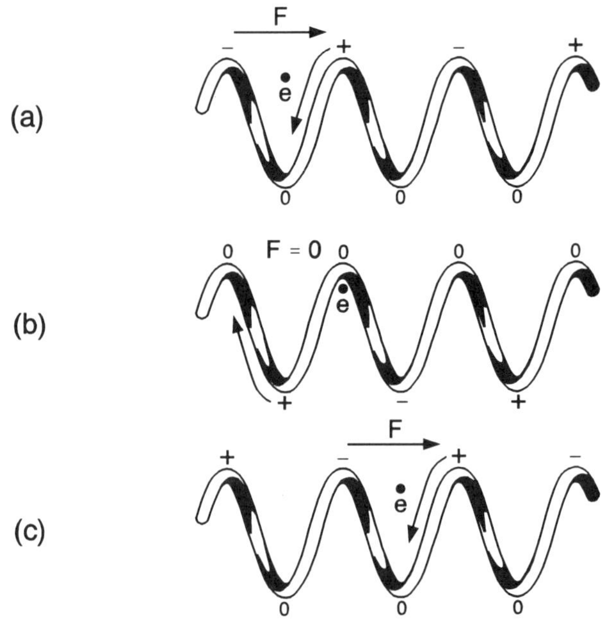

Figure 12-12 Backward wave interaction on a helix at two turns per wavelength.

Similarly, and not shown in Figure 12-12, electrons that are initially decelerated experience no forces 180° later and then are decelerated when they have moved another 180°. Thus, there is bunching even though the wave on the helix moves in the opposite direction from the electrons, and so backward wave interaction can take place.

If the backward wave interaction is sufficiently intense, oscillations occur. Whether or not a tube will oscillate in the BWO mode is dependent on a number of factors including beam-current level, beam focusing (and, in particular, how close the beam is to the helix), and the reflection coefficient at the collector end of the tube.

Because the frequency of backward wave oscillations is such that the beam has a 180° transit angle from turn to turn, the BWO frequency is tunable with beam voltage. In fact, tubes designed for operation in the BWO mode can be voltage tuned over a reasonably broad range of frequencies. Helix BWOs are discussed later in this chapter.

High power TWTs are especially prone to BWO, primarily because of the high beam currents. When these oscillations occur, there is a reduction in power at the desired frequency of operation and so the oscillations are said to produce in-band power holes. In addition, the BWO signal may induce oscillations or may intermodulate with the main signal to produce spurious signals inside the operating band of the tube.

12.1.4 BWO Suppression

One method of suppressing backward-wave oscillations is the introduction of frequency-sensitive attenuation into the helix circuit. A technique developed by Varian Associates for producing frequency-sensitive loss is illustrated in Figure 12-13 [2]. Meander-line circuits are placed on the support rods in the regions between turns of the helix. The circuits are fabricated using standard microwave integrated circuit technology. The total length of each line is equal to half the wavelength at the BWO frequency so that the lowest resonant frequency is the BWO frequency.

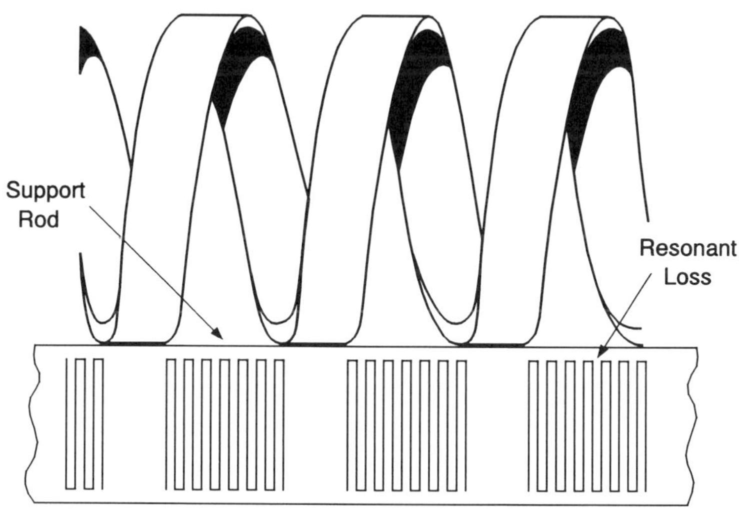

Figure 12-13 Use of resonant loss to produce frequency-sensitive attenuation. From A. Scott and M.J. Cascone, *Technical Digest*, 1978 IEDM. © 1978 IEEE.

The amount of loss that can be introduced with the resonant loss technique is very large, as shown by the example given in Figure 12-14. The loss pattern shown was used in a 10 kW tube designed for operation in the 3.0 to 6.0 GHz range. The (BWO) frequency was about 8.0 GHz. Power output as a function of frequency for this 10 kW tube is shown in Figure 12-15. Note the very sharp drop in power at the high end of the band where resonant loss becomes effective.

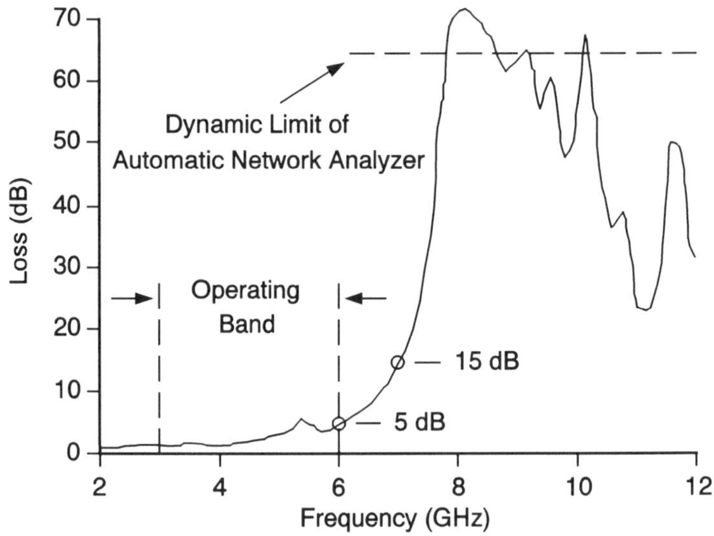

Figure 12-14 Example of insertion loss for resonant loss on three support rods. From Alan R. Jung, *Technical Digest*, 1978 IEDM. © 1978 IEEE.

In another technique for suppressing BWO, the output helix is constructed with a step change in the pitch and, thus, in the phase velocity. In addition, a step change in helix diameter is sometimes used. In Figure 12-16, estimates are shown of the peak output power that may be attainable using BWO suppression techniques. Note that peak power levels in excess of 100 kW appear to be possible in PPM-focused tubes using resonant loss.

12.2 TRANSITIONS

At the input to and output from a helix TWT, impedance transitions must be made to provide reasonable impedance matches. The impedance of the helix varies with frequency, as is shown in the example of Figure 12-17 [3]. In this example, the helix structure was designed for operation from 4.8 to 9.6 GHz. At the center of the band, the impedance is seen to be about 50 Ω. At the low end of the band, the impedance increases to over 80 Ω, and at the high end it is about 35 Ω.

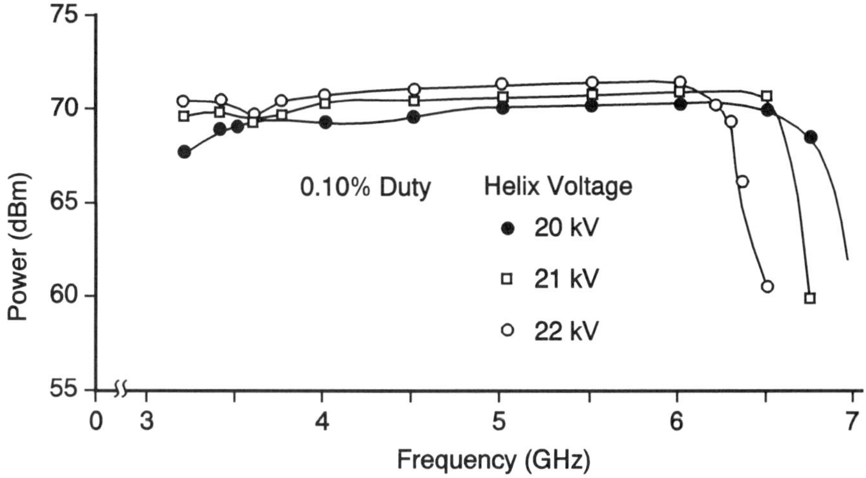

Figure 12-15 Saturated power output of 10 kW helix TWT with resonant loss. From Alan R. Jung, *Technical Digest*, 1978 IEDM. © 1978 IEEE.

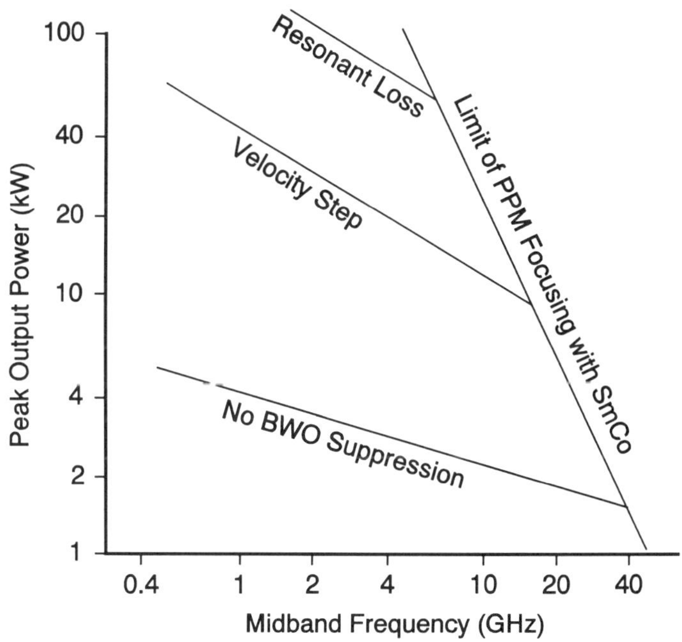

Figure 12-16 Obtainable peak power for helix TWTs with various oscillation-suppression techniques. From Alan R. Jung, *Technical Digest*, 1978 IEDM. © 1978 IEEE.

Ideally, a very broad band impedance match to the helix could be made with the transition shown in Figure 12-18. With this technique, a very gradual change in geometry is made so that the helix slowly changes into a coaxial line. As long as the overall transition is long compared with the wavelength of the signal being used, there will be an excellent impedance match.

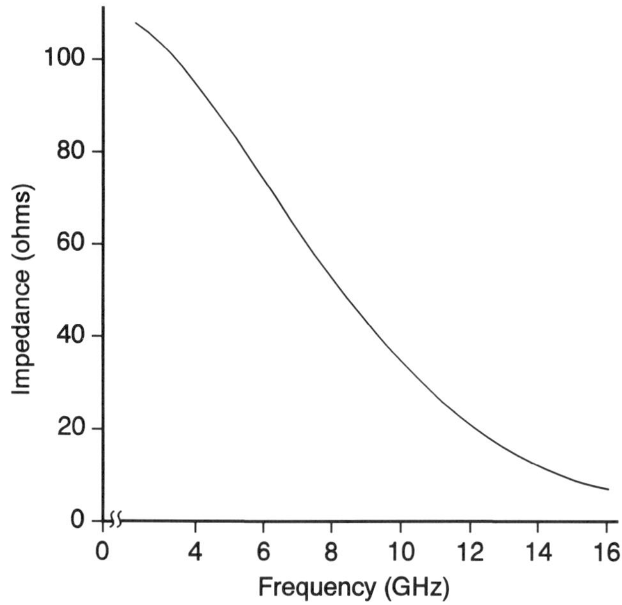

Figure 12-17 Typical helix impedance for an octave-bandwidth TWT. From A.S. Gilmour, Jr. *et al., IEEE Trans. Electron Devices*, Oct. 1979. © 1979 IEEE.

Unfortunately, transitions like that shown in Figure 12-18 have the very serious disadvantage of being long. If used, they would increase the length and weight of a TWT. In practice, transitions are normally made after the signal is removed from the helix. For example, the center conductor of the coaxial line can be tapered as shown in Figure 12-19. Another example of a taper transition was shown in the output waveguide in Figure 12-1. There, a stepped taper was used to match the tube impedance to the impedance of the waveguide.

A problem with the configuration in Figure 12-19 is that it requires a relatively long unsupported section of the center conductor of the coaxial line. Under severe shock and vibration conditions, the weld joint at the end of the helix may fail. Also, it should be noted that the last turn of the helix tends to be more poorly supported than other turns when the triangulation or pressure-insertion assembly techniques to be described in the following section are used. Thus, heat transfer from the last turn through the support rods may be relatively poor so that the last turn must be at least partially cooled by the center conductor of the coaxial line.

To provide adequate cooling and thereby help prevent failure of the weld, there should be a support for the center conductor located as close to the helix as possible. For example, in the output window of the TWT illustrated in Figure 12-1, the insulating support for the center conductor is only a fraction of a wavelength from the connection to the helix. The material of this support is usually beryllium oxide, which is a very good thermal conductor. The beryllium oxide is brazed to the center conductor and to the outer conductor and serves as part of the vacuum envelope.

Also, it should be pointed out that Teledyne Electronic Technologies (formerly Teledyne MEC) uses a flexible strap to connect the end of the helix to the center conductor. This prevents movements of the center conductor (because of thermal expansion and contraction, for example) from being transmitted to the helix.

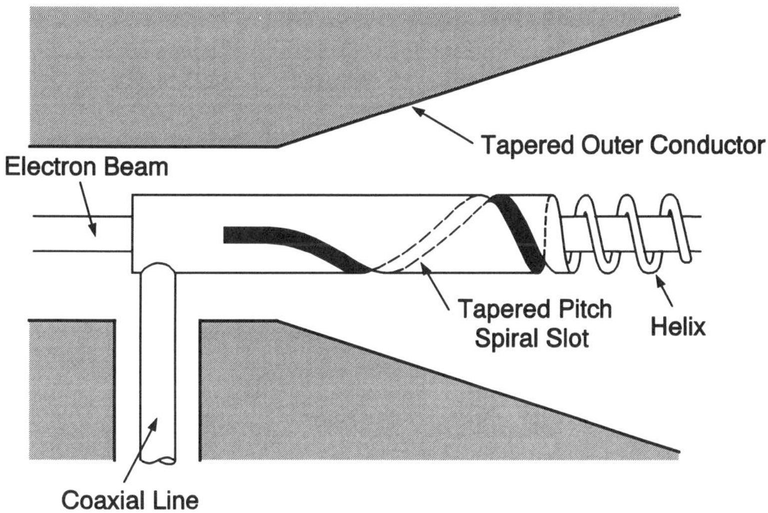

Figure 12-18 A very broadband transition from a helix to a coaxial line. From *Power Travelling Wave Tubes* by J.F. Gittins, published 1965 by American Elsevier, Inc.

12.3 HELIX SUPPORT TECHNIQUES

The helix of a TWT is a relatively delicate structure. As an example, a cross-sectional view is shown in Figure 12-20 for the helix of a 10 kW TWT designed to operate over the frequency range of 3.0 to 6.0 GHz. The helix is made of 0.010-in by 0.075-in tungsten tape. If the TWT is to be rugged and capable of operating at high average-power levels, this helix must be very well supported.

The use of metallic supports is out of the question because of their effect on the RF characteristics of the circuit. A dielectric cylinder completely surrounding the helix

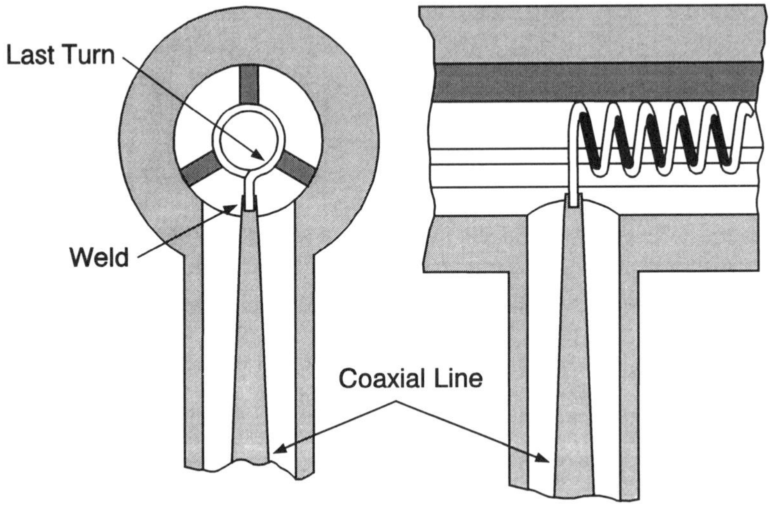

Figure 12-19 Transition using tapered coaxial line.

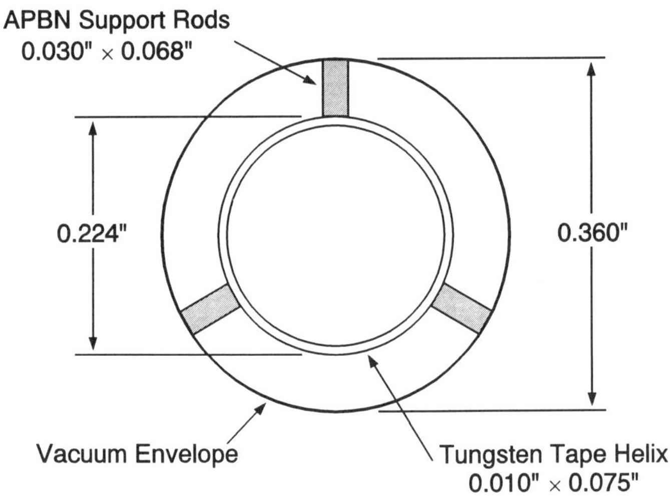

Figure 12-20 Cross section of helix of a 10 kW, 3.0 to 6.0 GHz TWT. From Alan R. Jung, *Technical Digest*, 1978 IEDM. © 1978 IEEE.

could conceivably provide excellent support and cooling. Unfortunately, the dielectric constants of suitable ceramic materials are high. As a result, their use in a cylinder would severely reduce the circuit interaction impedance of the helix. In fact, to minimize the loading effect of a ceramic support structure, it is necessary to minimize the amount of ceramic material used. As a result, thin, rectangular support rods like those shown in Figure 12-20 are typically used. Even these load the helix to some extent, as is indicated by the example given in Figure 12-21.

Only two dielectric materials are widely used for supporting the helices of TWTs: beryllium oxide (BeO) and APBN. These two materials have relatively high thermal conductivity (compared with other ceramics suitable for operation at high temperature in vacuum).

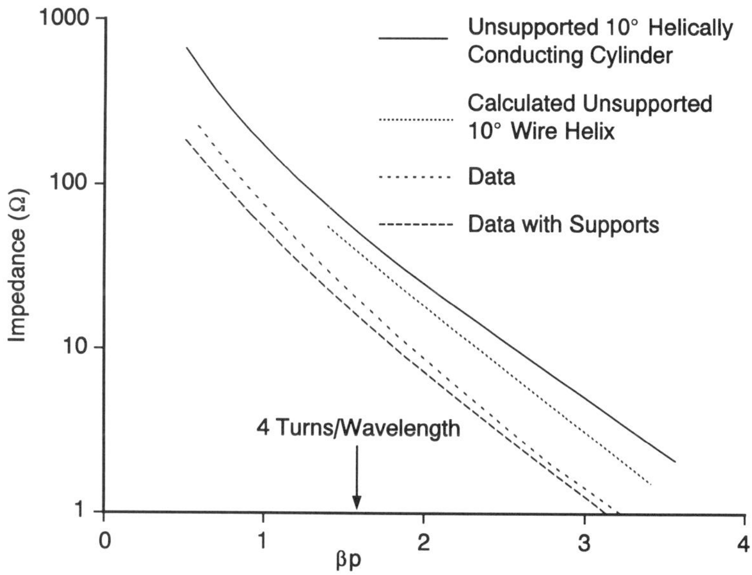

Figure 12-21 Coupled-interaction impedance of a helix with and without supports. From *Power Travelling Wave Tubes* by J.F. Gittins, published 1965 by American Elsevier, Inc.

With the use of high thermal conductivity ceramics, the temperature drop across the support rods is not a severe problem in high-power helix TWTs. Instead, the major problems are the thermal resistances of the interfaces between the helix and the support rods and between the support rods and the barrel. In the example shown in Figure 12-22, the thermal drops across these interfaces account for 70% of the total temperature difference between the hottest point on the helix and the barrel.

As a result of the large thermal drops at the support rod interfaces, the helix and support rod structures in high-power helix TWTs must be assembled using techniques that minimize the thermal interface resistances. Basically, three techniques are used:

- Triangulation,
- Pressure or hot insertion,
- Brazing.

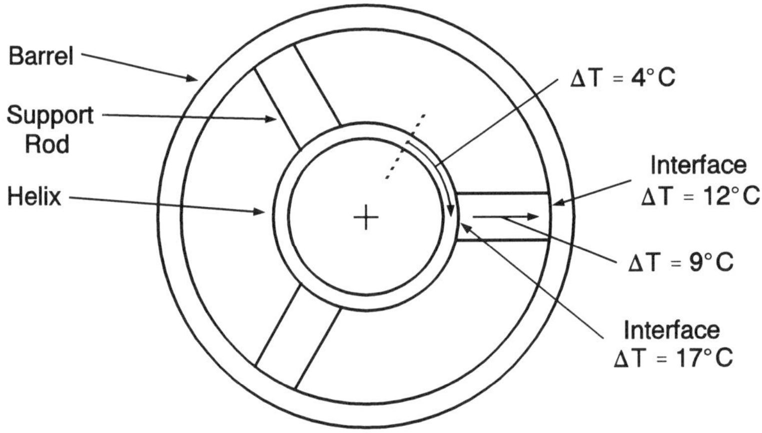

Figure 12-22 Temperature drops calculated for a tungsten helix and APBN rods dissipating 1 watt. From A. Scott and M.J. Cascone, *Technical Digest*, 1978 IEDM. © 1978 IEEE.

With the triangulation technique, the metal barrel that is used to surround the helix is slightly distorted into a triangular shape. The helix with three equally spaced support rods is then inserted into the barrel and the distorting forces on the barrel are removed. The higher the resulting forces are at the interfaces between the support rods and the helix or barrel, the lower are the thermal resistances. A limit is reached, however, when the forces are high enough to distort the helix excessively. For helix strength, the material normally used is tungsten or molybdenum.

It should be pointed out that with the triangulation technique the barrel is not a solid brazed structure (like that shown in Figure 12-11). Instead, it is a relatively thin metallic tube over which the magnetic pole pieces are slid after the helix has been inserted. With this technique, there is a relatively poor thermal interface between the barrel and the pole pieces. This problem is alleviated by making the pole pieces fit as tightly as possible on the barrel. Then, as the helix and barrel expand because of heating, the contact between the barrel and pole pieces improves and so heat transfer improves.

With the insertion technique, the helix and rod assembly is inserted into a barrel structure like that shown in Figure 12-11. In some cases, the barrel is slightly expanded by heating before the helix and rod assembly is inserted. This is normally referred to as *hot insertion*. In other cases, the helix and rod assembly are machined with tolerances for an interference fit and then inserted under high pressure into the cold barrel. The temperatures given in Figure 12-22 are for a pressure-inserted structure. As with the triangulation technique, the helix materials used are tungsten or molybdenum.

The most effective way of reducing the thermal interface resistances is to braze the interfaces. While brazing sounds attractive, it is difficult to implement. If rods that have been metallized on the surfaces to be brazed are used, there are difficulties with thermal expansion problems. Also, the support rods must be brazed to every turn of the helix. There may be several tens of contact points of one support rod with a helix and several support rods (usually three). Every contact point of every rod must be brazed for proper protection of the helix. After brazing, the metallization of the support rods must be removed in the regions between turns of the helix.

With the brazing technique, a wider choice of helix materials is possible than with the triangulation or pressure-insertion techniques. In addition to tungsten or molybdenum, copper can be used. While tungsten and molybdenum can be copper plated to reduce RF losses, only the use of pure copper throughout the helix leads to the maximum possible transfer of heat and the ultimate loss reduction,

Finally, it should be pointed out that in brazed structures the helix is mechanically bonded to the dielectric support material. There is reduced probability, therefore, that helix failure will result from overheating of the helix and the resulting possible loss of thermal contact between the helix and the dielectric.

An alternative to the use of brazed rods is the block-supported structure illustrated in Figure 12-23 [4]. Individual ceramic blocks are brazed to the helix and, usually, to the barrel. The use of ceramic blocks eliminates the thermal expansion problems associated with rods.

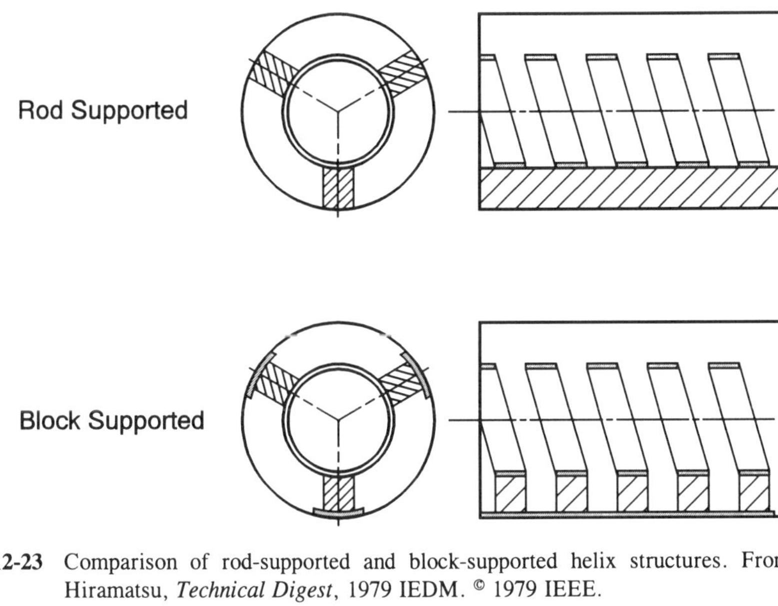

Figure 12-23 Comparison of rod-supported and block-supported helix structures. From Yukio Hiramatsu, *Technical Digest*, 1979 IEDM. © 1979 IEEE.

Additional advantages of the block-supported structure are reduced dielectric loading (and higher interaction impedance) and a reduced effect of the dielectric material on phase velocity. As an example of the capability made possible by the block-supporting technique, Thomson-Tubes Electroniques of France produces a 1.5 kW average power, 6 GHz TWT for communications service.

Comparisons of the advantages of various techniques are often misleading because test conditions differ. Still, because of the continuing emphasis on higher and higher powers at higher and higher frequencies, it is important to attempt to compare the triangulation, pressure-insertion and brazing techniques. This has been done, and the results are summarized in Figure 12-24 [5]. The base-plate temperature for the test fixture in these tests was 100°C. Thus, the temperature rise of the brazed helix (which was copper) was about two thirds of that for the insertion (shrunk) technique and about one half of that for the triangulation technique.

It is important to note that the results in Figure 12-24 were obtained by passing a dc current through the helices. RF losses and at least a small amount of additional heating are therefore, not included. The additional heating may be larger in the metallized, brazed helix than in the other two.

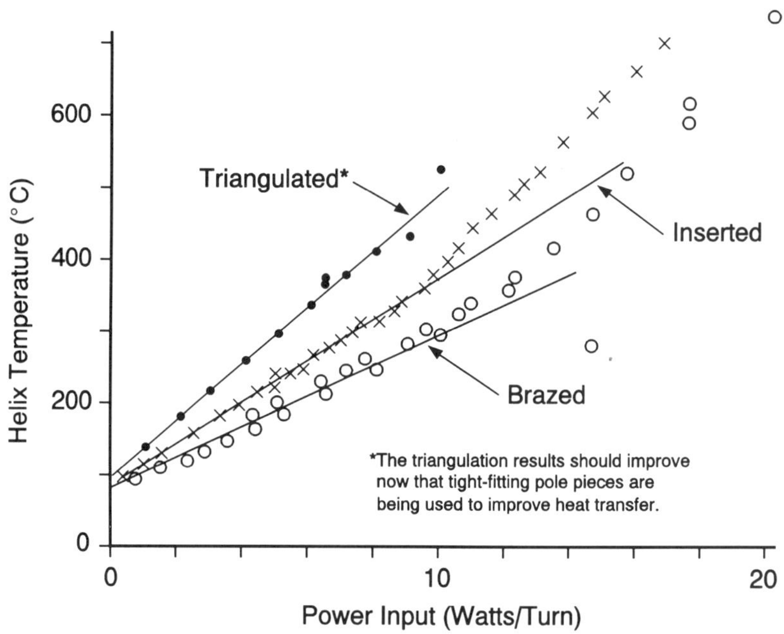

Figure 12-24 Comparison of thermal characteristics of triangulated, pressure-inserted (shrunk) and brazed helices. From T. Mosser *et al*, 1980 Microwave Power Tube Conference.

In general, TWT dimensions are scaled with wavelength. At frequencies in the millimeter-wave range, dimensions become extremely small and miniaturized helices and support rods are used. For example, rectangular support rods having cross-sectional dimensions of .010 in by .020 in, or smaller, are common. As helix assemblies become smaller, heat dissipation problems become more severe, and so the thermal management techniques described in the previous sections become even more important than they are at lower frequencies.

The brazing technique, which offers excellent thermal capabilities, is very difficult to implement in the millimeter-wave range. As a result, at present, pressure insertion is widely used and average power levels are limited to a few tens of watts at best.

The use of diamond may make possible significant improvements in average power capability. Type II-A diamond has a thermal conductivity several times that of copper. With diamond supports, it is predicted that average power levels as high as 200 W may be achievable at frequencies as high as 50 GHz [6].

12.4 ATTENUATORS AND SEVERS

In the analysis of traveling wave amplification, four waves are found to exist. One of these travels in the backward direction from the collector end of the tube to the gun end. This wave can carry power reflected from the load or output region backward through a TWT to the input. If there is a mismatch at the input, a portion of the signal will be reflected to provide a feedback signal. The results can be oscillations or severe variations in gain as a function of frequency.

If, as is indicated in Figure 12-25, the reflection coefficients at the output and input terminals are, respectively, ρ_o and ρ_i, then oscillations may occur when

$$G - L - \rho_o - \rho_i > 0 \qquad (12\text{-}3)$$

where G is the gain of the tube in dB, L is the loss of the circuit in dB and the reflection coefficients are expressed in dB.

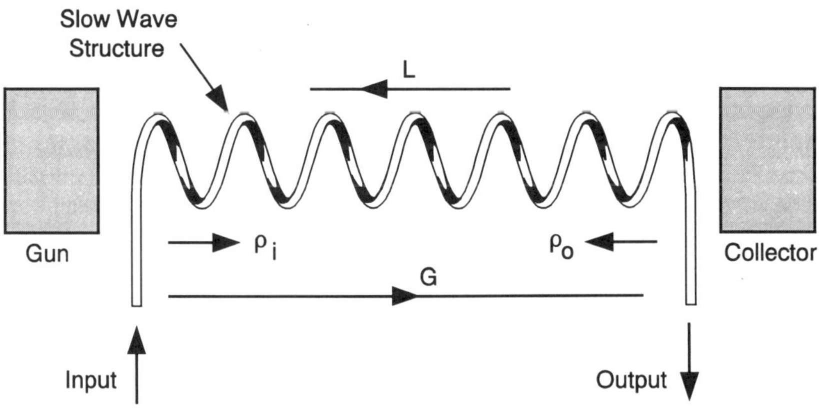

Figure 12-25 Quantities used to determine the oscillation condition.

Although the tube may be very well matched to the signal source and to the load, there will normally be reflections from the input and output terminals because of the difficulty in making impedance transitions from the RF structure. As a result, it is difficult to make ρ_o and ρ_i greater than 10 dB (10% reflections). Usually, the cold loss of the circuit cannot be made much more than 6 dB so that the limiting gain is on the order of 26 dB.

To prevent large variations in gain with frequency, the actual gain cannot be as high as 26 dB and, in fact, is limited to about 20 dB. To permit increased gain, attenuators and severs are used in TWTs.

A common form of attenuation is shown in Figure 12-26. Portions of the dielectric support rods in a helix TWT are coated with a lossy material. The thickness of the film is tapered to zero at the ends of the attenuation region to obtain a good impedance match over a wide range of frequencies.

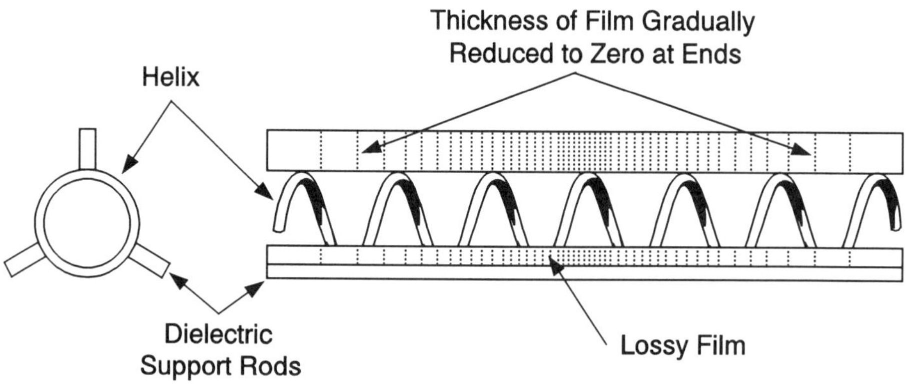

Figure 12-26 Lossy film attenuator for use with a helix.

The forward waves as well as the backward wave are attenuated by the lossy film. The loss in gain in the forward direction may be at least partially compensated for by an increase in length of the interaction region.

Unfortunately, if a good impedance match is to be realized, the attenuator must extend for a considerable length along the RF structure. On the other hand, the amplitude of the circuit wave is small in the region of significant attenuation. As a result, the electric fields of the circuit lose control of the bunching process and the velocity spread in the beam increases. The increase in beam velocity spread in the attenuation region results in less effective bunching after the attenuation and a reduction in efficiency. As a result, at high power and when efficiency is important, the lossy film technique for producing attenuation is not attractive.

In high-power tubes where efficiency is important, the most common technique for suppressing the backward wave is through the use of one or more severs. An illustration of a TWT with a severed helix is given in Figure 12-27. The sever prevents reflections from the load and the output terminal from reaching the input terminal.

Although the forward growing wave on the input section of the helix is lost at the sever, current and velocity modulation remain on the electron beam to carry the signal across the sever region for further amplification in the output section.

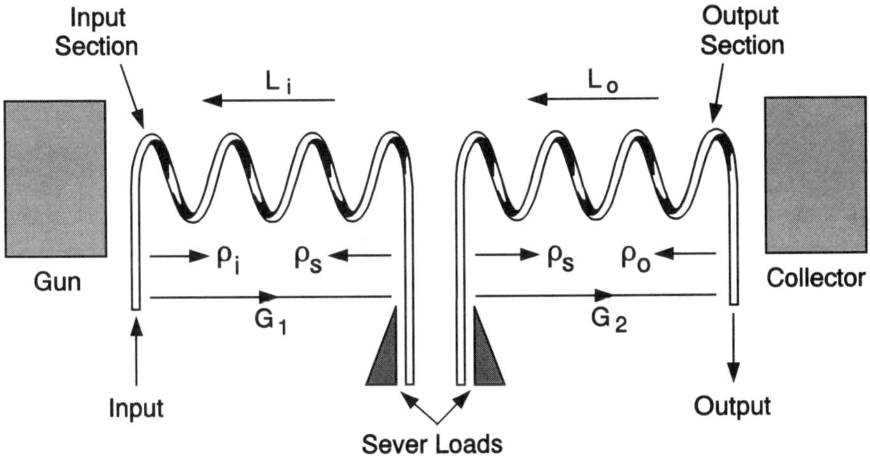

Figure 12-27 TWT containing a two-section severed helix. From *Power Travelling Wave Tubes* by J.F. Gittins, published 1965 by American Elsevier, Inc.

As the beam drifts across the sever region, bunching degrades and affects the final efficiency of the tube. As a result, the sever region should be as short as possible. Also, the amount of degradation in bunching in the sever region depends on the intensity of bunching that occurs in the input section of the slow wave structure. Thus, for maximum efficiency, the gain of the input section should be relatively low and the gain of the output section should be as high as possible.

Just as there is a limit of about 20 dB to the practical gain of a tube without a sever, there is a similar limit to the gain of each section of a tube with severs. Referring again to Figure 12 27, the reflection cocfficicnts at thc output, ρ_o, at the sever, ρ_s, and at the input, ρ_i, along with the circuit attenuations L_o and L_i, must be taken into consideration in determining practical gain levels. Because it is usually possible to keep the reflection coefficients at the sever terminations less than those for the input and output, the gain per section of a severed tube can be higher than that for a single-section tube. On the other hand, there is some signal loss in the sever region so that an overall gain from a two-section tube of about 40 dB is practical. With additional severs, higher gains are possible.

12.5 CIRCUIT EFFICIENCY [3]

Circuit efficiency is the efficiency with which the helix delivers the RF power generated in the TWT to the output port. As is shown in Figure 12-28, circuit efficiency can have a significant effect on the overall efficiency of a TWT. The collector efficiencies shown in Figure 12-28 are typical for multistage depressed collectors, which will be discussed in Chapter 14. Collector efficiency is the efficiency with which the collector recovers power from the electron beam leaving the interaction region of the TWT. For this figure it is assumed that the sum of the losses resulting from beam interception and from RF losses (power lost to the generation of unwanted signals such as harmonics) is 5% of the beam power.

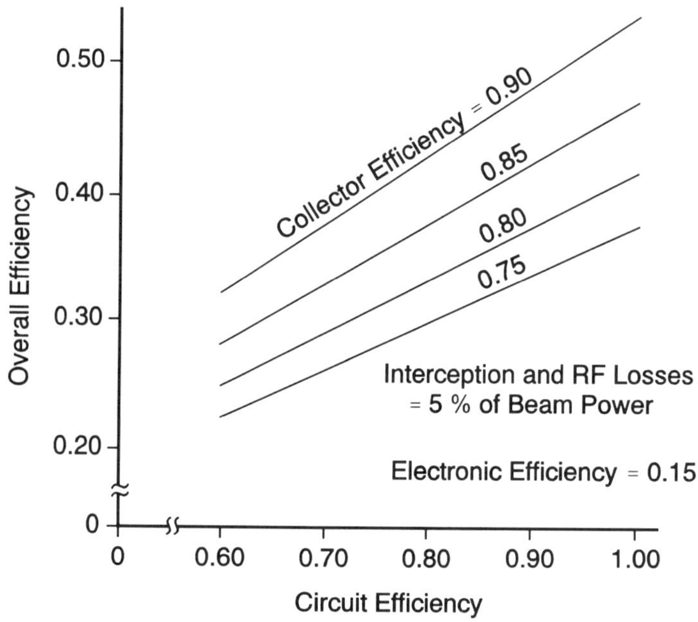

Figure 12-28 Effect of circuit efficiency on overall efficiency. Adapted from A.S. Gilmour, Jr. *et al., IEEE Trans. Electron Devices,* October 1979. © 1979 IEEE.

As an example of the importance of circuit efficiency, assume a collector efficiency of 85%. Assume further that a lossy helix such as rough tungsten (circuit efficiency ~ 70%) is replaced by a properly finished copper-plated helix (circuit efficiency ~ 90%). The increase in overall efficiency is 10%.

Circuit efficiency is dependent primarily on the effective resistance of the circuit material. At high frequencies, dielectric losses are also important. To illustrate the effects of circuit resistivity, calculations were performed of losses in the helix structure for a TWT. The model for the analysis is shown in Figure 12-29. A helically conducting sheath helix was assumed. Allowance was made for dielectric support rods, and the resistivity of the barrel as well as that of the helix were taken into consideration. Nominal values for the various helix parameters were as follows:

helix pitch	= 0.141 (18 T.P.I.)
helix radius	= 0.12 cm (0.047 in)
barrel-to-helix diameter ratio	= 2.0
helix resistivity	$= 2.5 \times 10^{-6}\ \Omega$ cm
barrel resistivity	$= 2.5 \times 10^{-6}\ \Omega$ cm
effective relative dielectric constant of support rods (ϵ_2)	= 1.5

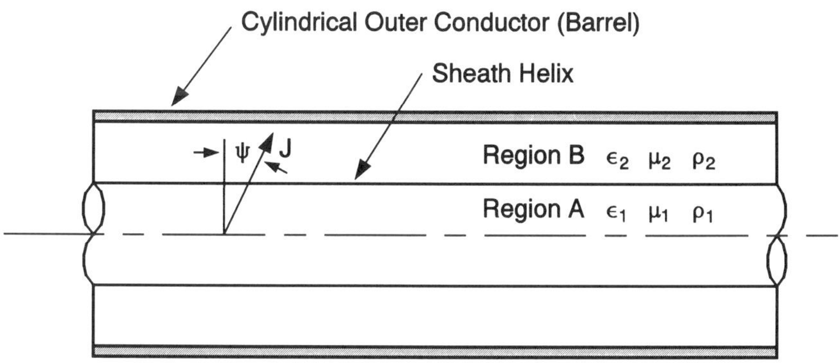

Figure 12-29 Helically conducting helix model used in analysis. Adapted from A.S. Gilmour, Jr. *et al., IEEE Trans. Electron Devices*, October 1979. © 1979 IEEE.

The effect of variations in helix resistivity with the other parameters is shown in Figure 12-30. Note that loss increases roughly as the square root of helix resistivity, as is expected for skin-effect resistance. Comparisons of measured and theoretical loss for a copper-plated helix are given in Figure 12-31.

The dashed curves of Figure 12-31 are calculated loss curves based on the bulk resistivity of copper ($1.724 \times 10^{-6}\ \Omega$-cm) and based on the best fit to experimental data at the frequencies where reflections in the measurement circuitry were small and a resistivity of $4 \times 10^{-6}\ \Omega$-cm was assumed. The circled data points are measured losses and the solid curve is the calculated loss for a resistivity of $4 \times 10^{-6}\ \Omega$-cm, taking into consideration reflections in the measurement circuitry.

The excellent fit of the theoretical curves to data when a resistivity of $4 \times 10^{-6}\ \Omega$-cm is assumed leads to the conclusion that $4 \times 10^{-6}\ \Omega$-cm was the effective value of the resistivity of the copper surface of the helix. This value is more than twice the bulk resistivity of copper.

The reason for the high effective resistivity is indicated in Figure 12-32, which

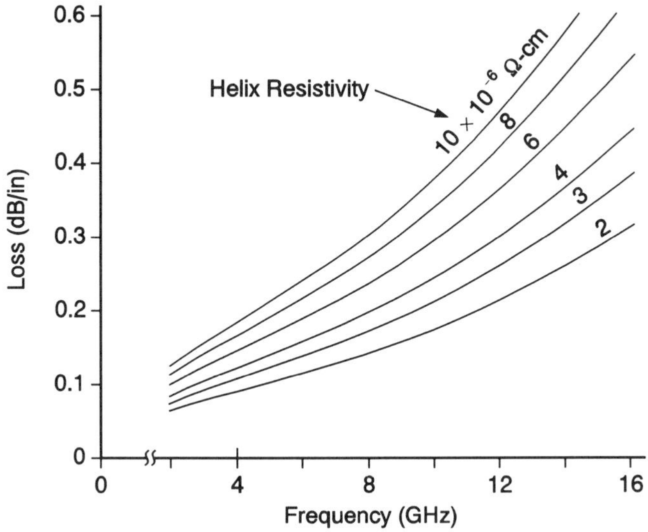

Figure 12-30 Loss of functions of frequency and helix resistivity. From A.S. Gilmour, Jr. *et al.*, *IEEE Trans. Electron Devices*, October 1979. © 1979 IEEE.

contains photomicrographs of the surface of the copper-plated helix [7]. The skin depth at 10 GHz was small compared with the sizes of the irregularities on the orange-peel textured copper surface. As a result, the path length for the RF current and the effective resistance were much larger than for a smooth surface. To minimize resistance and loss, it is essential to properly polish the surface of the helix.

The analysis showed that loss increased as helix pitch and radius increased. Furthermore, the amount of the increase scaled with the increase in current path length as expected. A similar effect caused by the relative effective dielectric constant of the helix support rods is illustrated in Figure 12-33. As the dielectric constant is increased, signal velocity decreases and loss increases (by the square root of dielectric constant).

Finally, the analysis showed that the resistivity of the metallic barrel surrounding the helix had essentially no effect on loss.

12.6 DUAL-MODE OPERATION

The purpose of a dual-mode TWT is to provide two different RF power levels (separated by as much as 10 dB) efficiently in a single device. This is highly desirable in airborne electronic countermeasures systems, for example, where volume and weight are of extreme importance. In these systems, the use of two tubes, each of which is designed for operation at one of the power levels, often results in excessive system weight and volume. There are basically three ways in which a single tube can be efficiently operated at two power levels:

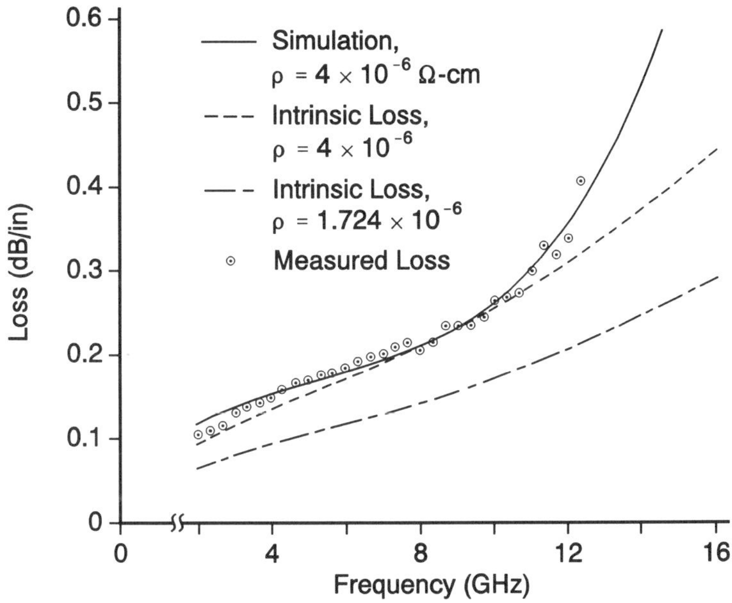

Figure 12-31 Measured and theoretical loss *versus* frequency for a copper-plated helix. From A.S. Gilmour, Jr. *et al., IEEE Trans. Electron Devices*, October 1979. © 1979 IEEE.

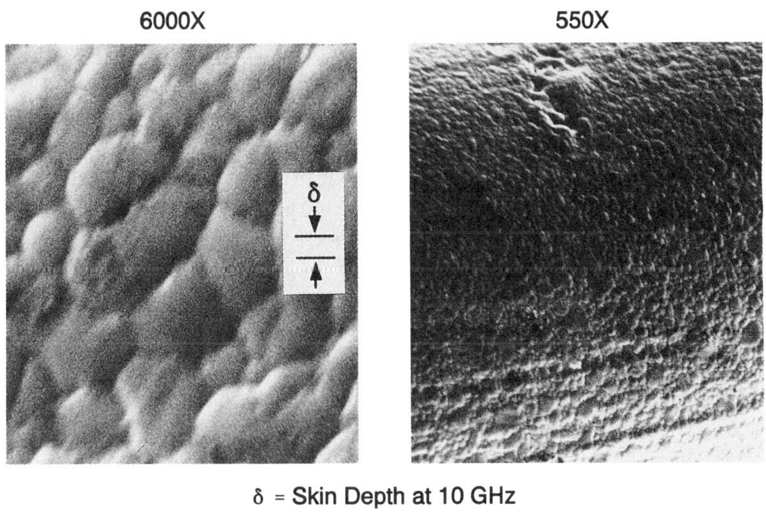

Figure 12-32 Photomicrographs of a copper-plated helix. From A.S. Gilmour, Jr. *et al., IEEE Trans. Electron Devices*, October 1979. © 1979 IEEE.

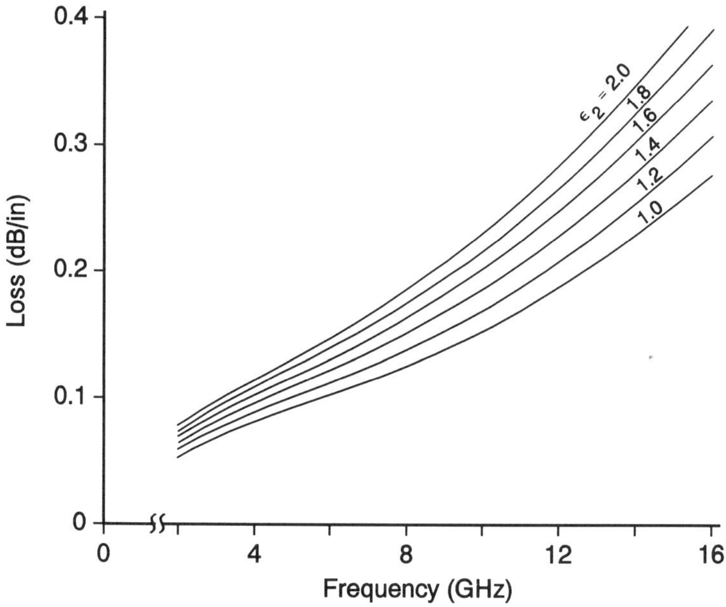

Figure 12-33 Effect on loss of effective value of ϵ of support rods. From A.S. Gilmour, Jr. *et al.*, *IEEE Trans. Electron Devices*, October 1979. © 1979 IEEE.

1. Beam current and voltage are held constant and RF drive level is changed. Efficiency is obtained by using a multistage depressed collector to recover spent-beam power. This technique is discussed further in Chapter 14.
2. A complex tube, basically consisting of two tubes in one envelope, is used. A drawing of a device of this type is shown in Figure 12-34. The tube consists of a conventional TWT in series with a very short, low gain (10 dB), nearly lossless device. In the low power mode, the beam in the 10 dB device (sometimes referred to as a *transparent tube*) is turned off and the signal from the conventional TWT passes through it with no amplification and little loss. In the high power mode, the transparent tube is turned on to provide an additional 10 dB of gain. This device has the advantage of providing a full 10 dB of pulse-up capability and the disadvantage of being more complex and heavier than the true single-tube device.
3. Beam current is varied with a gridded electron gun. (Beam voltage is not changed because of the weight and complexity of a modulator and because the beam velocity must remain near synchronism with the circuit phase velocity).

The third technique (changing beam current) for achieving dual-mode operation is very difficult to implement when the power separation between modes must be more than a few dB. One difficulty encountered is that of focusing both the low power and

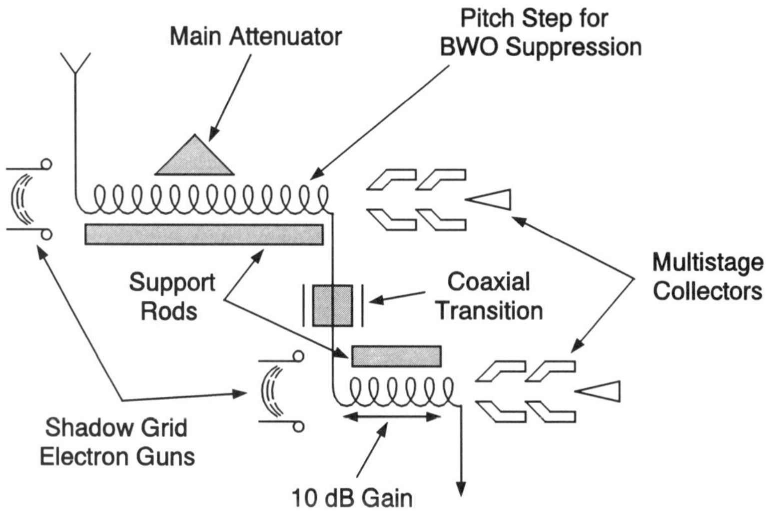

Figure 12-34 Dual-beam, dual-mode TWT. Courtesy of Varian Associates.

the high power beams with the same magnet structure (usually PPM because of the requirement for light weight).

Assuming constant efficiency, beam power and beam current must vary in direct proportion to RF power. If beam diameter is held constant, then beam charge density varies directly with beam current. Finally, the radial electric field in the beam is proportional to the charge density and, therefore, to the power level. If the PPM structure is designed for the low power mode, the magnetic field will be inadequate to offset the electric field in the high power mode. Of course, the opposite situation exists if the PPM structure is designed for the high-power mode. This problem may be alleviated by changing the beam diameter when the beam current is changed. An electron gun for changing beam diameter with beam current is shown in Chapter 6.

The other difficulty encountered when the beam current is changed is that the beam impedance changes and this changes Pierce's gain parameter, C. As a result, the RF characteristics of the tube change with beam current, making it extremely difficult to obtain the desired performance at two power levels that differ by a large amount.

12.7 HELIX BWOS

As was indicated earlier in this chapter, backward wave oscillations serve a useful purpose in voltage-tuned signal generators. Helix-type backward wave oscillators (BWOs) are rapidly tunable (in less than 1 μs) over a bandwidth up to an octave. Devices operating at frequencies up to 50 GHz are available.

Shown in Figure 12-35 are the essential elements of a BWO. Most significant, perhaps, is the electron-gun assembly with a ring-shaped cathode for generating a hollow electron beam. The use of a hollow electron beam traveling close to the helix structure enhances the backward wave interaction.

Power output is typically tens of milliwatts, as indicated by the power curve for a Varian 33-to-50 GHz BWO shown in Figure 12-36. The corresponding voltage tuning characteristic and helix modulation coefficient are given in Figure 12-37.

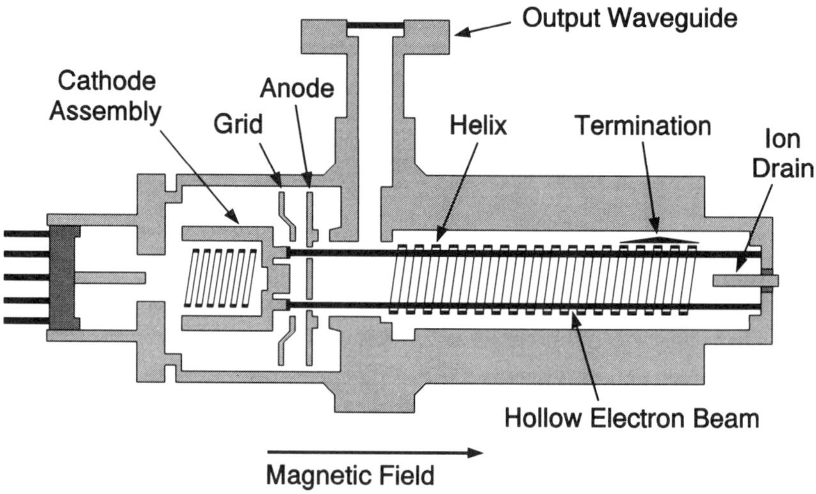

Figure 12-35 Helix-type backward wave oscillator. Courtesy of Varian Associates.

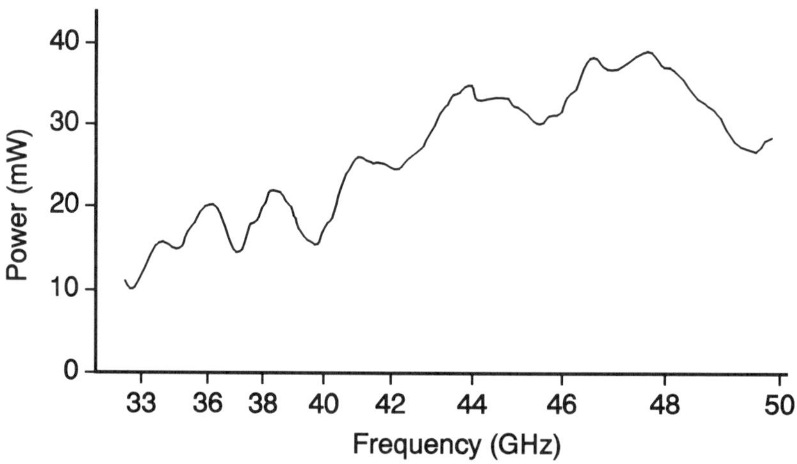

Figure 12-36 Power output for Varian helix BWO. Courtesy of Varian Associates.

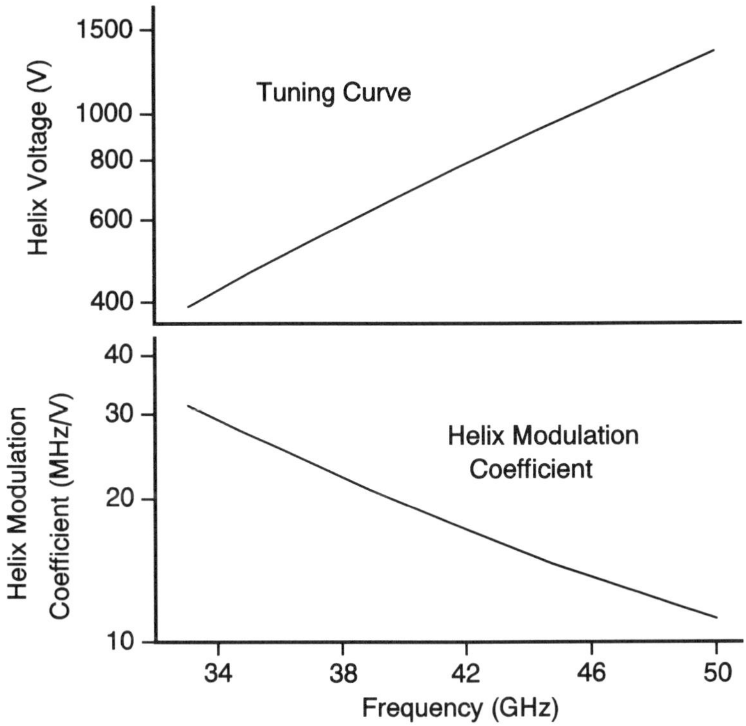

Figure 12-37 BWO voltage tuning characteristic and helix modulation coefficient. Courtesy of Varian Associates.

12.8 RING-BAR TWTS

The resonant loss and velocity step techniques described in the previous section, for suppression of backward wave oscillations, are relatively recent developments. Before these techniques were available, the ring-bar circuit illustrated in Figure 12-38(a) was developed to make possible high power operation of a helix-type circuit. Basically, the ring-bar circuit is derived from the contrawound helix circuit shown in Figure 12-38(b) and is far easier to fabricate.

To understand the operation of these circuits and how they suppress backward wave interactions, it is helpful to examine the bifilar helix shown in Figure 12-39 and to compare qualitatively its characteristics with those of a single helix. The pitch of the helices making up the bifilar helix is the same as that of the single helix. For the moment, interactions between the two helices of the bifilar and single helices are assumed to be the same. As a result, for a given frequency of operation, the voltage distributions indicated in Figure 12-39 for both the single helix and the bifilar helix are

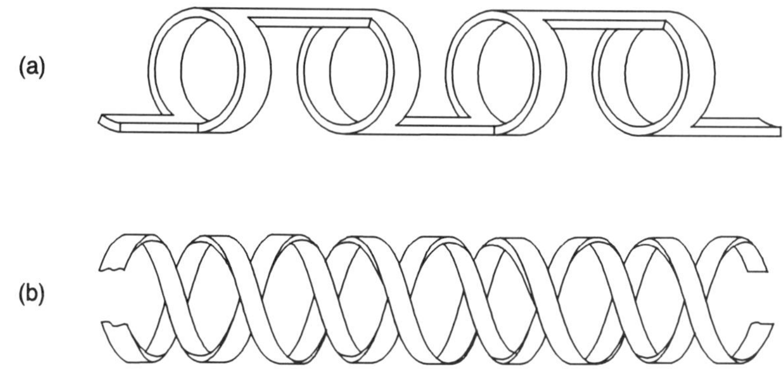

Figure 12-38 Ring-bar (a) and two-tape contrawound helix (b) circuits. From Hughes Aircraft Co., *TWT and TWTA Handbook*.

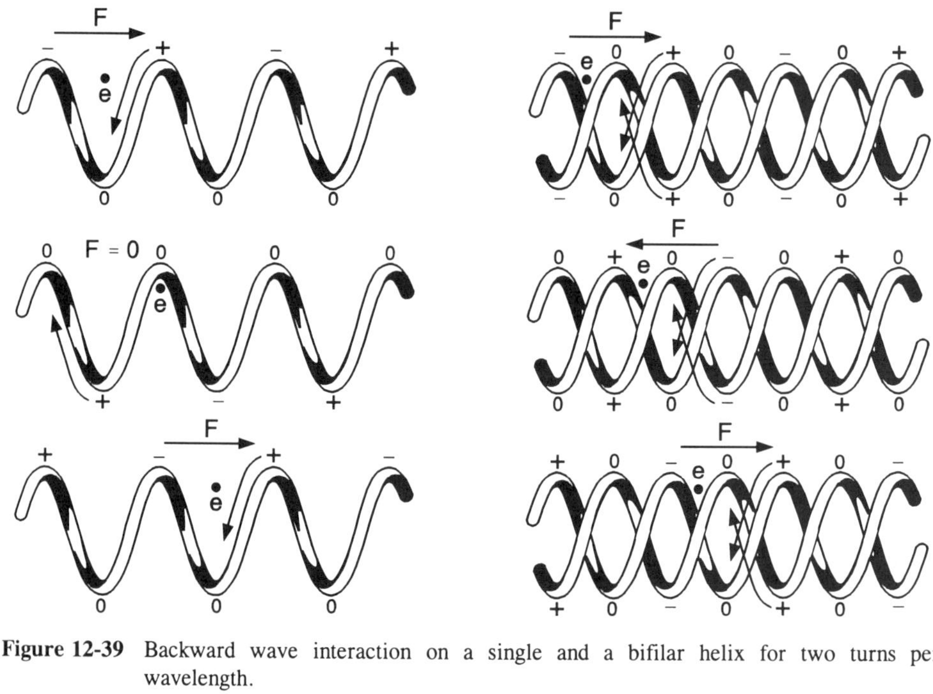

Figure 12-39 Backward wave interaction on a single and a bifilar helix for two turns per wavelength.

the same. For the BWO discussion, these have been chosen so as to be those associated with two turns per wavelength.

The interaction of electrons with a backward wave on a single helix has been discussed earlier in this chapter. To review, electrons that are initially accelerated (top left, Figure 12-39) experience essentially no forces 180° later (middle left) and then are accelerated again 360° later (bottom left). Similarly, and not shown in Figure 12-39, electrons that are initially decelerated experience no forces 180° later and then are decelerated 360° later. Thus, there is bunching and gain, even though the wave on the helix moves in the opposite direction from the electrons.

For the bifilar helix shown in Figure 12-39, it is assumed that the electrons and waves have the same velocities as for the single helix. In this case, however, electrons that are initially accelerated (top right) are decelerated 180° later (center right) and then accelerated again 360° later (bottom right). Because, on each cycle, the accelerating and decelerating forces tend to cancel each other, there is no net gain that can lead to backward wave oscillations. As a result, higher operating voltages, currents and RF power levels are possible with the bifilar helix than with the single helix (with no suppression techniques being used).

The same arguments apply to the contrawound helix and to the ring-bar circuit, so these circuits can operate without BWO at higher power levels than can the single helix with no BWO suppression. The ring-bar structure has been analyzed by Birdsall and Everhart [8]. Some of the parameters used in their analysis are shown in Figure 12-40. The interactions between the equivalent righthand-wound and lefthand-wound helices result in considerable dispersion, as is indicated by the ω-β diagram for a ring-bar circuit shown in Figure 12-41. Thus, while the operating power level is high for a tube using a ring-bar circuit, the bandwidth is generally limited to about 10% to 20%. Figure 12-42 contains a typical plot indicating the power and bandwidth capabilities of the ring-bar circuit.

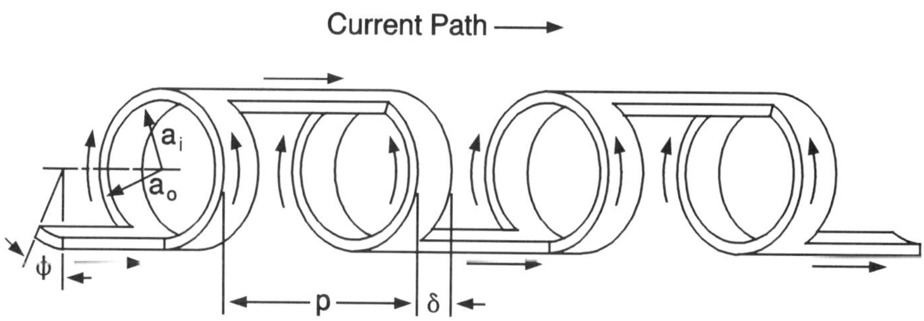

Figure 12-40 Quantities used in analysis of ring-bar structures. From *Power Travelling Wave Tubes* by J.F. Gittins, published 1965 by American Elsevier, Inc.

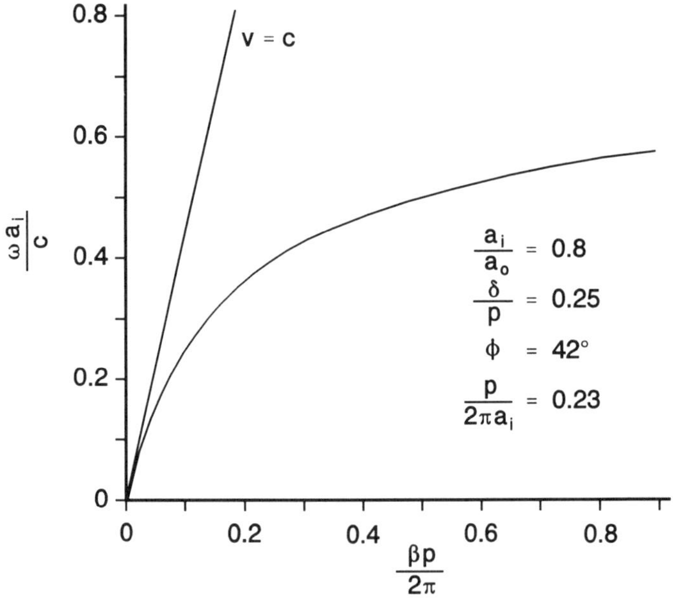

Figure 12-41 The ω-β diagram for a ring-bar structure. From *Power Travelling Wave Tubes* by J.F. Gittins, published 1965 by American Elsevier, Inc.

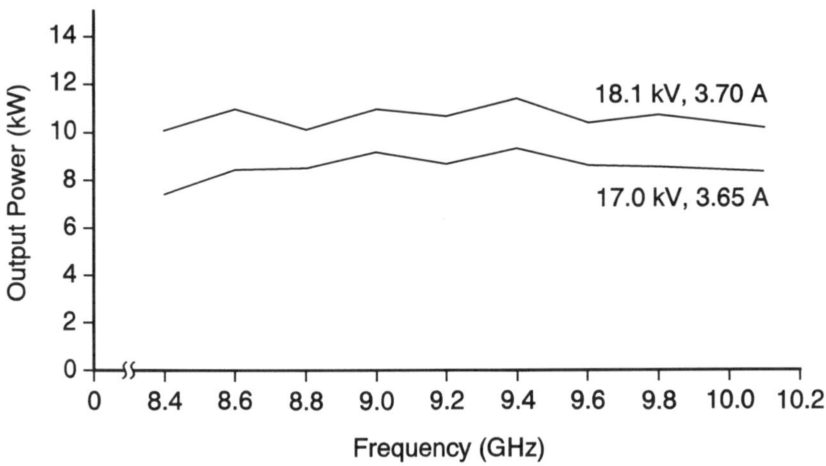

Figure 12-42 Output power characteristic for a ring-bar TWT. From Hughes Aircraft Co., *TWT and TWTA Handbook*.

REFERENCES

[1] J.L. Putz and M.J. Cascone, "Effective use of dispersion shaping in broadband helix TWT Circuits," *Technical Digest*, 1978 International Electron Devices Meeting, pp. 422–424.

[2] Alan R. Jung, "10 kW and up, from a helix TWT?" *Technical Digest*, 1978 International Electron Devices Meeting, pp. 530–553.

[3] A.S. Gilmour, Jr., M.R. Gillette and J.T. Chen, "Theoretical and experimental TWT helix loss determination," *IEEE Trans. Electron Devices*, vol. ED-26, Oct. 1979, pp. 1581–1588.

[4] Y. Hiramatsu, "Study of block-supported helix circuits for high-power TWTs," *Technical Digest*, 1979 International Electron Devices Meeting, pp. 416–417.

[5] T. Mosser, W. Pinger and D. Zavadil, "High-power brazed helix fabrication," 1980 Microwave Power Tube Conference.

[6] O. Sauseng, A.E. Manoly and A. Hall, "Thermal properties and power capability of helix structures for millimeter waves," *Technical Digest*, 1978 International Electron Devices Meeting, pp. 534–537.

[7] A.S. Gilmour, Jr., "SEM analysis of TWT helix samples," *IEEE Trans. Electron Devices*, vol. ED-24, June 1977, pp. 774–775.

[8] C.K. Birdsall and T.E. Everhart, *IRE Trans. Electron Devices*, vol. ED-3, 1956, p. 190.

Chapter 13

Coupled Cavity TWTs

In coupled cavity TWTs [1], several tens of klystronlike cavities are used for the slow wave structure. Except for sever regions, these cavities are electromagnetically coupled so that an RF signal can propagate along the structure. Coupling techniques take on widely differing forms and are a primary subject for discussion in this chapter. If an electron beam is passed through the structure and if it has a velocity near the phase velocity, then amplification can occur.

With appropriate choices of cavity dimensions and cavity coupling techniques, electron beam velocity can be high, and so operation at high peak-power levels is possible. In addition, coupled cavity structures are mostly metallic (copper and iron) and have geometries such that thermal resistances are very low. As a result, high average powers are possible.

Shown in Figure 13-1 are the major components of a coupled cavity TWT that is PPM focused. The operation of the electron gun was described in Chapter 6, and multistage collectors are discussed in Chapter 14. A qualitative explanation of the interaction of the electron beam with the coupled cavity circuit is given in the following section.

In the TWT shown in Figure 13-1, each cavity is coupled to the following cavity and to the preceding cavity with a kidney-shaped slot. The two slots in each cavity are 180° apart in rotational position and so this circuit can be viewed as a folded waveguide. In fact, many of the properties of this circuit are most easily understood by treating it as a folded waveguide. As a result, in the following discussions of ω-β characteristics, loaded waveguides and, in particular, folded waveguides are considered first.

While the waveguide approach is useful for a qualitative understanding of dispersion characteristics, a circuit model approach is most useful for a detailed examination of ω-β characteristics. The equivalent circuits described by Curnow [2, 3] and Gittins [4] are commonly used for modeling coupled cavity structures. The discussion of these circuits follows the waveguide discussion.

In the remainder of the chapter, examples of coupled cavity TWTs and their operating characteristics are described along with terminations, severs and transitions.

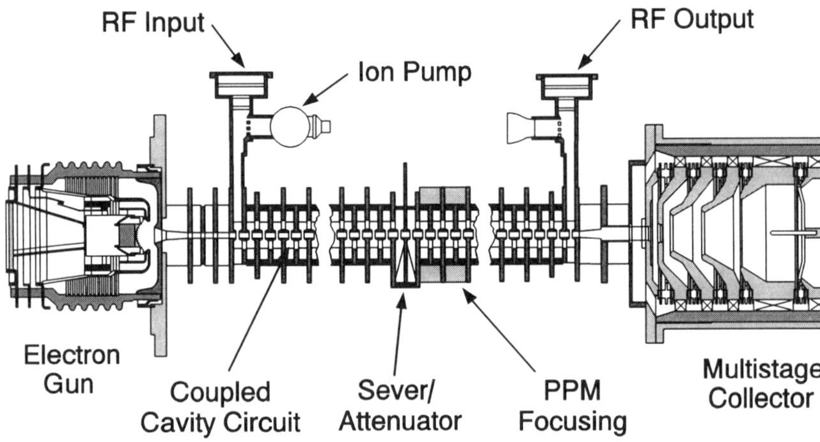

Figure 13-1 Coupled cavity TWT.

13.1 BASIC OPERATING PRINCIPLES

The operation of a coupled cavity TWT may be described with the aid of Figure 13-2. The input circuit supplies a signal to the first cavity through a coupling structure that often consists of either a magnetic coupling loop or an impedance matching section from a waveguide. The voltage, V_1, across the gap modulates the electron beam velocity so that an ac component of current develops as the beam moves toward the second cavity. The beam induces a signal into the second cavity, which splits into two equal wavelets, one traveling to the right and the other to the left. It will be assumed in this discussion that the phases of the reverse traveling wavelets from cavity 2 and from subsequent cavities are such that they cancel each other.

The voltage V_2 enhances the velocity modulation of the beam, thereby increasing the ac current in the beam. This process is repeated at subsequent gaps with each gap voltage being the combination of the voltage coupled from the previous gap and the sum of all the components induced by the ac beam current. The ac beam current at any point is the sum of all the currents produced by previous gaps.

The victor diagram used by Gittins [4] is helpful in illustrating how circuit voltages combine with beam induced voltages to produce signal amplification. The diagram in Figure 13-3 shows the relative phases of the voltages (and rotation of the vectors) as they would appear to an observer traveling at the beam velocity. In this case, the wave on the circuit is assumed to be traveling more slowly than the beam and so, with the convention that Gittins used, the voltage vectors rotate clockwise.

It will be noted in Figure 13-3 that the voltage resulting from the induced beam current is initially 90° out of phase with the cavity voltage. (For example, the very small voltage increment that is added to V_1 by the beam to obtain V_2 is perpendicular

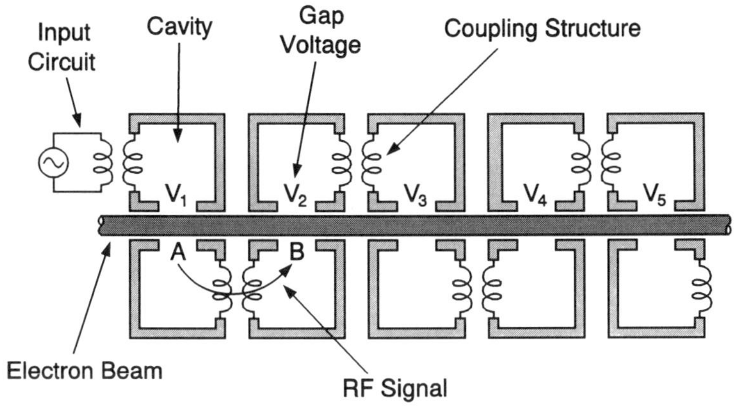

Figure 13-2 Basic coupled cavity circuit.

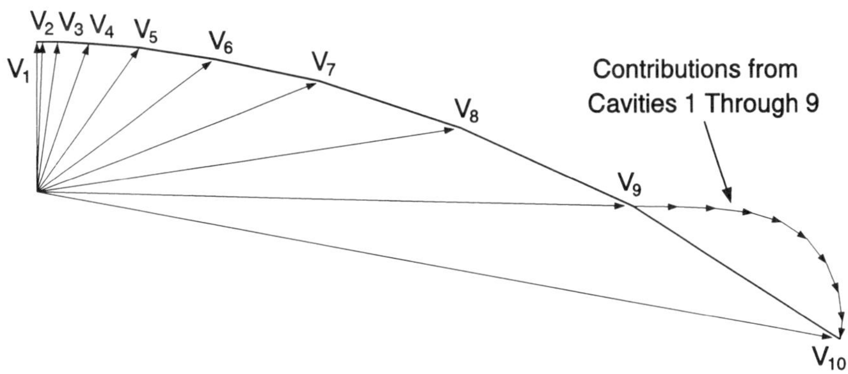

Figure 13-3 Vectors showing build-up of circuit voltage.

to V_1.) The reason for this is apparent from the Applegate diagram in Chapter 9, which showed the formation of electron bunches at alternate zeros of the modulating voltage. Thus, the ac current in the beam lags the modulating voltage by 90° (plus the electron transit angle).

To find the voltage at each gap, the increments of voltage induced by the beam current are added to the voltage coupled from the previous cavity. For example, in Figure 13-3, the voltage at cavity 10, V_{10}, is the voltage at cavity 9, V_9, plus beam-induced contributions produced by the modulation of the beam by cavities 1 through 9.

13.2 ω-β CHARACTERISTICS

13.2.1 Waveguide Approach

It is convenient to start the discussion of coupled cavity structures with the dispersion characteristics of waveguide. Consider the vane loaded waveguide from Chapter 11, which is shown again in Figure 13-4. To examine the interaction of waves in the guide with an electron beam, consider a single electron following the serpentine path shown in Figure 13-4.

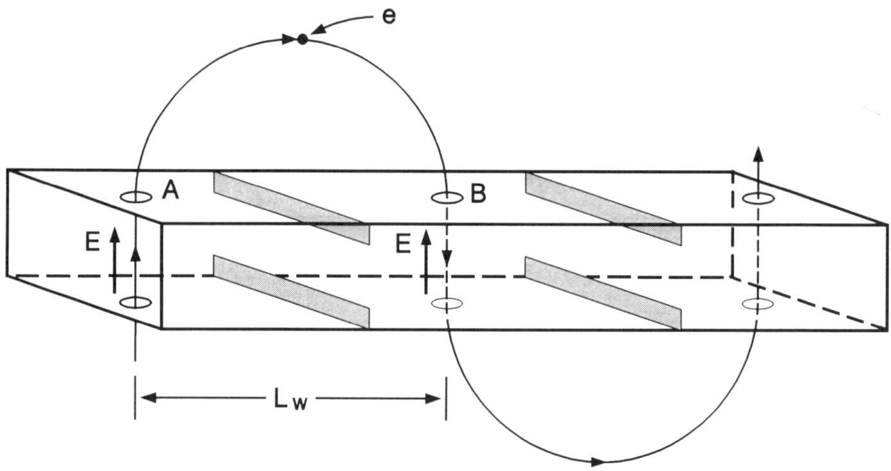

Figure 13-4 Serpentine electron path through vaned waveguide.

Assume first that the electron travels from A to B in the same time that the wave crest in the guide travels from A to B. With respect to the direction of motion of the electron, a 180° phase reversal takes place in the electric field. Thus, as "seen" by the electron, the ω-β diagram is shifted by π rad as shown in Figure 13-5. An obvious problem with the example given here is that the electron would have to travel faster than the speed of light to move from A to B in the time that the wave does. This problem can be alleviated by folding the waveguide as shown in Figure 13-6. Assuming that no additional impedance changes occur then the ω-β diagram for the wave is not changed. From the perspective of the electron, the phase velocity of the wave and the electron velocity from A to B along path L is important, and so βL_w must be replaced by βL on the abscissa of Figure 13-5 to consider electron interaction.

An electron accelerated at A can again be accelerated at B if one of the following two situations occurs:

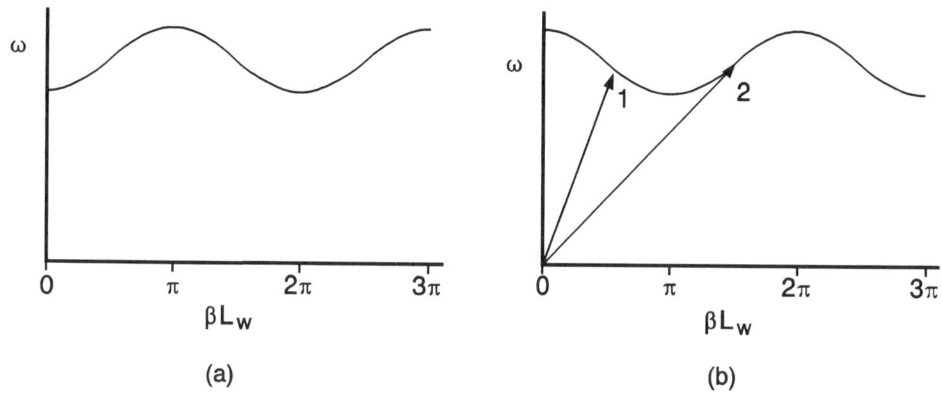

Figure 13-5 An ω-β diagram. (a) vane loaded waveguide, (b) as "seen" by an electron following the serpentine path in Figure 13-4.

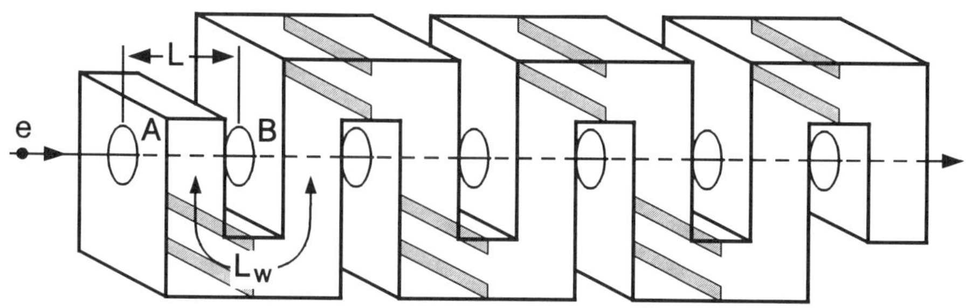

Figure 13-6 Folded waveguide that permits electron interaction below the velocity of light.

1. The direction of wave propagation is reversed, and the electron velocity is adjusted so that the electron arrives at B when an inverted wave crest traveling from the right does (see Figure 13-7(a)). This is a backward wave interaction and is indicated by arrow 1 on the ω-β diagram in Figure 13-5(b). The transit angle of the electron for this interaction is about $\pi/2$ rad.
2. The electron velocity is reduced so that the next inverted wave crest traveling to the right arrives at B when the electron does. This is a forward wave interaction and is indicated by arrow 2 in Figure 13-5(b). For this interaction to occur, the transit angle of the electron is about $3\pi/2$ rad.

The operation of the folded waveguide discussed in the preceding paragraphs is similar to that of a large class of coupled cavity TWTs, which includes the TWT shown in Figure 13-1. The circuits used in these TWTs are referred to as *fundamental backward wave circuits*.

Another class of coupled cavity TWTs uses structures referred to as fundamental

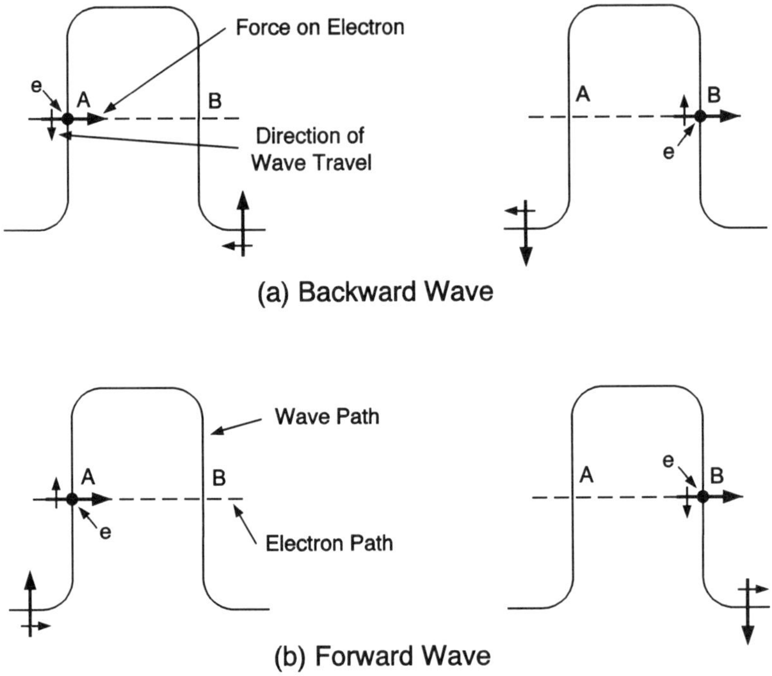

Figure 13-7 Wave propagation that reinforces acceleration of electron.

forward wave circuits. The operation of these circuits may be understood with the aid of Figures 13-8 and 13-9. In these circuits, there is a mechanism for shifting the phase of the RF signal by 180°. The 180° twist shown in Figure 13-8 is an obvious way to produce the desired phase shift but would be virtually impossible to implement in a folded waveguide structure.

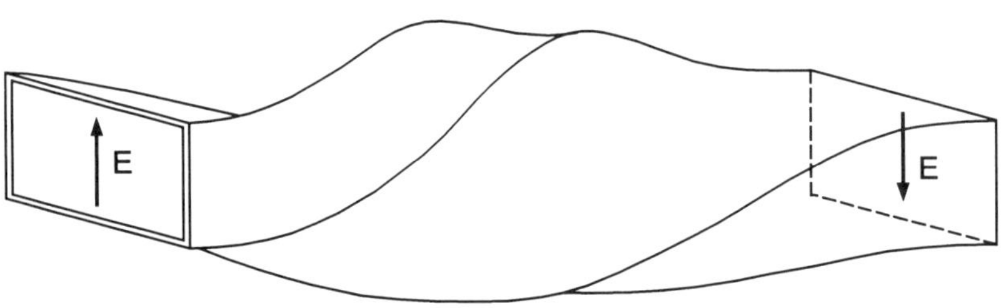

Figure 13-8 Waveguide with a 180° twist.

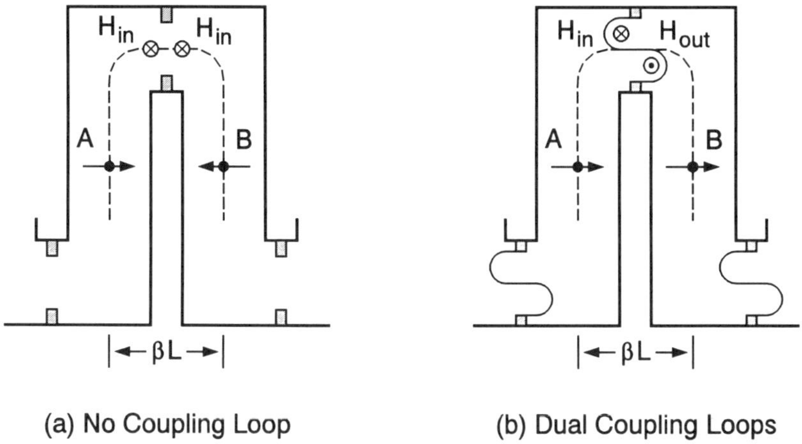

Figure 13-9 Use of a dual coupling loop to reverse the direction of the *H* field.

The dual coupling loop shown in Figure 13-9(b) reverses the magnetic field at the coupling aperture and, as a result, produces a 180° phase shift in the electric field. This action may be understood by considering, first, the magnetic field at the aperture in the folded guide with vanes. As indicated in Figure 13-9(a), the direction of the *H* field is the same on the two sides of the coupling aperture. In Figure 13-9(b), the magnetic field on the left side of the aperture causes current to flow in the coupling loops. This current, in turn, produces a magnetic field on the right side of the aperture that is the reverse of that on the left. This reversal of the magnetic field also produces a reversal of the electric field. An electron that was accelerated at A and travels from A to B in the same time that the wave does will again be accelerated at B.

The ω-β diagram for the loop coupled structure is shown in Figure 13-10. Interaction in this fundamental forward wave structure occurs with electrons at a transit angle near $\pi/2$ rad It should be noted that this transit angle is much smaller than for the fundamental backward wave structure ($-3\pi/2$ rad). As a result, tubes with fundamental forward wave structures operate with faster and, therefore, higher voltage, higher powered electron beams than fundamental backward wave tubes. Fundamental forward wave tubes can operate at power levels of several megawatts whereas fundamental backward wave tubes operate at kilowatts to tens of kilowatts.

13.2.2 Curnow-Gittins Equivalent Circuit Approach

The discussion of ω-β characteristics based on waveguide considerations is useful for a qualitative understanding of the dispersion characteristics of coupled cavity circuits. However, detailed analyses, which necessarily use electromagnetic field theory, are very difficult because of the complex boundary conditions. Thus, the waveguide

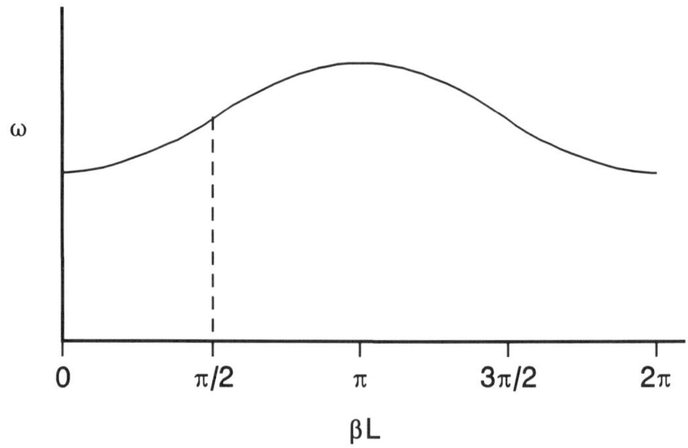

Figure 13-10 The ω-β diagram for a fundamental forward wave circuit.

approach does not provide much insight or guidance concerning the dependance of dispersion characteristics on variations in the geometrical characteristics of the circuits. Even computer modeling does not help much in developing an intuitive understanding of the operation of coupled cavity circuits. However, computer modeling, along with the use of equivalent circuits, can provide accurate predictions of the consequences of circuit changes along with an understanding of what is happening.

From the very early days of coupled cavity TWTs, equivalent circuits have been used to help explain circuit operation. The most successful of these equivalent circuits have been those described by Curnow [2, 3] and Gittins [4]. The development of these circuits is based on an examination of the current paths in a cavity. In a cavity without coupling apertures, current is distributed uniformly as is indicated in Figure 13-11(a). Current flows back and forth from one side of the cavity gap to the other through the cavity walls. The gap is the capacitive portion of the cavity and the walls are the inductive portion, and so the equivalent circuit is in Figure 13-11(b). The resonant frequency is f_c.

When coupling slots are placed in the cavity walls, some of the current paths are intercepted, as shown in Figure 13-12. In general, there can be four types of current paths:

1. Unbroken by any coupling slot (I_1),
2. Broken by the slot coupling to the preceding cavity (I_2),
3. Broken by the slot coupling to the following cavity (I_3),
4. Broken by both coupling slots (I_4).

The equivalent circuit to represent this distribution of currents can be constructed by dividing the cavity inductance, L_c, into four parts:

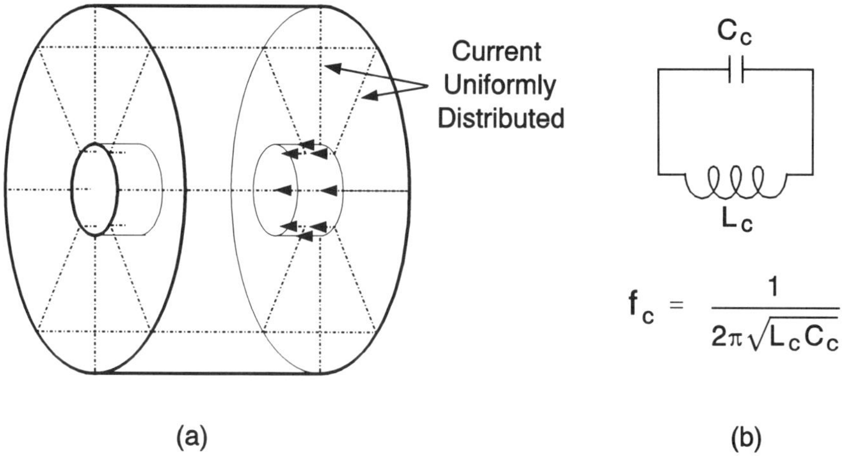

Figure 13-11 Current paths and the equivalent circuit for a cavity without coupling slots.

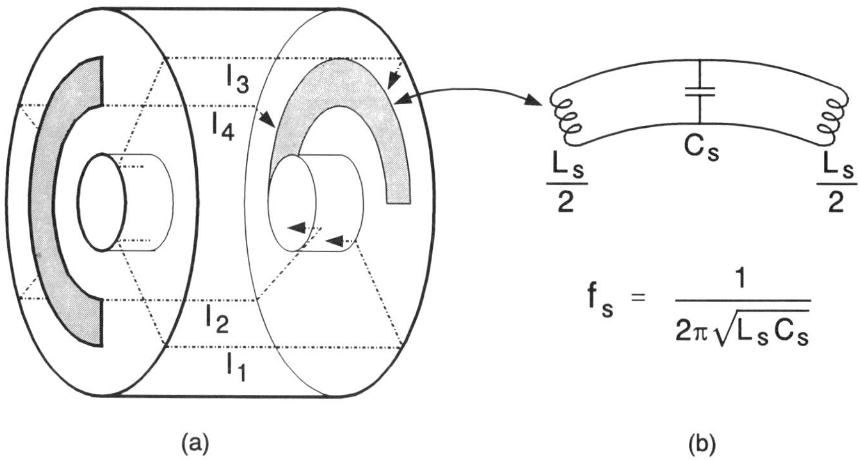

Figure 13-12 (a) Cavity currents used to define the inductance for the equivalent circuit, and (b) the equivalent circuit for a slot.

1. L_c/p, where p is the fraction of the current not involved in coupling;
2. $2L_c/k$, where k is the fraction involved in coupling to the following cavity;
3. $2L_c/k$, where k is the fraction involved in coupling to the preceding cavity;
4. L_c/n, where n is the fraction involved in coupling both ways.

Here, it has been assumed, as is usually the case, that the two coupling slots are the same. The total of the four currents is the total cavity current and so

$$p + k + n = 1 \tag{13-1}$$

The coupling slots can be represented as parallel resonant circuits with resonant frequency f_s, as shown in Figure 13-12(b). Thus, the cavity with all of its equivalent circuit elements is as shown in Figure 13-13. The generalized equivalent circuit derived from this cavity is in Figure 14-14.

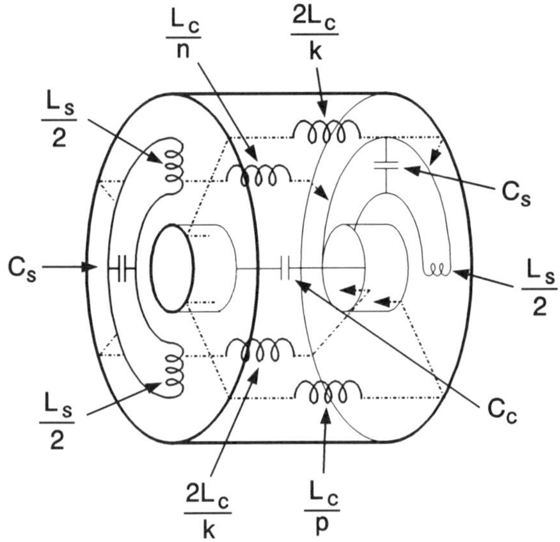

Figure 13-13 Equivalent circuit elements of a cavity.

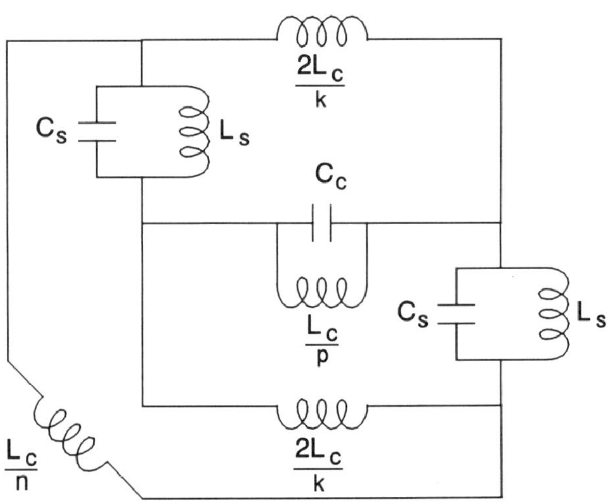

Figure 13-14 The equivalent circuit for a cavity.

Using the circuit in Figure 13-14 an expression can be developed for the phase shift, ϕ, from cavity to cavity. Using measured values for the resonant frequencies of the cavity and the slots and estimates for the current coupling fractions, the ω-β diagram can be constructed. Also, the impedance of the circuit can be calculated. Then, based on comparisons with experimental ω-β results and impedance measurements, the current fraction can be revised so that the equivalent circuit is a reasonably accurate model of the cavity.

One of the values of the Curnow-Gittins equivalent circuit is that it provides insight into the operation of the cavity. The other is that it provides a way to estimate the results in performance characteristics of changes to a coupled cavity structure. For example, if the coupling slots are made longer, the fraction of the current that they intercept is increased and so the change in dispersion and impedance can be estimated.

13.2.3 Example of an Application of the Curnow-Gittins Circuit

As an example of the use of the Curnow-Gittins equivalent circuit, the staggered-slot circuit will be examined. The most commonly used coupled-cavity circuit is the staggered-slot circuit used in the TWT shown in Figure 13-1, which is shown more clearly in Figure 13-15. In this circuit, there are no current paths that are intercepted by both slots. As a result, the fraction, n, used in the previous section is zero and so

$$p = 1 - k \tag{13-2}$$

where k is the total fraction of the current involved in coupling. For this case, the equivalent circuit is as shown in Figure 13-16. Circuits for adjacent cavities are also shown to facilitate with the analysis that follows.

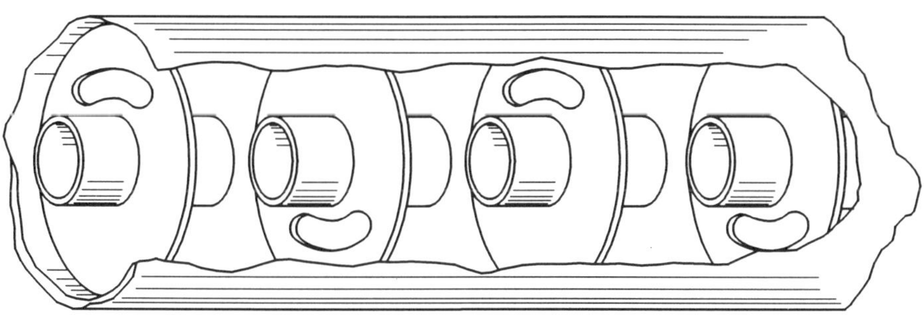

Figure 13-15 A staggered-slot fundamental backward wave circuit. From *Power Travelling Wave Tubes* by J.F. Gittins. Published 1965 by American Elsevier, Inc.

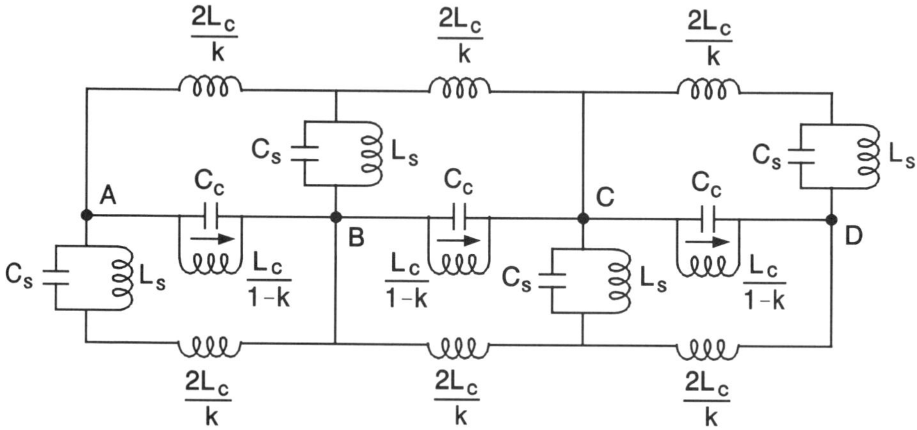

Figure 13-16 The equivalent circuit for the staggered-slot structure.

In manipulating the circuit so that it can be analyzed, it is important to keep track of the direction with which electrons pass through the circuit (across C_c). This direction is indicated by the arrow next to the circuit capacitance. By inspection, the circuit may be rearranged as is shown in Figure 13-17.

One cell (midcavity to midcavity) of the equivalent circuit is shown in Figure 13-18(a) and with equivalent impedances, in Figure 13-18(b). This circuit is in a form so that the phase shift ϕ from cavity to cavity can readily be determined. Also, the impedance looking in at X-X is the impedance "seen" by the electron beam, and this is the impedance needed for the calculation of Pierce's gain parameter.

From the theory of filter circuits the equation for the phase shift, ϕ, which is needed for constructing the ω-β diagram, is

$$\cos\phi = \left[\frac{Z_{oc}}{Z_{oc} - Z_{sc}}\right]^{1/2} \tag{13-3}$$

where Z_{oc} is the open circuit impedance looking in at X-X and Z_{sc} is the impedance at X-X with Y-Y short circuited. The phase shift "seen" by an electron, θ, is $\phi + \pi$ (note the direction of the arrow indicating electron flow) and so

$$\cos\theta = -\left[\frac{Z_{oc}}{Z_{oc} - Z_{sc}}\right]^{1/2} \tag{13-4}$$

The impedance, K, of the circuit is

$$K = (Z_{oc} Z_{sc})^{1/2} \tag{13-5}$$

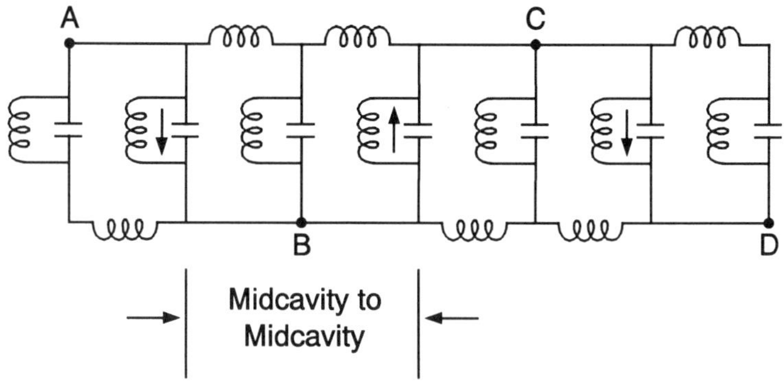

Figure 13-17 Rearranged circuit for staggered-slot structure.

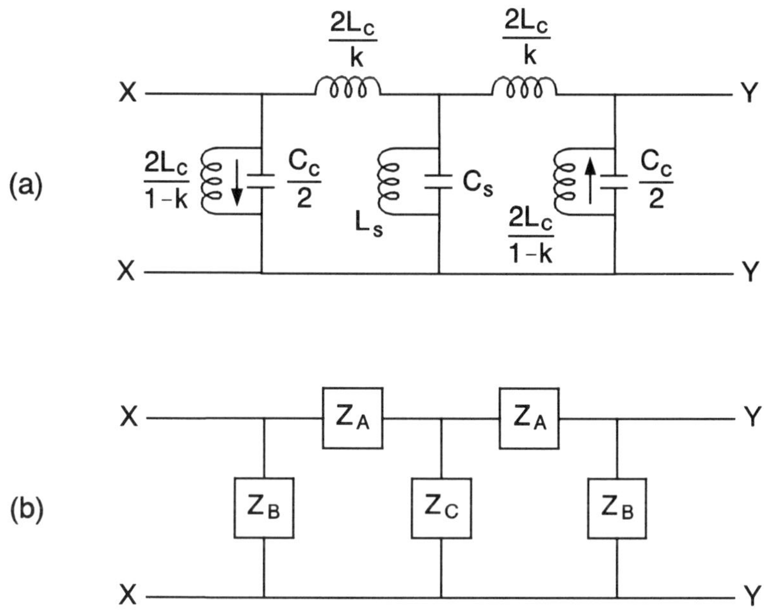

Figure 13-18 (a) One cell (midcavity to midcavity) of the equivalent circuit and (b) the equivalent impedances.

The calculation of Z_{oc} and Z_{sc} is tedious but straightforward. Referring to Figure 13-18(b),

$$Z_{oc} = \frac{Z_A^2 + 2Z_AZ_C + Z_AZ_B + Z_BZ_C}{Z_A^2 + Z_B^2 + 2Z_AZ_B + 2Z_AZ_C + 2Z_BZ_C} Z_B \tag{13-6}$$

and

$$Z_{sc} = \frac{Z_A^2 + 2Z_AZ_C}{Z_A^2 + 2Z_AZ_C + Z_AZ_B + Z_BZ_C} Z_B \tag{13-7}$$

Using these relations for Z_{oc} and Z_{sc} the equation for phase shift can be written as

$$\cos\theta = -\left[1 + \frac{2Z_A}{Z_B} + \frac{Z_A}{Z_C} + \frac{Z_A^2}{Z_BZ_C}\right] \tag{13-8}$$

From Figure 13-18,

$$Z_A = \frac{2j\omega L_c}{k} \tag{13-9}$$

$$Z_B = \frac{2j\omega L_c}{1 - k - \omega^2 L_c C_c} \tag{13-10}$$

and

$$Z_C = \frac{j\omega L_s}{1 - \omega^2 L_s C_s} \tag{13-11}$$

With the use of these impedance equations, the equation for phase shift becomes

$$\cos\theta = 1 - \frac{2L_c}{k^2 L_s}\left(1 - \omega^2 L_c C_c\right)\left[1 + \frac{kL_s}{L_c} - \omega^2 L_s C_s\right] \tag{13-12}$$

This equation may by put in a more usable form using the resonant frequencies

$$\omega_c = \frac{1}{\sqrt{L_c C_c}} \tag{13-13}$$

and

$$\omega_s = \frac{1}{\sqrt{L_s C_s}} \tag{13-14}$$

and, by defining a coupling factor, k_c, as follows:

$$k_c = \frac{kL_s}{L_c} \tag{13-15}$$

With these substitutions,

$$\cos\theta = 1 - \frac{2}{kk_c}\left[1 - \frac{\omega^2}{\omega_c^2}\right]\left[1 + k_c - \frac{\omega^2}{\omega_s^2}\right] \tag{13-16}$$

Now, it is noted that $\cos\theta = 1$ (and $\theta = 0$) for two conditions,

$$\omega = \omega_c \tag{13-17}$$

and

$$\omega = \omega_s\sqrt{1 + k_c} \tag{13-18}$$

These two conditions are "cutoff" conditions for reasons that will become clear in the following discussion. The first of these is, of course, just the cavity resonance condition. The second can be identified with the help of Figure 13-18. It is the series resonance of the T coupling elements made up of Z_A, Z_C and Z_A (Figure 13-18(b)), which is the resonance of the slot modified by the cavity inductance (Figure 13-18(a)).

To understand how phase shift varies with frequency, we will assume, first, that $\omega_s = \omega_c$. Then, since k_c is always positive, the modified slot resonant frequency is higher than ω_c. For frequencies between ω_c and the modified slot resonant frequency, $\cos\theta > 1$ and so there can be no signal propagation. Thus, a "stop" band occurs for these frequencies.

Continuing with the assumption that $\omega_s = \omega_c$, it is noted next that, as ω is either reduced or increased from one of the cutoff conditions; that is,

$$\omega < \omega_c \tag{13-19}$$

or

$$\omega > \omega_s\sqrt{1 + k_c} \tag{13-20}$$

the righthand side of (13-16) becomes less than unity, and this indicates that propagation through the circuit can occur. With assumptions for the values of k and k_c the phase shift as a function of frequency can be determined. For example, with assumptions that $k = 0.5$ and $k_c = 0.35$, the ω-β characteristic can be plotted as shown in Figure 13-19. The lower limit to the lower curve and the upper limit to the upper curve occur when $\theta = \pi$ or $\cos\theta = -1$. Then, from (13-16),

$$kk_c = \left[1 - \frac{\omega^2}{\omega_c^2}\right]\left[1 + k_c - \frac{\omega^2}{\omega_c^2}\right] \tag{13-21}$$

or

$$\frac{\omega^2}{\omega_c^2} = \frac{1}{2}\left[2 + k_c \pm \sqrt{k_c^2 + 4kk_c}\right] \tag{13-22}$$

At frequencies below the lower limit and above the upper limit, $\cos \theta < -1$, and so there can be no propagation.

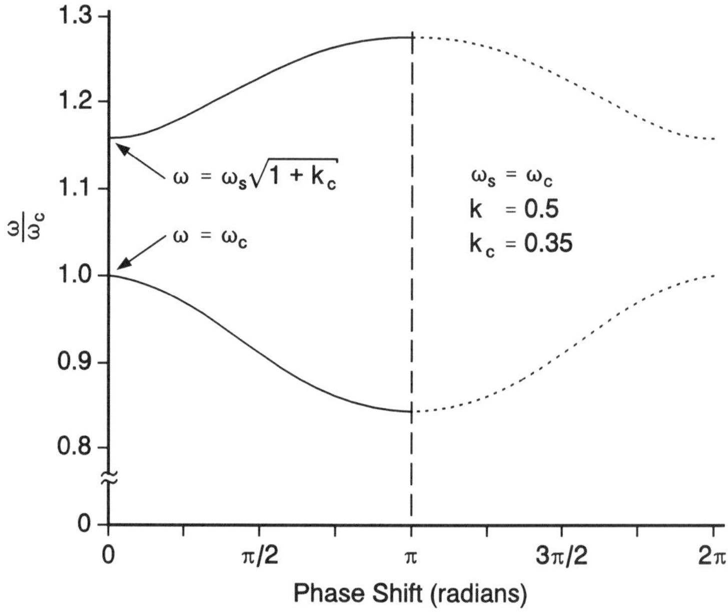

Figure 13-19 The ω-β diagram for the equivalent circuit for the staggered-slot structure.

Thus, there are two frequency bands in which propagation can occur. The lower one is referred to as the *cavity passband* because its characteristics are controlled largely by the cavity resonance condition. Similarly, the upper band is called the *slot passband* and its characteristics are controlled mostly by the slot resonance condition.

The solid curves in Figure 13-19 are the ω-β characteristics derived for one midcavity-to-midcavity cell of the equivalent circuit for the entire coupled cavity structure. Because only one cell is used, results for other portions of the ω-β characteristic, such as those shown by the dashed lines in Figure 13-19, are not obtained. However, these can be inferred from the discussion of ω-β characteristics contained in Chapter 11. Thus, the curves from π to 2π rad are simply mirror images of those from 0 to π rad. For the staggered-slot circuit, the lower dashed curve, which indicates forward wave interaction, is the ω-β characteristic that is needed.

There are several variables in the Curnow-Gittins equations. When changes are proposed for a coupled cavity structure, the effect of these changes on one or more of the variables can be estimated, and the change in dispersion characteristic (and interaction impedance) can be predicted. As an example, changes in the angle, θ, of the coupling slot shown in Figure 13-20 change the percentage of the cavity current involved in coupling (k), change the slot inductance (in k_c) and change the slot resonant frequency (ω_s). The effect of these changes on the dispersion characteristics is shown in Figure 13-21. These changes, in turn, affect the cold bandwidth (bandwidth of circuit with no electron beam) as shown in Figure 13-22. Changes in coupling slot angle also change the relative positions of the slot and cavity passbands as shown in Figure 13-23. Note that, for large coupling angles, the slot passband becomes the lower passband.

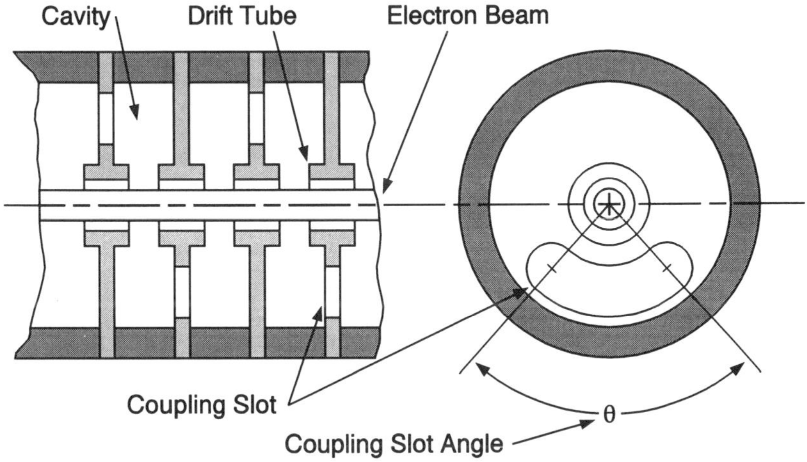

Figure 13-20 A coupling slot between cavities. From Hughes Aircraft Co., *TWT and TWTA Handbook.*

As noted previously in this section, the impedance as "seen" by the electron beam is just the characteristic impedance of the circuit in Figure 13-18, and that is given by $K = (Z_{oc}Z_{sc})^{1/2}$. Again it is tedious but straightforward to combine the expressions for Z_{oc} and Z_{sc}. In arriving at a useful expression for K, the loaded Q of the circuit must be taken into consideration. This is done by making the substitution

$$\left[\frac{L_c}{C_c}\right]^{1/2} = \frac{R}{Q} \tag{13-23}$$

where R/Q is found from circuit measurements. The final expression for impedance is

$$K = \frac{\omega R}{\omega_c Q}\left[\frac{8\left[1 + k_c - \dfrac{\omega^2}{\omega_s^2}\right]}{kk_c\left[1 - \dfrac{\omega^2}{\omega_c^2}\right]\left(1 + \cos\theta\right)}\right]^{1/2} \tag{13-24}$$

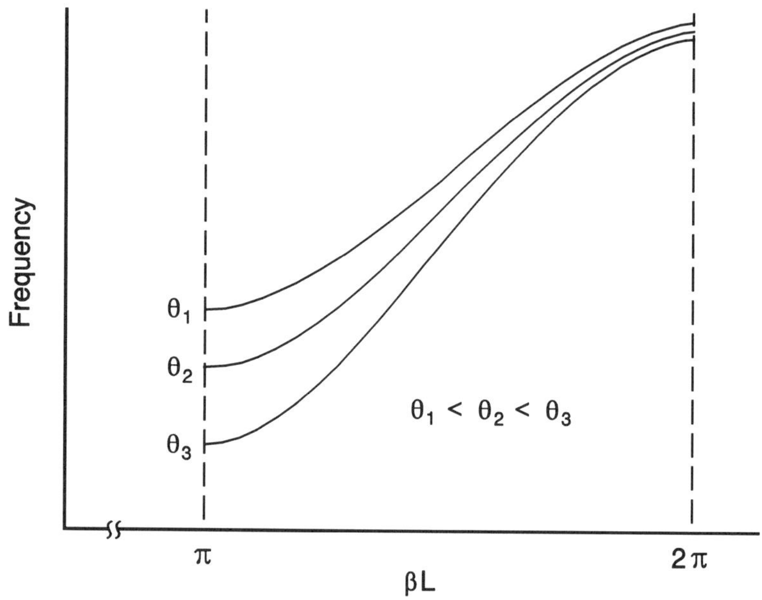

Figure 13-21 The effect of the coupling angle on the ω-β diagram. Courtesy of Hughes Electron Dynamics Division.

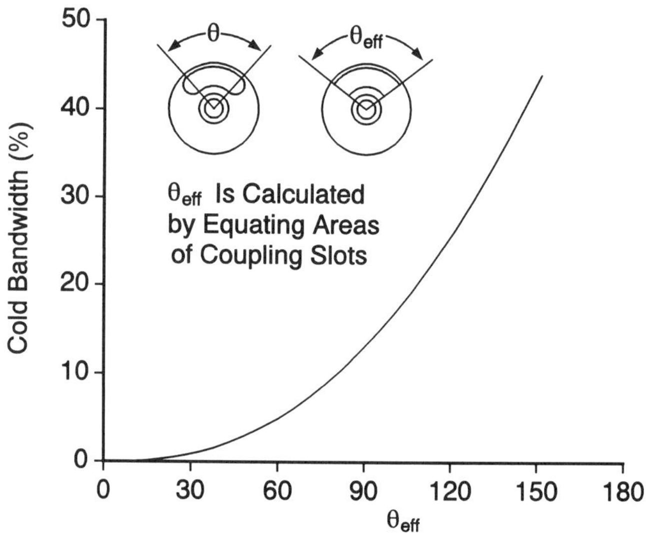

Figure 13-22 Cold bandwidth *versus* coupling slot angle. Courtesy of Hughes Electron Dynamics Division.

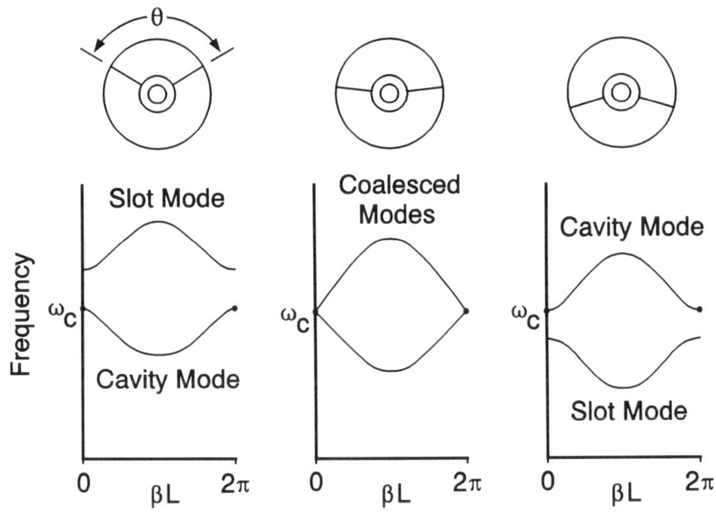

Figure 13-23 Change in passbands with increasing coupling slot angle. From J.R. Frey and I. Tammaru, *Technical Digest,* 1981 IEDM. © 1981 IEEE.

Again, using Gittins values of $k = 0.5$ and $k_c = 0.35$ and, also, assuming that $R/Q = 37$, impedance may be plotted as shown in Figure 13-24. Note that interaction impedance goes to infinity at all cutoff frequencies except for the lower slot-band cutoff frequency, where it is zero. For the values of constants used by Gittins, the impedance in the cavity band where normal operation occurs is a few hundred ohms, which is higher than normally encountered in staggered-slot TWTs.

Normally, the cavity passband is used for interaction with the electron beam, and such was the case for the results given in the previous section. It is possible for the slot pass band to be used, and tubes designed to operate in this mode have existed for many years [5]. An important property of the slot passband is that the interaction impedance at the upper cutoff frequency is zero (see Figure 13-24). This results from the vanishing axial electric field component on the axis, as is indicated in Figure 13-25. As a result, no special oscillation-suppression techniques are required at this frequency.

Attempts have been made in [6] to coalesce the cavity band and slot band as indicated in Figure 13-23 in an effort to achieve octave-bandwidth operation in staggered-slot TWTs. While the ω-β diagram indicates near synchronism of the circuit wave and the beam over the desired bandwidth, two difficulties are in-band backward wave instability and a large gain variation with frequency.

Once the equivalent circuit model with appropriate component values has been established for a coupled cavity structure, the analysis of the interaction with the electron beam can be carried out. The steps of the analysis are conceptually the same as those used by Gittins in the vector diagram description of coupled cavity TWT operation. That is, the electron beam induces current into the appropriate elements

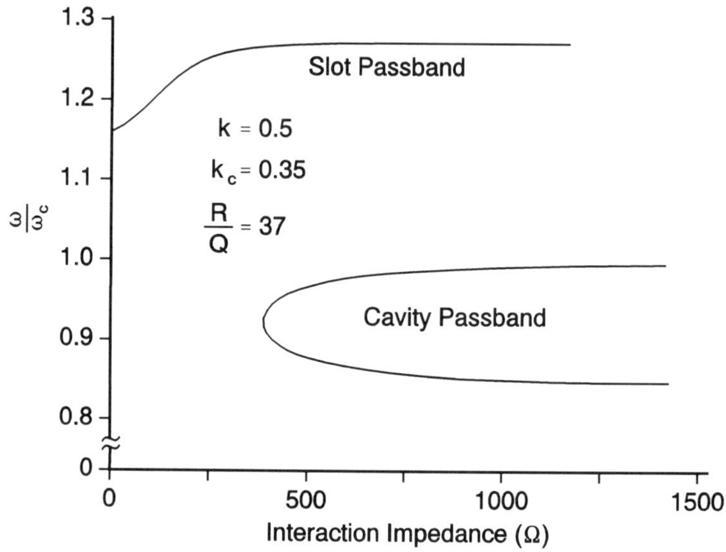

Figure 13-24 Interaction impedance derived from the circuit model for the staggered-slot structure.

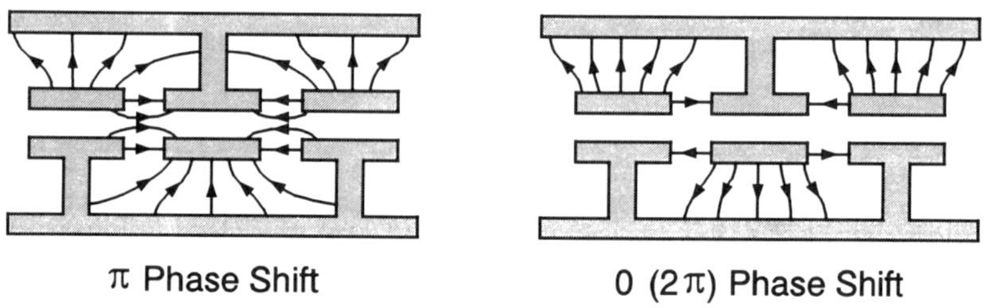

Figure 13-25 Electric field patterns for the slot-band cutoff frequencies. From J.R. Frey and I. Tammaru, *Technical Digest*, 1981 IEDM. © 1981 IEEE.

(cavity gaps) of the equivalent circuit, and the electric field of the circuit excites current and velocity modulation of the beam. The shapes of the electric fields of the circuit and the space charge fields in the beam must be accurately determined so that the analysis can properly predict tube performance.

13.3 FUNDAMENTAL BACKWARD WAVE OPERATION

Two commonly used fundamental backward wave circuits are illustrated in Figure 13-26. In each of these, apertures are used to couple the fields from one cavity to the next. The staggered-slot structure in Figure 13-26(a) was discussed in the previous section.

When the slots are aligned, as shown in Figure 13-26(b), all of the circulating cavity current that couples to the previous cavity also couples to the following cavity. As a result, the aligned-slot structure has properties that are quite different than those for the staggered-slot structure. The primary difference is that the aligned-slot structure has a smaller bandwidth. However, the shape of the ω-β characteristic is such that, for a given frequency of operation, larger cavity dimensions are possible with the inline structure and so higher powers are possible than with the staggered-slot structure.

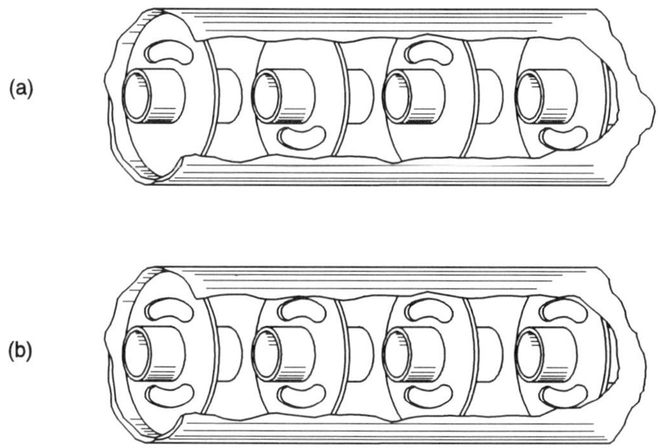

Figure 13-26 Staggered slot (a) and aligned slot (b) fundamental backward wave circuits. From *Power Travelling Wave Tubes* by J.F. Gittins. Published 1965 by American Elsevier, Inc.

An example of the capability of a device with a staggered-slot circuit is shown in Figure 13-27. Confined flow solenoid focusing and a depressed collector are used. The solid curves show the power output at various drive levels (referenced to the 0 dB saturation drive level of 40 mW). The overall efficiency is in the 20 to 30% range.

Coupled cavity structures are ideally suited for use in TWTs that are focused with periodic permanent magnets. As shown in Figure 13-28, the magnetic pole pieces can be made integral with the cavity structure. The pole pieces are copper plated, and copper spacers are used to form cavities of the required height. Liquid cooling channels may also be included in the coupled cavity structure. In some cases, the coolant is channeled around the drift tubes between cavities to carry away the heat resulting from interception of the electron beam.

The tube, for which the power output is given in Figure 13-29, is an example of what can be achieved in a staggered-slot circuit when PPM focusing is used. In this case, a confined flow version of PPM focusing is utilized to minimize beam interception. With 35 G at the cathode, the maximum interception current is less than 500 mA (out of a cathode current of 7.65 A) so the RF beam transmission is 93.4%

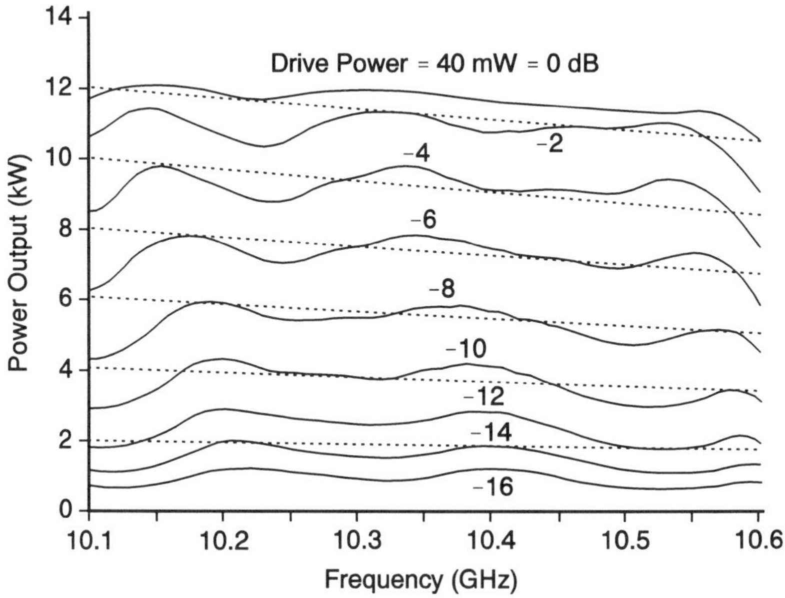

Figure 13-27 Example of the power capability of a cw coupled cavity TWT. From Armand Staprans *et al., Proc. IEEE*, March 1973. © 1973 IEEE.

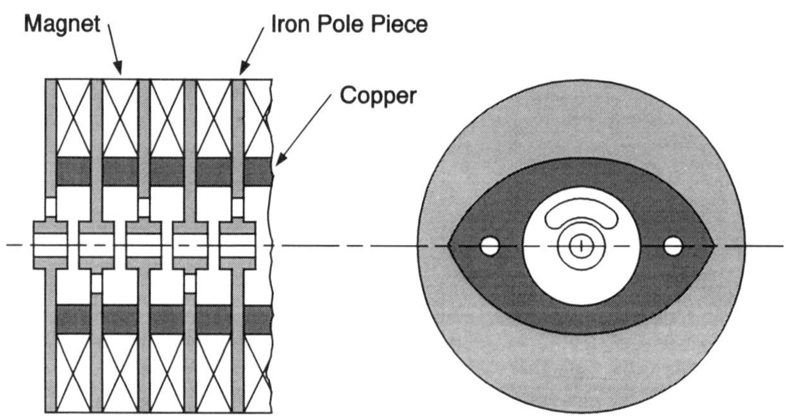

Figure 13-28 A coupled cavity circuit with integral periodic permanent-magnet focusing. From Hughes Aircraft Co., *TWT and TWTA Handbook*.

or better. A single-stage depressed collector is used, and the overall efficiency at midband is 33.7%. It is estimated that, with a five-stage collector, the efficiency would be 50%.

When periodic permanent magnets are used, the electron beam is not as well

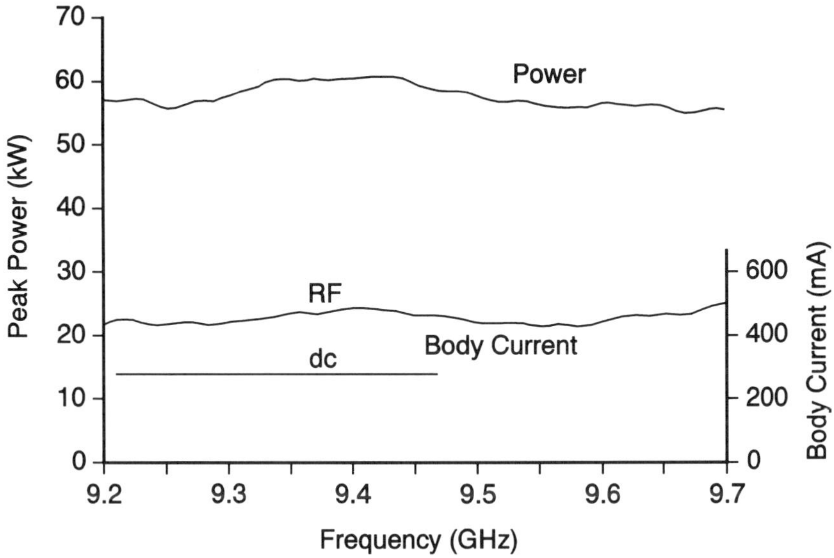

Figure 13-29 Performance of a pulsed PPM-focused coupled cavity TWT at 10% duty cycle. From Hughes Electron Dynamics Division, *Microwave Journal*, April 1989.

focused as it is when a solenoid is used. As a result, the percentage of the beam current intercepted by the circuit is relatively high, and so the average power capability is relatively low. The reduction in average power capability of PPM-focused tubes compared with solenoid-focused tubes is typically a factor of four or more.

When coupled cavity technology is extended to millimeter-wave frequencies, cavity dimensions become very small. The TWT circuit based on conventional cavity technology requires several dozen copper disks per section. Each disk contains a coupling slot and ferrules, which form the cavities. The disks are lathe turned, preferably by diamond-point machining. The kidney-shaped slots are normally cut by an EDM (electric discharge milling) operation. These disks are stacked and heated under pressure to braze or sinter them together.

At millimeter-wave frequencies, the problem with this conventional construction technique is that it requires the parts to be machined to extremely tight tolerances. One reason is that errors in cavity lengths may add cumulatively when the cavities are stacked. To maintain synchronism between the beam and circuit wave, extremely tight tolerances on the height of each cavity are necessary. A second reason is that the cavity resonant frequency is critically dependent on the cavity gap width. Machining tolerances on this dimension may be 70 millionths of an inch or smaller.

The problem of using a large number of parts machined to extremely tight tolerances can be circumvented by using the ladder-circuit designs to be described in following paragraphs [7]. With these designs, tolerances are relaxed to 200 millionths of an inch. The cost, in turn, is reduced by a factor of about 2.

The ladder circuit is fabricated by machining notches in a slab of copper to form a "ladder" or "comb" structure as is indicated in Figure 13-30. The beam tunnel is formed by cutting half-round grooves in the tips of the comb teeth. Two identical structures are then brazed together. The circuit is completed by brazing cover plates to the faces of the ladder assembly. Because the ferrules, which form the drift-tube tips in the cavities, are omitted with the ladder-assembly technique, gain per cavity is reduced and so the interaction circuit length is increased. Still, the advantage of needing only a few parts per section rather than several dozen is considerable.

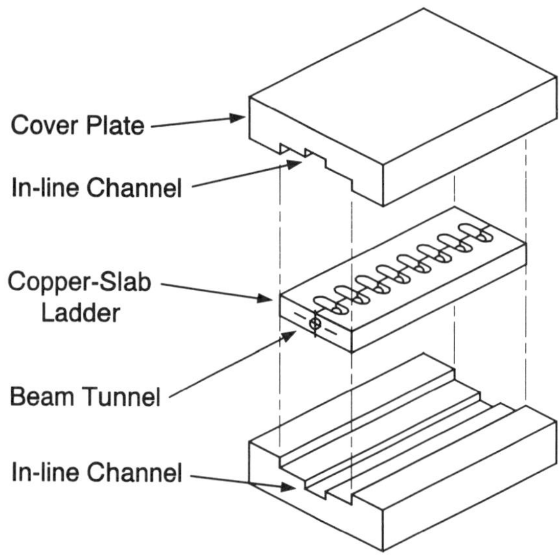

Figure 13-30 Ladder-core structure with dual in-line coupling channels. Adapted from Bill G. James, *MSN & CT*, September 1986.

In [7] three versions of ladder circuits are described:

1. In the aligned-slot structure shown in Figure 13-30, the coupling slots from cavity to cavity are formed by machining channels in the cover plates.
2. In the staggered-slot structure shown in Figure 13-31, slots are machined into the edge of every other ladder rung, with the skipped rungs notched on the other side. The cover plates are flat slabs of copper.
3. In the double-staggered ladder structure, two coupling slots spaced 180° apart are used in each cavity wall. The pair of coupling slots is rotated 90° from cavity to cavity.

As shown in Figure 13-32, there are significant differences between the ω-β characteristics for these structures. These differences lead to advantages (or disadvantages) that govern the applications of the tubes.

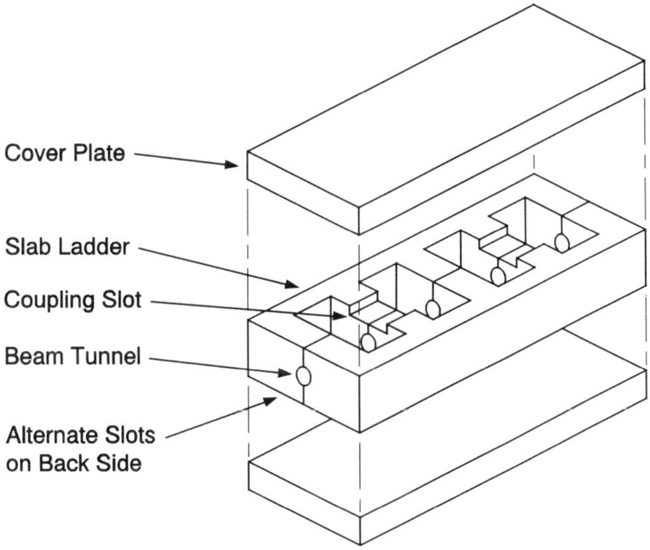

Figure 13-31 Ladder-core structure with staggered coupling slots. Adapted from Bill G. James, *MSN & CT*, September 1986.

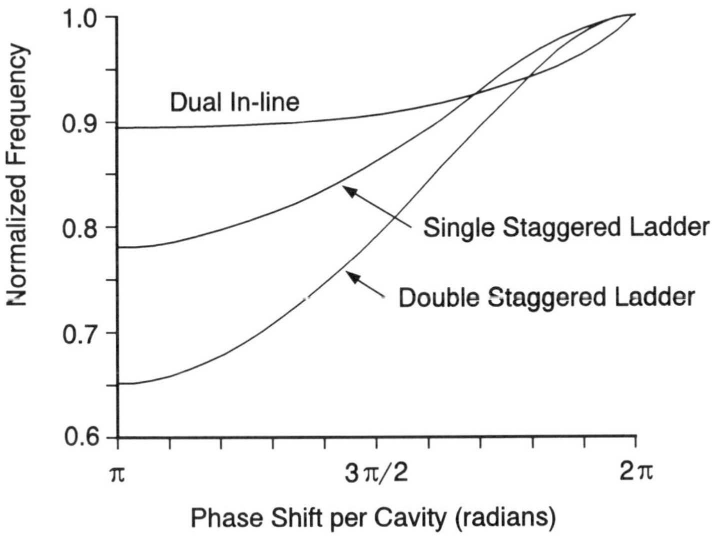

Figure 13-32 ω-β characteristics for ladder structures. Adapted from Bill G. James, *MSN & CT*, September 1986.

With the aligned-slot design, the slot and cavity modes are coalesced (see the discussion of ω-β characteristics previously in this chapter), which allows the use of a reduced cavity resonant frequency and larger cavity dimensions than for the staggered-slot circuit. In addition, because of the nature of the ω-β characteristic, a larger value of βL can be used and so the cavities can be longer than for the staggered-slot circuit. These two properties essentially double the power-handling capability of the circuit, making it one of the highest power linear beam millimeter-wave circuits available. Figure 13-33 contains power and efficiency curves for a tube with a dual aligned-slot circuit. The bandwidth for this tube is approximately 1%. Theoretically, it is possible to achieve bandwidths of over 3%, depending on operating voltage and perveance.

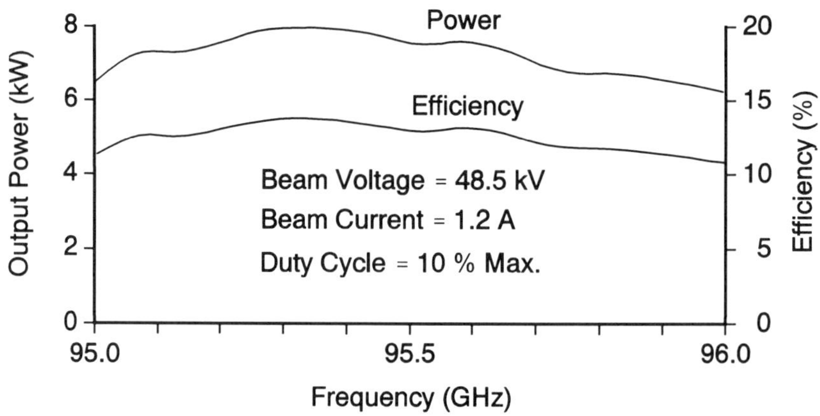

Figure 13-33 Power output for a tube with a dual aligned-slot ladder assembly. Adapted from Bill G. James, *MSN & CT*, September 1986.

The double-staggered ladder is the broadest bandwidth circuit of the three ladder structures discussed here. This circuit has demonstrated 20% operating bandwidth over the 80 to 100 GHz range with a power output of 100 W CW at 90 GHz.

The estimated power capability of millimeter-wave TWTs using solenoid-focused ladder structures is given in Figure 13-34. Results for the VTA-5700 and VTW-5795 are shown. When PPM focusing is used, the estimated power capability for ladder structures is about an order of magnitude lower than shown in Figure 13-34.

The 1989 status of power capabilities of fundamental backward wave coupled cavity TWTs with moderate bandwidth (few percent) are summarized for pulsed tubes in Figure 13-35 and for CW tubes in Figure 13-36 [8]. For larger bandwidths the maximum power capability is lower than shown here. In Figure 13-35, E_k is the electron gun voltage and A_c is the area compression ratio of the electron gun. Note that, in these plots, the power capabilities of PPM tubes are about a factor of three below those for solenoid-focused tubes.

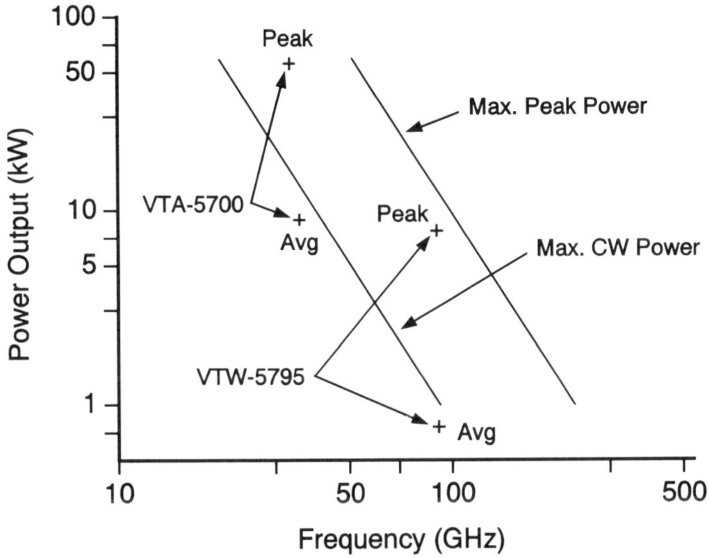

Figure 13-34 Practical power limits for millimeter-wave TWTs with solenoid-focused ladder structures. Adapted from Bill G. James, *MSN & CT*, September 1986.

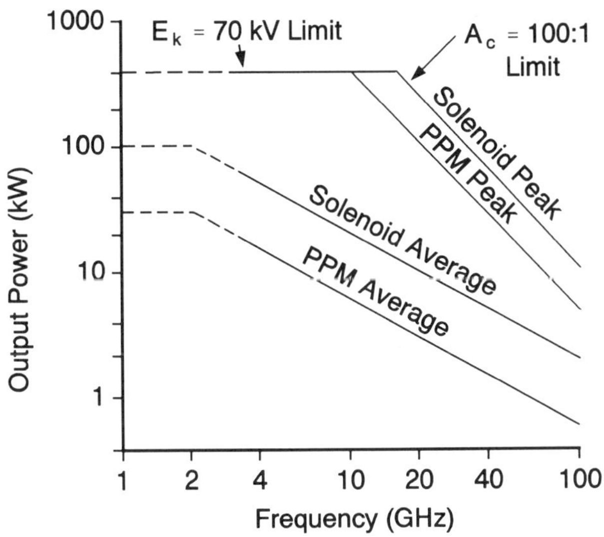

Figure 13-35 Output power for pulsed coupled-cavity TWTs with moderate bandwidths. From James W. Hansen, *Microwave Journal*, 1989 State of the Art Reference.

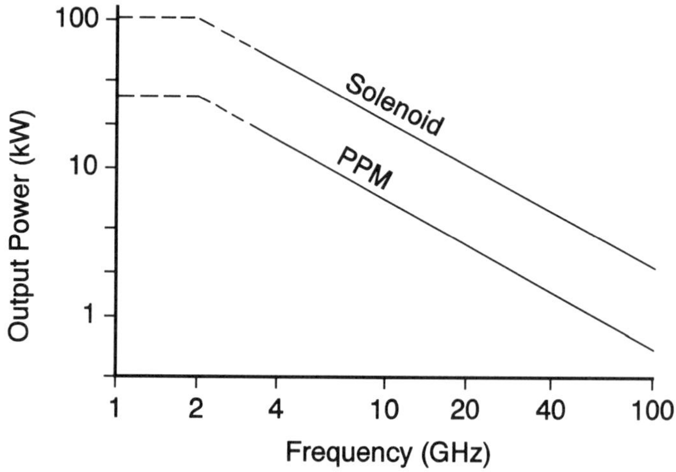

Figure 13-36 Output power for cw coupled cavity TWTs with moderate bandwidths. From James W. Hansen, *Microwave Journal*, 1989 State of the Art Reference.

13.4 FUNDAMENTAL FORWARD WAVE OPERATION

The centipede (interlaced) structure shown in Figure 13-37 is an implementation of the coupling loop scheme discussed at the beginning of the chapter for achieving fundamental forward wave operation [9]. Many coupling loops are used to maximize coupling and operating bandwidth. As shown in Figure 13-37(b), the height of each cavity must be adequate to accommodate the large coupling loops required to maximize coupling. Since the beam transit angle from cavity to cavity is only about 90°, the electron beam velocity must be high, so the centipede structure is suitable for high power operation.

The cloverleaf structure shown in Figure 13-38 is a commonly used fundamental forward wave circuit [9]. The four-lobed cavity distorts the magnetic field lines into a cloverleaf pattern. Radial components of the field are coupled from cavity to cavity by radial slots. By rotating each cavity 45° with respect to adjacent cavities, the fields in adjacent cavities are in phase except for a delay of approximately 90° associated with coupling the signal from cavity to cavity.

The cloverleaf structure is used in high power TWTs. An example, shown in Figure 13-39, is the Varian VA-125 for which the power output is given in Figure 13-40. Note that output powers in excess of 4 MW are obtained.

The cloverleaf structure is also used for the output circuit in twystrons. In a twystron, as shown in Figure 13-41, a klystronlike driver section is used in conjunction with (usually) a cloverleaf output section. Also, a high-perveance beam is used to provide strong beam loading, which helps increase bandwidth. With this combination,

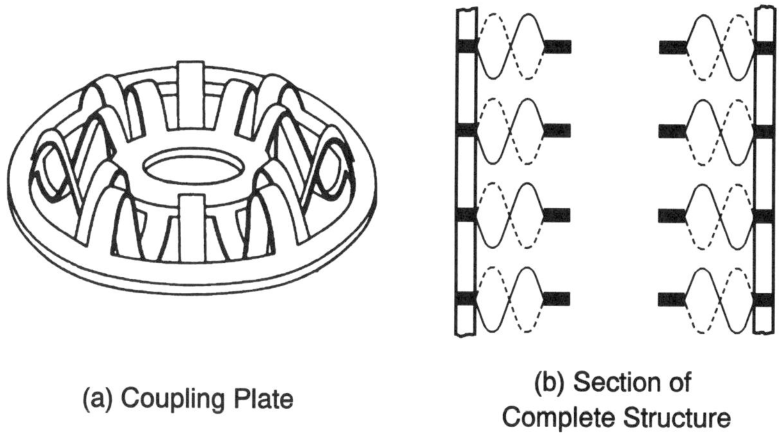

Figure 13-37 Centipede (interlaced) structure. From *Power Travelling Wave Tubes* by J.F. Gittins. Published 1965 by American Elsevier, Inc.

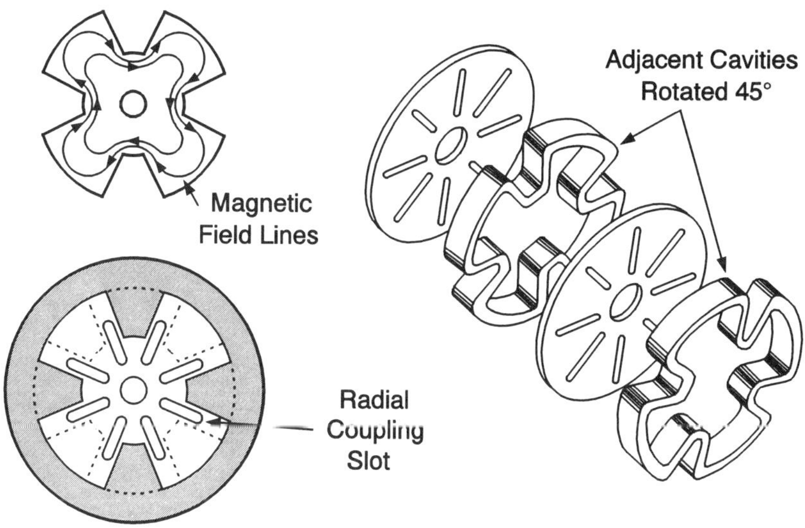

Figure 13-38 Cloverleaf structure. Adapted from M. Chodorow and R.A. Craig, *Proc. IRE*, August 1957. © 1957 IRE (now IEEE).

the low gain of the TWT section at band edges (for an example of typical effect on power output, see Figure 13-40) is offset by the relatively high gain of the stagger-tuned klystron driver section. The overall result, as Figure 13-42 indicates, can be a very flat gain characteristic over a bandwidth of up to 15%. The resulting power output as a function of frequency is also very flat, as shown in Figure 13-43.

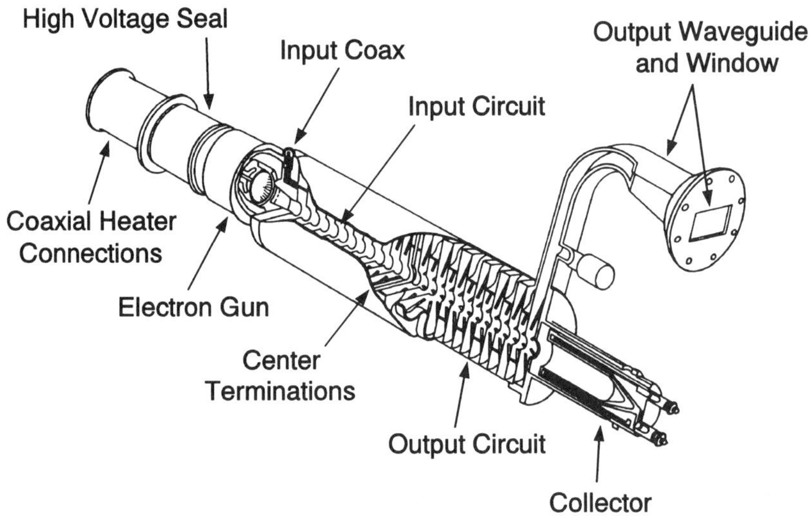

Figure 13-39 The use of a cloverleaf structure in the Varian VA-125 high power TWT. From *Power Travelling Wave Tubes* by J.F. Gittins. Published 1965 by American Elsevier, Inc.

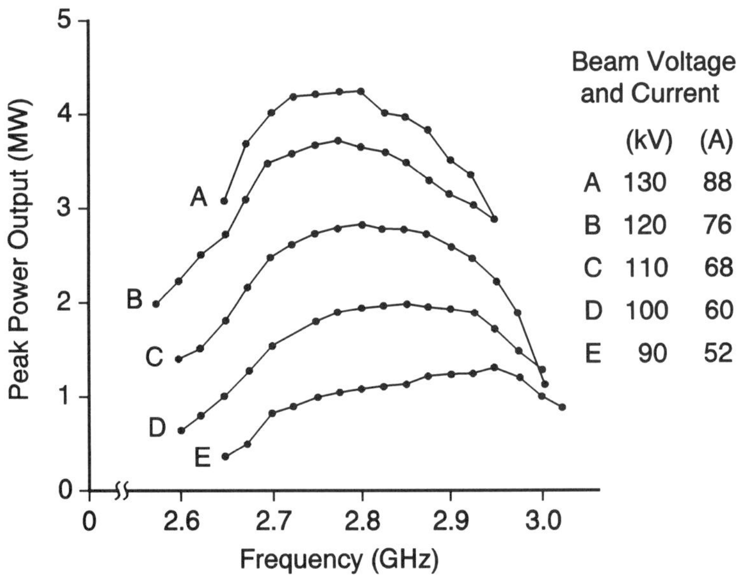

Figure 13-40 The power output for a VA-125A TWT. From Armand Staprans *et al., Proc. IEEE*, March 1973. © 1973 IEEE.

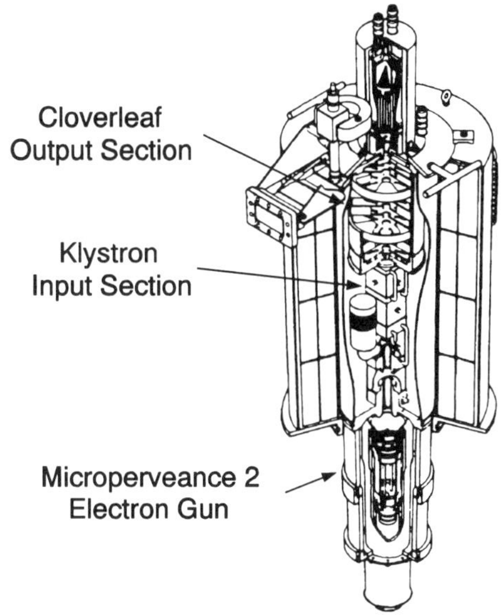

Figure 13-41 A typical multimegawatt twystron. From *Microwave Tube Manual* by Varian, Air Force Publication No. T.O.00-25-251, 1979.

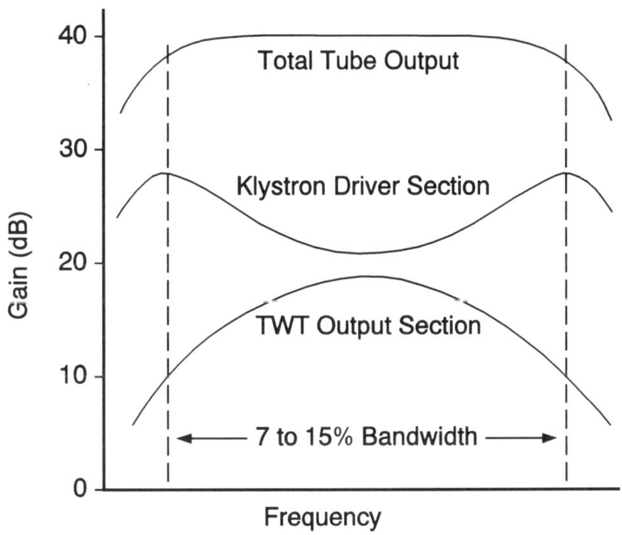

Figure 13-42 The gain of a twystron amplifier. From *Microwave Tube Manual* by Varian, Air Force Publication No. T.O.00-25-251, 1979.

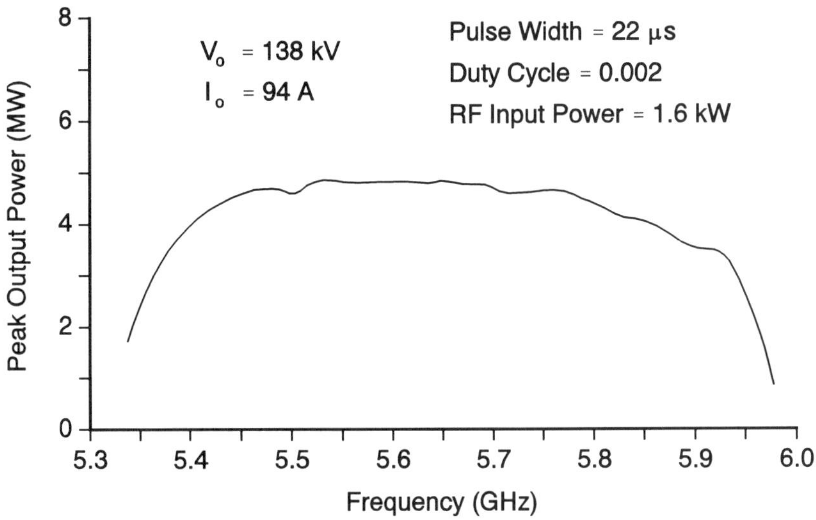

Figure 13-43 The power output for twystron amplifier VA-146. From *Microwave Tube Manual* by Varian, Air Force Publication No. T.O.00-25-251, 1979.

Twystrons with peak power outputs up to at least 10 MW have been built. Average powers over 30 kW are available and can be increased about an order of magnitude before circuit losses and interception present fundamental limitations.

13.5 TERMINATIONS AND TRANSITIONS

The need for severs in TWTs has been discussed in previous chapters. Terminations are required at the severs to minimize reflections. The RF power that a termination must dissipate can be appreciable in coupled-cavity TWTs. As a result, the termination is normally a lossy ceramic material capable of high temperature operation and appropriately bonded to the surrounding metallic structures to achieve the required heat dissipation capability. Examples of terminations for a cloverleaf structure and a staggered-slot structure are shown in Figures 13-44 and 13-45, respectively.

Two types of transitions are shown for the high-power coupled-cavity TWT in Figure 13-39. At the input, where the power level is low, a coaxial line is used to couple to a cavity, and at the output, a transition from the cavity to a waveguide is used. The use of a coaxial line and a coupling loop is common at low power levels. A coaxial line and coupling loop for a coupled-cavity structure are illustrated in Figure 13-46. Also shown is the side view of a waveguide transition. In coupling a cavity to a waveguide, it is often necessary to use inductive impedance-matching elements as shown in Figure 13-47. These elements offset the capacitive reactance of the cavity gap.

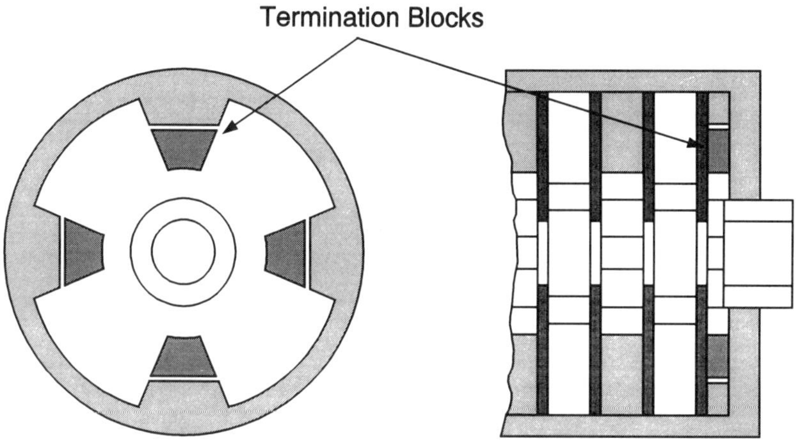

Figure 13-44 Termination for a cloverleaf structure. From *Power Travelling Wave Tubes* by J.F. Gittins. Published 1965 by American Elsevier, Inc.

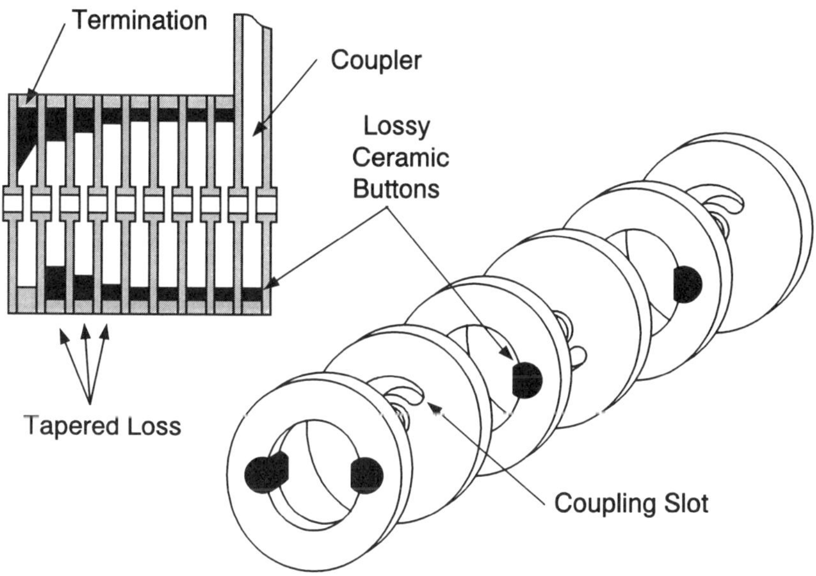

Figure 13-45 Termination and lossy ceramic buttons (for producing tapered loss pattern). From Hughes Aircraft Co., *TWT and TWTA Handbook*.

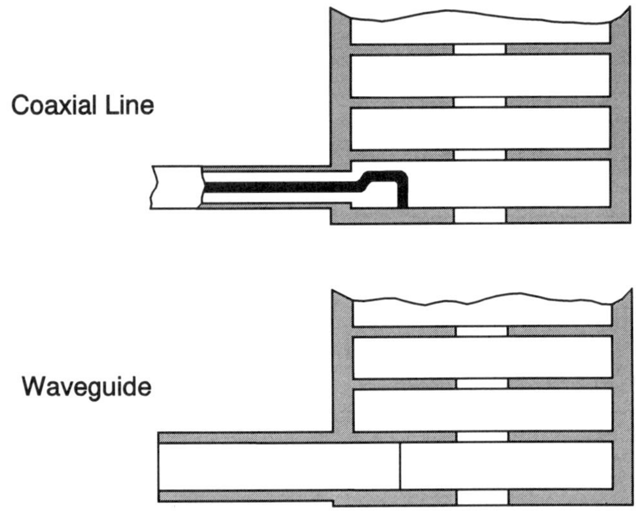

Figure 13-46 Coaxial-line and waveguide transition to coupled-cavity structures. From *Power Travelling Wave Tubes* by J.F. Gittins. Published 1965 by American Elsevier, Inc.

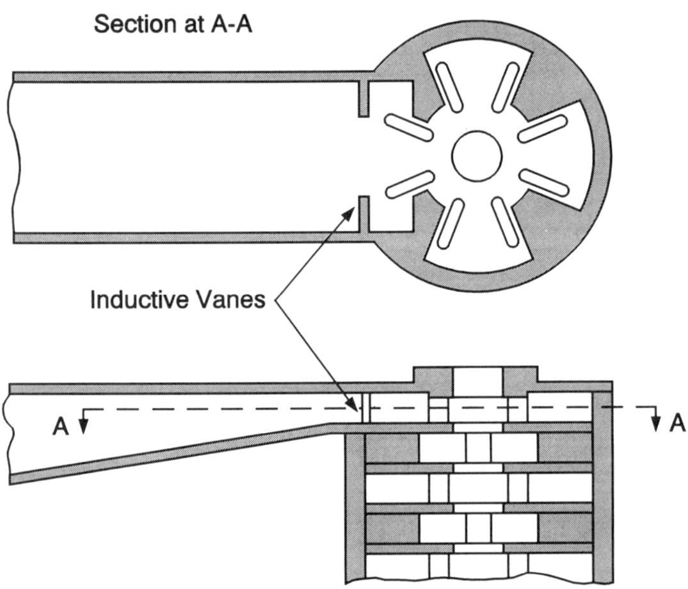

Figure 13-47 Transition from waveguide to clover-leaf structure. From *Power Travelling Wave Tubes* by J.F. Gittins. Published 1965 by American Elsevier, Inc.

REFERENCES

[1] Recommended for further reading: John T.Mendel, "Helix and Coupled-Cavity Traveling-Wave Tubes," *Proc. IEEE*, vol. 61, no. 3, March 1973, pp. 280–298; Armand Staprans, Earl William McCune and Jack A. Ruetz, "High-power linear-beam tubes," *Proc. IEEE,* vol. 61, no. 3, March 1973, pp. 299–327.

[2] H.J. Curnow, "A general equivalent circuit for coupled-cavity slow-wave structures," *IEEE Trans. Microwave Theory Tech.*, vol. MTT-13, Sept. 1965, pp. 671–675.

[3] H.J. Curnow, " A new equivalent circuit for coupled cavity structures," *Proc. 5th International Congress on Microwave Tubes*, conference held in Paris, France, September 14–18, 1964, published by Academic Press, New York and London, 1964.

[4] J.F. Gittins, *Power Travelling-Wave Tubes*, New York: American Elsevier, 1965.

[5] J.R. Frey and I. Tammaru, "A coupled-cavity TWT operating in the inverted slot mode," *Technical Digest*, 1981 International Electron Devices Meeting, pp. 504–506.

[6] Arthur Karp and W. Revis Ayers, "Design concepts for an octave-bandwidth coupled-cavity TWT," *Technical Digest*, 1978 International Electron Devices Meeting, pp. 546–549.

[7] Bill G. James, "Coupled-cavity TWT designed for future mm-wave systems," *MSN & CT*, September 1986, pp. 105–116.

[8] James W. Hansen, "US TWTs from 1 to 100 GHz," *Microwave Journal*, 1989 State of the Art Reference, pp. 179–193.

[9] M. Chodorow and R.A. Craig, "Some new circuits for high-power traveling-wave tubes," *Proc. IEEE*, August 1957, pp. 1106-1118.

Chapter 14

Depressed Collectors

A primary consideration in selecting microwave tubes for many applications is the overall efficiency in converting dc input power to RF output power at the desired frequency. One very important part of overall efficiency is electronic efficiency, which is the efficiency of conversion of power in the electron beam to RF power. Electronic efficiency has received considerable attention since the invention of microwave tubes. In this chapter, it will be assumed that the electronic efficiency has been established in the design of the RF characteristics of a tube. Then, with electronic efficiency fixed, the enhancement of the overall efficiency with the collector will be discussed.

14.1 POWER FLOW

The various elements of power flow in a linear beam tube are illustrated in Figure 14-1. Power is supplied for the heater, the electron beam and the solenoid. The RF input power is normally at least 30 dB below the RF output power and is assumed to be negligible.

Power is lost when electrons are intercepted by the circuit and by RF losses within the tube. In addition, some of the RF power that is generated is lost because it is not at the desired frequency (harmonics, intermodulation products, *et cetera*).

Finally, after the RF output is removed from the tube, the beam leaving the RF circuit (the spent beam) contains power that is delivered to the collector. Some of this power is converted to heat by electrons striking the collector. If the collector is designed properly, the electrons can be slowed down before they strike the collector surfaces, and the collector can thereby recover power from the spent beam [1].

Now, using the various elements of power identified in Figure 14-1, it is straightforward to derive power flow and efficiency relationships. The overall tube efficiency, η_{ov}, is simply the ratio of the RF output power to the total input power, P_{in}. That is,

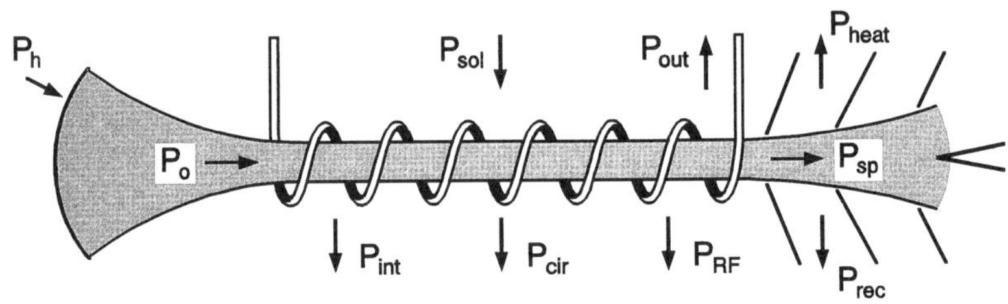

P_h = Heater Power
P_o = Beam Power from Gun
P_{int} = Interception Loss
P_{cir} = Circuit Loss
P_{RF} = Harmonics, Intermods, etc.
P_{sol} = Solenoid Power
P_{out} = RF Output
P_{sp} = Power in Spent Beam
P_{heat} = Heat in Collector
P_{rec} = Power Recovered by Collector

Figure 14-1 Power flow in a linear beam tube.

$$\eta_{ov} = \frac{P_{out}}{P_{in}} \tag{14-1}$$

where

$$P_{in} = P_o + P_h + P_{sol} - P_{rec} \tag{14-2}$$

or

$$P_{in} = P_{ot} - P_{rec} \tag{14-3}$$

where

$$P_{ot} = P_o + P_h + P_{sol} \tag{14-4}$$

is the total power associated with generating and focusing the electron beam. As a result,

$$\eta_{ov} = \frac{P_{out}}{P_{ot} - P_{rec}} \tag{14-5}$$

The power recovered by the collector may be written as

$$P_{rec} = \eta_{coll} P_{sp} \tag{14-6}$$

where η_{coll} is the efficiency of the collector. The spent beam power is beam power from the gun less the power removed from the beam as it passes through the tube. That is,

$$P_{sp} = P_o - P_{int} - P_{cir} - P_{RF} - P_{out} \tag{14-7}$$

so

$$P_{rec} = \eta_{coll}\left(P_o - P_{int} - P_{cir} - P_{RF} - P_{out}\right) \tag{14-8}$$

Two additional efficiency terms are useful in completing the analysis. One is the electronic efficiency, η_e, which, as pointed out earlier, is the efficiency of conversion of beam power to RF power at the desired frequency, that is,

$$P_{out} + P_{cir} = \eta_e P_o \tag{14-9}$$

Values for η_e range from a few percent to well over 50%.

The other efficiency is the circuit efficiency, η_{cir}, which is the efficiency of the RF circuit in delivering the generated RF power at the desired frequency to the output connector of the tube. Thus,

$$P_{out} = \eta_{cir}\left(P_{out} + P_{cir}\right) = \eta_{cir}\eta_e P_o \tag{14-10}$$

Values for η_{cir} are usually in the 75 to 90% range.

Now, by combining (14-10) and (14-8) with (14-5), the overall efficiency may be written as

$$\eta_{ov} = \frac{\eta_{cir}\eta_e}{\dfrac{P_{ot}}{P_o} - \eta_{coll}\left[1 - \eta_e - \dfrac{P_{int} + P_{RF}}{P_o}\right]} \tag{14-11}$$

The plot of this equation in Figure 14-2 is useful in illustrating the relative significances of the various efficiencies and powers.

In Figure 14-2, the effect of P_{ot} is shown on the 90% collector efficiency plot. This same range of variations would apply to other η_{coll} plots, but is shown for only the 90% plot. Similarly, the effects of variations in $P_{int} + P_{RF}$ are shown on the 70% η_{coll} plot, but would apply to other collector efficiency plots. The effects of variations in circuit efficiency are shown on the 50% η_{coll} plot but would apply to other η_{coll} plots.

Perhaps the most striking aspect of Figure 14-2 is the strong dependence of overall efficiency on collector efficiency, especially for high values of collector efficiency. This leads directly to the next section of this chapter, which deals with the design of efficient collectors.

The importance of circuit efficiency is also evident in Figure 14-2. Note that a 10% increase in η_{cir} increases η_{ov} about five percentage points. This leads to considerations of circuit efficiency, which were discussed in Chapter 12.

Finally, in considering (14-11) and Figure 14-2, heater power and RF losses (harmonics and intermodulation products) are normally small. Thus, P_{ot} contains primarily the power penalty resulting from the use of a solenoid for focusing. Increases in η_{ov} that result from decreases in solenoid power (and P_{ot}) can be readily be traded

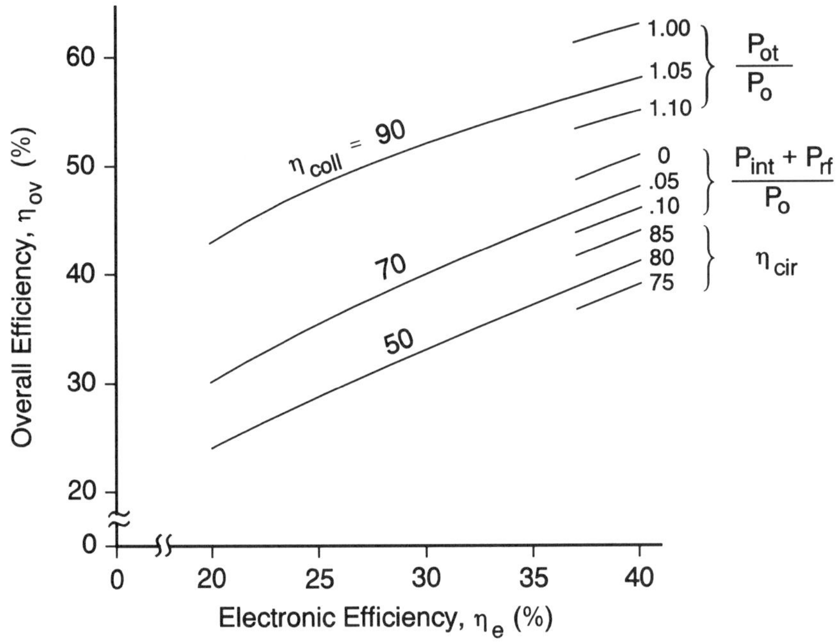

Figure 14-2 Variations of overall efficiency with the various elements of efficiency and power.

off against the decrease in η_{ov} that results from an increase in P_{int}, (interception power), which, in turn, normally results from a decrease in solenoid power.

14.2 POWER RECOVERY WITH A DEPRESSED COLLECTOR

A collector for a linear beam tube is shown in Figure 14-3. As the beam passes through the pole piece, the magnetic focusing field is removed, and so space charge forces cause the beam to expand as it enters the collector.

With the collector at the same potential as the body of the tube, electrons strike the collector at relatively high velocity. Electron energy is converted to heat at the collector surface. With a collector like that shown in Figure 14-3, a relatively large percentage (normally 50% or more depending on electronic efficiency) of the power input to the tube is dissipated in heat in the collector.

By reducing (depressing) the voltage on the collector below the body potential, the velocities of electrons striking the collector and the heat generated are reduced. As a result, the collector recovers some of the power in the spent electron beam.

The concept of power recovery by a collector is sometimes confusing. The following example should help with the understanding of power recovery. Assume that a TWT with a 10 kV, 1 A electron gun produces an RF power of 2 kW as is indicated

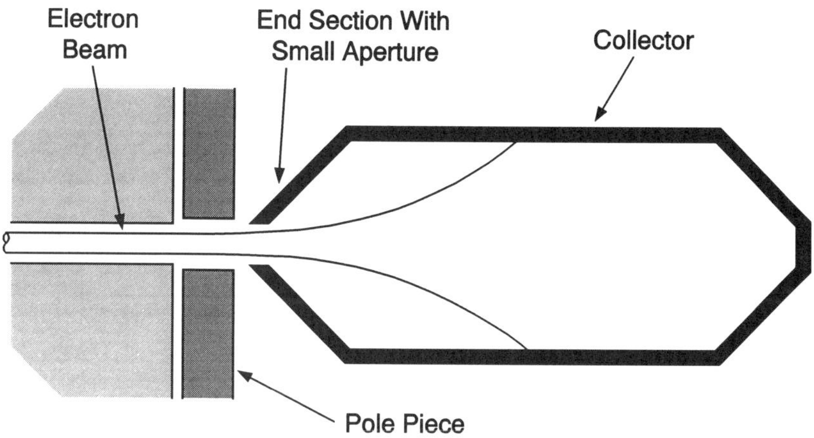

Figure 14-3 Collector for a linear beam tube.

in Figure 14-4. To simplify the example, we will assume that, other than interception, there are no losses, that heater power is negligible and that PPM focusing is used. With the collector at the same potential as the body (and the helix) of the tube (ground potential) as shown in Figure 14-4(a), the overall efficiency is simply

$$\eta_{ov} = \frac{2 \text{ kW}}{10 \text{ kV} \times 1 \text{ A}} = 20\,\% \tag{14-12}$$

All of the current intercepted by the circuit and collected by the collector returns through the cathode power supply. It should also be noted that, to regulate the RF output, the full 10 kW of input power must be regulated.

In Figure 14-4(b), a second 10 kV power supply has been provided for the collector. The total power supplied to the tube is the same as in (a) and so the efficiency is still 20%. It should be noticed that, to regulate the RF output power, only the cathode supply must be regulated and the power provided by that supply is now only 100 W. The regulation requirement for the collector supply of 9900 W is normally far less severe than for the cathode supply. Thus, while no improvement in efficiency has been gained in Figure 14-4(b), a substantial reduction in power supply complexity may have been achieved.

In Figure 14-5, the voltage of the collector supply has been reduced to 6000 V. The current provided by this supply is still the full collector current of 0.99 A; however, the power provided is only 5940 W. The RF output power is still 2 kW and so the overall efficiency is now

$$\eta_{ov} = \frac{2000 \text{ W}}{100 \text{ W} + 5940 \text{ W}} = 33\,\% \tag{14-13}$$

which is a substantial improvement over the 20% obtained in Figure 14-4(a) and (b).

(a)

(b)

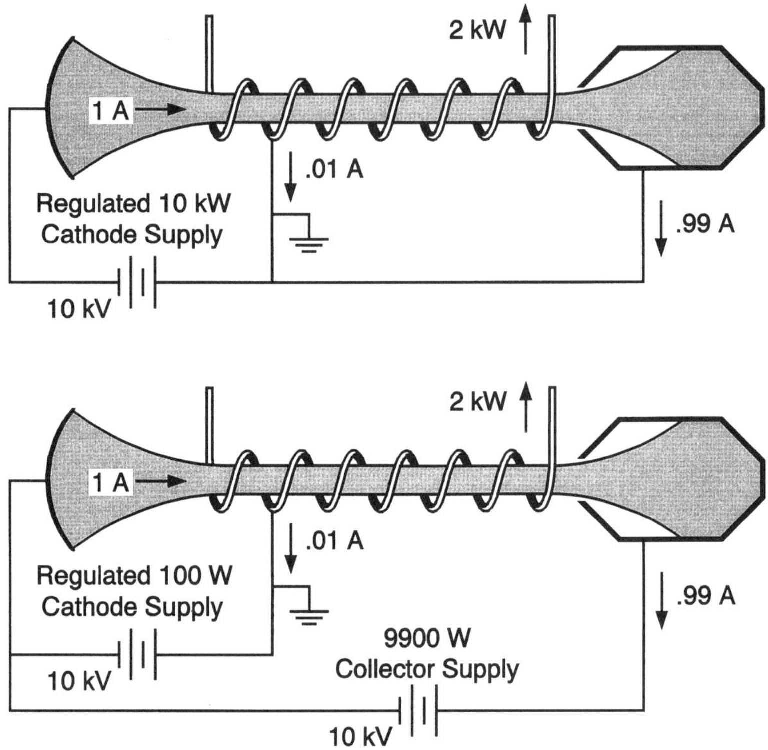

Figure 14-4 Power supply connections for collectors at ground potential; (a) common power supply and (b) separate power supplies.

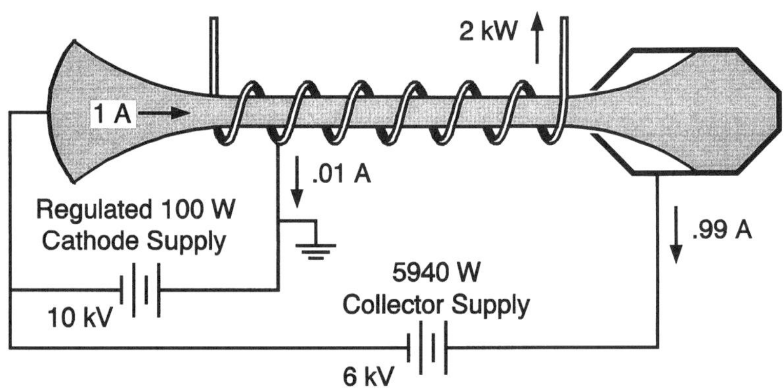

Figure 14-5 Depressed collector circuit for recovering beam power.

The improvement in efficiency in Figure 14-5 has been achieved by reducing (depressing) the voltage of the collector below the body and helix voltage and, thereby, slowing down the electrons before collecting them. Thus, we say that the collector has "recovered" power from the electron beam.

It should be noted here that the power supply engineer considers collector depression to be a technique for reducing power input rather than for recovering power because the power was not supplied to the tube in the first place. In fact, of course, both points of view are correct. At the risk of belaboring this point, consider the example shown in Figure 14-6. Here, the collector is at cathode potential and there is no RF circuit to extract power from the electron beam. It is also assumed that no current is intercepted by the anode. With 10 kV applied to the anode, a 10 kW electron beam is generated. However, as this beam enters the collector, it is slowed down and the electron velocity drops to zero as the electrons land on the collector. No heat is generated in the collector and there are no other losses and so no power is dissipated. Thus, a 10 kW beam has been generated by the electron gun and all of this power has been recovered by the collector. No power is provided by the 10 kV supply because there is no current flowing to it.

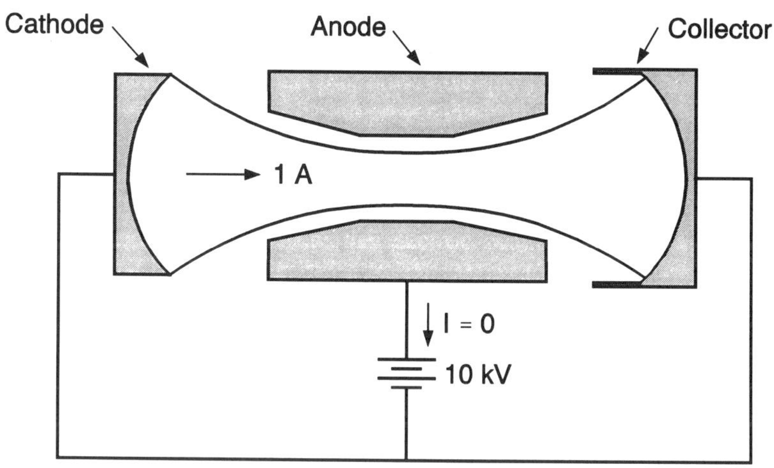

Figure 14-6 Electron beam device with a 10 kW beam and full beam power recovery.

14.2.1 Electron Energy Distribution

In the example shown in Figure 14-5, the collector voltage was 6 kV, or 4 kV below the body potential. Thus, the collector was "depressed" 4 kV, and this raised the efficiency from 20% to 33%. The next question that might be asked is: Why not depress the collector further and improve the efficiency more? The problem that arises is that the operating potential of a depressed collector power supply like that shown in

Figure 14-5 is highly dependent on the distribution of electron velocities in the electron beam. The slower electrons will be reflected by the collector if the potential is depressed too far. With increasing collector voltage, as shown in Figure 14-7, collector current will decrease and body current will increase. Excessive body current will result in damage to the RF structure. Also, reflected electron current traveling back through the tube induces noise in the RF circuit and, in addition, provides a signal feedback path. Single-element collectors, like that illustrated in Figures 14-5, are capable of recovering roughly 30 to 40% of the power in a spent beam (highly dependent on the tube type).

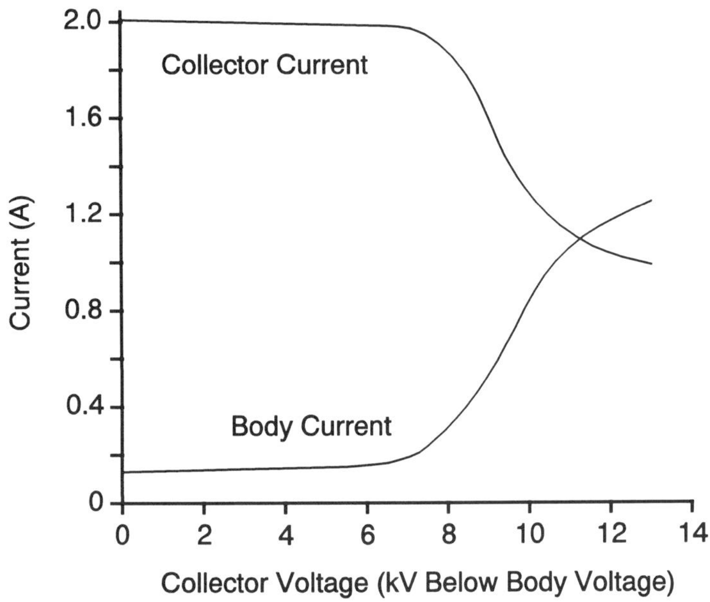

Figure 14-7 Effect of collector voltage on body current.

To recover a large fraction (over 80%) of the power in a spent beam and, at the same time, prevent electrons from being reflected, a collector must contain electric fields shaped so that all electrons are nearly stopped, regardless of velocity, just as they strike the collector electrodes. In addition, the fields must be shaped to prevent secondary electrons from being accelerated (thus wasting energy) or leaving the collector.

To better understand the approach to the problem of building a highly efficient collector, it is necessary to examine the energies of the electrons in the spent beam entering the collector. For purposes of illustration, it will be assumed that the collector power supply is connected as shown in Figure 14-8 and that the collector voltage can be varied. (Note that this is not a practical way to apply voltage to the collector

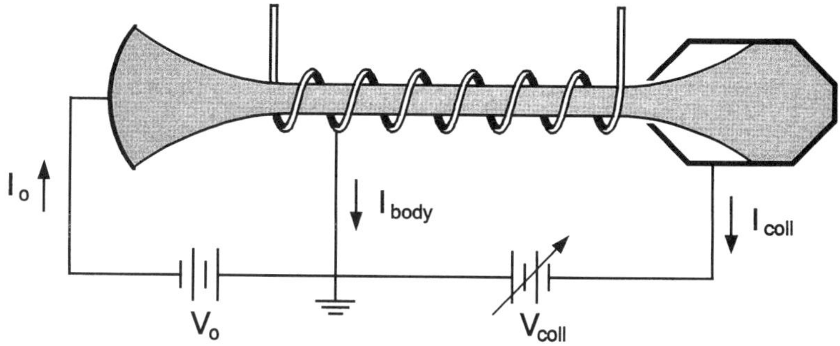

Figure 14-8 Power supply configuration for consideration of energy distribution of spent beam.

because power is always flowing *into* the supply. Thus, the supply is acting like a battery that is continually being charged.)

In Figure 14-9, the current to the collector is plotted as a function of the amount by which the collector voltage is depressed, V_{coll}. It is assumed here that the body current is zero as long as the collector is not depressed too far. Also, the secondary electron current is ignored. When $V_{coll} = 0$, all of the cathode current reaches the collector and so $I_{coll} = I_o$. With no RF drive, no energy is extracted from the beam and so *all* of the electrons entering the collector have the same energy that they had when leaving the electron gun. As a result, as the collector voltage is depressed, all of the electrons can reach the collector until $V_{coll} = V_o$. As V_{coll} becomes larger than V_o, suddenly none of the electrons can reach the collector and so I_{coll} drops to zero.

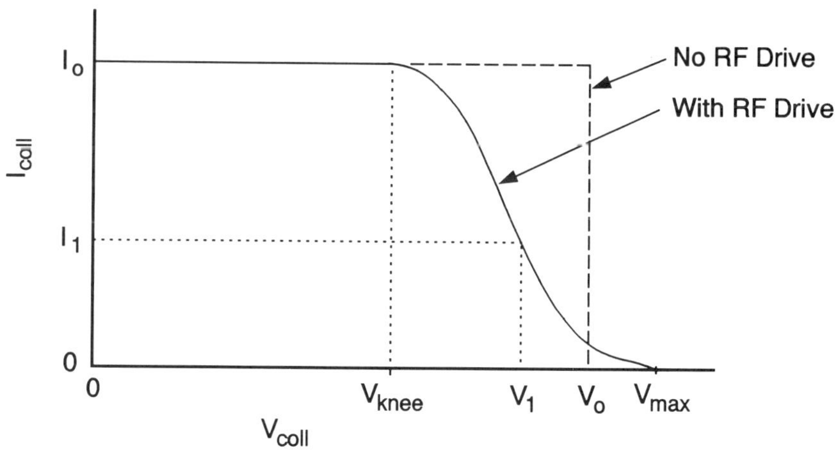

Figure 14-9 Collector current *versus* collector voltage.

When the RF drive is applied to the tube, energy in varying amounts is extracted from (or in some cases, added to) electrons during the interaction of the RF wave on the circuit with the beam. In a helix TWT, some electrons lose as much as 30 to 40% of their initial energy (when they left the gun) and some may gain as much as 30%.

With the RF drive, as the collector voltage is depressed from the body potential, all of the cathode current reaches the collector until the collector voltage reaches the point where the slowest electrons can no longer reach the collector surface. This voltage is indicated by V_{knee} in Figure 14-9. As the collector voltage is depressed below V_{knee}, more and more electrons are reflected and so I_{coll} decreases. At some voltage V_{max}, which is actually more negative that the cathode potential, V_o, the collector current drops to zero. The small "tail" of the curve at voltages between V_o and V_{max} is caused by those electrons that were accelerated rather than decelerated as the beam interacted with the RF wave.

In general, any current, such as I_1 in Figure 14-9, consists of electrons with energy eV_1 or greater. Thus, with the collector at potential V_1, the maximum current that could be collected would be I_1. The remaining current, I_0 - I_1, would be reflected and returned to the body of the tube.

The shape of the curve in Figure 14-9 depends a great deal on the nature of the interaction between the beam and the RF circuit. As an example, Figure 14-10 contains curves for a helix TWT and for a high efficiency klystron. (For these plots, the current and voltage have been normalized to the cathode values. More will be said about the reasons for this normalization in following paragraphs.) For the klystron, because the interaction is so intense, some electrons lose nearly all of their energy while others are accelerated so much that their energy is nearly doubled. Thus, a collector that was depressed even a small amount would reflect some current.

14.2.2 Spent Beam Power

In Figure 14-9, current is plotted as a function of voltage and so the area under a curve represents power. For example, the area under the *No RF Drive* plot is simply $I_o \times V_o$, which is the dc beam power, P_o. When RF drive is applied, as shown by the shaded area in Figure 14-11, the difference between the no-drive rectangle and the curve with drive is nearly the beam power that is converted to RF power. Some power is lost to the $V_{max}-V_o$ tail (cross-hatched area) and there are RF losses, and so the RF power out of the tube is somewhat smaller than is shown by the shaded area.

In Figure 14-11, the area under the solid curve is the power remaining in the spent beam. This is the maximum power that could possibly be recovered by an ideal collector. To recover part of that power, the power supply configuration must be as shown in Figure 14-12 (and as was shown in Figure 14-5). The collector supply voltage, V_c, is the cathode voltage, V_o, less the amount by which the collector voltage is depressed, V_{coll}. The maximum power that could be recovered by a single-electrode collector is shown by the shaded area in Figure 14-13. (The collector with a single

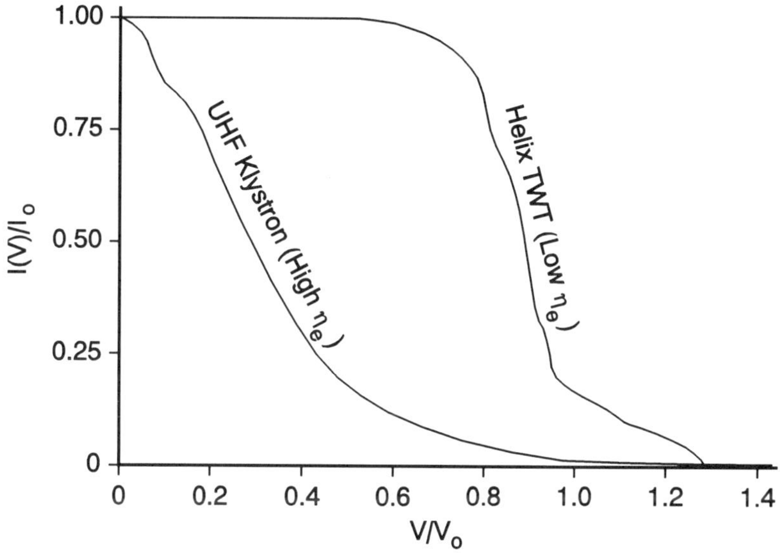

Figure 14-10 Beam energy distributions for tubes with high and low electronic efficiency.

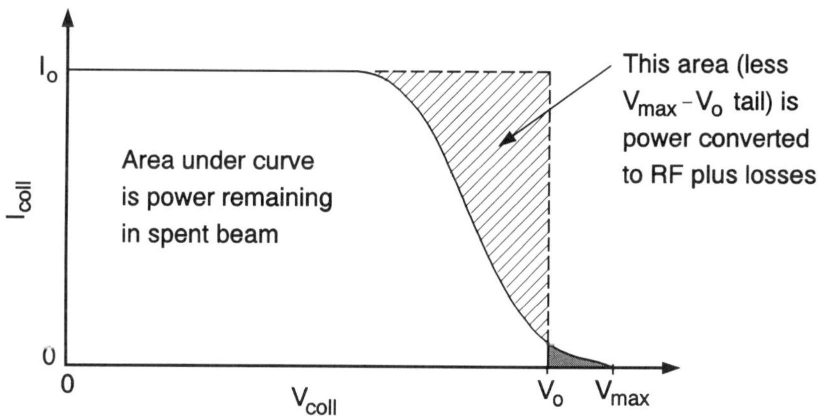

Figure 14-11 Spent beam power.

electrode is often referred to as a *single-stage collector*. Collectors with several electrodes, to be discussed in following paragraphs, are called *multistage collectors*.)

When V_c is smaller than shown in Figure 14-13, electrons are reflected by the collector because the amount by which the collector is depressed, V_{coll}, is greater than the knee voltage. When V_c is greater that shown in Figure 14-13, all electrons are collected. In practice, it is not possible to operate a single-stage collector as close to the knee of the curve as is indicated in Figure 14-13. One reason for this is that V_c is

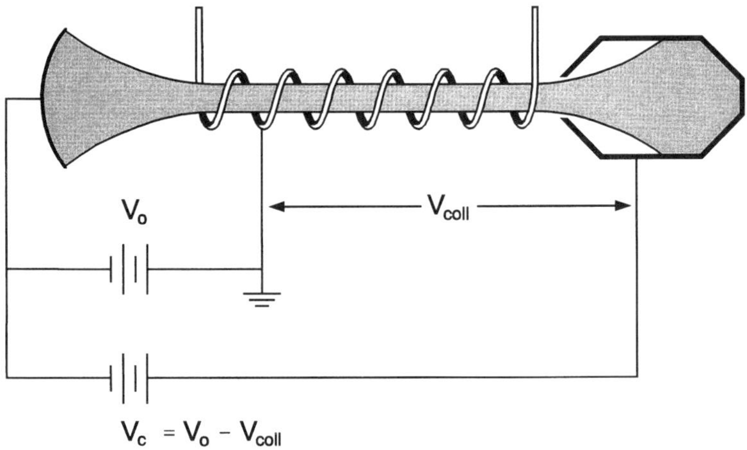

gure 14-12 Power supply configuration for depressed collector operation.

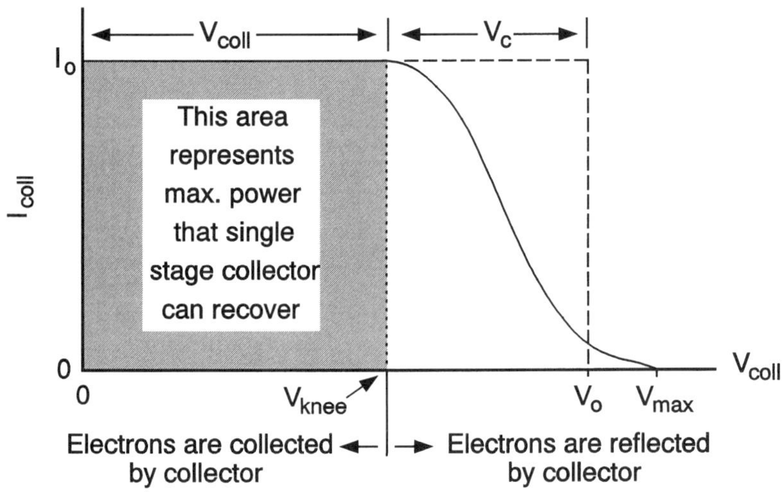

Figure 14-13 Maximum power recovered by a single-stage collector.

not normally highly regulated and so V_c must be high enough to avoid electron reflections under all operating conditions. Another reason is that the knee of the curve is normally not as sharply defined as is shown in Figures 14-9, 14-11 and 14-13 and so excessive electron reflection (and body current) occurs with collector voltages significantly above that indicated in Figure 14-13.

14.2.3 Effect of Body Current [2]

The effect of body current on the energy distribution curves is shown in Figure 14-14. Here, currents and voltages have been normalized with respect to the cathode values. This normalization is often used because body current and collector currents are often expressed as percentages of the cathode values.

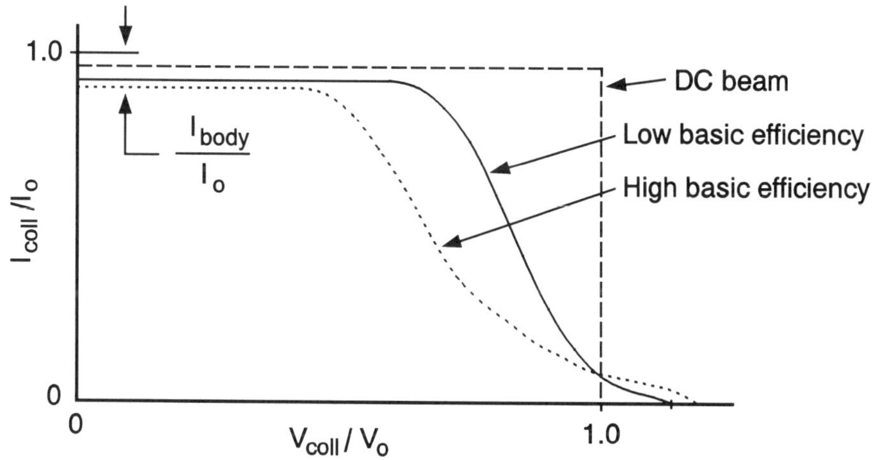

Figure 14-14 Effect of body current on energy distribution plots.

The three plots given in Figure 14-14 show how increases in body current with drive level affect the energy distribution curves as the RF drive is increased from zero (dc beam) to a low level (low basic efficiency) and then to a high level (high basic efficiency). Body current reduces the areas under the curves, and this reduces the power available to be recovered by a collector.

14.3 MULTISTAGE DEPRESSED COLLECTORS

The preceding discussion provides the basis for the design of efficient collectors. Modern, highly efficient collectors use several electrodes at different potentials to selectively collect electrons at low energy levels. These collectors are called *multistage depressed collectors* (MDCs).

In the previous section, energy distribution curves were shown for the electron beam entering the collector. It was pointed out that the area under a curve represents the power available in the spent beam for recovery by the collector. To recover a large fraction of this power, the collector must be designed to sort the electrons in the spent beam into various energy classes. Then, electrons in each energy class must be collected on an electrode at a voltage that recovers as much of that energy as possible.

Figure 14-15 illustrates the recovery of a substantial amount of spent beam power by a four-stage collector. Here, as in previous discussions in this chapter, the effects of secondary electrons are not yet being considered. A collector electrode depressed to the cathode potential, V_o, can collect that current, I_4, consisting of electrons with energies in the range $eV_{max}-eV_o$. An electrode depressed to potential, V_3, can collect current, I_3, but not I_4 because that current goes on to the electrode at V_o.

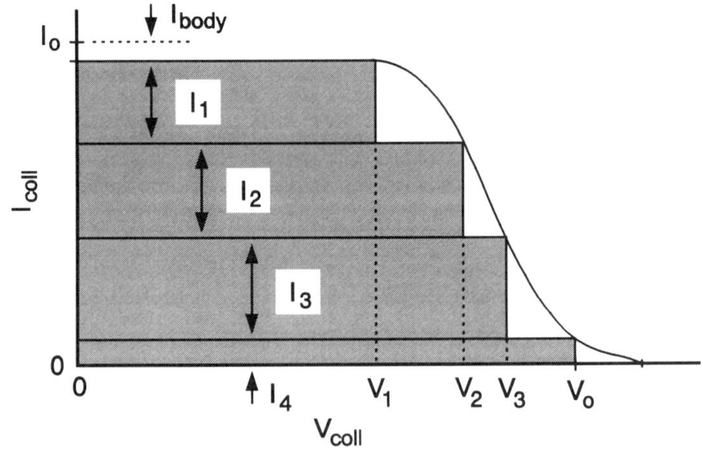

Figure 14-15 Power recovered by a four-stage depressed collector.

The total power recovered by the four-stage collector is represented by the shaded area in Figure 14-15 and is

$$P_{rec} = I_1 V_1 + I_2 V_2 + I_3 V_3 + I_4 V_o \tag{14-14}$$

The research and development for the collectors that most efficiently operate as indicated in Figure 14-15 has been the work of the NASA Lewis Research Center in Cleveland, Ohio. The original impetus for the NASA work was the extremely high premium on efficiency resulting from the high cost of power generation and heat reflection on satellites using microwave tubes. Now, the NASA collector technology is being put to use in a broad range of applications where high microwave power is required and prime power is severely limited. An example is airborne electronic countermeasures systems [3].

The collector-electrode configurations for the NASA MDCs are illustrated in Figures 14-16 and 14-17. The first NASA collectors were derived from mathematical solutions to the problem of both slowing down and dispersing the electron beam entering the collector. As shown in Figure 14-16, the equipotential surfaces were designed to be concave toward the aperture where the beam entered the collector. This forced electrons to move radially outward. Collector electrodes were designed to coincide with the equipotential surfaces that were derived mathematically. The aperture sizes in the electrodes were chosen so that electrons would land on electrodes at the appropriate voltage. Since most beam electrons landed on the back sides of the electrodes, secondary electrons were emitted into a negative potential region and, therefore, were immediately returned to the electrode surfaces. More will be said about secondary electrons in subsequent paragraphs.

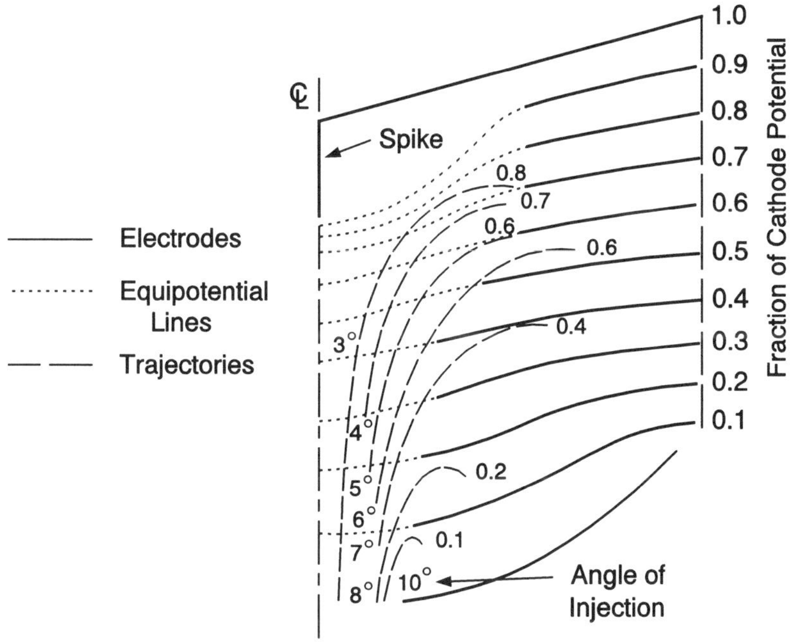

Figure 14-16 Axisymmetric MDC with equipotential surfaces determined mathematically. From H.G. Kosmahl, *Proc. IEEE*, November 1982.

As has been pointed out in previous paragraphs, in all microwave tubes, some electrons are out of phase with the RF voltage and are accelerated, rather than decelerated, by the interaction process. These electrons leave the interaction region with energies greater than the original beam energy. In the NASA collector in Figure 14-16, a sharp spike is shown at the back of the collector. This spike helps shape the fields at the back of the collector to disperse high energy electrons. In addition, the surface area presented to high energy electrons entering the collector on axis is minimized. Thus, the generation of reflected electrons is minimized.

The efficiencies of collectors patterned after the design in Figure 14-16 is very high (> 85%) and the current reflected from the collector is very small. Complexity and size are disadvantages.

Collectors of the type shown in Figure 14-17 were originally developed during a joint NASA/USAF program. Collector size was reduced significantly with a small compromise in efficiency so that MDCs could be used in airborne applications, where

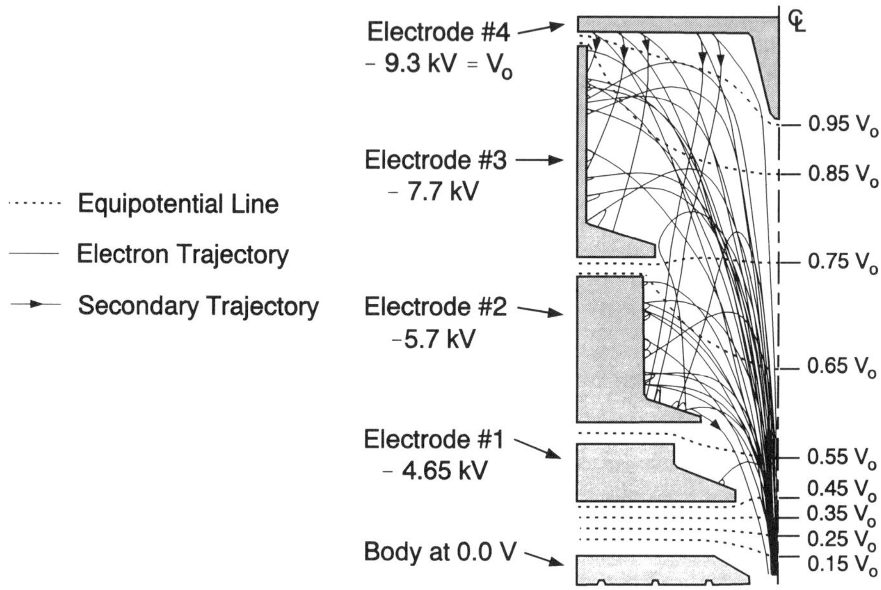

Figure 14-17 Equipotential surfaces and electron trajectories in "individual lens" collector. From J.A. Dayton, Jr., AIAA 12th International Communication Satellite Systems Conference.

size and weight are important considerations. In these collectors, called *individual lens collectors*, each electrode and its associated fields were optimized to produce maximum collector efficiency while minimizing the current resulting from reflected electrons.

In order for an MDC (either the mathematically designed version or the individual lens version) to operate efficiently, the electron beam entering it must not diverge rapidly. Rapid divergence leads to large radial components of electron velocities and to energy components that cannot be recovered by the collector.

To reduce the divergence of a beam it is sometimes necessary to expand its diameter so that space charge forces are reduced. A magnetic refocusing system, one version of which is shown in Figure 14-18, is used for this purpose. An additional benefit of the refocusing system is an improvement in the laminarity of the beam entering the collector.

Improvements in overall efficiency that have been demonstrated through the use of a MDC on a dual-mode helix TWT for electronic countermeasures applications are illustrated in Figure 14-19. An overall efficiency improvement by a factor of three or more was demonstrated across a bandwidth of an octave. The collector efficiency was nearly constant at about 85%, as shown in Figure 14-20.

A very interesting application for MDCs is in tubes where efficient operation is necessary in the linear (small-signal) region. As shown in Figure 14-21, the efficiency of the MDC increases significantly as the input signal level is reduced. For the example

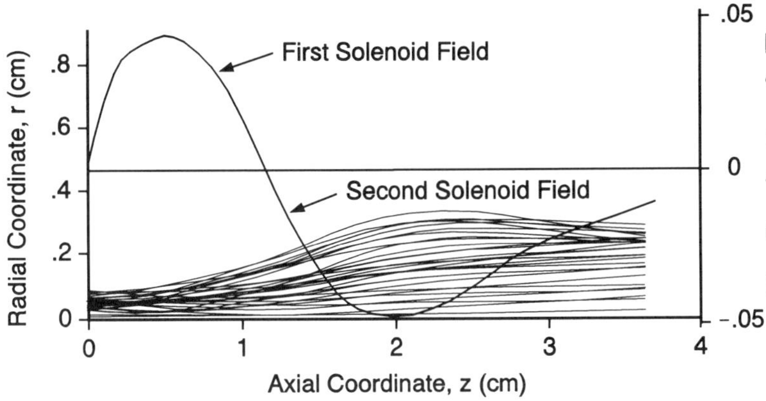

Figure 14-18 Magnetic refocusing system for reducing beam divergence and improving laminarity. Adapted from H.G. Kosmahl, *Proc. IEEE*, November 1982.

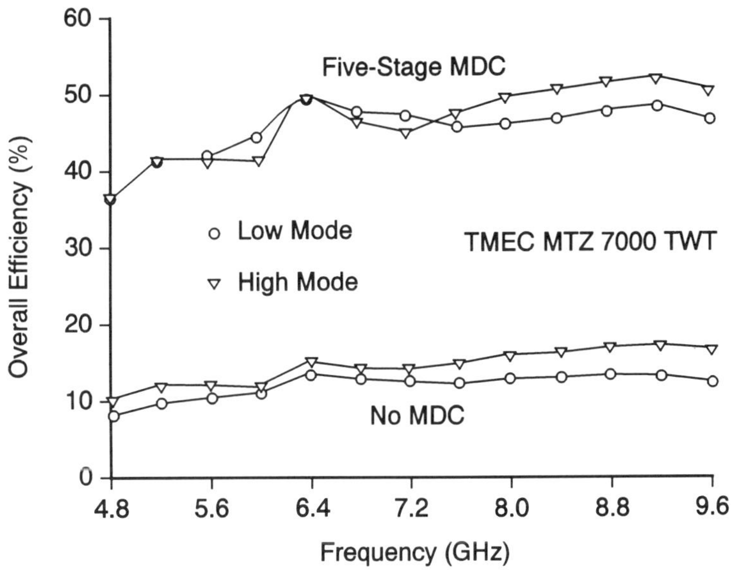

Figure 14-19 Overall efficiency of a dual-mode ECM helix TWT with and without a MDC. Adapted from H.G. Kosmahl, *Proc. IEEE*, November 1982.

in Figure 14-21, the MDC efficiency exceeds 97% when the drive signal level (and basic efficiency) are reduced to zero.

At a drive level low enough so that the basic (electronic) efficiency and the overall efficiency (without MDC) is 1%, use of an MDC increases the overall efficiency to 14%. At a basic efficiency of 4%, the MDC increases the efficiency to 30%. Thus,

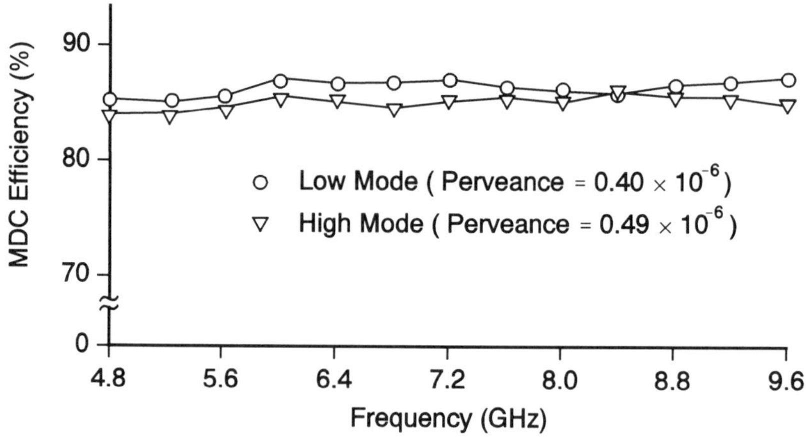

Figure 14-20 Collector efficiency at saturation. Adapted from H.G. Kosmahl, *Proc. IEEE*, November 1982.

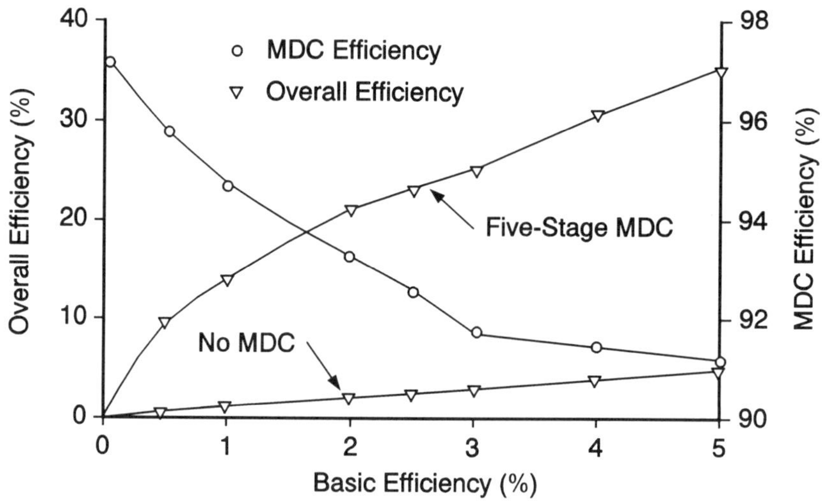

Figure 14-21 Overall and MDC efficiency *versus* basic (electronic) efficiency in the small-signal region. Adapted from H.G. Kosmahl, *Proc. IEEE*, November 1982.

using a MDC, efficient operation over a wide range of drive levels is possible.

A typical power supply configuration for a MDC is shown in Figure 14-22. As was the case for the single-element depressed collector, the cathode-to-anode (and helix) supply must be regulated, but it must supply only the power resulting from interception current. The tapped power supply for the collector is essentially unregulated. This is because variations in voltages to collector elements may cause some electrons to move from one collector element to another, but the overall effect on efficiency is negligible.

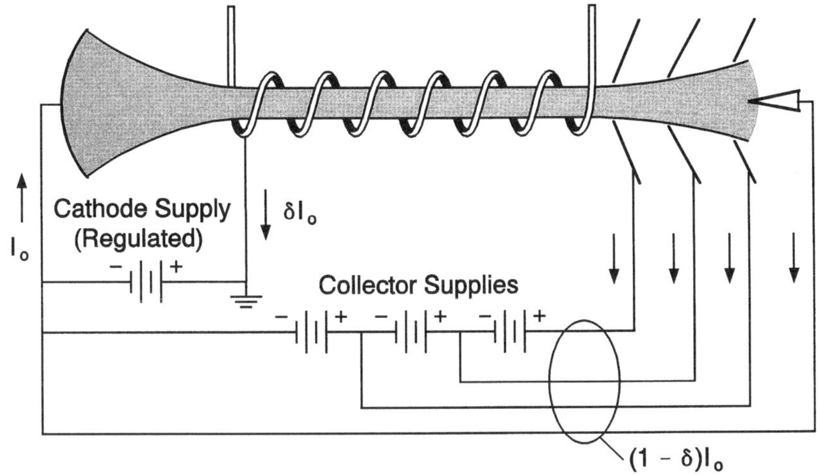

Figure 14-22 Power supply connection with a multistage depressed collector.

In Figure 14-16, the number of stages in the MDC was ten and, in Figure 14-17, it was four. The optimum number of stages for any application is determined by considering the increase in tube efficiency that can be achieved with a large number of electrodes and trading this off against the increased complexity of tube construction and the power supply. As Figure 14-23 shows, the increase in efficiency achieved with more than about three or four stages is small and so, for most applications, no more than three or four stages are used. The efficiencies in Figure 14-23 were computed and are based on a knowledge of the spent beam characteristics of each type of tube.

14.4 SECONDARY ELECTRONS IN DEPRESSED COLLECTORS

Excessive secondary electron emission in depressed collectors can cause two problems:

1. Escape of secondaries from the collector, which may cause excessive noise, signal distortions and heating of the circuit;
2. Current flow between collector electrodes and excessive heating of electrodes which, in turn, results in the reduction of collector efficiency.

The first of these problems is discussed elsewhere in this book (Chapters 15 and 16). The second problem arises when secondary electrons released from the front surface of a collector electrode are accelerated toward and then collected by a more positive electrode. An example of this situation was shown in Figure 14-17 where trajectories for five secondary electrons released from electrode #4 are plotted. Three of these electrons strike electrode #3 after being accelerated through 1.6 kV

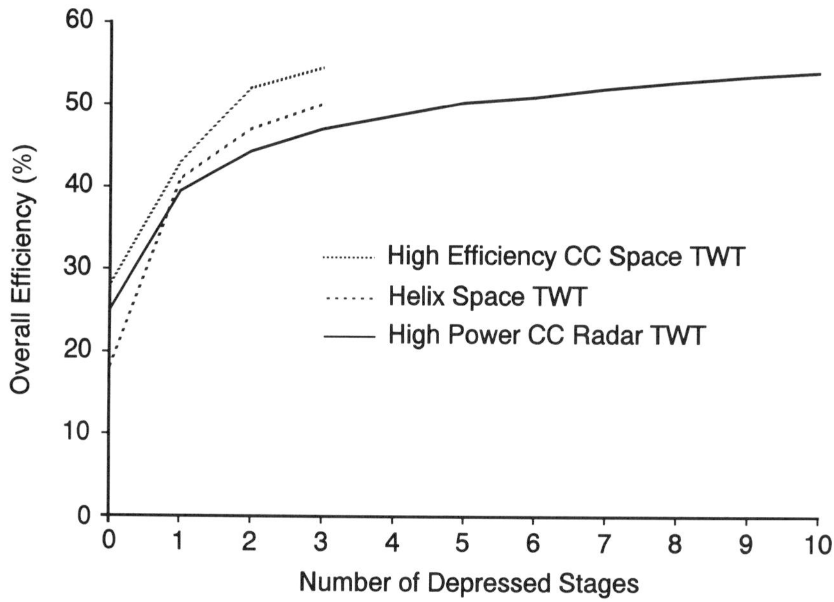

Figure 14-23 Efficiency *versus* the number of depressed collector stages. Courtesy Hughes Electron Dynamics Division.

(9.3 kV–7.7 kV). The other two strike electrode #2 after being accelerated through 3.6 kV.

The effect of secondary electrons on collector efficiency and overall efficiency is shown in Figure 14-24 [4]. These data were obtained from a helix TWT with an electronic efficiency of 16%. Tests were performed with a four stage MDC having various electrode surface treatments. The collector efficiency was increased about 10% by using electrodes with textured isotropic graphite surfaces rather than untreated copper surfaces. The overall efficiency was increased over 5%.

The technique of texturing materials for collector surfaces has been developed by the NASA Lewis Research Center [5]. The texturing is accomplished by bombarding the surface with argon ions. For the textured graphite surfaces described in [5], the ion accelerating potential was 1000–1500 V dc, the surface current density of the sample was 3–5 mA/cm^2 and the texturing period was 1–4 hrs. A scanning electron microscope photomicrograph of a fully developed textured carbon surface on a copper substrate is shown in Figure 14-25. The surface appearance in this figure is also typical of fully developed ion-textured pyrolytic and high purity isotropic graphite. The larger spires have average heights of 6–12 μm and average spacings of 3–6 μm.

The secondary emission characteristics of textured carbon on a copper substrate are shown in Figure 14-26. The true secondary emission coefficient for normal incidence of the primary beam is from 0.2 to about 0.25. This is significantly below the value

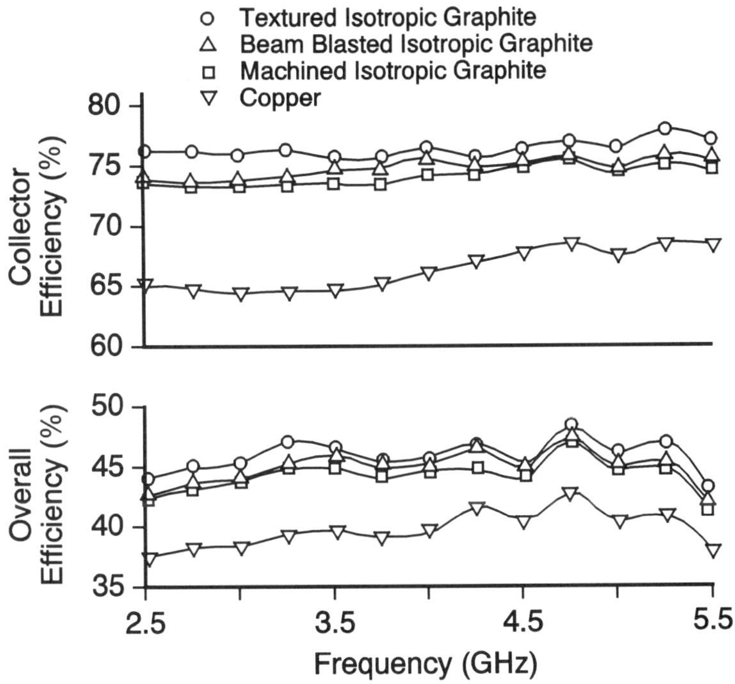

Figure 14-24 Efficiencies for various surfaces in a four-stage MDC on a TWT with $\eta_e = 16\%$. From P. Ramins and B.T. Ebihara, *IEEE Trans. on Electron Devices*, November 1986.

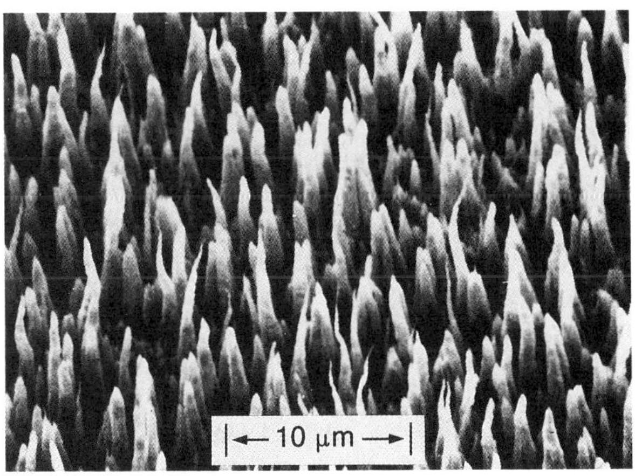

Figure 14-25 SEM photomicrograph of textured carbon on a copper substrate. From A.N. Curren, *IEEE Trans. on Electron Devices*, November 1986.

for soot, which has, in the past, been the lowest yield surface available. Of course, soot has not been used in microwave tubes because of its poor physical characteristics. The reflected primary yield index, π (shown in Figure 14-26(b)) for normal incidence ranges from about 0.5 to about 1.5. The basis for this index is the primary yield for soot with a primary electron energy of 1 keV and normal incidence. For these conditions, the value of π is unity.

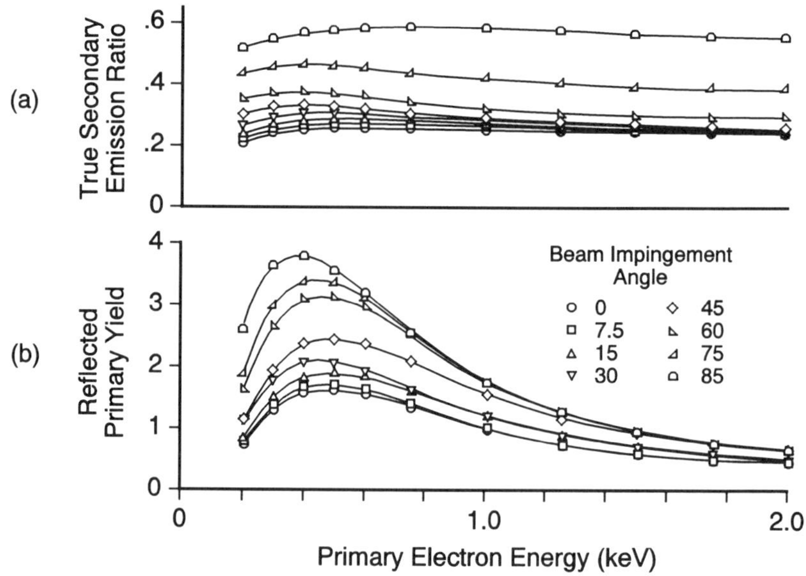

Figure 14-26 Secondary emission characteristics of textured carbon on a copper substrate. From A.N. Curren, *IEEE Trans. on Electron Devices*, November 1986.

In addition to texturing various carbon surfaces, NASA has developed techniques for texturing copper [6]. With one technique, sputtered molybdenum is used as a texture-inducing masking film. The true secondary electron emission ratios for these surfaces range from 0.3 to 0.5 and are a factor of three below ratios for untreated copper surfaces. The reflected primary yield index is a factor of about three to seven below that for an untreated copper surface. The results reported in [6] were for three samples of ion textured copper. One conclusion of the NASA work is "with further development, the molybdenum-masked, ion-textured OFHC copper surface will be a promising material for high efficiency MDC electrodes."

REFERENCES

[1] H.G. Kosmahl, "Modern multistage depressed collectors—A review," *Proc. IEEE*, vol. 70, Nov. 1982, pp. 1325–1334.

[2] Lecture notes from the Hughes Electron Dynamics Division prepared by I. Tammaru were of considerable help in the preparation of some of the material in this chapter.

[3] H.G. Kosmahl and P. Ramins, "Small-size 81 to 83.5 percent efficient 2- and 4-stage depressed collectors for octave-bandwidth high performance TWT's," *IEEE Trans. Electron Devices*, vol. ED-24, Jan. 1977, pp. 36–44.

[4] P. Ramins and B.T. Ebihara, "Improvements in MDC and TWT overall efficiency through the application of carbon electrode surfaces," *IEEE Trans. Electron Devices*, vol. ED-33, no. 11, Nov. 1986, pp. 1915–1924.

[5] A.N. Curren, "Carbon and carbon-coated electrodes for multistage depressed collectors for electron-beam devices—A technology review," *IEEE Trans. Electron Devices*, vol. ED-33, no. 11, Nov. 1986, pp. 1902–1914.

[6] A.N. Curren, K.A. Jensen and R.F. Roman, "Secondary electron emission characteristics of molybdenum-masked, ion-textured OFHC copper," *NASA Technical Paper 2967*, January 1990.

Chapter 15

Noise

There are many sources of noise in a TWT. Some are unavoidable, but most can be minimized or eliminated through proper design and construction. In this chapter, an attempt has been made to provide discussions of all of the important noise sources and techniques that can be used to minimize their effects. The chapter starts with a description of thermal agitation noise, which is the noise basis against which all other noise is compared, and a review of the definitions of *noise figure*. The chapter ends with a discussion of the minimum noise figure obtainable in a TWT (or any other beam-type device).

15.1 THERMAL AGITATION NOISE

Thermal agitation noise results from the minute voltage generated by the random motion of free electrons in conductors. The voltage produced by thermal agitation has energy distributed throughout the entire frequency spectrum. In the frequency band f_2-f_1, the effective voltage produced, e, is given by

$$e^2 = 4kT \int_{f_1}^{f_2} R\,df \tag{15-1}$$

where k = Boltzmann's constant
T = absolute temperature
R = resistive component of impedance

For R independent of frequency

$$\begin{aligned} e^2 &= 4kTR(f_2 - f_1) \\ &= 4kTRB \end{aligned} \tag{15-2}$$

where B is bandwidth.

Thermal agitation noise sets the ultimate limit to the smallest signal that can be amplified without being lost in a background of noise. For example, for R = 1/2 MΩ,

$T = 300$K and $B = 5$ kHz, the effective noise voltage, e, is 6.4 μV. Signal amplitudes on the order of this voltage level (6.4 μV) or smaller would be distorted or lost in the noise.

The thermal power, N_g, generated in a resistance is

$$N_g = \frac{e^2}{R} = 4kTB \tag{15-3}$$

If R is connected to a matched load resistance, R_L, maximum noise transfer occurs. Then, the noise voltage, e, divides equally between R and R_L and so the noise power, N_d, delivered to R_L is

$$N_d = \frac{\left(\frac{e}{2}\right)^2}{R} = kTB \tag{15-4}$$

This is the theoretical minimum noise level. At room temperature, its value is

$$N_d(\text{room temp}) = -144 \text{ dBW/MHz} \tag{15-5}$$

That is, for a 1 MHz bandwidth, N_d is 144 dB below 1 W (i.e., 4×10^{-15} W).

15.2 DEFINITIONS OF NOISE FIGURE

For an amplifier or receiver that

1. is noise free,
2. has gain G,
3. is perfectly matched to its input termination,

the output noise power is $kTBG$. The noise figure, F, for a real amplifier or receiver that generates noise is

$$F = \frac{\text{actual output noise power at 290K}}{\text{theoretical minimum noise power at 290K}} \tag{15-6}$$

or

$$F = \frac{N_1}{kT_oBG} \tag{15-7}$$

where N_1 is the measured output noise power at temperature $T_o = 290$K. The amount of noise added, N_a, by an amplifier is

$$N_a = N_1 - kT_oBG = (F - 1)kT_oBG \tag{15-8}$$

For a noise-free amplifier, $N_a = 0$ and $F = 1$.

Noise figure can be expressed in dB as follows

$$F_{dB} = 10 \log_{10} F \tag{15-9}$$

For a noise-free amplifier, $F_{dB} = 0$.

The equivalent noise temperature, T_e, of an amplifier is defined using

$$N_a = kT_e BG \tag{15-10}$$

and is

$$T_e = \frac{N_a}{kBG} = (F - 1)T_o \tag{15-11}$$

15.3 OVERVIEW OF NOISE PHENOMENA IN TWTS

There are three unavoidable sources of noise in a TWT:

- Shot noise
- Velocity noise
- Thermal agitation noise

Shot noise results from the discrete nature of the electron and the fact that electron emission from the cathode is a random process. Velocity noise occurs because of the wide distribution (Maxwellian) in velocities of electrons emitted by the hot cathode. This noise is often referred to as *Rack velocity noise*, named after A.J. Rack, who first described a technique for analyzing the effects of this noise. Thermal agitation noise, discussed at the beginning of this chapter, is present because the RF circuit, which has loss, is normally at room temperature or above.

In addition to these three unavoidable noise sources, there are many other sources of noise that depend largely on some aspect of the design and construction of a TWT. Most of these can be made insignificant through the use of careful design and construction procedures. Some of these noise sources result from

- Partition effect,
- Flicker effect,
- Collision ionization,
- Secondary and reflected electrons,
- Lens effects,
- Noise growth,
- Poor insulation,
- Varying insulator charges,

- Microphonics resulting from vibration,
- Multipactor,
- Power-supply induced noise, such as hum.

The most significant noises in a TWT are usually those produced by the cathode (shot and velocity). These noises are modified in significant ways as the electron beam travels through the electron gun and into the RF region of a the tube.

An extremely important characteristic of noise impressed on the electron beam in a TWT is the fact that noise propagates like any other RF signal on the beam. That is, noise travels as fast and slow space charge waves. The two sources that produce these waves are the current resulting from the shot effect and the velocity noise. In the general case, the two sources are assumed to be uncorrelated, and so each source produces fast and slow waves. As these waves propagate along the beam, noise current and velocity standing waves are independently produced.

In low power, low noise TWTs, some degree of correlation between the current and velocity noise sources is produced with a special low velocity drift region in the electron gun. Then, care is taken in accelerating the electron beam in the gun so that the amplitudes of the standing waves that develop are minimized. The design of the acceleration region in the gun is analogous to the design of an impedance matching section of a transmission line. Finally, as the beam leaves the gun, the axial position of the entrance of the beam into the helix is adjusted (by properly positioning the helix) to coincide with a noise minimum.

In the absence of correlation between current and velocity noise, the minimum noise figure for a TWT is predicted to be 6.5 dB. With correlation (produced by the low velocity drift region in the gun) noise figures below 3 dB have been achieved. It has also been found that the use of a large axial magnetic field in the acceleration region of the gun and in the input section of the RF structure further inhibits noise. With large magnetic fields in these regions, noise figures as low as 1 dB have been achieved and it has been predicted that, with additional refinements, a noise figure of 0 dB could be approached.

Some of the discussions in this chapter relate primarily to low power, low noise TWTs that are no longer used in modern systems. However, the noise phenomena in these TWTs exist, to some extent, in all TWTs and so it is of value to discuss them.

15.4 NOISE IN ELECTRON GUNS

For a discussion of noise properties, it is convenient to divide the electron gun into four regions as shown in Figure 15-1. The unavoidable sources of noise at the cathode surface (Region I) are velocity noise as described by Rack and shot noise. There may also be flicker noise but this is a low frequency noise that is negligible at microwave frequencies. Flicker noise will be discussed at the end of this section.

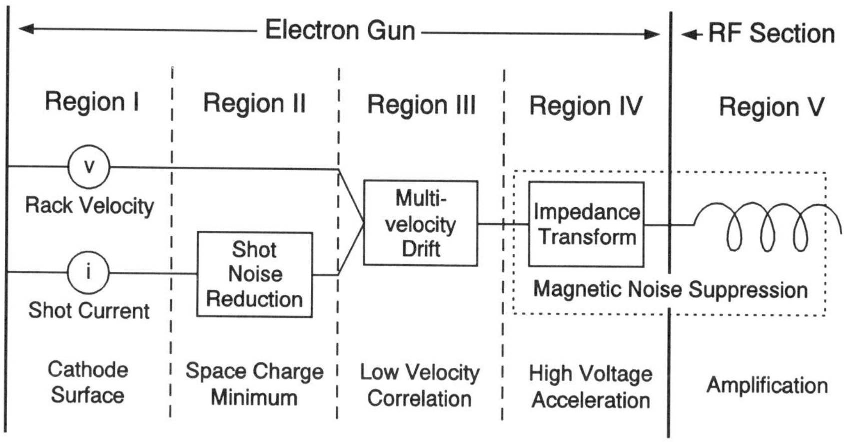

Figure 15-1 Noise regions in an electron gun.

Because of the initial emission velocities of the electrons, the potential of the space charge cloud (Region II) near the cathode surface is slightly below cathode potential. Low velocity electrons are returned to the cathode by this potential minimum. The amount by which the potential is depressed fluctuates as the cathode current fluctuates (shot effect). Thus, the number of electrons returned to the cathode fluctuates with time and this, in turn, changes (reduces at low microwave frequencies) the shot noise current passing the potential minimum. Additional noise may be introduced at the potential minimum by ions and reflected electrons from the collector. It turns out that the velocity distribution of the electrons is unaffected by the potential minimum.

Region III in the electron gun exists in low noise TWTs through careful design of the cathode configuration and the electrodes near the cathode. In the electron gun of a high power TWT this region, if it exists at all, is so short that it has negligible effect on noise performance.

In low noise TWTs, electrons drift at low velocities (energy < 1 eV) for an extended distance in Region III. The result is that the noise current and velocity become correlated to some degree, and this reduces noise figure by about 3 dB. The physical reason for this correlation is not clear.

When the electron beam emerges from Region III and undergoes acceleration in Region IV, space charge waves are launched and a standing wave pattern is produced. As is the case in an RF transmission line, a sudden change in impedance produces an increase in standing wave ratio. In a low noise TWT, an increase standing wave ratio results in an increase in noise figure. To avoid this problem in a low noise TWT, several electrodes are used in Region IV to control the rate of voltage rise (and the rate of impedance change). In high power TWTs, the function of thc clcctron gun is to produce a well-focused high perveance beam and no attempt is made to control the rate of impedance change.

As the electron beam emerges from the electron gun into Region V, care is taken in low noise tubes to position the entrance of the RF structure at the noise minimum. A strong magnetic field in Regions IV and V has been found to reduce noise.

In the sections that follow, the phenomena occurring in Regions I–IV are described in greater detail.

15.5 NOISE GENERATION AT THE CATHODE

15.5.1 Shot Noise

Shot noise [1] results from the random emission of electrons from the cathode. The randomly emitted electrons are a component of the noise current in the electron beam. This noise current is coupled to nearby RF circuits as was discussed in Chapter 8 and is illustrated in Figure 15-2. The circuit current can be found from a Fourier analysis of the induced current pulses. The mean square of the current fluctuations, i^2, depends only on emitted current, I_o, and bandwidth, B, as follows:

$$i^2 = 2eI_oB \tag{15-12}$$

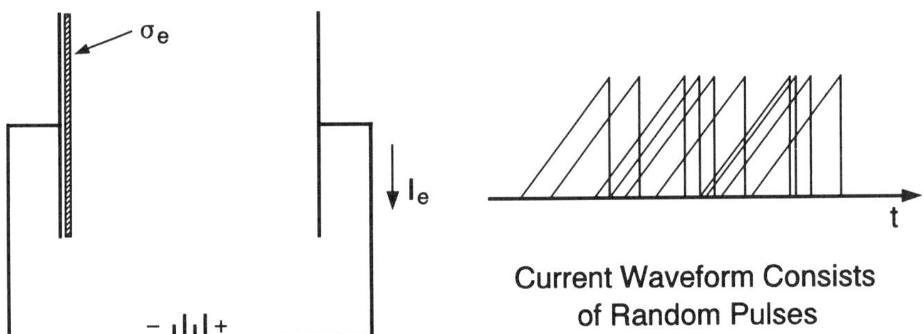

Figure 15-2 Random emission of electrons from cathode produces shot noise.

If this current is passed through a resistance, R, the noise power generated is

$$i^2R = 2eI_oBR \tag{15-13}$$

As an example, assuming $R = 100\ \Omega$, $I_o = 1$ A and $B = 1$ MHz the noise power generated is

$$i^2R - 3.2 \times 10^{-11}\ \text{W/MHz} = \quad 105\ \text{dB/MHz} \tag{15-14}$$

The resulting noise figure is

$$F = 39 \text{ dB} \tag{15-15}$$

Thus, shot noise can be a major source of noise in a TWT.

15.5.2 Velocity Noise

Velocity noise arises from the distribution in velocities of the electrons emitted from a hot cathode. The distribution is Maxwellian and, at 888°C, is as shown in Figure 15-3.

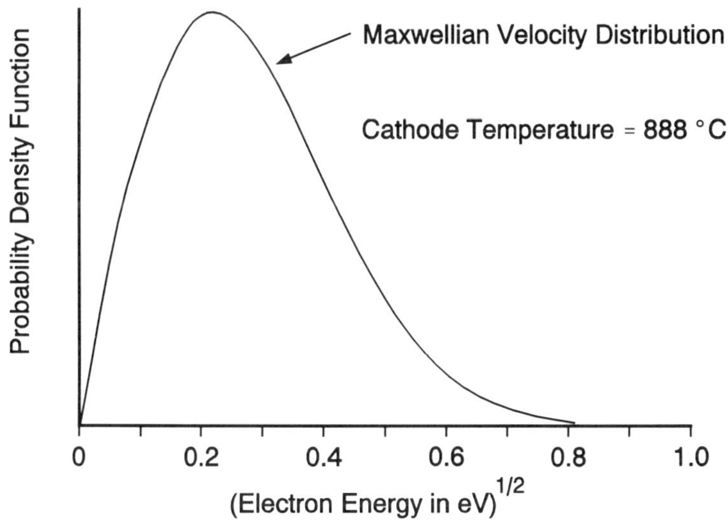

Figure 15-3 Velocity distribution of electrons from a cathode at a temperature of 888°C.

The multivelocity flow of electrons from the cathode that results from the Maxwellian velocity distribution is difficult, if not impossible, to treat analytically. Rack [2] showed that, at low frequencies, this multivelocity flow can be replaced by a single-velocity flow with the same mean square fluctuation in velocity, which is

$$v^2 = (4 - \pi)\eta \frac{kT_c}{I_o} B \tag{15-16}$$

where T_c is the temperature of the cathode in Kelvins.

We will discuss in the following paragraphs how Peterson and Pierce used the Rack velocity expression to predict noise behavior at microwave frequencies. Cutler and Quate found surprisingly good experimental agreement with the Pierce prediction.

15.5.3 Other Noise Generation Mechanisms

At low frequencies, two other (other than shot and velocity) noise generation mechanisms occur. These are hum produced by the filament and flicker noise. The reason for the hum was discussed in Chapter 5 on cathodes. Magnetic flux from the filament linking through the cathode surface disturbs the trajectories of electrons and results in modulation of the beam shape.

The Flicker effect produces a large amount of noise over and above shot noise. Figure 15-4 contains representative plots of flicker noise for directly and indirectly heated cathodes [3]. Note that flicker noise can be orders of magnitude above shot noise. Note also that flicker noise drops rapidly as frequency is increased. There appear to be many possible causes for flicker noise [4], depending on the type of cathode being used. For dispenser cathodes of the types normally used in TWTs, the original theory of flicker noise proposed by Schottky may apply.

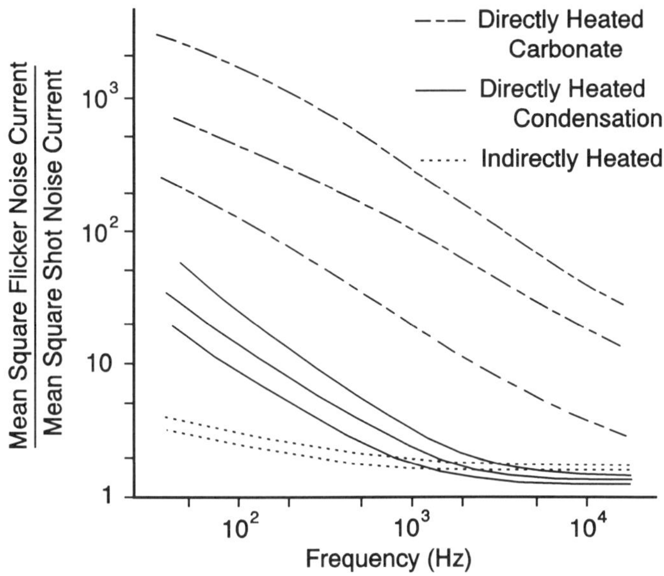

Figure 15-4 Flicker noise for some directly and indirectly heated cathodes. Adapted from *The Oxide-Coated Cathode* by G. Herrman and P.S. Wagener, Chapman and Hall, 1951.

Schottky attributed the flicker effect to foreign atoms arriving and departing from the cathode surface at random. For temperature limited cathodes, he predicted that the mean square noise current should be proportional to the square of emission current density and, at high frequencies, inversely proportional to the square of frequency. It should also be pointed out that this flicker noise was predicted to be even more strongly suppressed by space charge than shot noise.

A dispenser cathode consists of a monoatomic layer of barium on oxygen on the substrate metal (tungsten in a B cathode, osmium in an M cathode, *et cetera*). During operation of the cathode there is continual evaporation of atoms from the surface and migration of additional atoms from the tungsten substrate to the surface. The evaporation of atoms from the surface is a random thermal process that produces fluctuations in work function and, thereby, in emission current. These fluctuations are thought to be the primary source of flicker noise in a dispenser cathode.

15.6 THE SPACE CHARGE MINIMUM REGION

In the analysis of space charge limited flow in Chapter 5 it was assumed that the electric field at the surface of the cathode is zero. This assumption is valid when the emission velocities of the electrons are ignored. In the analysis of noise behavior, emission velocities are considered. The result is that there is a small depression in the potential near the cathode surface, as shown in Figure 15-5. This potential depression reflects low energy ($< \sim 0.2$ eV) electrons and returns them to the cathode. At the potential minimum, the energies of all other electrons are reduced by ~ 0.2 eV.

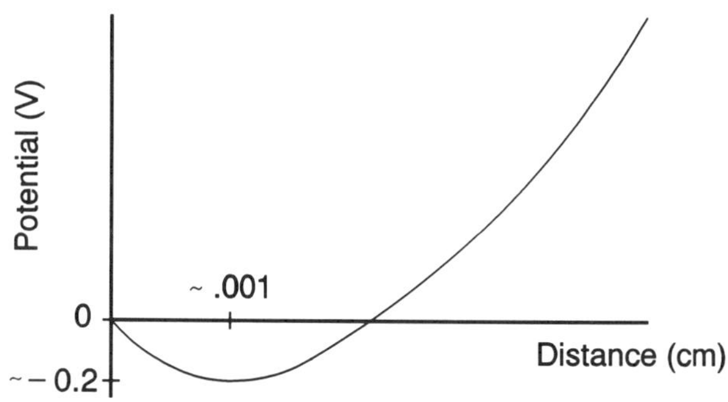

Figure 15-5 Initial emission velocities of electrons result in depression of potential near cathode.

Plots of typical electron positions as functions of time are shown in Figure 15-6 [5]. Each plot also shows the distance from the cathode of an electron at time *t* and the time at which the electron must have left the cathode surface (depending on its initial velocity) to arrive at that position. The effects of the potential minimum on electron velocities and positions are evident. Note that the time scale is in nanoseconds and so the time required for electrons to travel from the cathode to the potential minimum (located at 0.0075 mm) is only ~ 0.02 ns.

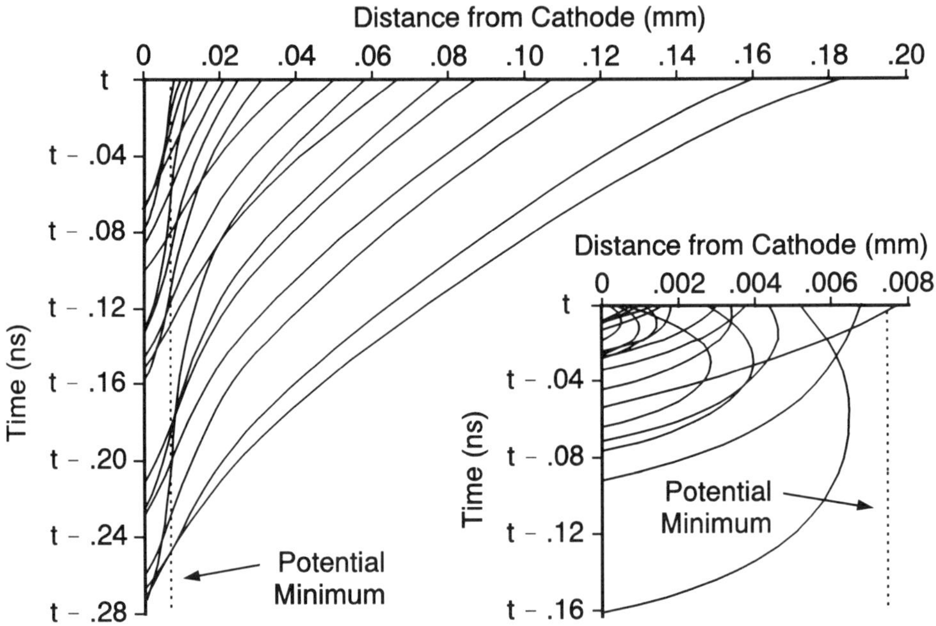

Figure 15-6 Typical electron velocity plots resulting from the Maxwellian velocity distribution in Figure 15-2. From P.K. Tien and J. Moshman, *Jour. Appl. Phys.*, September 1956.

15.6.1 Rack Noise Invariance

Because the initial velocity distribution is Maxwellian and because the velocities of all electrons that reach the potential minimum are reduced by the same amount, the velocity distribution at the potential minimum is still Maxwellian. The velocity distributions at the cathode surface and at the potential minimum are shown in Figure 15-7.

15.6.2 Shot Noise Reduction

Because the emission of electrons from the cathode is random (shot effect) the potential minimum produced by the electrons fluctuates with time. The perturbation (increased depression) of the potential minimum produced by a burst of electrons is shown in Figure 15-8. The additional depression increases the number of reflected electrons, thereby reducing the number of electrons in the burst that passes the minimum. Thus, the potential minimum reduces the shot noise current. The amount of reduction depends on frequency and approaches 100% at low frequencies.

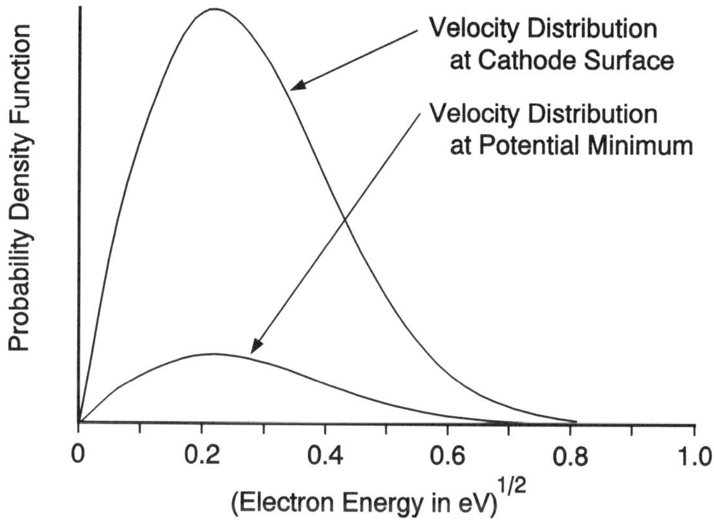

Figure 15-7 Electron velocity distributions at the cathode surface and at the potential minimum.

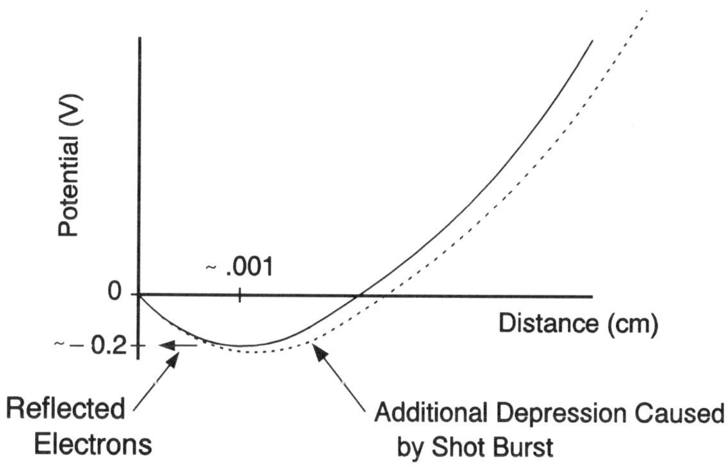

Figure 15-8 Perturbation of potential minimum by shot burst.

At frequencies approaching the plasma frequency of the electrons a resonance conditions exists that causes an enhancement of shot noise. One explanation for this resonance may be understood with the aid of Figure 15-9 [6]. The equivalent circuit for the cathode-to-potential-minimum region may be thought of as consisting of the capacitance of this region in parallel with an inductance that represents the inertial effect of the space charge cloud at the minimum. The quantity, G, is the conductance of the retarding field diode (cathode to potential minimum). The plasma frequency at the potential minimum for the current densities used in some low noise TWTs is in the low microwave frequency range (3–4 GHz). Some predictions of the shot noise reduction factor Γ^2 (the amount by which the shot current is reduced) are shown in Figure 15-10. For these predictions, the ratio of emitted current, I_s, to the current passing the potential minimum, I_o, was five. (The remainder of the current consisted of low velocity electrons that were reflected by the potential minimum and returned to the cathode.) For frequencies near the plasma frequency, shot noise current is increased by a factor of about two.

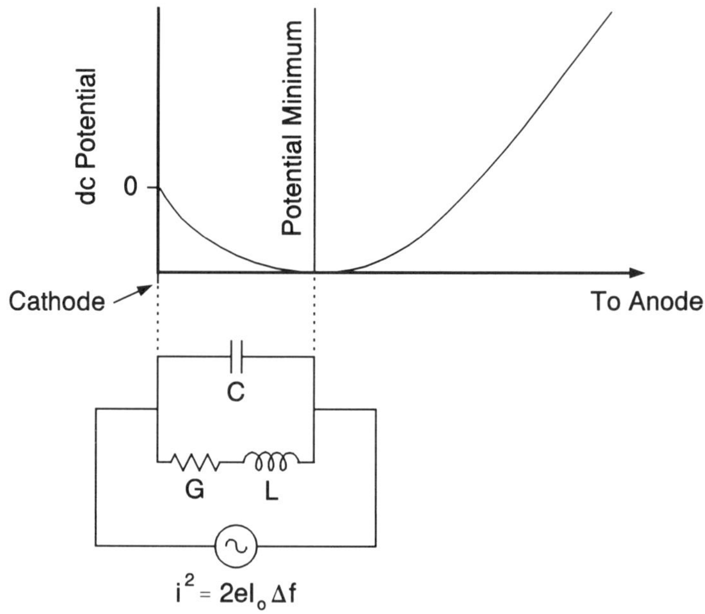

Figure 15-9 Circuit analog for potential minimum region. From R.W. Peter, *Noise in Electron Devices*, published by MIT and John Wiley & Sons, 1959.

It is important to understand that fluctuations in the potential minimum do not change the average velocity or the velocity distribution of the electrons.

15.6.3 Other Noise Effects

We have seen in previous paragraphs how fluctuations in the charge entering the potential minimum region from the cathode cause the potential minimum voltage to change. Other sources of the charge will have the same effect on the potential minimum. For example, positive ions generated in or near the electron gun by collisions of the beam electrons with gas molecules will travel toward the cathode. As

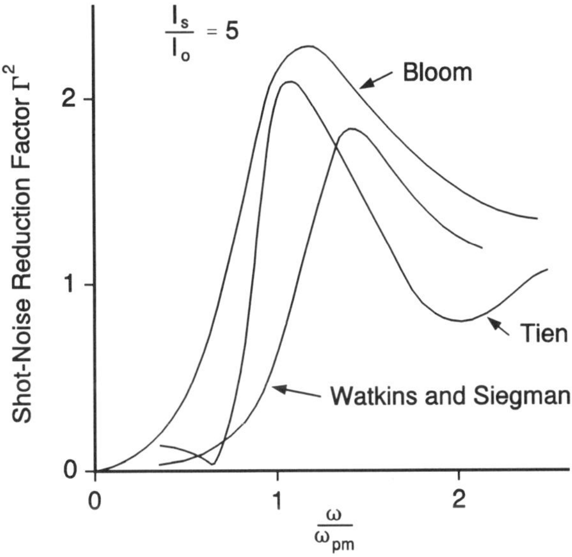

Figure 15-10 Predictions of shot noise reduction factor. From R.W. Peter, *Noise in Electron Devices*, published by MIT and John Wiley & Sons, 1959.

they pass through the potential minimum they cause the voltage to rise, and so additional current from the cathode can pass the minimum region. Since the ion production process is random, the fluctuations of the potential minimum that they produce are random and the resulting current fluctuations are random.

Another source of charge that can affect the potential minimum is high velocity secondary electrons, primarily from the collector. These electrons are guided through the tube in the reverse direction by the magnetic focusing field. If their interaction with the RF fields is such that they gain (or at least do not lose) energy, they will be able to reach the potential minimum region. The secondary emission process is a noisy process and so secondary electrons in the potential minimum region produce random fluctuations of the potential minimum, which, in turn, produce an additional component of noise modulation on the emission current passing the potential minimum.

Finally, in considering other possible noise effects in the potential minimum region, it is noted that there is evidence that noiselike fluctuations of the space charge cloud may occur. These fluctuations occur on a large scale in the vircator (virtual cathode oscillator). In this device a dense beam of electrons is used to produce a fluctuating potential minimum. Reasonably efficient conversion of beam energy to generally noisy RF power takes place.

15.7 LOW VELOCITY CORRELATION REGION

The biggest breakthrough on low noise tubes was made by Currie [7] when the possibility of producing correlation between shot current and velocity noise was made apparent. Currie's low noise electron gun is shown in Figure 15-11. The cathode was an annulus surrounded by an electrode, called the *profile shaping electrode*. In Currie's early experiments, there was also a portion of the profile shaping electrode in the center of the annular cathode. This was later shown to not be needed. Anode 1, which played an important role in shaping the potential profiles near the cathode, contained an annular opening through which the hollow beam from the cathode could pass. Anodes 2, 3 and 4 were used to provide the voltages for the desired impedance transformation to minimize the standing wave ratio of the of the space charge waves (to be described in following paragraphs).

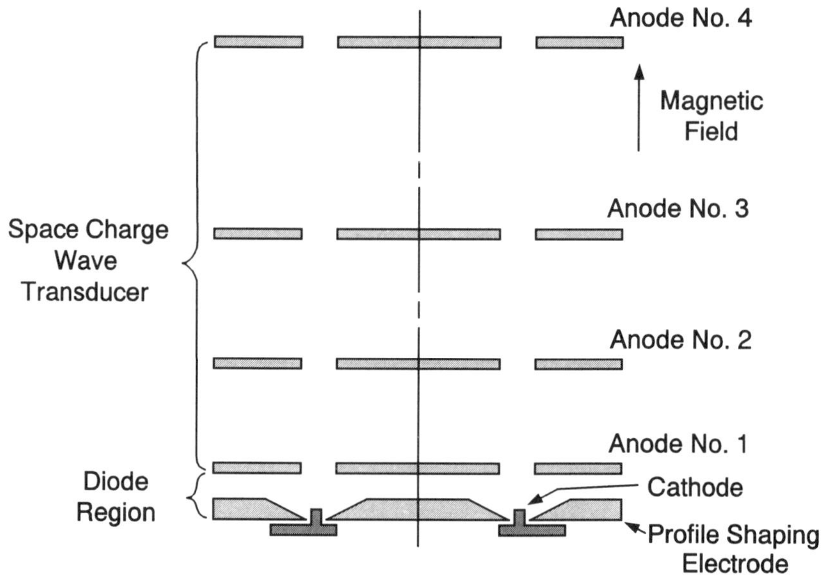

Figure 15-11 Currie's low noise electron gun. From M.R. Currie and D.C. Forster, *Jour. Appl. Phys.*, January 1959.

The excess noise (noise produced by the beam) with beam conditions adjusted for minimum beam noisiness is shown in Figure 15-12. By comparison, the minimum excess noise predicted by theory with no correlation is shown by the dashed line. The noise figure for the TWT at 2.54 GHz was 3.6 dB. Noise figures as low as 2.9 dB were obtained.

The potential profiles occurring in the cathode region when voltages were adjusted for near-optimum conditions are shown in Figure 15-13. These profiles were obtained

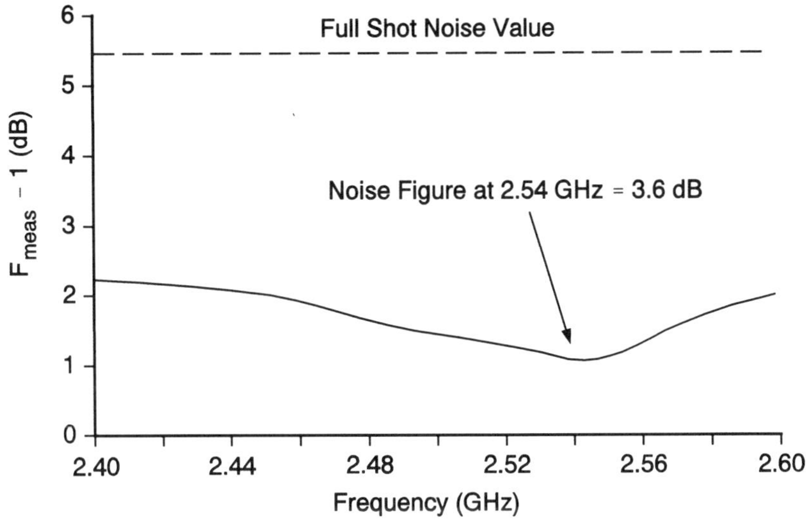

Figure 15-12 Excess noise with beam conditions adjusted for minimum beam noisiness. From M.R. Currie and D.C. Forster, *Jour. Appl. Phys.*, January 1959.

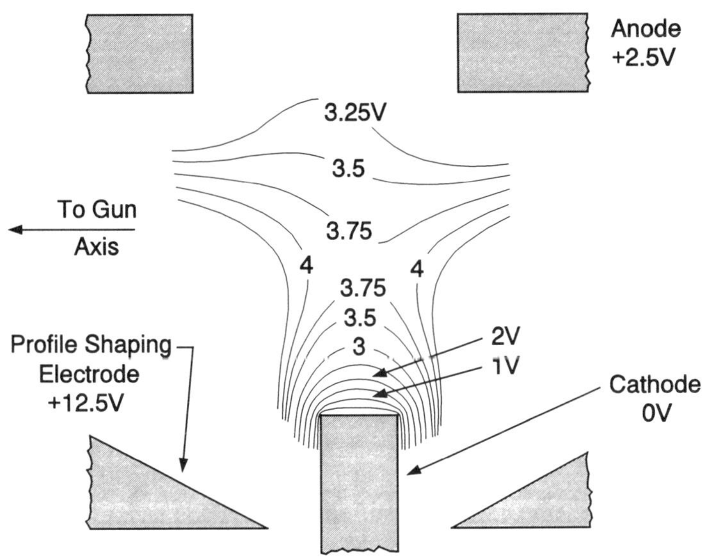

Figure 15-13 Near optimum potential distribution for Currie's low noise gun. From M.R. Currie and D.C. Forster, *Jour. Appl. Phys.*, January 1959.

using an electrolytic tank without charge injection, and so the effects of space charge are not included. The resulting current profile is given in Figure 15-14. Most of the electron emission was from the edges of the cathode. A strong axial magnetic field was used to form the electrons into a very thin hollow beam. (Only emission from one edge of the cathode is shown in Figure 15-14. Emission from the other edge was the same.)

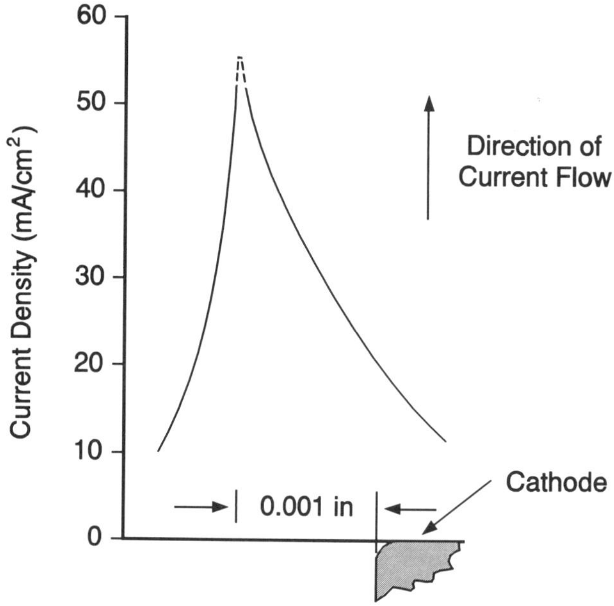

Figure 15-14 Current density *versus* radial distance for Currie's gun. From M.R. Currie and D.C. Forster, *Jour. Appl. Phys.*, January 1959.

From Figure 15-13, it is seen that after the electrons left the region near the edge of the cathode and moved toward the anode, they drifted through a region of very slowly varying potentials. This variation of potential with distance, including the simulated effects of space charge, is shown in Figure 15-15 (Poisson tank solution). The Fry-Langmuir solutions show potential profiles for space charge limited flow corrected for the effects of thermal velocities. Note that, for the Currie profile, in a distance of 0.010 in., the potential was still less than 1 V. Thus, the situation created by Currie in his electron gun forced the electrons to drift for a considerable distance at a low average velocity that was comparable to the noise velocities of the electrons. As a result of this multivelocity flow, the basic noise quantities in the beam were transformed and some degree of correlation between noise current and velocity was achieved. The physical mechanism(s) for this transformation have not been described.

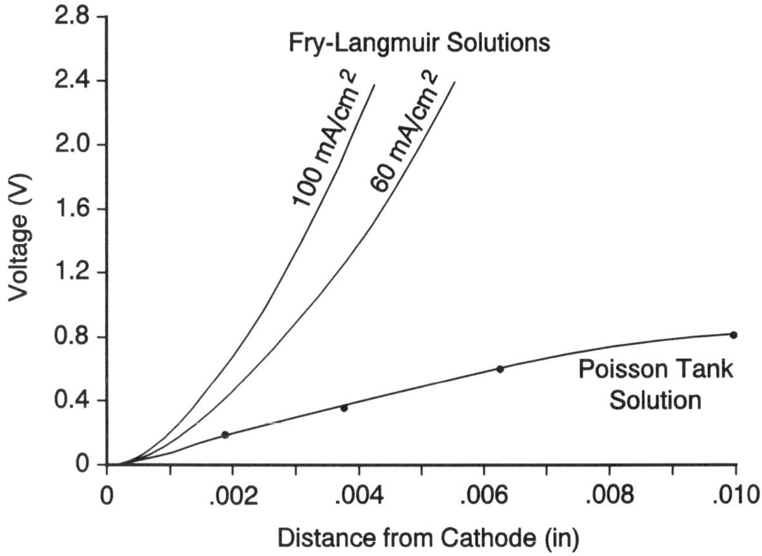

Figure 15-15 Currie (Poisson tank) potential profile and Fry-Langmuir profiles. From M.R. Currie and D.C. Forster, *Jour. Appl. Phys.*, January 1959.

15.8 HIGH VOLTAGE ACCELERATION REGION

To understand the rational for the design of the high voltage acceleration region in a low noise TWT it is necessary to understand the nature of the noise space charge waves that propagate on the beam. In this section we will first discuss the propagation of noise space charge waves and then will discuss the impedance transformation in an electron gun necessary to avoid large noise standing wave ratios. At the end of this section, a discussion of lens effects on noise is included. Lens effects cause radial components of noise to be converted to axial noise and can thereby significantly increase the noise figure of a TWT.

15.8.1 Noise Space Charge Waves

The propagation of noise as fast and slow space charge waves was predicted by Pierce [8] and by Peterson [9]. They carried out analyses of the noise resulting in an electron beam after acceleration of the electrons in the gun. They used the equations of Llewellyn and Peterson [10] for a space charge limited diode. We will not reproduce the Llewellyn-Peterson equations here but will point out that two of the initial conditions for these equations are the noise current and velocity at the potential minimum. Because the dc velocity of the electrons is zero at the potential minimum, the equations predict that the effect of shot noise current vanishes. This is equivalent to using a shot-noise reduction factor of zero, which in turn, is equivalent to assuming that fluctuations of the potential minimum produced by the shot current completely eliminate shot noise current beyond the potential minimum. Thus, only Rack velocity noise was used to predict noise in an electron beam.

During acceleration of the electrons in the gun, some of the Rack velocity fluctuations result in current noise, and so both current and velocity noise exist in the electron beam. Pierce used the noise current and velocity at the exit from the gun as input conditions to a TWT and used his TWT theory to predict fast and slow noise space charge waves on an electron beam.

The theories of Peterson and Pierce (and of Rack) met with considerable skepticism. However, the December 1950 issue of *Physical Review* [11] published C.C. Cutler and C.F. Quate's paper "Experimental Verification of Space Charge and Transit Time Reduction of Noise in Electron Beams." The noise measuring apparatus used by Cutler and Quate is shown in Figure 15-16. A cavity was used to detect a portion of the noise current in the electron beam. A shutter could be moved across the beam to investigate partition current (to be discussed later in this chapter).

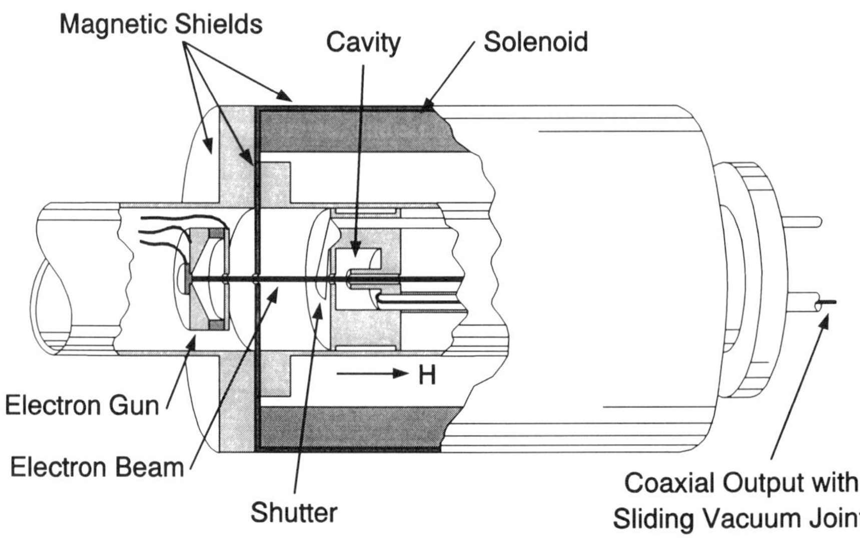

Figure 15-16 Cutler and Quate's noise measuring apparatus. From C.C. Cutler and C.F. Quate, *Phys. Rev.*, Dec. 1950.

The key result of the Cutler-Quate experiment is shown in Figure 15-17. Measured noise power was compared with the predictions of Pierce. The existence of noise space charge waves was clearly verified. In addition, the variation of the position of the minimum with voltage and with the transit angle of the electrons in the gun agreed with the theoretical prediction. The reason that the position of the minimum differs from the predicted minimum by 0.4 in was thought to be that calculations were based on the infinite plane diode configuration rather that a cylindrical beam configuration in the gun region.

The absolute values of the noise levels in Figure 15-17 were determined by comparing them with the temperature limited shot noise. This was done by reducing the cathode temperature and measuring the noise power as a function of beam current as shown in Figure 15-18.

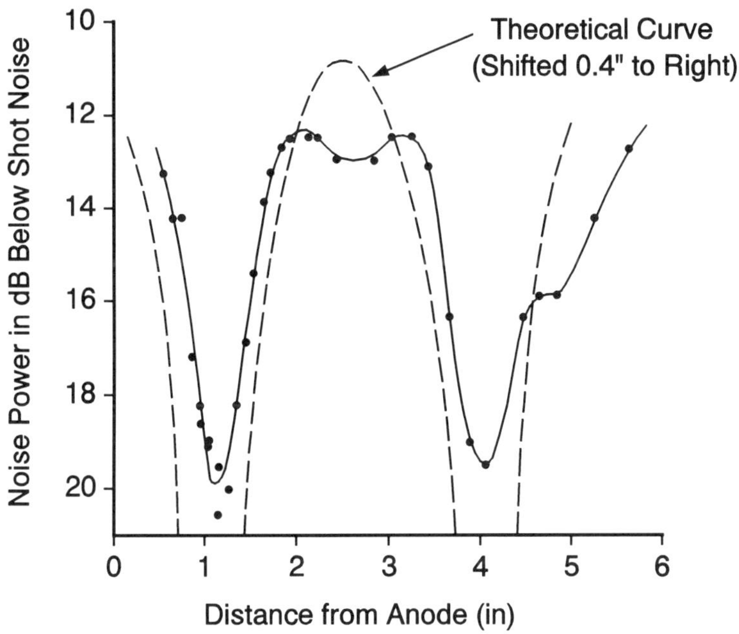

Figure 15-17 Noise power *versus* distance from the anode. From C.C. Cutler and C.F. Quate, *Phys. Rev.*, Dec. 1950.

The slope of the curve at low beam currents shows the shot noise power to be proportional to current as expected (15-12). At larger beam currents, the noise power was reduced by space charge. The greatest total reduction of noise power occurred at the first minimum and was 20 dB below the shot noise level.

Pierce predicted a minimum noise figure for a TWT of ~6–7 dB. By comparison, the noise figure for a conventional TWT is 30–40 dB. In an early low noise TWT developed by Watkins [12], the Cutler and Quate results were used along with a velocity jump to reduce the noise figure to ~10 dB. The reason for using the velocity jump is that the axial component of RF velocity on an electron beam is inversely proportional to the "jump" in dc beam velocity. Figure 15-19 shows sketches of the TWT analyzed by Pierce (a) and the velocity jump, low noise TWT of Watkins. The

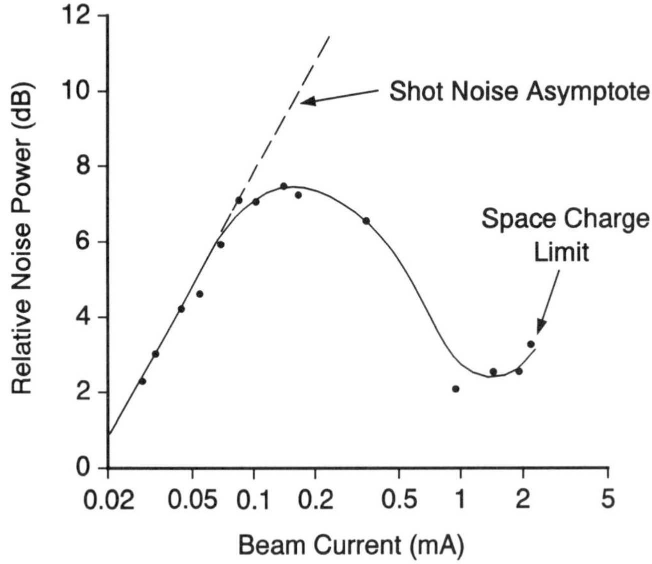

Figure 15-18 Noise power *versus* beam current for temperature limited emission. From C.C. Cutler and C.F. Quate, *Phys. Rev.* Dec. 1950.

difference between the two was the short gap across which a voltage difference, $V_2 - V_1$, could be applied to reduce the noise velocity of the beam.

15.8.2 Impedance Transformation for Low Noise Tubes

The electron beam acts as a transmission line to carry space charge waves launched by shot current and Rack velocity noise generators in the potential minimum region. From the shot current source, fast and slow current waves emerge and produce a current standing wave. Associated with this current wave is a velocity standing wave that is displaced one quarter of a plasma wavelength from the current wave.

The Rack velocity noise generator produces fast and slow velocity waves that form a velocity standing wave. Associated with this velocity wave, is a current standing wave that is displaced one quarter of a plasma wavelength from the velocity wave. These standing waves produced by the velocity generator are shown for a space charge limited planar diode in Figure 15-20 [13]. Velocity and current waves for two space charge conditions are shown. For a high space charge a velocity minimum and current maximum occur near the potential minimum (where the relative distance through the diode is zero). Subsequently, as the electrons are accelerated through the diode, the noise velocity wave grows and the noise current wave decreases. For the low space charge condition, the velocity minimum and noise current maximum occur near the anode.

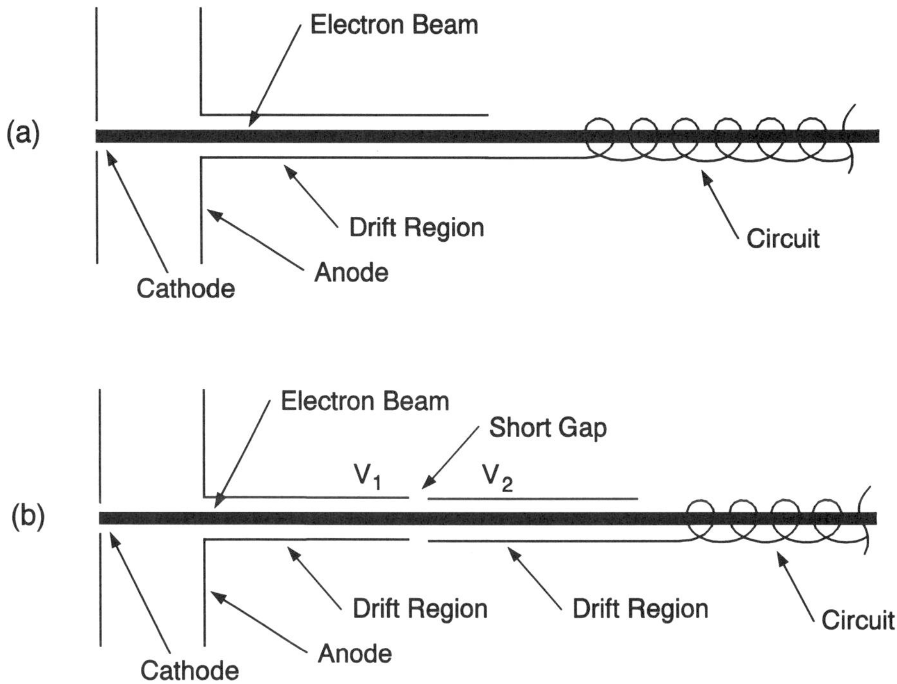

Figure 15-19 TWTs analyzed by Pierce (a) and by Watkins (b).

The two current standing waves (one from the shot noise current source and the other from the Rack velocity noise source) are separated by one-quarter plasma wavelength (unless correlation has been introduced by the multivelocity drift region). As a result, the current distributions along these two waves at the exit from the gun are as shown in Figure 15-21. Each of these waves contributes to the noise figure of a TWT. Thus, in the design of a low noise TWT, it is necessary to

1. Control the amplitudes of the standing waves,
2. Locate the entrance to the helix at a position that minimizes the noise induced by the two waves.

The amplitudes of the standing waves in the electron gun are governed by the impedance of the electron beam. This impedance is very low at the potential minimum where the electron velocity is low. As the beam is accelerated and the velocity increases, the impedance increases (see Figure 15-21). The impedance at the input to the helix may be over two orders of magnitude greater than the impedance at the potential minimum.

To control the current standing waves so that minimum noise is introduced in the RF section, it is necessary to avoid rapid changes in beam impedance. This may be

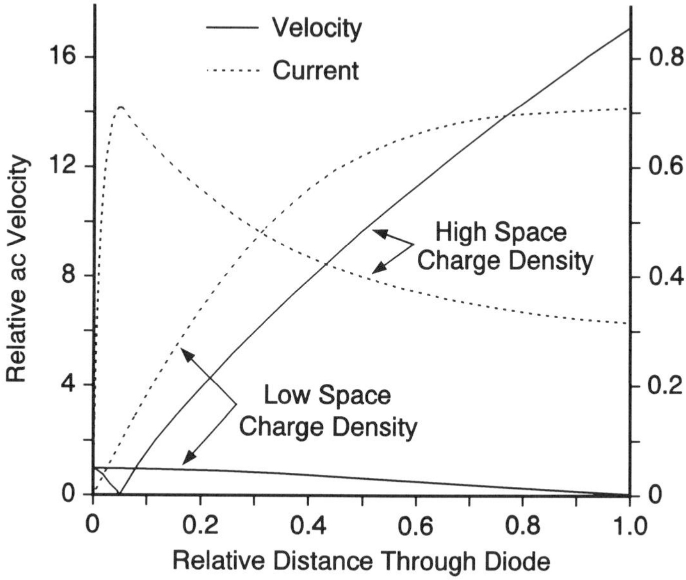

Figure 15-20 Noise velocity and current in a space charge limited diode. From S. Bloom and R.W. Peter, *RCA Review*, March 1954, copyright 1954 by RCA, reproduced with permission of General Electric Company.

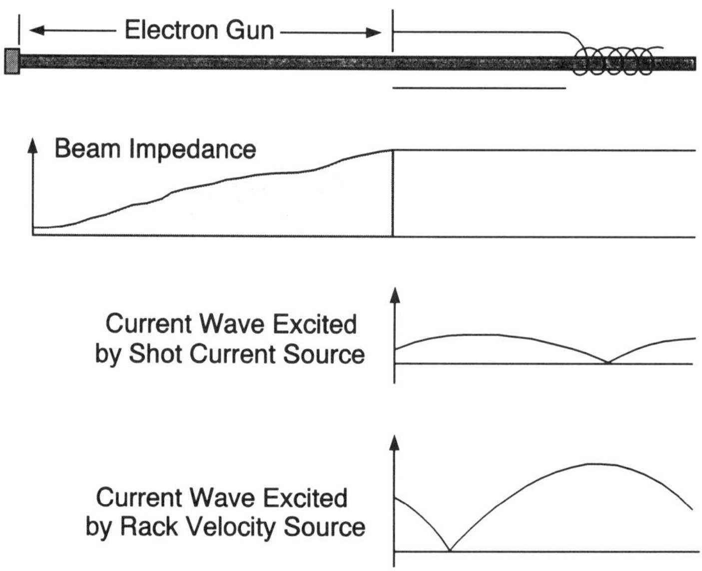

Figure 15-21 Beam impedance in gun and current waves that emerge from gun.

accomplished with an impedance transformer like that shown in Figure 15-22 [6]. Several electrodes are used in the electron gun to control the rate of acceleration of the beam and, thereby, to control its impedance. In the example shown in Figure 15-22, the impedance variation as a function of distance from the potential minimum to the helix in nearly exponential. This is analogous to the situation in acoustics or RF transmission lines, where an exponential variation in impedance is used to match a low impedance source to a high impedance load.

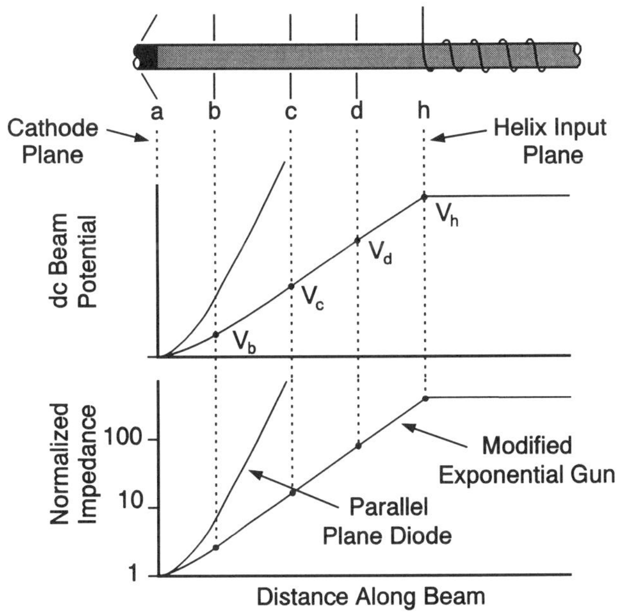

Figure 15-22 Potential and impedance for a low noise gun. Adapted from R.W. Peter, *Noise in Electron Devices*, published by MIT and John Wiley & Sons, 1959.

15.8.3 Lens Effects

The transverse and axial velocities of electrons emitted from a thermionic cathode are assumed to be the same. In a low noise gun, as electrons are accelerated, the axial velocity fluctuations decrease (see the low space charge curve in Figure 15-19, for example). However, the transverse components of velocity are not appreciably affected. Thus, after acceleration, the transverse velocity components are much larger than the axial components.

As electrons pass through a lens, their transverse and axial velocities are altered. If, as the electrons enter the lens, the fluctuations in their transverse velocities are larger that those in the axial velocities, then the axial velocity fluctuations are increased by the lens. This was proven theoretically and experimentally by Knechtli [14]. The situation illustrated in Figure 15-23 will help in providing a qualitative understanding of the effect of a lens on transverse velocities. The trajectories of three electrons with different transverse velocities are shown. To simplify the illustration, it is assumed that the electrons have no axial velocity fluctuations when entering the lens and that they all pass some plane, z_o, at the same time. The electrons all pass through the lens at the same time and the trajectory of each is deflected through an angle, ϕ. At some later time, the electrons arrive at positions 1, 2 and 3, which are separated in the axial direction. For this to happen, the electrons must have different axial velocities. Thus, in passing through the lens, transverse velocity components have been converted to axial velocity components.

It is clear, therefore, that discontinuities in accelerating or decelerating voltage increase beam noisiness (as long as the transverse velocity fluctuations are larger than the axial ones), and so these discontinuities, which act as electrostatic lenses should be avoided in low noise TWTs. A limiting factor in the noise figure achieved by Watkins (discussed previously in this section) is thought to have been the conversion of transverse noise velocities to axial noise in the velocity jump regions.

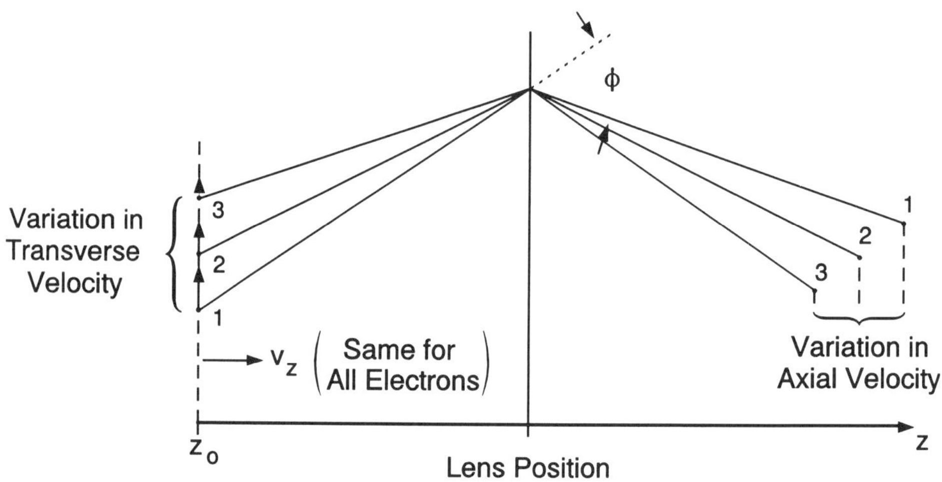

Figure 15-23 Conversion of transverse electron velocities to axial velocities by a lens.

15.9 RF SECTION NOISE PHENOMENA

15.9.1 Circuit Loss

There are two effects of circuit loss on the noise figure of a TWT. The smaller of the two is the thermal agitation noise that is generated by the resistance of the helix. The larger affect occurs because noise figure depends on the gain of a TWT and gain, in turn, is affected by circuit loss. The affect of circuit loss on noise figure amounts to, at most, a few dB as is shown in Figure 15-24 [15] and is important in low noise TWTs. As an example of the effect of helix material on the loss and noise figure, we will assume that the Pierce loss parameter, *d*, is 0.35 and so the contribution to noise figure is about 2.3 dB. Compared with tungsten, if copper with an excellent surface finish was used for the helix, the value of *d* would drop to about 0.2 and the improvement in noise figure would be 0.9 dB.

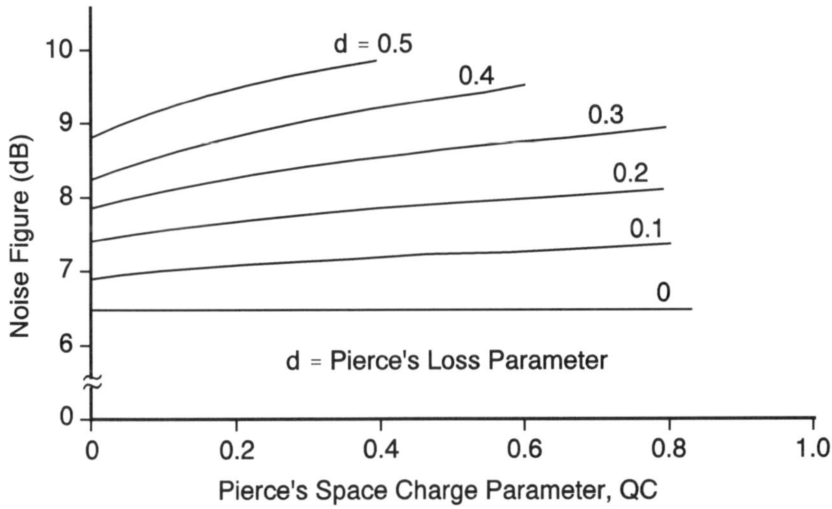

Figure 15-24 Predicted noise figure for low noise TWTs including the effect of circuit loss.

15.9.2 Partition Noise

Partition noise arises because transverse components of thermal velocity cause fluctuations in the current intercepted by electrodes other than the collector. Transverse components of velocity may induce electron motion in a helix and produce a noise analogous to partition current even though no electrons actually strike the helix. A similar effect can occur in the grid cathode region, where space charge fluctuations induce grid signals. The noise added to the beam by partition current equals the shot noise contained in an equal amount of temperature limited current [8 and 16]. Also, the partition current adds a fully random velocity component to the beam [16].

In the experiments of Cutler and Quate [11] a shutter was used (see Figure 15-16) to intercept a small fraction of the beam current and, thus, control the partition current. The resulting noise data given in Figure 15-25 show a direct correlation between noise power and partition current.

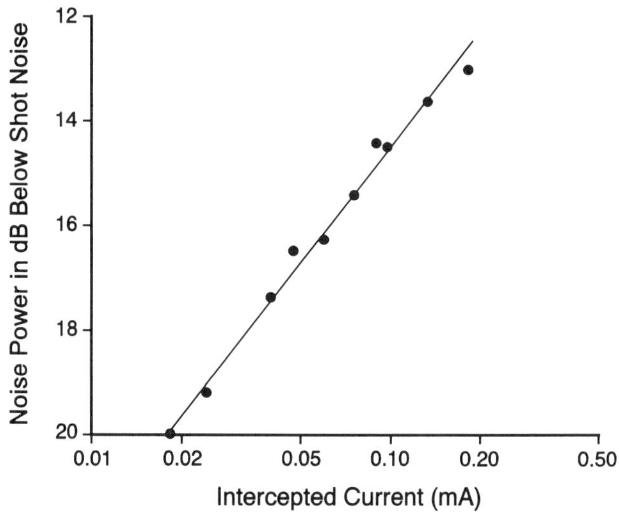

Figure 15-25 Noise power resulting from partition effect in Cutler and Quate's experiment. From C.C. Cutler and C.F. Quate, *Phys. Rev.*, Dec. 1950.

15.9.3 Secondary Electron Interactions

The electron beam striking the collector generates a large number of slow secondary electrons and a few percent of fast secondaries (often referred to as *reflected electrons*) that have energies nearly equal to the incident electron energy. The secondary electron current contains noise of the primary beam plus shot noise from the secondary emission process. If a portion of the secondary current escapes from the collector, it can introduce noise on the RF circuit through interception or by interacting with reverse traveling waves on the RF circuit. This noise is amplified and, at the input of the RF circuit, may be partially reflected and thus contribute to the total noise output.

In modern multistage depressed collectors (discussed in Chapter 14) great care is taken to minimize the generation of secondary electrons and to prevent their escape from the collector. Still, some secondaries are generated and those that do escape from the collector have a wide range of energies. This is because those electrons were generated from electrodes at depressed potentials. Thus, those electrons are accelerated as they travel toward the exit from the collector (and into the interaction region). The problem of noise generation by secondaries from the collector is greatly reduced with multistage depressed collectors as long as they are properly designed.

15.9.4 Noise Growth

Measurements of noise on a beam usually show a standing wave pattern like that found by Cutler and Quate [11]. Sometimes, however, noise increases very rapidly up to very large saturation values, as shown by results obtained at RCA and given in Figure 15-26 [6]. The results of measurements first made by Gilmour and reported in [17] show (see Figure 15-27) that a large increase in peak velocity of the electrons may also occur. Voltage equivalents of electron velocities of the order of 100 V above the dc beam voltage were observed. In both the RCA and Cornell experiments, the electron gun was shielded and the magnetic field at the cathode was near zero. In measurements made at MIT [18] an auxiliary magnetic field was applied to the cathode region. The growing noise could be eliminated by making the auxiliary field have the same direction as the axial component of the main focusing field that leaked through the cathode. The opposite polarity of the auxiliary field increased the noise.

As yet no theory adequately explains the noise growth phenomena. Scalloped beam amplification has been proposed but the variation of noise growth with magnetic field that is predicted is far different from the measured results.

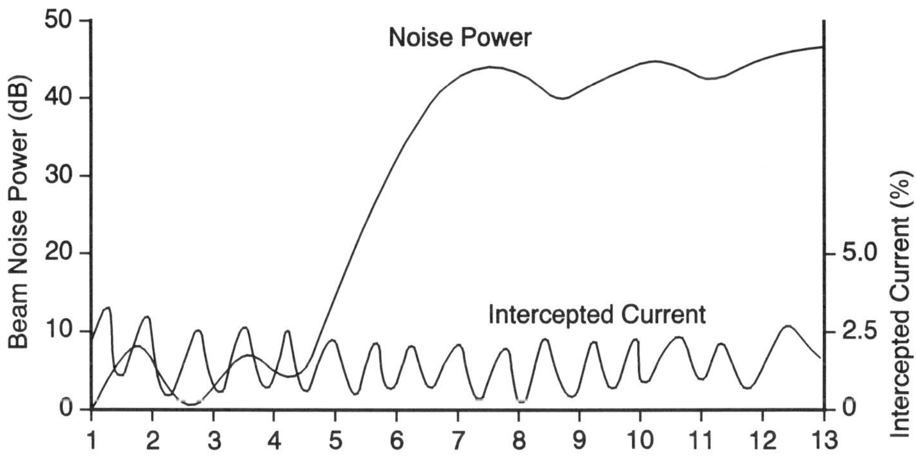

Figure 15-26 Noise growth on an electron beam. From R.W. Peter, *Noise in Electron Devices*, published by MIT and John Wiley & Sons, 1959.

15.9.5 Magnetic Noise Suppression

A high magnetic field applied to the helix input region of a low noise TWT can reduce the noise figure to a very low value. For example, in [19], Hammer and Thomas reported TWT noise figures of 1.0 dB at S-band. The equivalent beam noise

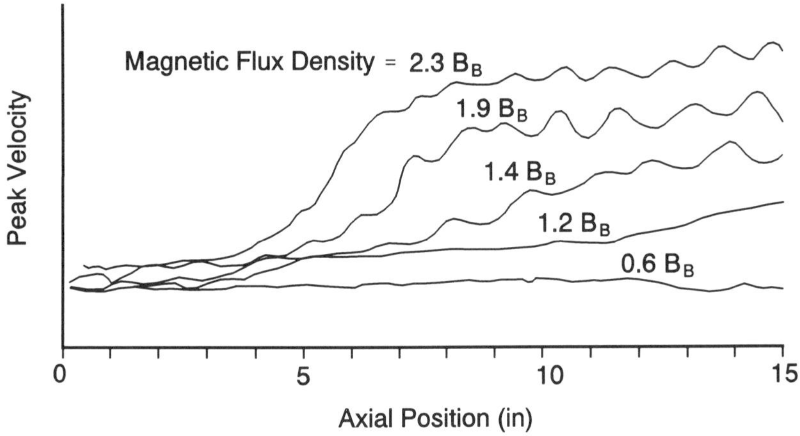

Figure 15-27 Peak electron velocity for various beam focusing conditions. From *Technical Report* No. RADC-TR-64-563, April 1965.

temperature was only 57K, and the ratio of this noise temperature to room temperature was 0.19. In a subsequent paper [20], Hammer and Wen showed that the beam noise temperature varied inversely with magnetic field as shown in Figure 15-28. They stated that these results were obtained after some degradation in noise performance. Before degradation, the straight line through the data passed through the origin, indicating a beam noise temperature of 0K and a noise figure of 0 dB. The Hammer and Wen study indicated that the magnetic field suppresses transverse noise fluctuations that couple to the helix. The transverse velocity and position fluctuation originate in the current emitted from a thermionic cathode, which are statistically independent of the axial fluctuations.

15.10 OTHER NOISE SOURCES

There are several other sources of noise in a TWT. Some of these are the following:

1. One or more grids is often used in an electron gun to control current. Some interception of emitted current occurs, and this introduces partition noise. In addition, field distortions occur near the grid, and these introduce transverse velocity components to electron trajectories. The overall affect of the grid is to raise the apparent temperature of the cathode. In one report, in which a computer code was used to simulate a shadow gridded gun with a fictitious cathode temperature, the best correlation between computed and measured beam interception was obtained with an apparent cathode temperature of 10,750K.

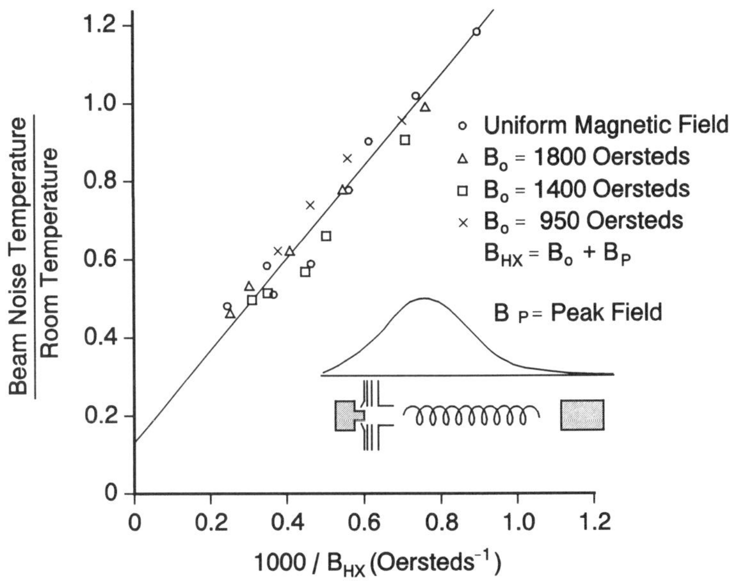

Figure 15-28 Reduction of noise with a high magnetic field. From J.M. Hammer and C.P. Wen, *RCA Review*, December 1964, copyright 1964 by RCA, reproduced with permission of General Electric Company.

2. Hum results from inadequate filtering of power supply voltages and can result in RF modulation at power frequencies and their harmonics. This will be discussed further in Chapter 16. The hum produced by an ac filament magnetic field that links the cathode surface was discussed previously in this chapter.
3. Faulty insulation can result in breakdown, corona or partial discharges that produce fluctuations in voltages and modulation of the RF signal.
4. Multipactor is a form of RF vacuum breakdown that occurs across vacuum gaps or on insulator surfaces. Noiselike impedance fluctuations occur that modulate the RF signal.
5. Time varying charge accumulations on insulator surfaces, such as the support rods for the helix can modulate the RF signal.
6. Movement of the cathode, grid, helix, *et cetera* as the result of shock and vibration produces modulation of the RF signal.
7. "Gas noise" is generated under poor vacuum conditions. As has previously been mentioned in this chapter, ions produced by electron interaction with gas molecules can produce fluctuations of the potential minimum at the cathode surface and, thereby, produce noise. In addition, ions can produce oscillations, beam compression and reduced power output, RF phase shift and phase noise and reduced cathode activity, which in turn results in patchy emission and increased flicker and shot noise.

15.11 MINIMUM NOISE FIGURE OF A TWT

An expression for the minimum noise figure of a TWT has been derived by several researchers [21–23] with the assumption that the noise at the potential minimum is uncorrelated shot noise current and Rack velocity noise. With the additional assumptions that the gain is high and that circuit losses are negligible, the minimum noise figure is

$$F_{\min} = 1 + \sqrt{4 - \pi}\,\frac{T_c}{T_o} \tag{15-17}$$

where T_c is the cathode temperature and T_o is room temperature. For typical cathode temperatures, F_{min} is in the range of 6.4 to 7.3 dB. This is the minimum noise figure if noise reduction (but not correlation) is provided to minimize noise figure. When helix losses are considered, the minimum noise figure is somewhat higher, as was shown in Figure 15-24.

The derivation of (15-17) is reasonably straightforward but does not provide a means for taking into account correlation between the shot current and Rack velocity noise. However, the elegant noise theory for linear beam devices developed by Haus [24 and 25] and Robinson [25] takes correlation into consideration. Their theory contains four basic quantities, all related to noise power on a beam. In terms of the noise current, i_1, and the noise voltage, V_1, (where V_1 is the kinetic voltage or voltage equivalent of the velocity fluctuation, that is $V_1 = v_1 u_o$) these quantities are as follows:

1. The self-power density spectrum of current fluctuations,

$$\Psi = \frac{\overline{i_1^2}}{4\pi} \tag{15-18}$$

2. The self-power density spectrum of voltage fluctuations,

$$\Phi = \frac{\overline{V_1^2}}{4\pi} \tag{15-19}$$

3. The real part of the cross-power density spectrum,

$$\Pi = \mathrm{R_e}\left[\frac{\overline{i_1 V_1}}{4\pi}\right] \tag{15-20}$$

4. The imaginary part of the cross-power density spectrum,

$$\Lambda = \mathrm{I_m}\left[\frac{\overline{i_1 V_1}}{4\pi}\right] \tag{15-21}$$

The factor 4π in these expressions arises from conventions used for frequencies (ordinary or angular) and negative frequencies. The quantity Π has special meaning because it is a direct measure of the noise power carried by the beam. It is, therefore, invariant through conservative beam regions. For the case of uncorrelated shot noise current and Rack velocity noise, Π is zero and is assumed to remain zero throughout the gun when there is no region that introduces correlation (such as Currie's multivelocity region).

Another quantity used by Haus, which is invariant in conservative regions, is usually denoted by S and defined by

$$S^2 = \Phi\Psi - \Lambda^2 \tag{15-22}$$

The minimum noise figure found by Haus and Robinson for linear beam devices (TWTs as well as klystrons and other linear beam devices) is given by

$$F_{min} = 1 + \left[\frac{2\pi}{kT_o}\right](S - \Pi) \tag{15-23}$$

where k is Boltzmann's constant.

When there is no correlation, Π and Λ are zero and S has the value $2kT_c/4\pi$ and so the expression for minimum noise figure becomes

$$F_{min} = 1 + \frac{T_c}{T_o} \tag{15-24}$$

which is nearly identical to that given in (15-17).

With the correlation produced by Currie's low noise gun, the value of $2\pi/kT_o(S - \Pi)$ was only about 27% of that corresponding to full uncorrelated shot noise. In the work of Hammer and Wen, it was predicted that $(S - \Pi)$ could approach zero (and result in F_{min} = 0 dB) if very large magnetic fields were used in a low noise TWT.

REFERENCES

[1] W. Schottky, "Spontaneous current fluctuations in electron streams," *Ann. Physik*, vol. 57, December 1918, pp. 541–567.

[2] A.J. Rack, "Effect of space charge and transit time on the shot noise in diodes," *Bell Sys. Tech. Jour.*, vol. 17, October 1938, pp. 592–619.

[3] G. Herrman and P.S. Wagener, *The Oxide-Coated Cathode*, London: Chapman and Hall, 1951.

[4] A. van der Ziel, "Low-frequency noise in vacuum tubes (flicker effect)," chapter in *Noise in Electron Devices*, published jointly by The Technology Press of MIT and John Wiley & Sons, New York, 1959.

[5] P.K. Tien and J. Moshman, "Monte Carlo calculation of noise near the potential minimum of a high-frequency diode," *J.A.P.*, vol. 27, no. 9, September 1956, pp. 1067–1078.

[6] R.W. Peter, "Low-noise traveling-wave tubes," chapter in *Noise in Electron Devices*, published jointly by The Technology Press of MIT and John Wiley & Sons, Inc. New York, 1959.
[7] M.R. Currie and D.C. Forster, "New mechanism of noise reduction in electron beams," *J.A.P.*, vol. 30, no. 1, January 1959, pp. 94–103.
[8] J.R. Pierce, *Traveling Wave Tubes*, Princeton, NJ: Van Nostrand, 1950.
[9] L.C. Peterson, "Space-charge and transit-time effects on signal and noise in microwave tetrodes," *Proc. IRE*, November 1947, pp. 1264–1272.
[10] F.B. Llewellyn and L.C. Peterson, "Vacuum-tube networks," *Proc. IRE*, March 1944, pp. 144–166.
[11] C.C. Cutler and C.F. Quate, "Experimental verification of space charge and transit time reduction of noise in electron beams," *Phys. Rev.*, vol. 80, no. 5, December 1950, pp. 875–878.
[12] D.A. Watkins, "Traveling-wave tube noise figure," *Proc. IRE*, January 1952, pp. 65–70.
[13] S. Bloom and R.W. Peter, "Transmission-line analog of a modulated electron beam," *RCA Review*, March 1954, pp. 95–112.
[14] R.C. Knechtli, "Effect of electron lenses on beam noise," *IRE Trans. on Electron Devices*, April 1958, pp. 84–88.
[15] S. Bloom, "The effect of initial noise current and velocity correlation on the noise figure of traveling-wave tubes," *RCA Review*, vol. 16, no. 2, June 1955, pp. 179–196.
[16] W.R. Beam, "Interception noise in electron beams at microwave frequencies, *RCA Review*, vol. 16, no. 4, December 1955, pp. 551–579.
[17] T.G. Mihran, B.K. Andal, A.S. Gilmour, Jr., L.A. Mackenzie and M. Roberts, "The growth of peak velocity, noise, and signal along O-type beams from shielded guns," 1964 Conference on Electron Device Research, Cornell University.
[18] L.D. Smullin and C. Fried, "Microwave noise measurements on electron beams," *IRE Trans. on Electron Devices*, vol. ED-1, 1954, pp. 168–183.
[19] J.M. Hammer and E.E. Thomas, "Traveling-wave tube noise figures of 1.0 dB at S-band," *Proc. IEEE*, February 1964, p. 207.
[20] J.M. Hammer and C.P. Wen, "Effect of high magnetic field on electron-beam noise," *RCA Review*, December 1964, pp. 785–789.
[21] F.N.H. Robinson, "Space-charge smoothing of microwave shot noise in electron beams," *Phil. Mag.*, vol. 43, 1952, p. 51.
[22] S. Bloom and R.W. Peter, "A minimum noise factor for the traveling-wave tube," *RCA Review*, vol. 15, 1954, p. 252.
[23] J.R. Pierce and W.E. Danielson, "Minimum noise figure of traveling-wave tubes with uniform helices," *J.A.P.*, vol. 25, 1954, p. 1163.
[24] H.A. Haus, "Noise in one-dimensional electron beams," *J.A.P.*, vol. 26, no. 5, May 1955, pp. 560–571.
[25] H.A. Haus and F.N.H. Robinson, "The minimum noise figure of microwave beam amplifiers," *Proc. IRE*, August 1955, pp. 981–991.

Chapter 16

Nonlinearities and Distortion [1]

At small signal levels, the power output *versus* power input curve for a TWT is reasonably linear and signal distortion produced by the tube is very small. As a TWT is driven toward saturation, the transfer characteristic becomes more and more nonlinear. This nonlinearity results in distortions of amplitude modulated (AM) or phase modulated (PM) signals for several reasons, including

- AM/AM conversion,
- AM/PM conversion,
- Harmonic generation,
- Intermodulation distortion.

In addition, other signal distortions occur, which are not related to saturation effects. Noise is one example, and this was discussed in Chapter 15. In addition, the following are often of concern:

- Variations of gain and power output with frequency,
- Phase nonlinearities or time delay distortion,
- Spurious AM and PM outputs.

16.1 DISTORTION RESULTING FROM SATURATION EFFECTS

16.1.1 AM/AM Conversion

A typical RF power output *versus* RF power input curve for a TWT is shown in Figure 16-1. Also, gain and AM/AM conversion are shown as functions of RF power input. Gain is defined as the ratio of RF power output to RF power input. Two values of gain are usually of importance. These are the small signal gain and the saturated gain. Saturated gain is usually 5 to 6 dB less than small signal gain.

AM/AM conversion is a measure of the change in RF power output that results from a change in RF power input and is usually defined by the slope of the curve of RF power output *versus* RF power input expressed in dB/dB. Thus,

$$\text{AM/AM Conversion} = \frac{\Delta \text{ RF power output (in dB)}}{\text{1 dB change in input power}} \tag{16-1}$$

In the linear (small signal) region, the AM/AM conversion is very nearly 1 dB/dB. At saturation, where no change in output occurs as the input is varied, the AM/AM conversion is 0 dB/dB.

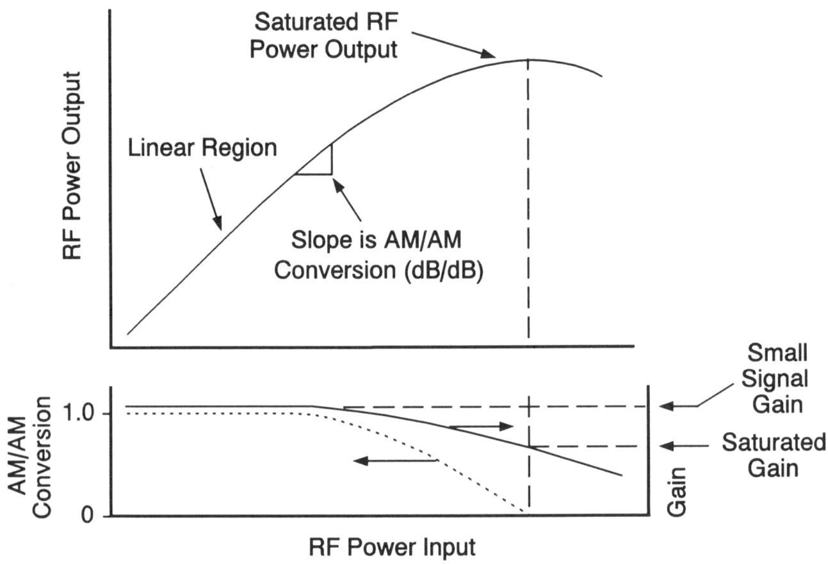

Figure 16-1 Gain and AM/AM conversion *versus* RF power input.

AM/AM conversion is important in communication applications of TWTs. The specification of AM/AM conversion depends on the type of modulation used. For example, if amplitude modulation is employed and if amplitude distortion must be avoided, then it may be necessary to operate the TWT at a maximum power level well below saturation. For this operating condition, we say that the power level must be "backed off" from saturation. The amount of "back off" depends on the amplitude distortion that can be tolerated. Most amplitude modulation communication systems can tolerate no less than 0.9 dB/dB of AM/AM conversion [2]. If no provisions are made for linearizing the TWT, it may be necessary to "back off" as much as 6–8 dB to meet this AM/AM conversion requirement. Through the use of a pitch taper in the TWT [3, 4] and an external linearizing circuit, distortion is significantly reduced and it is possible to operate much closer to the saturation power level.

16.1.2 AM/PM Conversion

AM/PM conversion is a measure of the change in TWT phase length that results from a change in RF drive level. This occurs because, as drive level is increased, more power is extracted from the electron beam and the velocity of the beam is reduced. As beam velocity decreases, the velocity of the RF wave on the circuit is reduced and this increases the phase length of the TWT. It should be noted here that the cathode voltage supply may aggravate the phase delay. As drive increases, body current usually increases, which in turn increases the load on the cathode supply causing a reduction in voltage and an increase in phase length.

A typical curve of the variation of phase length with RF input power is shown in Figure 16-2. In the linear region of the RF power output curve, phase shift is relatively insensitive to drive level. As drive is increased toward saturation, the rate of phase change increases rapidly and then decreases as the TWT saturates. The slope of the phase length curve in degrees/dB is the AM/PM conversion. An example is plotted in Figure 16-3.

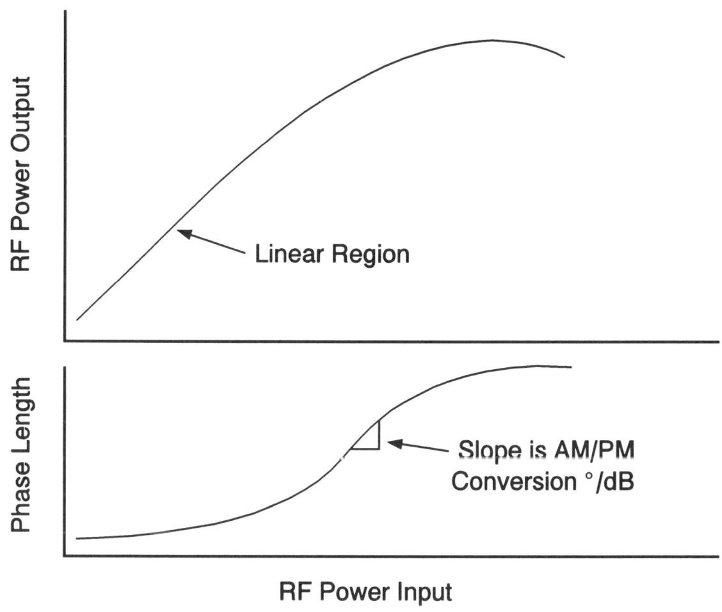

Figure 16-2 Change of phase length with drive power.

The maximum value of AM/PM conversion is usually the value that is specified. This occurs at 3 to 10 dB below the input power that produces saturation. The type of modulation employed determines the need for specifying AM/PM conversion. For example, in amplitude modulated systems, AM/PM conversion can result in amplitude distortions. In multiple carrier systems, in which both AM and PM signals are present,

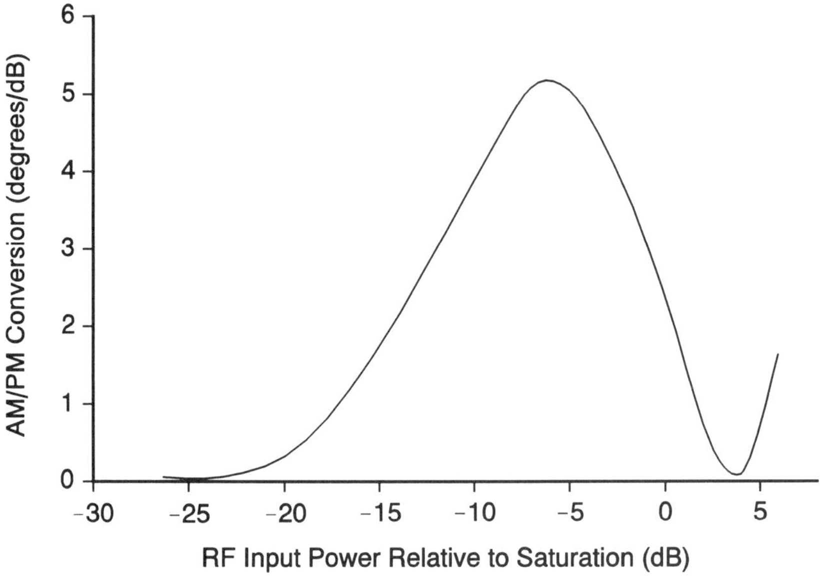

Figure 16-3 Example of variation of AM/PM conversion with RF drive. From Hughes TWT/TWTA Handbook, 1992.

AM/PM conversion can cause distortion or interference with the phase or frequency modulated signals. Finally, amplitude variations in phase or frequency modulated systems can cause distortions or interference.

16.1.3 Harmonic Generation

Near saturation, the electron bunches in the electron beam of a TWT become sharply defined, and so the RF current in the beam is rich in harmonics. At saturation, helix TWTs used for communication applications usually produce second harmonic power that is 8 to 10 dB below the fundamental power. Third harmonic power at saturation is usually 16 to 18 dB below the fundamental power [2].

Harmonic generation is a by-product of the bunching process in a TWT, and little can be done in the design of the TWT to reduce harmonics. The system must be able to tolerate harmonics or else filters must be used to reduce the level of the harmonics.

In some extremely broadband ECM applications, TWTs are used that operate with bandwidths greater than an octave. There have been cases where the second harmonic power output is equal to or greater than the fundamental power when one of these TWTs was driven to saturation at the low end of the frequency band. One of the reasons for this high harmonic output is that, at the low end of the band, the harmonic frequency is within the amplifying band of the TWT.

In coupled cavity TWTs, the slow-wave structure is a filter that normally does not propagate harmonics. Even if the structure does contain a higher order mode at a harmonic frequency, the field configuration is such that the interaction with the beam is weak. As a result, the harmonic output from a coupled cavity TWT may be 30 dB or more below the fundamental power.

16.1.4 Intermodulation Products

When two or more carrier signals are being amplified by a TWT, intermodulation occurs and signals (intermodulation products) are produced that are displaced from the carriers by multiples of the frequency separation of the carriers. This intermodulation occurs because of the non linearities in the amplification process. The third order intermodulation products are the most important because they are closest to the signal frequencies and largest in amplitude. These third order intermodulation products result from the second order distortion products of one signal mixing with the first order products of another signal. Figure 16-4 indicates relative frequencies and amplitudes of third and fifth order intermodulation products.

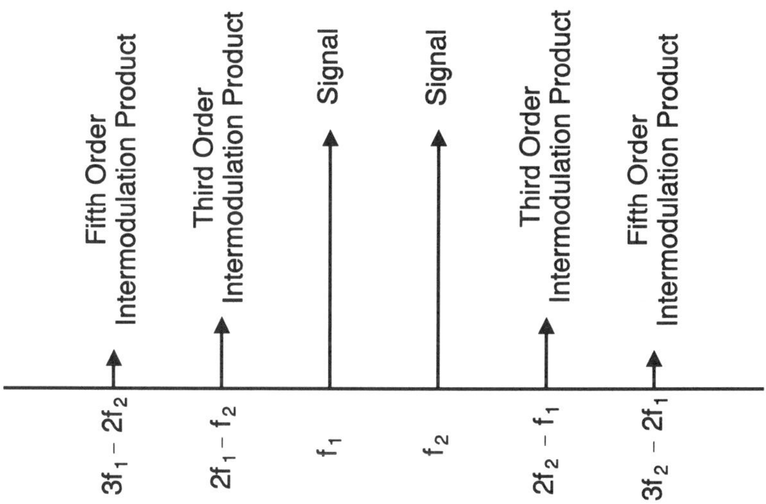

Figure 16-4 Intermodulation products.

As was the case with harmonic generation, it is usually not possible to control the amplitudes of the intermodulation products with the design of a TWT. As indicated in Figure 16-5, when a TWT is driven to saturation with two signals of equal amplitude, the third order intermodulation products are each 8 to 10 dB below either of the signals. By reducing the amplitudes of the drive signals so that the TWT operates well below saturated power output, the amplitudes of the intermodulation products can be

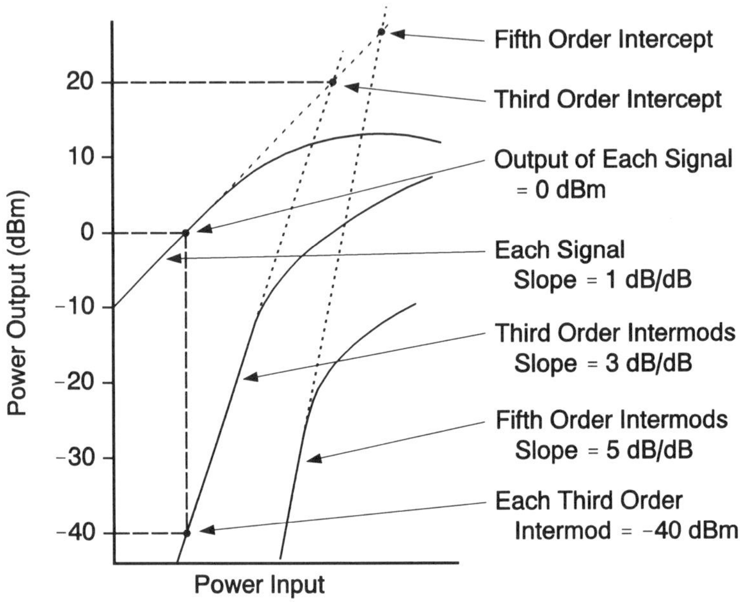

Figure 16-5 "Intercept point" method for specifying intermodulation distortion. From *Introduction to Microwaves*, by J.W. Hansen, to be published.

greatly reduced. In the example in Figure 16-5, with the TWT operated about 10 dB below saturation, the third order intermodulation products drop to about 40 dB below the signal output level.

The specification of intermodulation products is not straightforward. Usually, the amplitudes of the products are specified as the maximum allowable relative to the carrier signal levels, when these signals have equal amplitudes. The amplitude of the carrier signals is also usually specified relative to the saturated output power when the TWT is amplifying two signals of equal amplitude.

An alternative method illustrated in Figure 16-5 is the called the *intercept point* method [5]. This is based on an understanding of the rate at which the levels of the intermodulation products vary as drive level is varied. If one point is known, such as the intercept of the projection from the linear regions of the power curve for the carrier signals with the projection of the power curve for the third order intermodulation products, then the intermodulation product amplitudes can be calculated for various drive levels.

16.2 VARIATIONS WITH FREQUENCY

When the gain *versus* frequency characteristics of a TWT are examined, two distinct types of gain variation are usually observed. One is a broadband variation for which the gain varies slowly as a function of frequency. The other is a relatively rapid variation of gain in narrow bandwidths. These two types of gain variation are discussed in the following paragraphs.

16.2.1 Broadband Gain Variations

An example of the broadband variation in gain is shown in Figure 16-6. This gain characteristic was computed for small signal operation of a coupled cavity TWT [6]. The variation in gain with frequency results from the dispersion and impedance characteristics of the circuit along with the fact that gain varies with the electrical length of the circuit, which, in turn, varies with frequency. The imposition of a tight specification on gain flatness results in compromises in circuit design and beam conditions that reduce output power and efficiency. In an example [6], output power was a factor of 2.5 lower for the TWT with the required gain flatness than for a TWT designed for maximum efficiency.

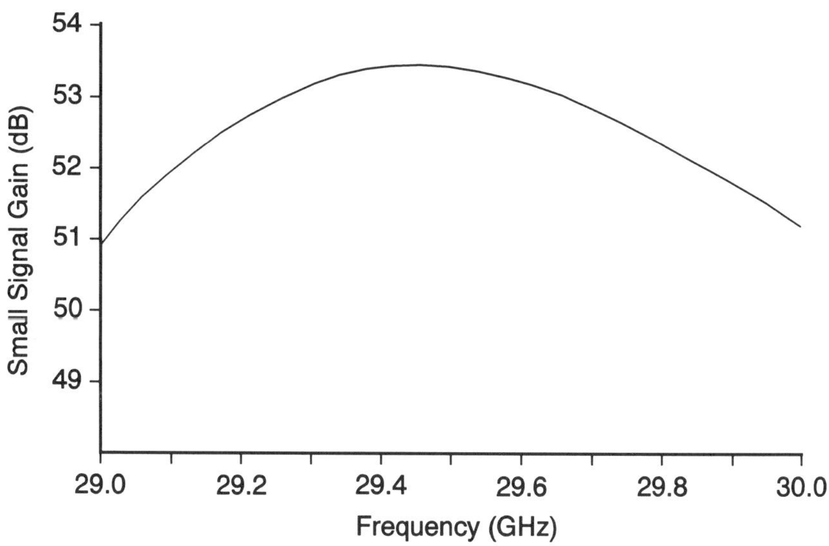

Figure 16-6 Small signal gain with no internal reflections. From M.L. Cabrales *et al., NASA Contractor Report 185146*, January 1990.

16.2.2 Narrowband Gain Variations

In narrow bandwidths, the dominant effect on gain is nearly always some feedback path, usually within the TWT. As an example, internal mismatches, which can never be completely eliminated, cause a periodic (with frequency) variation in gain. Figure 16-7 illustrates the way in which this can occur. A mismatch at the output end of the slow wave structure causes the reflection of a small portion of the signal. This portion is then reflected by a mismatch at the input end of the structure. It then combines, in various phase relationships, with the desired growing wave. the result is that a "gain ripple" is produced as is shown in Figure 16-8. This particular curve was calculated for the millimeter-wave coupled cavity TWT mentioned in a previous paragraph [6]. The frequency difference between gain peaks depends on the frequency of operation and the length of the RF circuit (which produce a closed loop phase change of 360°) and is typically in the 100–300 MHz range.

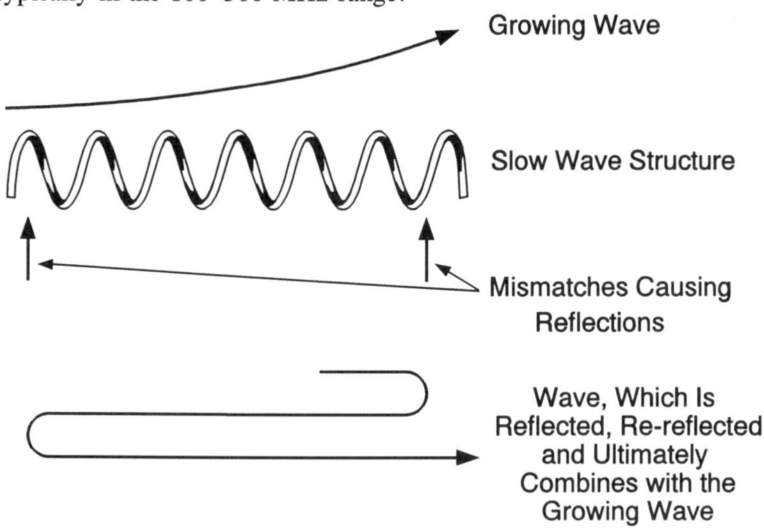

Figure 16-7 TWT section with two internal mismatches.

Internal reflections within a TWT are the primary cause for gain ripple. These reflections can occur at the following locations:

1. At the attenuator at the end of the input section of the TWT. These reflections are then re-reflected at the input port to the TWT,
2. Within the center section of a TWT with two severs,
3. At the output port. These reflections are then re-reflected at the sever or attenuator at the input end of the output section. This feedback path is usually the dominant path because gain is high and reflections from the output port vary with load characteristics. An isolator at the output helps control load reflections,

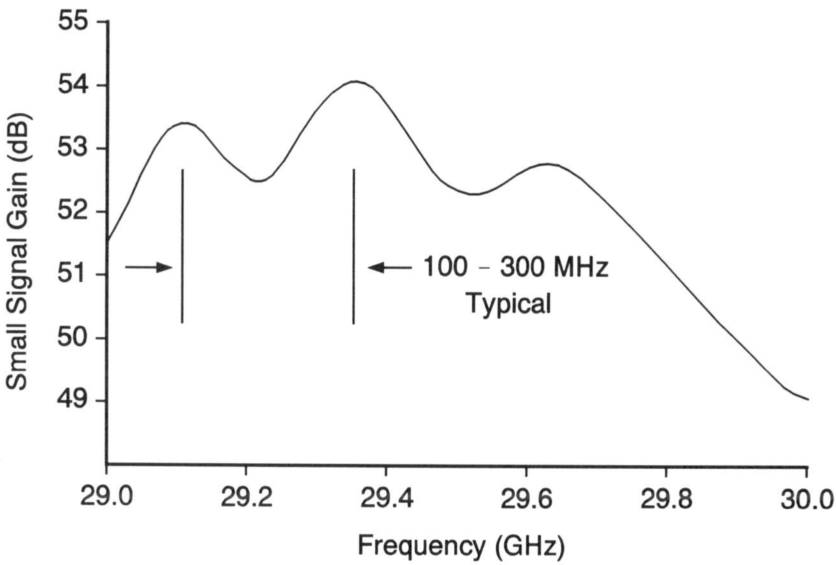

Figure 16-8 Gain ripple with 7% voltage reflections. From M.L. Cabrales *et al.*, *NASA Contractor Report 185146*, January 1990.

4. At imperfections in the slow wave structure. Periodically varying imperfections, for example at helix support rods, can be particularly detrimental in this respect.

It is also possible for returning electrons from the collector to provide a feedback path that produces gain ripple.

16.2.3 Phase Nonlinearities or Time Delay Distortion

Shown in Figure 16-9 are phase and time delay as functions of frequency. A linear increase of phase as a function of frequency is unavoidable because the phase length of a TWT depends on wavelength. The ripple-portion of the phase curve depends on the same internal reflection phenomena that produce gain ripple. Time delay is the total time that it takes for a signal to pass through a TWT and is the derivative of phase delay. Thus, the same mechanisms that cause phase nonlinearities are responsible for time delay distortion.

The frequency band of interest is indicated in Figure 16-9 to illustrate the fact that, for most TWTs, the narrowband effects dominate the broadband effects. This is true because most TWTs operate over only a small portion of the passband of the circuit and so band edge effects are far from the frequency band of interest.

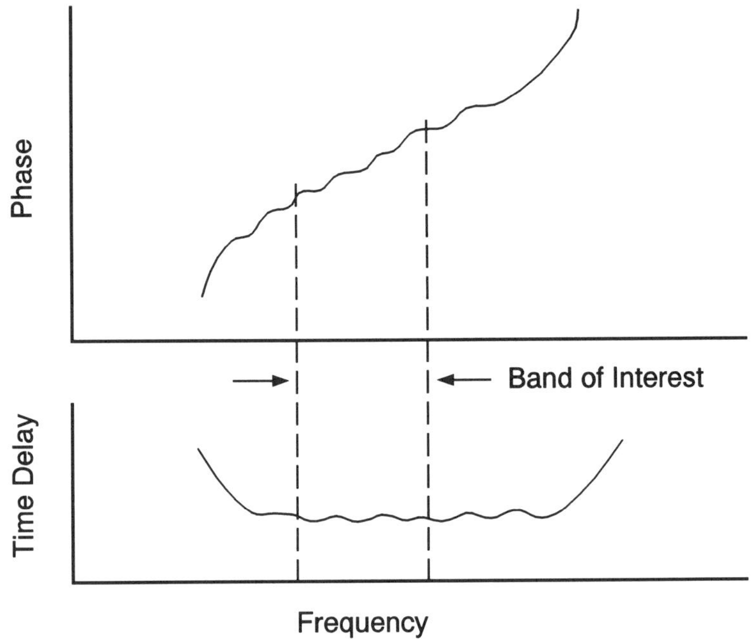

Figure 16-9 Phase nonlinearity and time delay distortion.

16.3 SPURIOUS AM AND PM OUTPUTS

The principle spurious output is phase modulation caused by ripple on the TWT cathode voltage. Ripple is usually small because most TWT power supplies use dc–dc converters operating at ~5 to 30 kHz, and the ripple voltages are easily filtered. Still, ac components of voltage may be present on the TWT cathode voltage because filtering is never perfect and because ripple or transients on the input voltage pass through the converter.

The pushing factor of a TWT is a measure of the sensitivity of phase shift to changes in electrode voltage. For the cathode voltage, it is [2]

$$\frac{\Delta\phi}{\Delta V} = -\frac{1}{3}\frac{\phi}{V} \tag{16-2}$$

where ϕ is the total phase length of the TWT and V is the cathode voltage. For $\phi = 10{,}000°$ and $V = 10{,}000$ V, $\Delta\phi/\Delta V = 1/3$ degree/V.

REFERENCES

[1] The author is indebted to J.W. Hansen of the Hughes Electron Dynamics Division for much of the material in this chapter.

[2] J.W. Hansen, G.A. Lange, A.S. Rostad and R.L. Woods, "Systems aspects of communications TWTAs or how to deal with the tube manufacturer to your best advantage," Hughes Aircraft Company Electron Dynamics Division Applications Note, August 1982.

[3] H.G. Kosmahl and J.C. Peterson, "A TWT amplifier with a linear power transfer characteristic and improved efficiency," NASA Technical Memorandum 83590, prepared for the Tenth Communications Satellite Systems Conference, Orlando, FL, March 18–22 1984.

[4] J.W. Hansen, "Linearized TWTAs for low-distortion communications applications," presented at the 1992 Microwave Power Tube Conference, Monterey, CA, May 11–13 1992.

[5] F.C. McVay, "Don't guess the spurious level," *Electronic Design*, February 1967.

[6] M.L. Cabrales, H.A. Davis, H.C. Limburg, I. Tammaru and J.P. Vaszari, "Study of methods for the reduction of distortion in high power traveling-wave tubes," *NASA Contractor Report 185146*, January 1990.

Chapter 17

Breakdown and Protection

There are many ways in which traveling wave tubes can fail. A reasonably thorough discussion of their failure mechanisms is well beyond the scope of this book and, in fact, would require another book of size comparable to this one.

One of the most common and, perhaps, most poorly understood classes of failure mechanisms centers on electrical breakdown in one form or another. In the short courses from which this book was originally developed, electrical breakdown was, at first, only briefly discussed. Because of the strong interest in breakdown phenomena expressed by those taking the course, that section was developed to cover major aspects of breakdown phenomena of concern to those who must use traveling wave tubes.

Breakdown in microwave tubes may occur inside or outside the vacuum envelope at any of several locations as is indicated in the following and in Figure 17-1.

- In the electron gun (between electrodes, between leads or from electrodes or leads to ground),
- In the collector (between electrodes, between leads or from electrodes or leads to ground),
- In the high power portions of the RF structure.

Often, it is not the tube that is initially at fault. For example, there may be a failure at any one of several locations in the RF system attached to a tube. Shown in Figure 17-2 are some of the common problem areas in these RF systems. If excessive RF power is reflected, the tube may be damaged for any one of several reasons. All too often, the tube is thought to be at fault and is replaced before the original cause of the difficulty is corrected.

Whether or not breakdown will occur depends on two factors:

1. The applied field level and local field enhancement effects,
2. The breakdown field of the medium (gas, vacuum, liquid or solid).

Very approximately, the breakdown fields for various media are as follows:

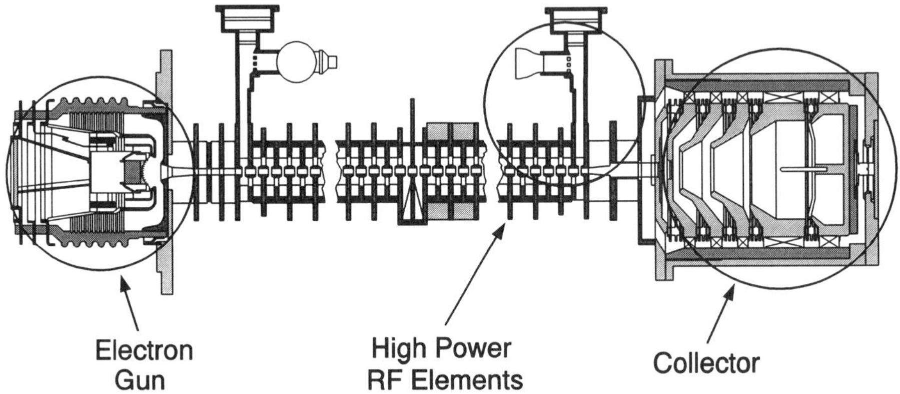

Figure 17-1 Regions in a microwave tube where electrical breakdown may occur.

Gas	~ tens of V/cm to 10^5 V/cm depending on pressure and type
Vacuum	~ $0.5\text{–}3 \times 10^5$ V/cm
Liquid	~ $0.5\text{–}1.0 \times 10^6$ V/cm
Solid	~ $0.5\text{–}1.0 \times 10^6$ V/cm

At interfaces between solids, liquids and vacuum or gas, the breakdown field may be significantly degraded. In addition, in a void in a solid or a liquid, the field is normally enhanced. The amount of enhancement depends on the shape and orientation of the void and the dielectric constants of the media, as indicated in Figure 17-3.

The low breakdown fields of gas or vacuum relative to those of solids or liquids coupled with various field enhancement effects result in breakdown occurring, in most cases, in the gas or vacuum rather than in the solid or liquid. As a result, most of the breakdown discussions in this chapter will concentrate on gas and vacuum. First, a section is devoted to various field enhancement effects that reduce the value of the applied field at which breakdown occurs. Then, dc and RF breakdown in gas and in vacuum are covered. Finally, the special case of breakdown on the surface of an insulator is discussed.

17.1 FIELD ENHANCEMENT

The electric field between parallel plane electrodes is the voltage applied from one electrode to the other divided by the distance by which the electrodes are separated. In practice, it is common to calculate field levels in all sorts of configurations simply by dividing the applied voltage by the electrode separation distance. Fields calculated in this manner are often in error by a factor of from two to ten or more. Worse yet, because the power varies as the square of the field, the calculated power handling capability is often in error by a factor ranging up to two orders of magnitude or more.

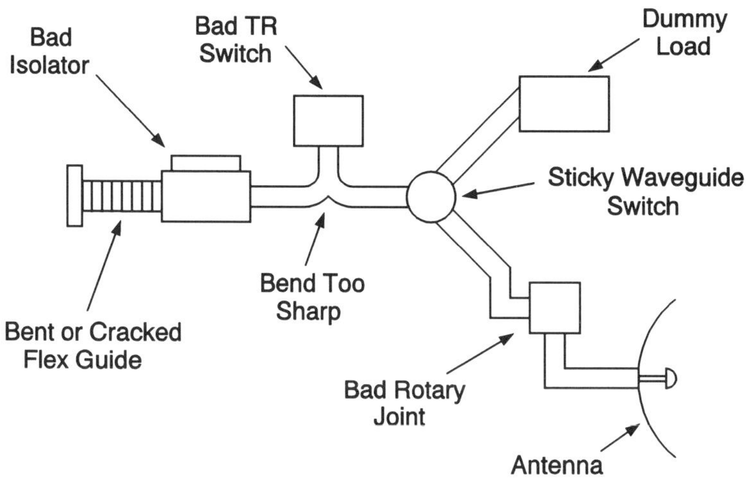

Figure 17-2 Common problem areas in RF systems. From *Microwave Tube Manual*, by Varian, Air Force Publication No. T.O.00-25-251, October 1979.

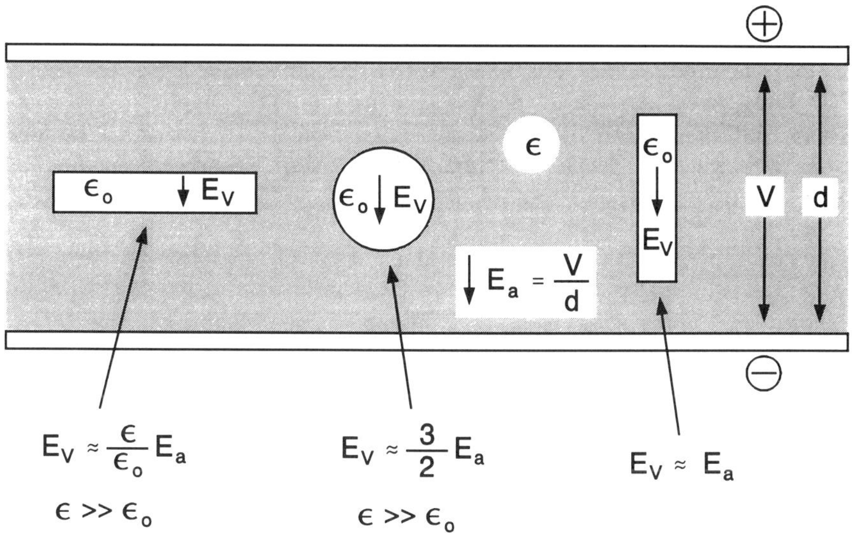

Figure 17-3 Electric field enhancement in voids in dielectric media.

The parallel plane and coaxial conductors shown in Figure 17-4 provide a simple case for the comparison of field levels. A potential of 1000 V between electrodes is assumed, and equipotential profiles at 250 V, 500 V, and 750 V are shown. For the parallel-plane electrode case, the potential profiles are equally spaced and the electric field is 1000 V/cm everywhere between the electrodes.

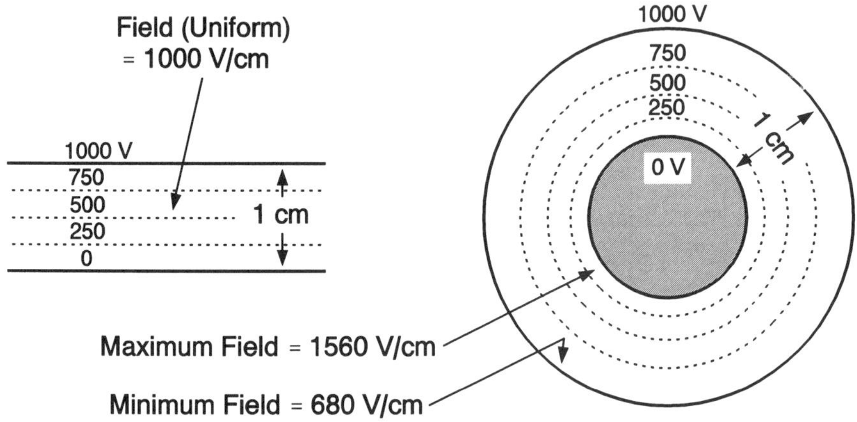

Figure 17-4 Potential distributions in parallel plane and 50 Ω coaxial electrode configurations.

For the coaxial electrode case, the equipotential profiles near the center conductor are more closely spaced than in the parallel plane electrode case. The electric field at the surface of the center conductor is easy to calculate for this configuration and is 1560 V/cm, or over 50% above that for parallel plane electrodes.

If the maximum field was limited to 1000 V/cm because of breakdown considerations, then the maximum voltage that could be applied to the coaxial electrodes would be 1000/1.56, or 641 V. In the TE_{11} mode (for which the potential distributions in Figure 17-4 are correct), the maximum RF power capability of the coaxial configuration would be $(1/1.56)^2$, or 40% of that for the parallel plane electrodes.

The factor by which the electric field exceeds the field for the parallel plane electrode case is normally called the *field enhancement factor*, denoted β. At the center conductor of the 50 Ω coaxial line, the field enhancement factor is 1.56.

The magnitude of β is highly dependent on the radius of an electrode and its distance to another electrode. These factors are illustrated in Figure 17-5 for electrode pairs consisting of a sphere and a plane and a hemisphere on a plane and a plane.

Field enhancement factors have been calculated for many electrode configurations. Some of these are given in Figure 17-6. In most practical cases of interest, the electrode geometries are close enough to those shown in Figure 17-6 to permit a reasonable estimate of the value of β to be made. In some cases, values of β for two or more configurations can be combined to obtain an estimate for a configuration in question. For example, the value of β at the surface of a drop of condensation on the center conductor of a 50 Ω coaxial line is approximately 3 (β for a hemisphere on a plane) times 1.56 (β at the center conductor of a 50 Ω coaxial line), or about 4.7.

It is also possible to estimate field enhancement factors for some configurations by using a simple manual iterative technique. This is based on Laplace's equation, from which it can be shown that the potential at the center of a square is the average of the

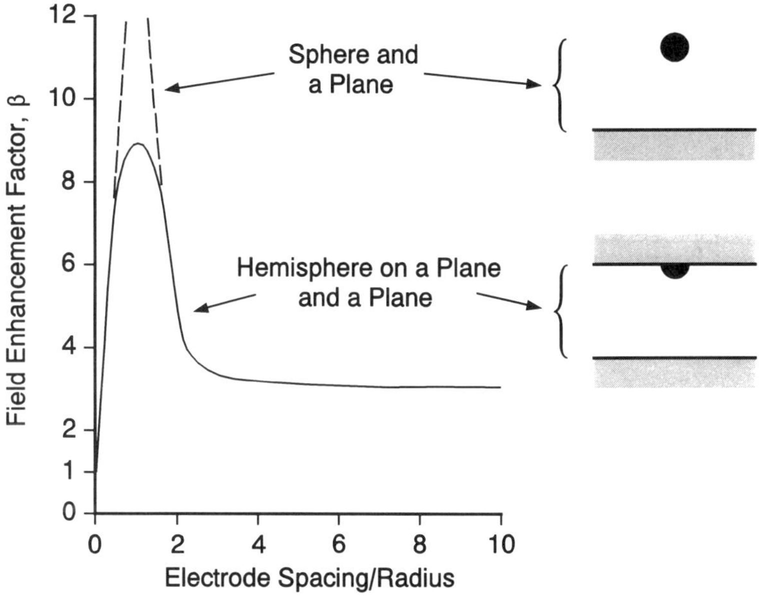

Figure 17-5 Value of β for two electrode configurations. From V.N. Glazanov, *Elektrichestvo*, vol. 3, 1958.

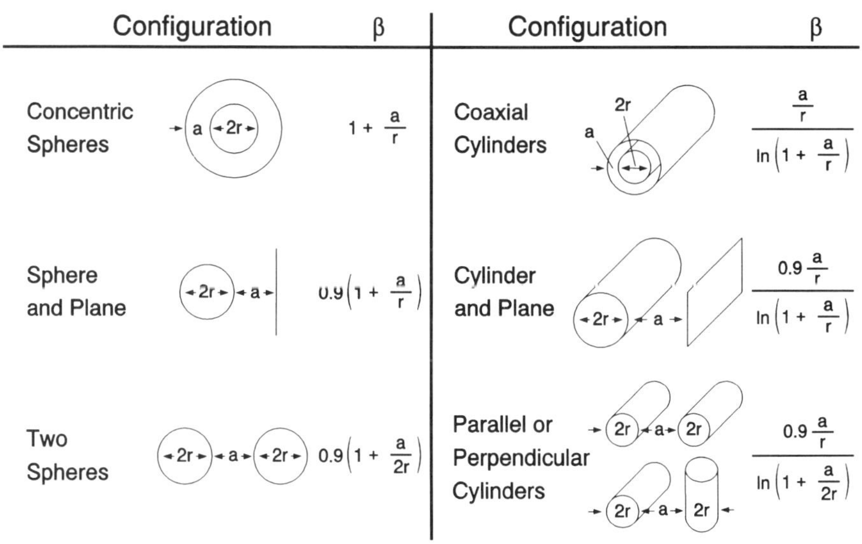

Figure 17-6 Maximum field enhancement factor, β, for various electrode configurations.

potentials at the four corners of the square. That is, referring to Figure 17-7,

$$V = \frac{V_m + V_n + V_o + V_p}{4} \tag{17-1}$$

The use of this technique in estimating field enhancement factors is, perhaps, best explained with the aid of an example. Consider the cross section of the electrode configuration shown in Figure 17-8. The outer conductor is a waveguide-like box and the center electrode is a flat strip at potential V with respect to the outer conductor. The problem is to find the field enhancement factor at the edge of the strip. The radius of the edge is 1 unit, and the spacing between the strip and the outer conductor is 32 units.

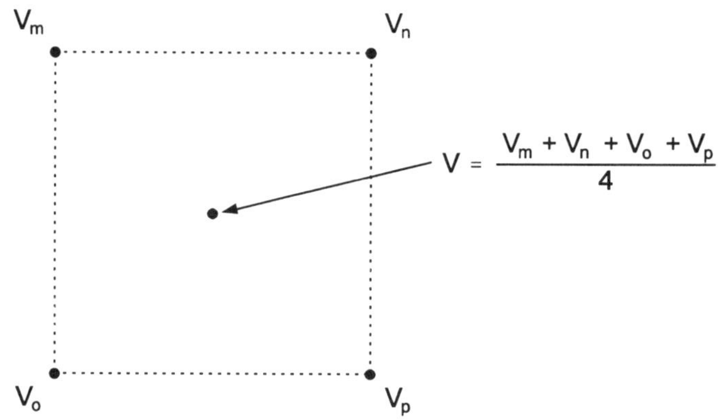

Figure 17-7 Potential at the center of a square.

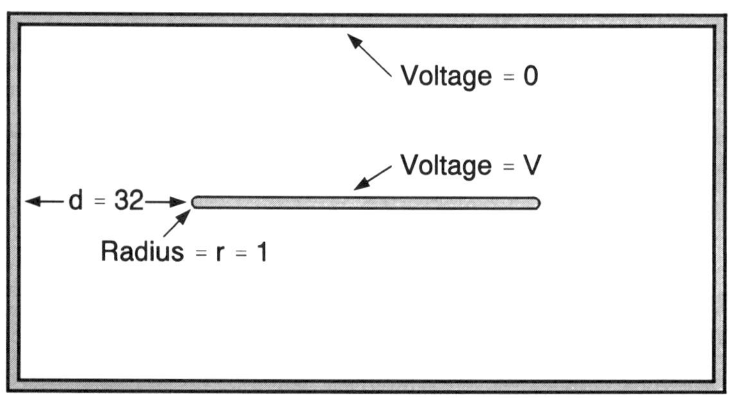

Figure 17-8 Electrode configuration for illustrating use of iterative technique for determining β.

The potential distribution between the electrodes may be estimated with (17-1) as indicated in Figure 17-9. Referring to Figure 17-9(a), estimates of potentials are shown as fractions of the voltage on the strip. The first estimate of the potential at a point is preceded by a (1). The second estimate is preceded by a (2), *et cetera*. The average of the potentials at m, n, o and p is

$$\frac{0 + 0 + 0 + 1}{4} = .25 \tag{17-2}$$

and is shown at the center of the mnop square. For the first round of estimates, known potentials on the boundaries are used for calculations. For the second round of calculations, a combination of boundary values plus estimates from the first round are used. For example, the value of .38 at the point midway between o and p is the average of .25 + 1 + .25 + 0. Additional calculations are made until (usually) after about four or five iterations, little change occurs in the estimates.

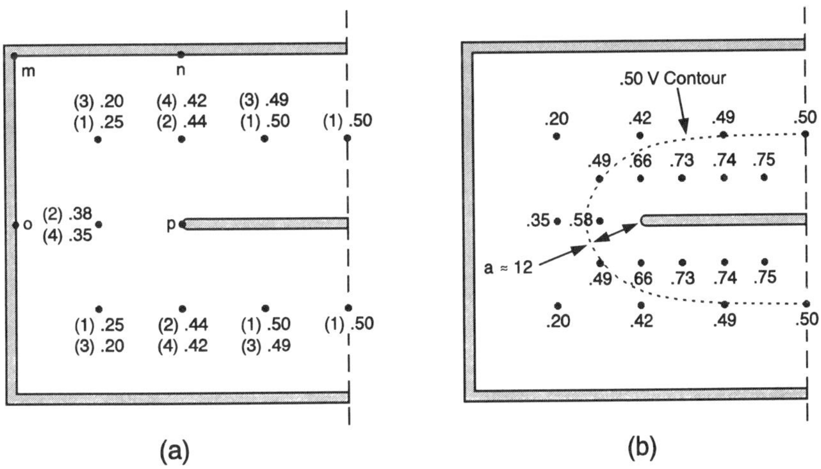

Figure 17-9 Potential calculations for determining β.

For the problem at hand, it is noted that the configuration of the edge of the strip electrode and the nearby equipotential contours is similar to that for coaxial cylinders given in Figure 17-6. So, in this case as shown in Figure 17-9(b), a contour is selected (the .50 V contour) and the distance from that contour to the electrode of interest is determined. Finally, the field enhancement factor is found by dividing the field at the edge of the strip electrode by the nominal overall field ($V/32$ from Figure 17-8). The field at the strip conductor edge is .50 $V/12$ multiplied by the field enhancement factor for coaxial cylinders (from Figure 17-6). The result is

$$\beta = \frac{\dfrac{.50\,V}{12}\,\dfrac{a/r}{\ln\left(1 + a/r\right)}}{\dfrac{V}{32}} \approx 6 \tag{17-3}$$

Obstacles of various shapes are often purposely placed in waveguides and coaxial lines. This may be for impedance matching purposes, for mounting dielectric supports, *et cetera*. Values of β and power handling degradation are given in Figure 17-10 for some of these obstacle shapes in rectangular waveguide. Note that the reduction in power capability can easily be an order of magnitude or more.

The importance of field enhancement on electrical breakdown cannot be overemphasized. The most troublesome aspect of field enhancement is that, in many cases, its magnitude is either not known or is badly underestimated. The value of β of 1.56 calculated for the center conductor of a 50 Ω coaxial line is virtually meaningless if there are irregularities, such as scratches or machining marks, and so forth on the center conductor. These can easily have a β of 10 or more.

Finally, it should be noted here that the breakdown characteristics of electrodes in a vacuum may, in some cases, be almost completely controlled by tiny protuberances on the cathode surface having values of β approaching 100. This is discussed in Section 17.4.

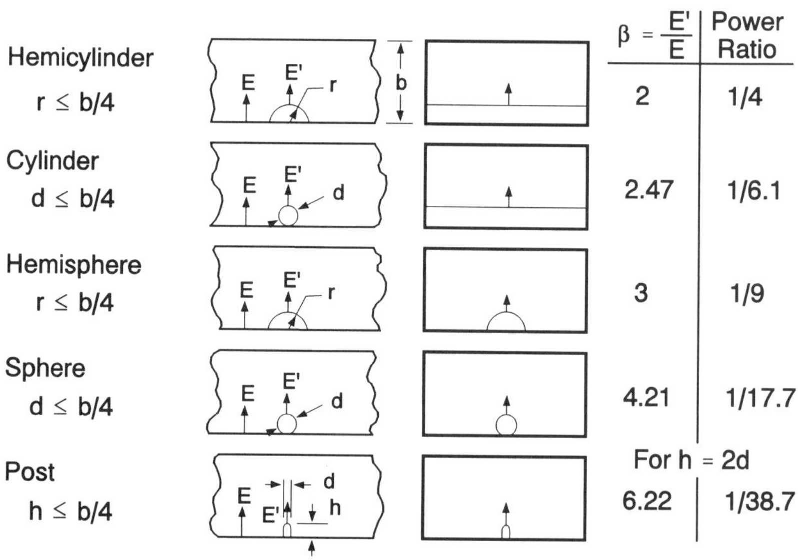

Figure 17-10 Values of β and power handling degradation for various obstacles in rectangular waveguides. From R.D. Wegenroth, Wheeler Laboratories Report 953P, June 1960.

17.2 DC BREAKDOWN IN GAS [1–3]

Electrical discharges in gases are extremely complex. Several different types of discharges can exist, depending on the current and voltage applied to the gas. These various discharge types are indicated in Figure 17-11. Current and relative voltages shown in Figure 17-11 are representative values only and depend on factors such as gas type and area of the electrodes used during testing.

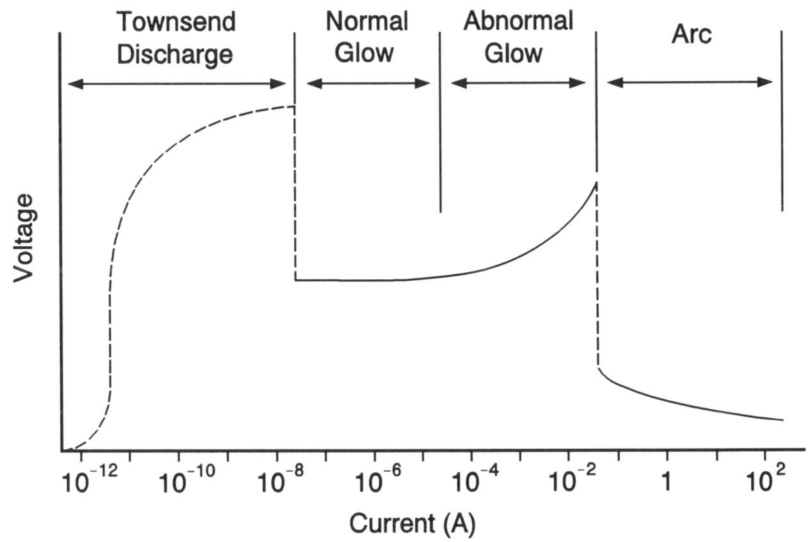

Figure 17-11 Typical volt-ampere characteristics of a gas discharge. Adapted from *Principles and Applications of Electron Devices* by P.D. Ankrum.

By far, the most damaging discharge is the arc. In the following paragraphs, various phenomena leading up to an arc are discussed and then arc characteristics are described.

The discharges in Figure 17-11 are roughly categorized as non-self-sustaining and self-sustaining. The non-self-sustained discharges occur at very low current levels. The various phases of the non-self-sustained discharge are indicated in Figure 17-12, which is an expansion of the lefthand side of Figure 17-11 with the axes reversed. The curve in Figure 17-12 shows the variation of current with voltage between electrodes in a gas at room temperature (the exact temperature is not important as long as it is not absolute zero). It is also assumed that the radiation (cosmic, ultraviolet, *et cetera*) that normally exists is present. Because of the temperature and the radiation, electrons are emitted by the electrodes and are released by ionization of gas atoms and molecules. (At room temperature there is about 1 electron/cm^3 in atmosphere air).

When voltage is first applied to the electrodes, the current increases slowly and then saturates. That is, all free electrons are drawn to the positive electrode. Further

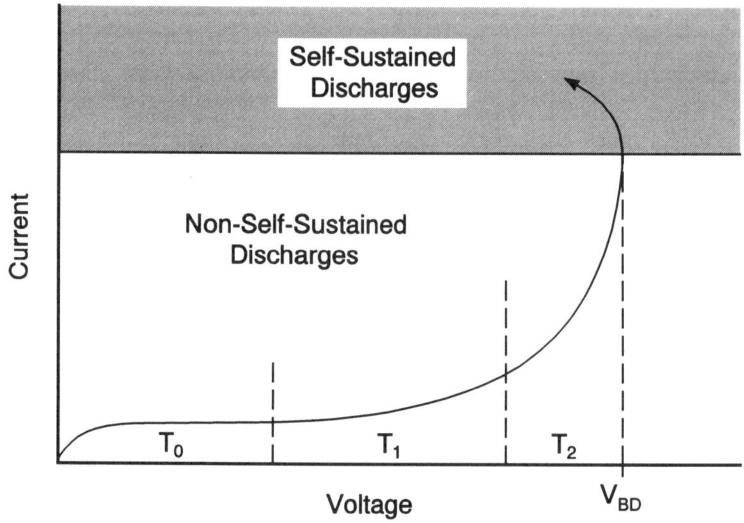

Figure 17-12 Current-voltage relations in prebreakdown regions. Adapted from J.D. Cobine, *Gaseous Electronics*, Published 1958 by Dover Publications.

increases in voltage produce no further increase in current (region T_0 in Figure 17-12). When the voltage is increased to region T_1, the current increases slowly again. In this region, electrons gain sufficient energy from the electric field to produce ionization of the gas and the resulting additional electrons (and ions) contribute to the current level.

In region T_2, the rate of increase of current accelerates because positive ions produced by electron collisions gain sufficient energy from the field to produce secondary emission of electrons from the negative electrode as well as additional ionization.

The discharges in regions T_1 and T_2 are called Townsend discharges, after J.S. Townsend, who conducted early investigations in this field. These discharges (and that in region T_0) are referred to as *non-self-sustaining discharges* because the current ceases as soon as the external radiation (including the heat to bring the electrodes and gas to room temperature) is removed.

When the voltage reaches a critical level, V_{BD}, current increases very rapidly and is accompanied by a decrease in voltage. The voltage, V_{BD}, is normally called the *breakdown voltage*, but is sometimes referred to as the firing voltage, ignition voltage, or sparking voltage. The resulting discharge is self-sustaining because each electron leaving the negative electrode establishes secondary processes (ionization, *et cetera*), which, in turn, cause one or more electrons to be emitted. Removal of external radiation does not cause the discharge to extinguish.

The breakdown voltage of a gas depends on the ease with which the gas is ionized and on the effectiveness of the ions in producing electrons when they strike the negative electrode. Thus, the breakdown voltage is a function of gas type and pressure, and electrode spacing and material.

The breakdown voltage is normally plotted in the manner shown in Figure 17-13. Breakdown voltage plots of this type are referred to as Paschen characteristics because of Paschen's law, which states that, for parallel plane electrodes in a particular gas at a particular temperature, the breakdown voltage is a function of the product of the gas pressure and the electrode spacing.

The shapes of the Paschen characteristics may be explained qualitatively as follows: to the left of the minimum in the Paschen curve, the breakdown voltage increases as pressure and density decrease (with fixed electrode spacing). This is because the gas molecules are farther apart, and more of the electrons travel through the gas and are collected by the positive electrode without having produced ionization. In the limit, under high-vacuum conditions, there is essentially no ionization of the gas between electrodes and breakdown occurs because of effects occurring at the surfaces of the electrodes. Direct current (dc) breakdown in a vacuum is discussed in Section 17.4.

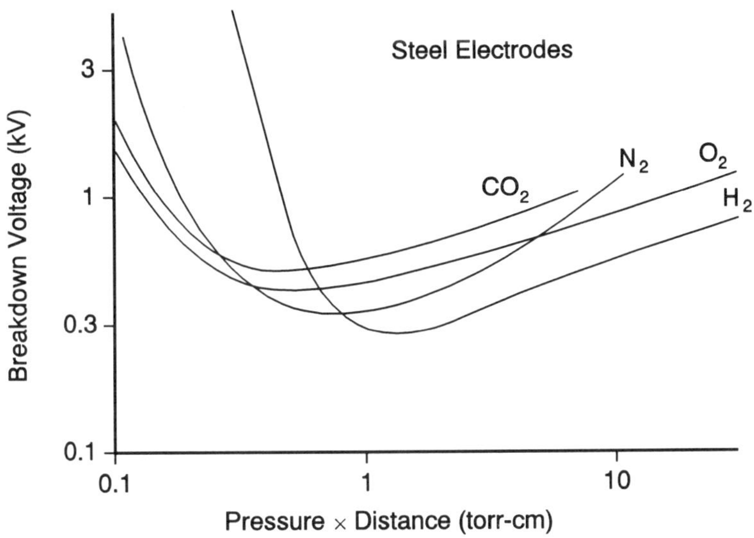

Figure 17-13 Paschen characteristics for some molecular gases. From Max J. Schonhuber, *IEEE Trans. P.A.S.* February 1969. © 1969 IEEE.

To the right of the Paschen minimum, the breakdown voltage increases as pressure and gas density increase (with fixed spacing between the electrodes). In this case, gas molecules are closer together, and so a large electric field is required to give the electrons enough energy in the short distances between collisions to produce ionization. Thus, the breakdown voltage is high. At atmospheric pressure and above, the breakdown voltage for a 1 cm gap can be several tens and even hundreds of thousands of volts.

At intermediate pressures and electrode spacings, optimum conditions for ionization occur, and so there is a minimum in the Paschen curve. The minimum breakdown voltage for most gases is a few hundred volts and occurs at a *Pd* (pressure times distance in torr-cm) near unity.

Breakdown voltages for several commonly used gases are listed in Table 17-1. The values given were obtained with a 1 cm gap at the Paschen minimum (at the critical pressure-spacing product) and at atmospheric pressure. In particular, it should be noted that the freons and sulphur hexafluoride have relatively high breakdown voltages. These gases are often referred to as electronegative gases. Their molecules require one or two electrons to fill the outer shell. As a result, they can easily capture free electrons and this results in the high breakdown voltages for these gases.

TABLE 17.1

Breakdown Voltages for Several Commonly Used Gases (From W.G. Dunbar, "High Voltage Design Guide: Aircraft," Vol. IV, AFWAL-TR-82-2057.)

Gas	*Volts (dc) (Paschen Min)*	*Kilovolts/cm (dc) (Atm. Pressure)*
Air	315	33
Ammonia	—	26
Argon	280	4.8
Carbon Dioxide	430	28
Freon 14	480	32
Freon 114	420	90
Freon 115	430	90
Freon 116	500	—
Freon C 138	450	—
Helium	189	1.63
Hydrogen	292	17
Nitrogen	265	32
Oxygen	440	—
Sulfur Hexafluoride	520	89

Electronegative gases are expensive and tend to be somewhat hazardous, and so they are often mixed with more common gases such as air or dry nitrogen. As shown in Figure 17-14, even a small amount of sulphur hexafluoride (SF_6) in nitrogen significantly increases breakdown voltage.

When breakdown occurs, the discharge increases in intensity very rapidly. Depending on circuit conditions (voltage capability and impedance), either a glow discharge or an arc discharge may develop. The characteristics of these discharges are indicated in Figure 17-11.

The power supply or modulator for a microwave tube is normally capable of delivering a short-circuit current of at least several amperes and sometimes several kiloamperes. As a result, when breakdown occurs, an arc is established rather than a glow discharge.

Still, glow discharges cannot be ignored because they can occur under special circumstances and go unnoticed until they lead to an arc. For example, a glow discharge can exist in a void at the interface of a potting compound and a conductor. Sputtering and heat from the discharge can enlarge the void and eventually lead to an arc.

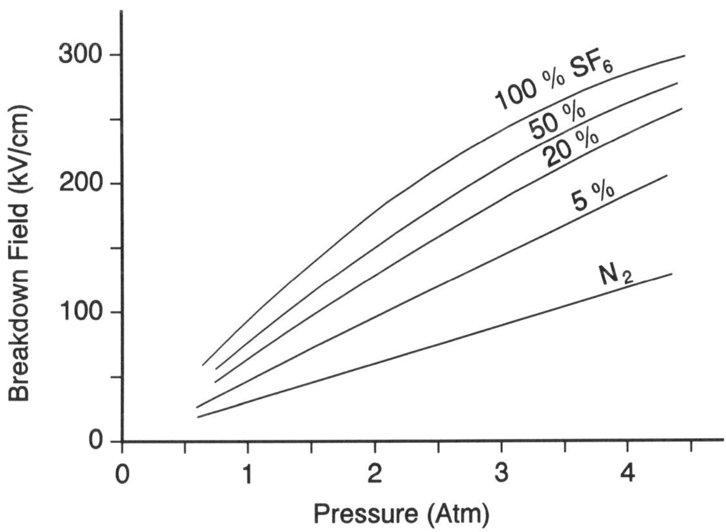

Figure 17-14 Effect on breakdown voltage of mixing SF_6 with dry nitrogen. From N.H. Malik and A.H. Qureshi, *IEEE Trans. on Elec. Ins.*, October 1980. © 1980 IEEE.

The voltage required to maintain a glow discharge is on the order of the minimum breakdown voltage for the gas. That is, from roughly 100 V to a few hundred volts, depending on gas type. The glow discharge is characterized by a glow in the gas and on the cathode surface and by a low current density, on the order of 1 to 10 mA/cm^2. The discharge is maintained by secondary emission of electrons from the cathode as a result of positive-ion bombardment.

The potential distribution between two electrodes is altered significantly when a glow discharge occurs. For parallel plane electrodes, for example, the potential distribution before breakdown is linear. As shown in Figure 17-15, after a glow discharge has been established, the potential is close to the anode potential throughout most of the interelectrode region. Near the cathode, the potential drops rapidly to that of the cathode in the cathode fall region.

The large number of positive ions formed in the interelectrode space is the primary reason for the shape of the potential profile. These ions neutralize the negative space charge due to the electrons. Near the cathode a positive-ion space charge cloud forms that is called the *cathode sheath*. Ions accelerated through the cathode sheath produce secondary electron emission and sputtering at the cathode surface. The sputtered cathode material may be deposited on nearby insulating surfaces and result in undesirable conduction. Electrons accelerated through the cathode sheath may cause heating and decomposition of materials other than metals.

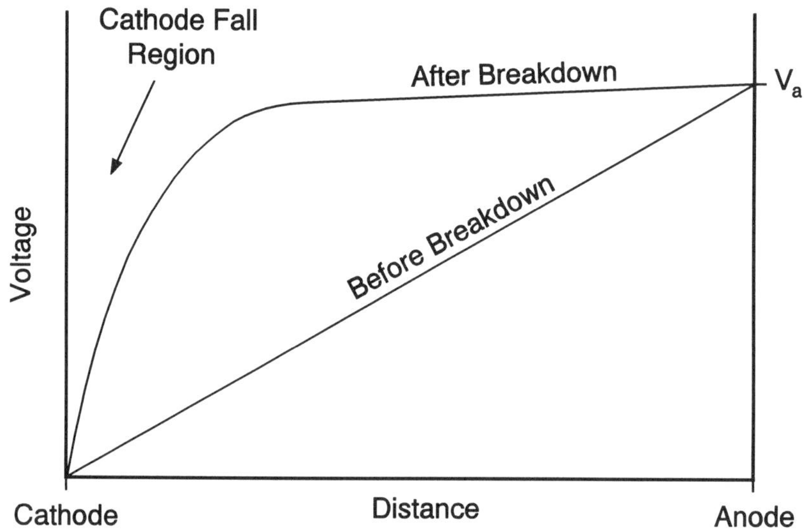

Figure 17-15 Potential distribution in a glow discharge between parallel plane electrodes.

If the current through a glow discharge is increased beyond the point where the cathode is covered with the glow, the discharge voltage will increase to produce the additional ionization required for the current flow. Under these conditions, the discharge is called an *abnormal glow*.

Continued increases in current result in more intense ion bombardment and heating of the cathode surface. Finally, a sudden transition to an arc discharge will occur. An arc discharge concentrates on a small spot on the cathode and the voltage across the discharge drops to a value comparable to the ionizing voltage of the gas. Because of the concentrated discharge, the current density and electric fields are very high compared with those in the glow discharge. With high melting temperature metals, the cathode spot is so hot that the dominant emission mechanism is thermionic emission. In other cases, field emission may dominate, or emission may result from a combination of thermionic and field emission.

At this point, it should be noted that, in spite of intensive research, the cathode and anode regions in an arc are still poorly understood. This is because of the complexities caused by the interactions of electrical, magnetic, thermal and fluid effects in addition to surface effects.

The voltage profile between electrodes during an arc is somewhat like that of a glow discharge (Figure 17-15) and is shown in Figure 17-16. The primary difference in shape is a marked increase in voltage at the anode.

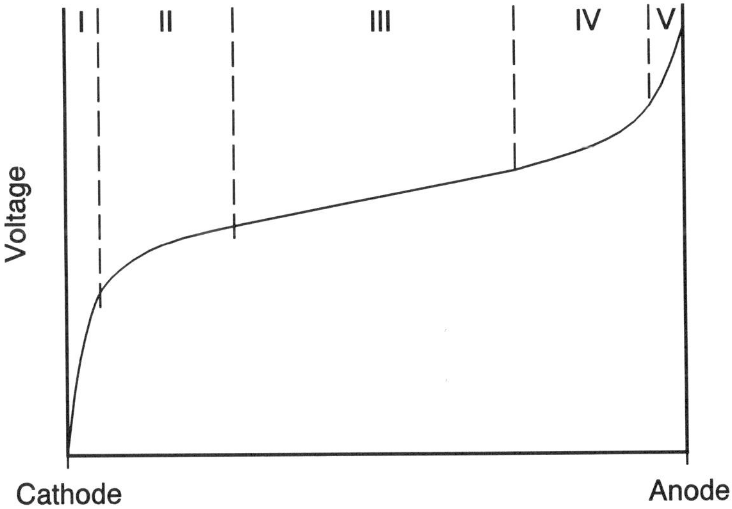

Figure 17-16 The regions of an arc. Adapted from *Gaseous Electronics* by M.H. Hirsh and H.J. Oskam, copyright 1978 by Academic Press.

The thicknesses of the electrode regions are exaggerated in Figure 17-16. At atmospheric pressure, the thicknesses of the cathode fall (I) and anode fall (V) regions are 10^{-4} to 10^{-3} cm. The thicknesses of the transition regions (II and IV) are on the order of 10^{-1} cm under the same conditions. In the transition regions, the arc diameter changes from the size at the electrode to the size of the arc column (III) and strong gradients exist in the plasma properties.

The voltage drops in the cathode region (I and II) and in the anode region (IV and V) are dependent on gas type, pressure and electrode material. At atmospheric pressure, the voltage drop in the cathode region is typically on the order of 10 V. The voltage drop in the anode region may be anywhere from 1 or 2 V to 10 V or more.

The voltage gradient in the arc column is largely dependent on gas type and arc current as well as pressure. The arc-column voltage gradients for several common gases at atmospheric pressure are given in Figure 17-17. The combination of the column gradient of a few tens of volts/cm with the electrode drops results in arc voltages in roughly the 50 to 100 V range for arcs of average length (a few centimeters).

At the end of Section 17.4, estimates are made of the energy and charge levels at which excessive damage may occur during a dc arc inside the vacuum envelope of a tube. Unfortunately, it is not so easy to make comparable estimates for dc arcs outside the vacuum envelope. If the arc occurs between bare electrodes, it is possible that a substantial amount of energy (hundreds of joules or more) can be dissipated without causing excessive damage.

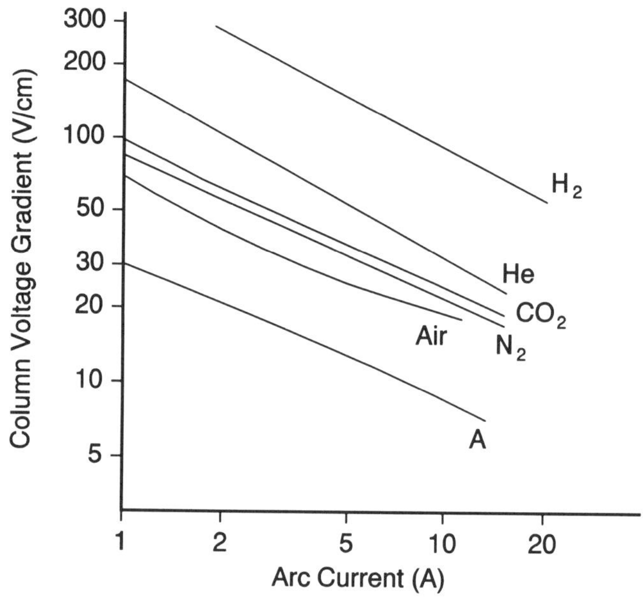

Figure 17-17 Arc-column voltage gradient at atmospheric pressure. Adapted from *Gaseous Conductors* by J.D. Cobine. Published 1958 by Dover Publications.

If, however, the arc occurs in the vicinity of an insulating material, it is possible that the material may be damaged with only a few joules of energy dissipation.

17.3 RF BREAKDOWN IN GAS [2]

In the previous section, it was shown that the dc breakdown field for air at atmospheric pressure is about 30 kV/cm. At low frequencies (up to tens of Hz), the breakdown field remains at 30 kV/cm. As frequency is increased to the audio range and then into the RF range, the breakdown field drops and becomes highly dependent on frequency and electrode spacing (as well as pressure, of course).

The unified plot given in Figure 17-18 was developed by Woo [4] for 50 Ω coaxial transmission lines and gives breakdown power as functions of frequency, pressure and spacing under laboratory conditions. This plot appears complex at first glance, but it is relatively straightforward once a person becomes familiar with it.

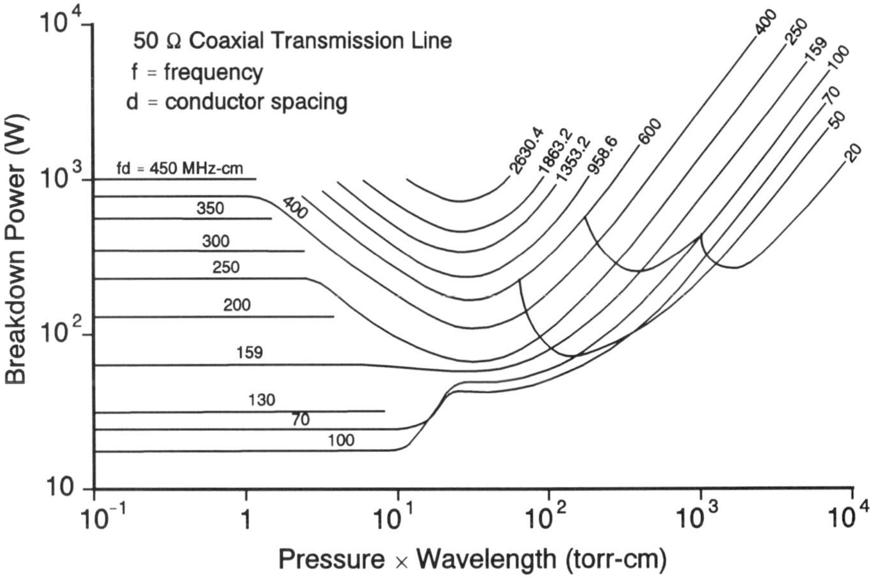

Figure 17-18 Unified plot for RF breakdown of air in 50 Ω coaxial transmission lines. From R. Woo, *Proc. IEEE*, February 1969. © 1969 IEEE.

Two primary breakdown processes are responsible for the shapes of the plots in Figure 17-18. These breakdown processes and the ranges of frequency, pressure and spacing in which they occur are given in Figure 17-19. Descriptions of the breakdown processes identified in Figure 17-19 are as follows:

1. Multipacting breakdown occurs when the electron's mean free path length is greater than the electrode spacing. Electrons travel back and forth across a gap between electrodes in synchronism with an oscillating RF field. If the secondary emission coefficient of the electrodes is greater than unity, a discharge develops. Multipacting will be discussed in Section 17.5.

2. In the region identified as the ionization breakdown region, the frequency is high enough and the separation distance large enough so that electrons are not swept out of the discharge by the field. If the field level is high enough, the electrons gain sufficient energy between collisions to cause ionization.

The minimum breakdown power occurs when the frequency of collision of electrons

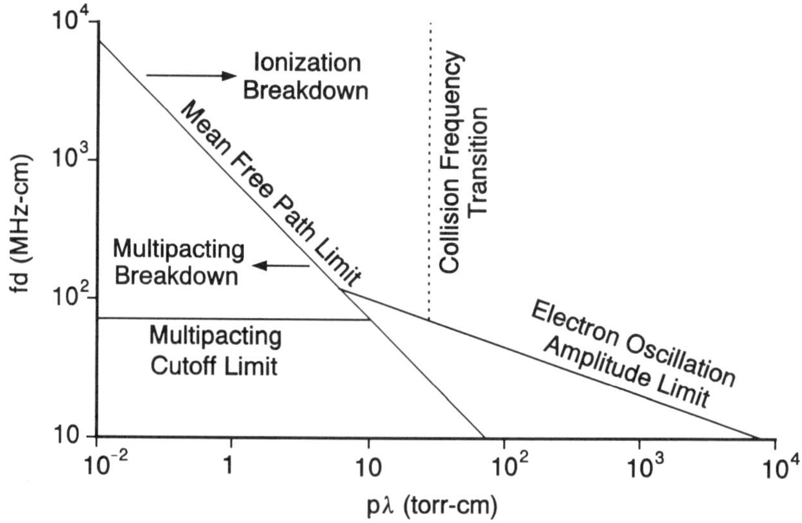

Figure 17-19 The *fd*-*p*λ plane showing ranges of breakdown processes. From R. Woo, *Proc. IEEE*, February 1969. © 1969 IEEE.

with neutral molecules equals the applied RF frequency. This minimum occurs at a value of *p*λ of about 30 torr-cm, indicated by the dashed line identified as the collision frequency transition.

When the frequency-distance (*fd*) product is small, the frequency is low enough and the separation between electrodes is short enough so that the amplitude of the electron oscillations exceeds the electrode separation distance. As a result, at the electron oscillation amplitude limit, electrons are lost to the conductor surfaces and the power required for breakdown increases rapidly.

The actual value of electric field at which breakdown occurs has been determined for selected frequencies in the microwave range. Two such selected frequencies are shown in Figures 17-20 and 17-21. In both of these cases (992 MHz and 9.4 GHz), extrapolations to atmospheric pressure indicate that the breakdown field is about 20 kV/cm, or comparable to the dc breakdown field level of about 30 kV/cm. The characteristic diffusion length in Figures 17-20 and 17-21 is a measure of the size and shape of the cavity in which the measurements were made.

The information in Figures 17-18, 17-20 and 17-21 would lead one to believe that extremely high power levels can be transmitted through waveguides and coaxial lines of relatively modest size. For example, an extrapolation of Woo's unified plots to atmospheric pressure in a 7/8-in coaxial line and a frequency of 1 GHz indicates that the breakdown power is about 200 kW. Still, as shown in Figure 17-22, the power ratings for a standard 7/8-in coaxial line are 43 kW peak and only about 1250 W average. Some of the degradation results from field enhancement associated with supporting elements for the center conductor. More important, for the average power

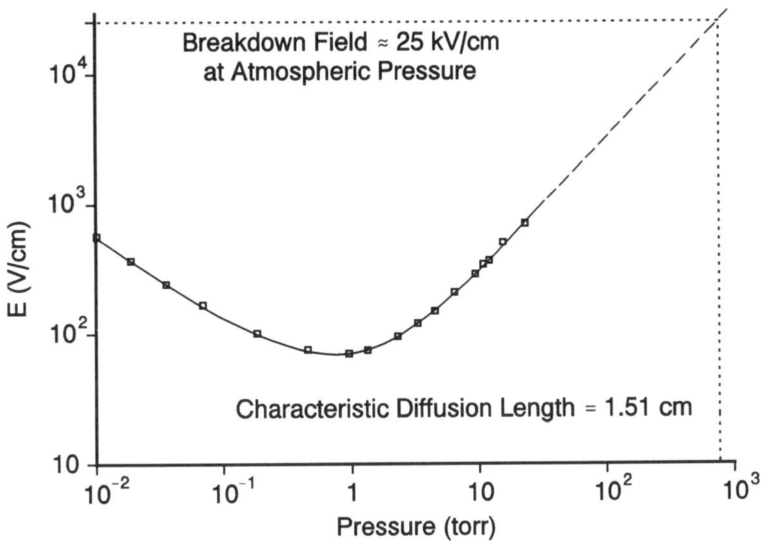

Figure 17-20 Breakdown in air at 992 MHz. Adapted from *Gaseous Electronics*, Volume I, "Electric Discharge," edited by M.H. Hirsh and H.J. Oskam, copyright 1978 by Academic Press.

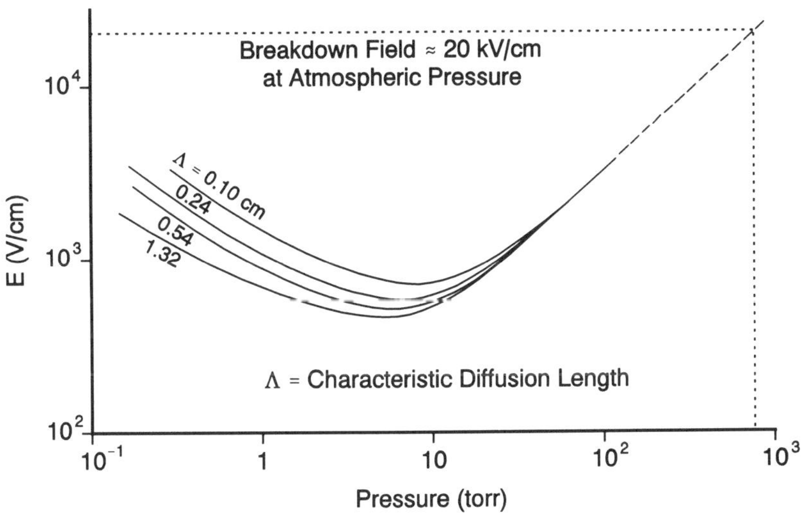

Figure 17-21 Breakdown in air at 9.4 GHz. Adapted from *Gaseous Electronics*, Volume I, "Electric Discharge," edited by M.H. Hirsh and H.J. Oskam, copyright 1978 by Academic Press.

case, heating, especially of the center conductor and gas, is one of the major factors limiting the power rating.

A situation similar to that in coaxial lines exists in waveguides as shown in Figure 17-23. The ratings in the lefthand plot are for a conductor temperature rise of 42.4°C, and those for the righthand plot are for a conductor temperature rise of 110°C. Based on a breakdown field of 20 kV/cm, it would be expected that WR 90, which is used at X band and has a small dimension of about 1 cm, should be able to transmit megawatts of power. Instead, the rating is in the kilowatt range.

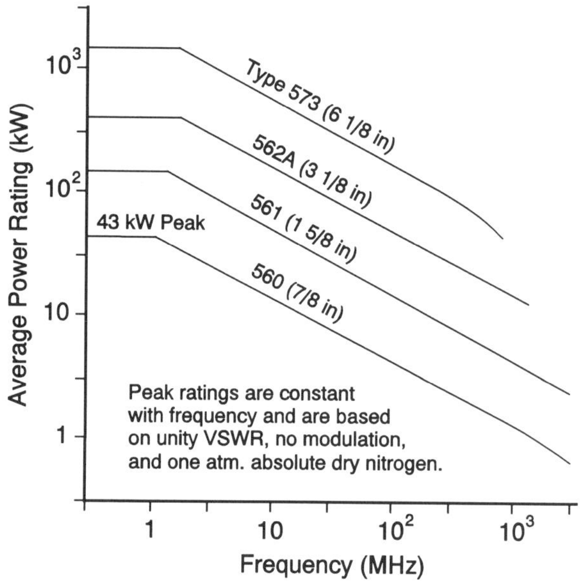

Figure 17-22 Average power ratings for standard rigid lines. Adapted from *Microwave Engineers Handbook*, edited by T. Saad, copyright 1971 by Artech House.

The power ratings for the waveguides in Figure 17-23 decrease by a factor of about three when the conductor temperature rise increases from 42.4°C to 110°C. Part of this decrease in rating may be attributed to the decrease in gas density in the waveguide as temperature is increased while pressure is held constant. This is equivalent to moving to the left in the high pressure region on the Paschen curve toward lower breakdown voltages. The reduction in pressure at constant temperature in moving to the left on the Paschen curve is equivalent to a reduction in gas density. This reduction in power capability with temperature is illustrated more clearly in Figure 17-24.

As would be expected, a change in pressure in waveguide or a coaxial line has an extremely large effect on power handling capability. As an example, the change in pressure from sea level to about 50,000 ft is given in Figure 17-25, and the corresponding reduction in waveguide power capacity is shown in Figure 17-26. Note

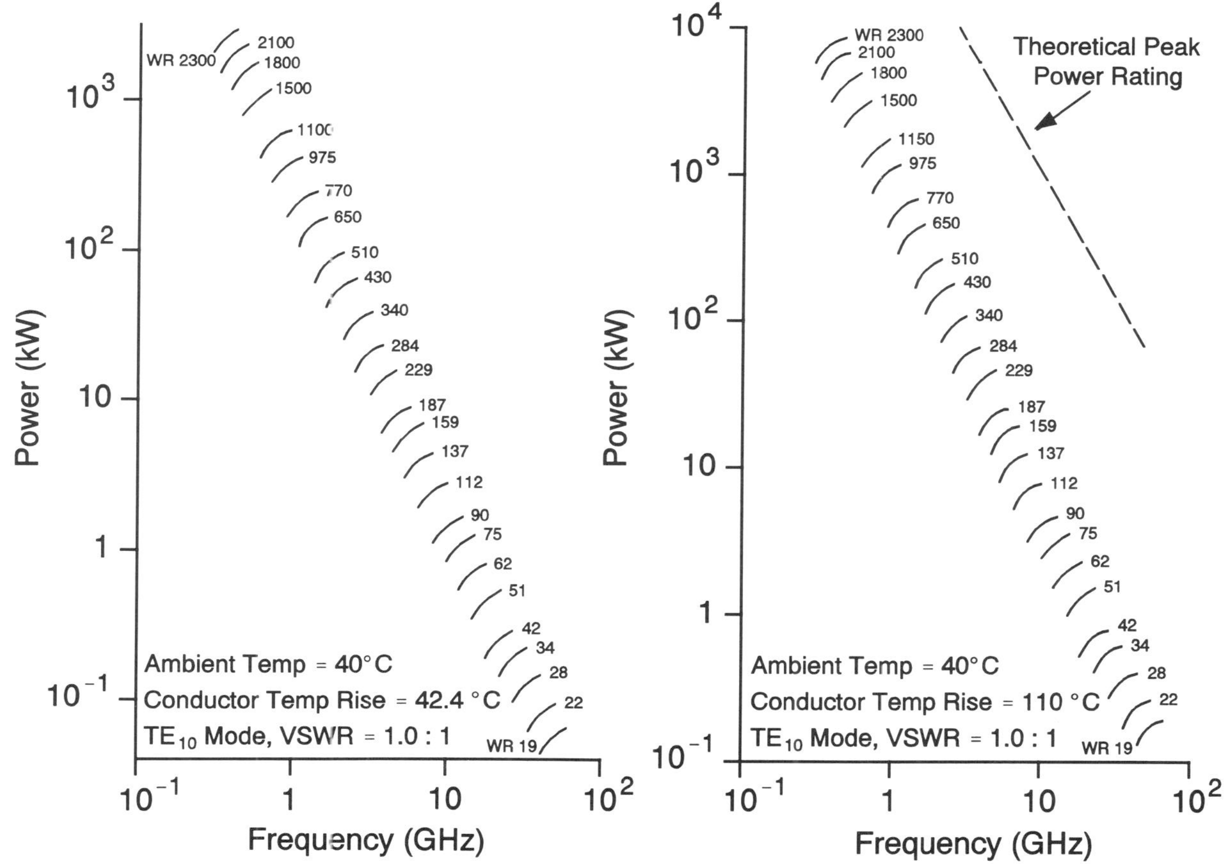

Figure 17-23 Average power rating for copper rectangular waveguide. Adapted from *Microwave Engineers Handbook*, edited by T. Saad, copyright 1971 by Artech House.

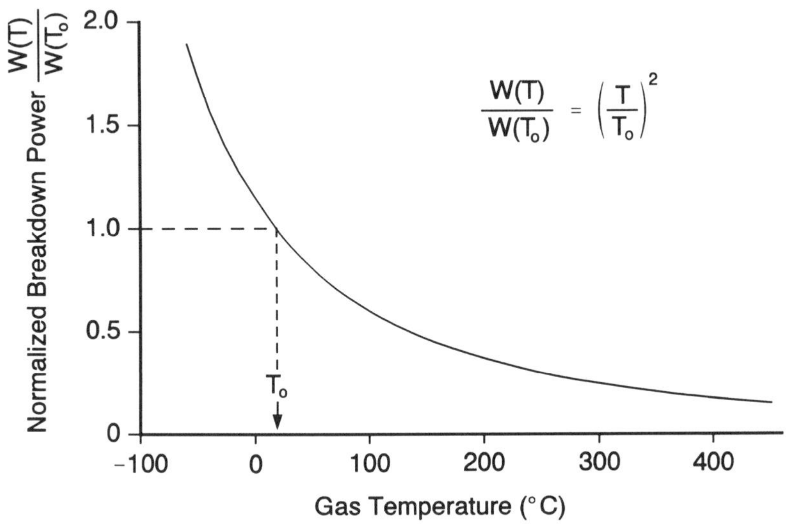

Figure 17-24 Waveguide breakdown power as a function of temperature at fixed pressure. From R.H. Stone and R.M. White, Final Report TIS R61 ELM 205 General Electric Co. January 1961.

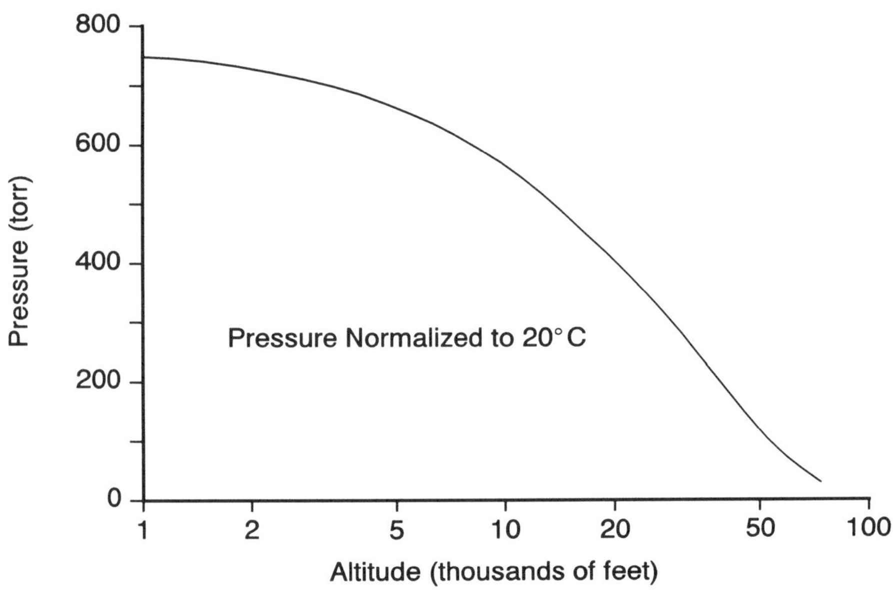

Figure 17-25 Pressure as a function of altitude.

that at 50,000 ft, the power capacity drops to about 3% of the sea level capacity. If an RF system is pressurized, a significant increase in power handling capability can be realized as shown in Figure 17-27.

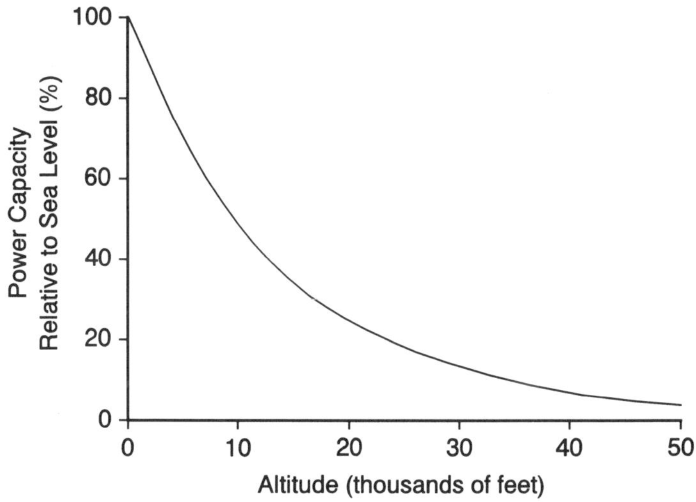

Figure 17-26 Waveguide power capacity *versus* altitude. From *Microwave Tube Manual*, by Varian, Air Force Publication No. T.O.00-25-251, October 1979.

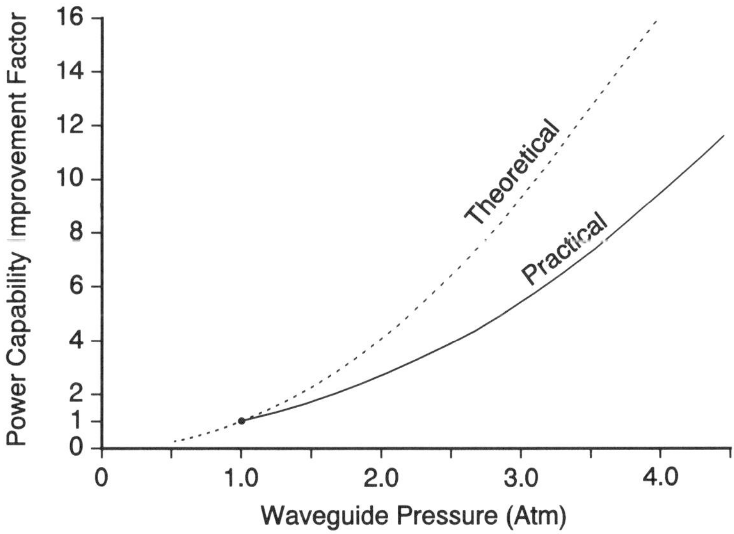

Figure 17-27 Improvement in power capability with pressurization. From *Microwave Tube Manual*, by Varian, Air Force Publication No. T.O.00-25-251, October 1979.

In Section 17.2, the use of electronegative gases for increasing breakdown voltage was discussed. As shown in Figure 17-28, similar improvements are possible in RF components. The improvement in power handling capability with pure sulphur hexafluoride (SF_6) at a pressure of 20 psi is a factor of about 16. Even a small percentage of SF_6 mixed with air produces a significant increase in power capability. Other electronegative gases such as freon 12 and freon 114 produce similar improvements in power handling capability.

In pulsed RF systems, the peak power handling capability increases for at least two reasons. The first has already been pointed out in connection with Figure 17-22. That is, the heating effect is decreased. The other reason is based on the fact that it takes time for a gas to break down. First of all, there is the time required for the breakdown process to start. This is called the *formative time*. Second, there is the time required for the discharge to develop to the point where appreciable RF power is absorbed or reflected.

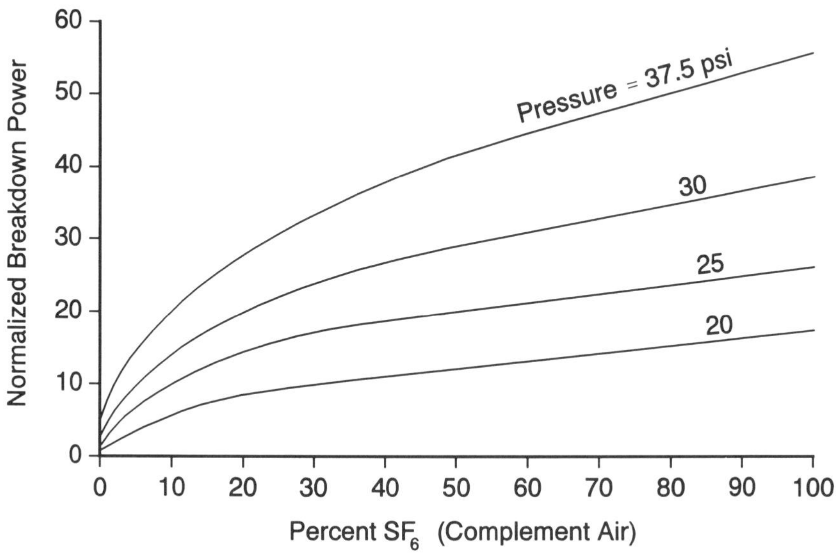

Figure 17-28 Breakdown power for mixtures of SF_6 and air normalized to air at 20 psi. From V.C. Vannicola and L.L. Stevens, Air Force Report No. RADC-TDR-64-67, March 1964.

The formative time for air can be determined from Figure 17-29. In this figure, P is gas pressure, E is electric field and τ is the formative time. For a 1 μs pulse, and atmospheric pressure, the $P\tau$ product is slightly less than 10^{-3} torr-sec, which is near the right side of Figure 17-29. Thus, at pulse lengths substantially shorter than 1 μs, the formative time is long enough so that the breakdown field increases.

An example of the change in breakdown power resulting from pulsewidth considerations is given in Figure 17-30. Note that the reduction in power capability

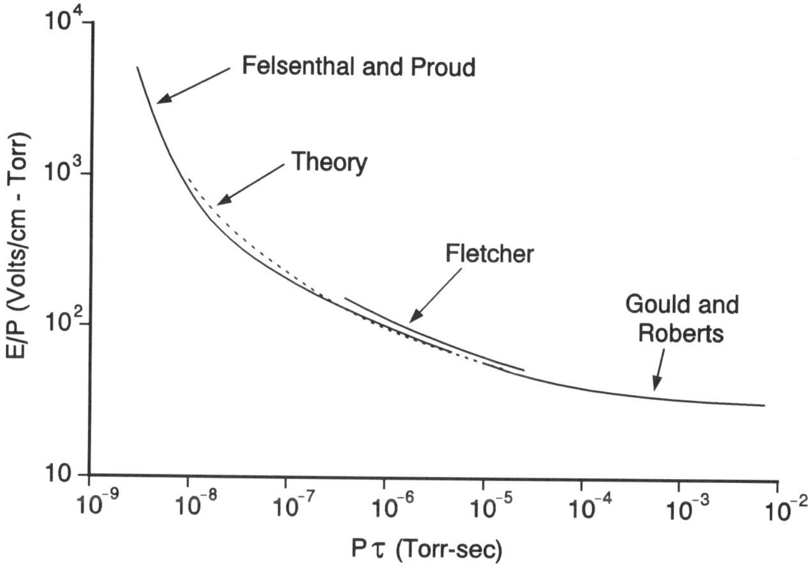

Figure 17-29 Formative time in air. From P. Felsenthal and J.M. Proud, *Phys. Rev.*, 1965.

when the pulse length is increased from 1 to 3 μs is about 30%.

The voltage standing wave ratio (VSWR) in the RF output system of a tube may cause breakdown to occur. The reason for this is, of course, that the reflected wave adds to the forward wave to produce regions of increased field level. A VSWR of 1.5:1 reduces the breakdown power by 35%, and 2:1 reduces breakdown power by 50%. If a system must be designed for operation with a significant VSWR, then the normal waveguide or coaxial line rating must be reduced to prevent breakdown as indicated by Figure 17-31.

If breakdown does occur in the RF output system connected to a microwave tube, it is probable that the resulting arc will reflect most of the incident power back toward the tube. As a result, the VSWR will be very high and may produce breakdown at other points along the RF system, as is indicated in Figure 17-32. In this case, an arc at plane I produces a standing wave and causes the electric field to exceed the breakdown level of a critical element such as the output window of a tube located at plane IV.

An isolator is often used to protect a microwave tube from excessive reflected RF power. The isolator may also impede or even stop a traveling arc. The effect of a ferrite isolator on VSWR is shown in Figure 17-33. While the isolator is effective in protecting the tube, it will not prevent arcing in the remainder of the RF system.

In some microwave systems, an attempt is made to reduce arcing problems and arc damage by using a detector of some type. Reflected power can be sensed with a directional coupler and measured with a threshold detector that generates a signal to

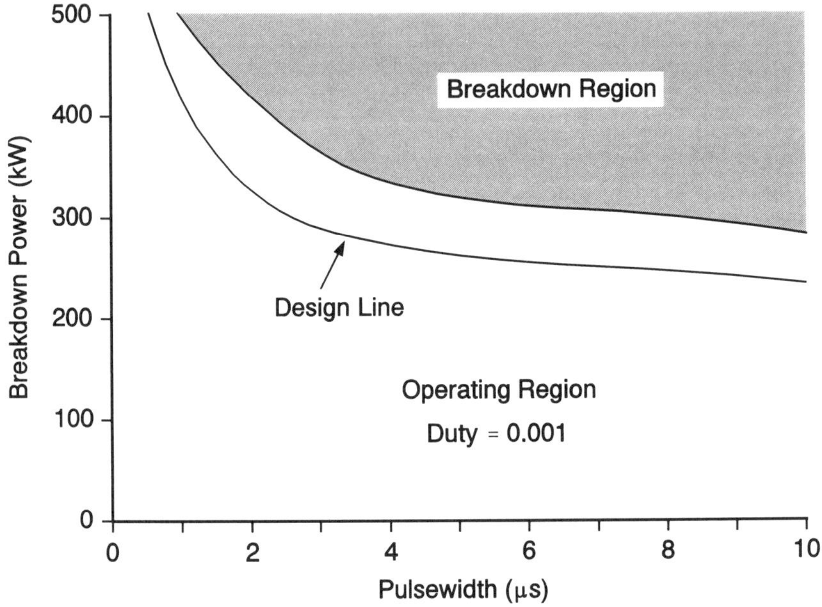

Figure 17-30 Effect of pulsewidth on breakdown power in waveguide. From *Microwave Tube Manual*, by Varian, Air Force Publication No. T.O.00-25-251, October 1979.

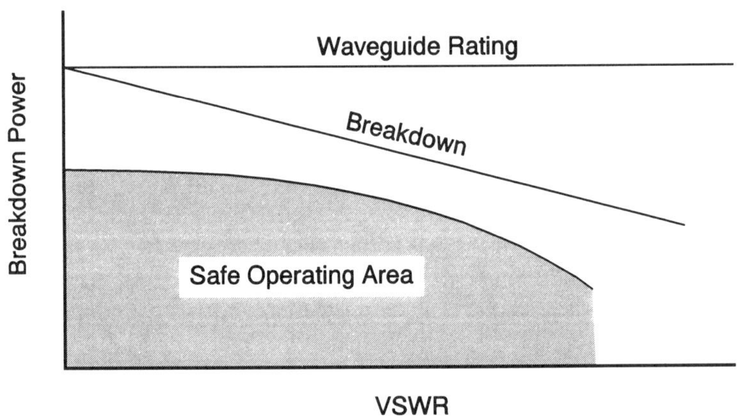

Figure 17-31 Effect of VSWR on RF power system rating. From *Microwave Tube Manual*, by Varian, Air Force Publication No. T.O.00-25-251, October 1979.

turn off the microwave tube if excessive reflections occur.

A waveguide arc may be sensed with an optical device for detecting visible arcs. The sensing portion of the device is usually mounted in a 90° waveguide bend located between the tube output flange and the waveguide transmission line. The output signal

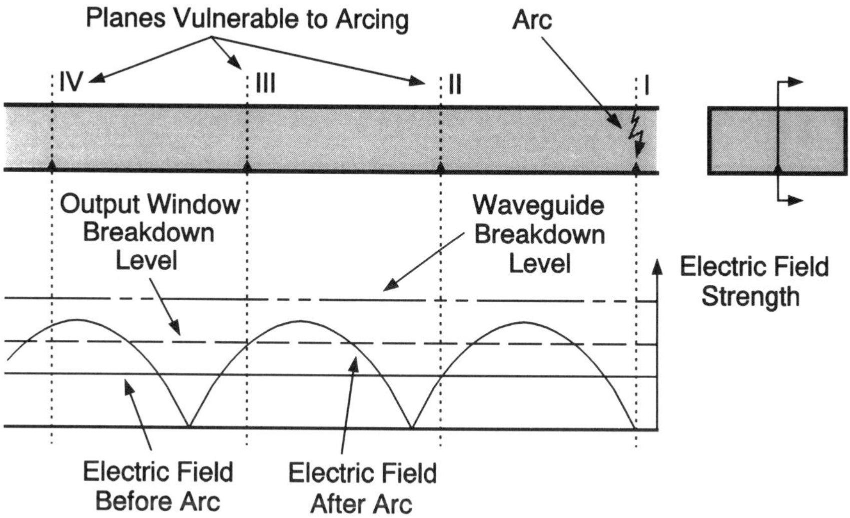

Figure 17-32 Arc propagation in a waveguide. From R.H. Stone and R.M. White, Final Report TIS R61 ELM 205 General Electric Co., January 1961.

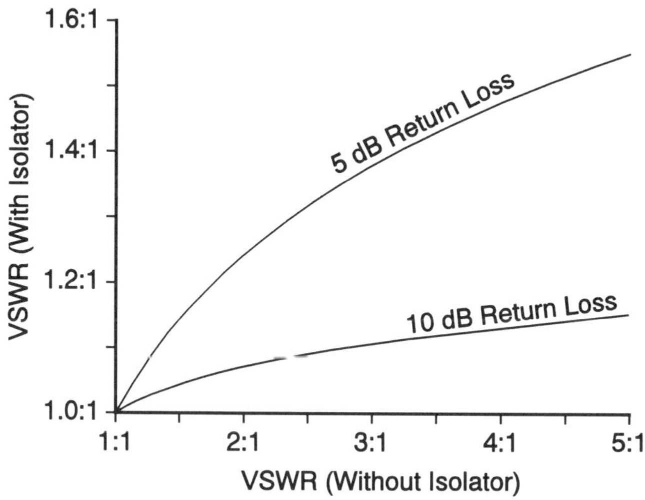

Figure 17-33 Effect of ferrite isolator on VSWR. From *Microwave Tube Manual*, by Varian, Air Force Publication No. T.O.00-25-251, October 1979.

from the sensing device is normally used to turn off the high power tube.

In another technique sometimes used to protect the output window of a tube, a metal probe is inserted through, but insulated from, the waveguide in the region near the output window. When an arc occurs and moves toward the window, the electrical circuit between the probe and the waveguide is closed, with the result that a relay can be activated to turn off the microwave tube.

17.4 DC BREAKDOWN IN VACUUM [5]

Under ideal conditions, the breakdown field level for vacuum exceeds that of all media including liquids and solids, and may be 10 MV/cm or higher. The upper limit results from field emission (see Section 5.1.2) from the negative electrode. In practice, the field level at which breakdown occurs is found to be two to three orders of magnitude below that occurring under ideal conditions. For example, for average gap lengths (~ 1 cm) breakdown normally occurs at a field level ranging from a few tens to a few hundreds of kV/cm.

For many vacuum tube applications, "Kilpatrick's criterion" [6] is used as the guideline to the maximum electric field that can be used in vacuum if breakdown is to be avoided. According to Kilpatrick, breakdown in vacuum varies as electrode spacing raised to the 0.75 power. Kilpatrick refers to field levels below this criterion as the *no-spark region* and to higher field levels as the *possible-spark region*.

Figure 17-34 contains Kilpatrick's criterion and some results from the tube industry [7, 8]. The shaded areas indicate roughly the values of plate hold-off voltages for gridded tubes with thoriated tungsten (ThW) and oxide-coated cathodes, and cylindrical configurations. The data points represent the experimental breakdown characteristics of Pierce guns. The data points represented by circles are for dc operation. The triangles and crosses represent long pulse and short pulse operation, respectively. Recent results presented by Staprans [9] are in basic agreement with those given in Figure 17-34. Note that, for pulse operation, field levels somewhat above the Kilpatrick curve are sometimes used.

The Kilpatrick criterion is based on results for parallel plane electrodes while the tube results in Figure 17-34 are for various electrode configurations. As a result, varying degrees of field enhancement contribute to the scatter in the data. In addition, apparently, when oxide coated cathodes are used, decomposition products form surface layers of reduced work function, greater roughness, different hardness, *et cetera*, that result in a reduction of the breakdown voltage.

The mechanisms responsible for breakdown in vacuum at field levels that are orders of magnitude below the field emission limit have been under investigation for decades. Still, there is often controversy over which mechanism dominates in any circumstance. Those mechanisms thought most likely to occur in microwave tubes are discussed in this section.

Conductive whiskers form in high field regions on electrodes in a vacuum. These whiskers can result from the migration of debris such as the decomposition products of a cathode or loose particles left from the manufacturing process. Whiskers can also grow from metallic surfaces subjected to high electric fields. The whisker shown in Figure 17-35 was produced by subjecting an optically flat stainless steel surface to a

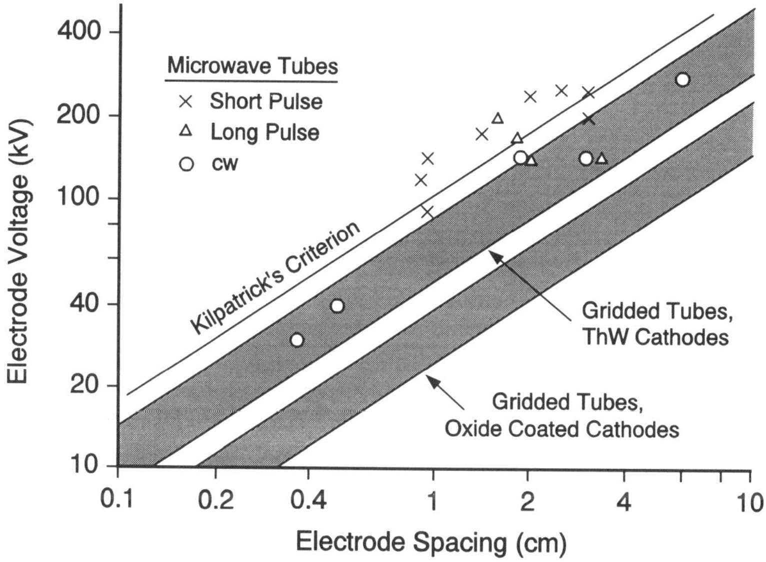

Figure 17-34 Kilpatrick's criterion and experimental breakdown characteristics of tubes.

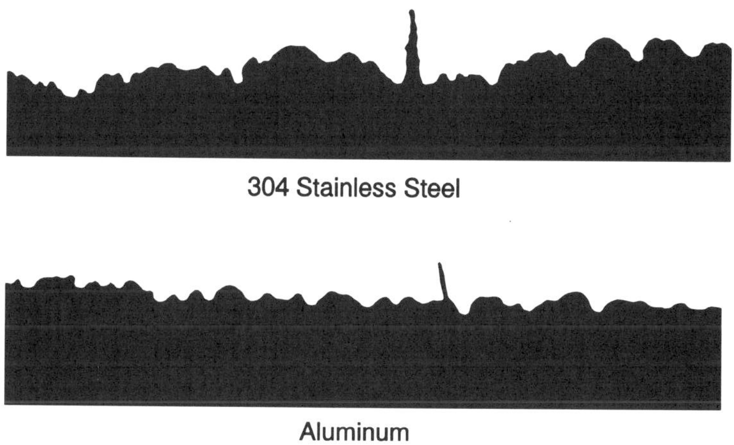

Figure 17-35 Whiskers on stainless steel and aluminum produced by electric field. From R.P. Little and W.T. Whitney, *Jour. Appl. Phys.*, August 1963.

field of somewhat less than 200 kV/cm [10]. This field level is above the nominal field levels indicated by Figure 17-34. It must be kept in mind, however, that the field enhancement factors resulting from elcctrodc curvature, machining irregularities, *et cetera*, will produce fields on electrode surfaces well above those indicated in Figure 17-34.

The field-enhancement factor at the tip of a microprotrusion or a whisker, like that shown in Figure 17-35, can be extremely high. Values as high as 100 have been estimated. The resulting electric field at the tip of a microprotrusion can produce field emission of electrons. The current flow through the tip coupled with ion bombardment (ions produced when the field emitted electrons strike gas atoms) produces heating of the tip, gas evolution and additional electron emission. Further heating occurs, which may produce an explosion of the tip. A microdischarge occurs, which may or may not grow into a full scale breakdown between electrodes.

An example of the time profiles of current and voltage for a microdischarge that does not result in breakdown is shown in Figure 17-36. At this point, it is appropriate to point out that microdischarges of the type indicated by Figure 17-36 are common occurrences during the voltage conditioning of electrodes in high voltage devices. During this process, high voltage is gradually applied to the electrodes, as shown in Figure 17-37. Microdischarges are produced that "clean up " the electrodes. That is, whiskers and other microprotrusions present on the electrode surfaces for any of many reasons are removed to permit high voltage operation. In a high voltage device, this process must be carried out when voltage is first applied and then must normally be repeated in an accelerated manner each time the device is turned on after being off for even relatively short periods (hours in some cases).

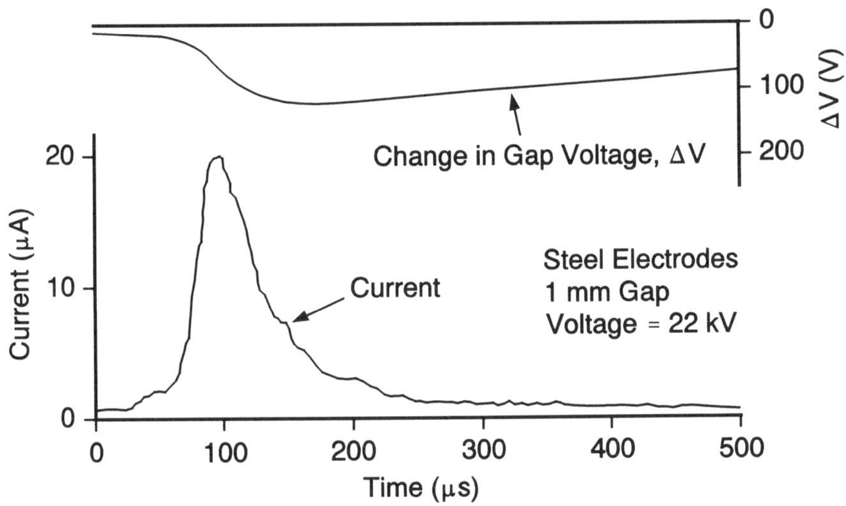

Figure 17-36 Microdischarge for 22 kV applied to a 1 mm gap between steel electrodes. From High Voltage Technology Seminar, Ion Physics Corporation, 1969.

It should also be noted that, even for a newly conditioned gap with voltage applied, the voltage hold-off capability degrades with time, as is shown by Figure 17-38.

Several mechanisms have been proposed by which microprotrusions may produce full scale breakdown in a vacuum. Two are illustrated in Figure 17-39. For small gaps

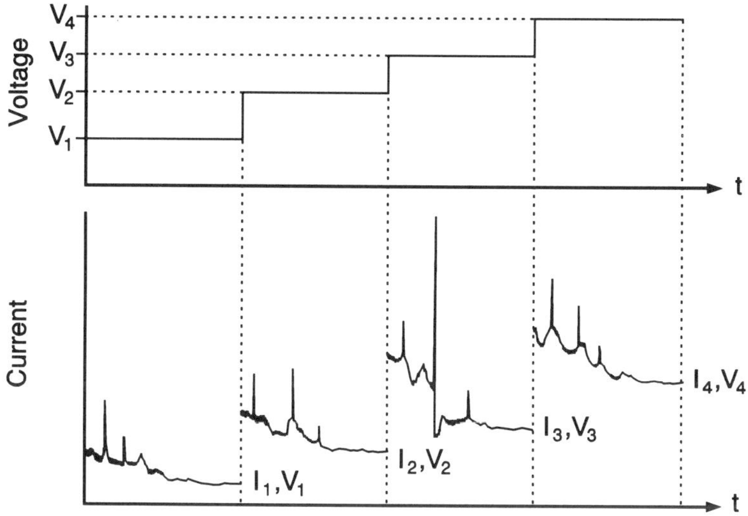

Figure 17-37 Voltage and current during conditioning. From *High Voltage Vacuum Insulation: The Physical Basis*, by R.V. Latham, copyright 1981 by Academic Press. Reproduced with permission.

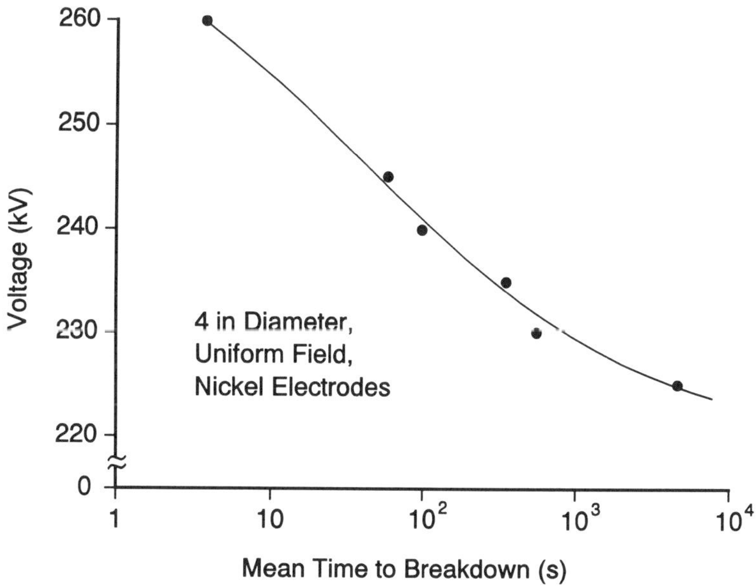

Figure 17-38 Breakdown characteristics for a conditioned vacuum gap. From High Voltage Technology Seminar, Ion Physics Corporation, 1969.

(<1 mm) the microprotrusion may explode and produce a plasma cloud of sufficient density between the electrodes to cause a vacuum arc (also called a metal vapor plasma arc) to form. For the relatively modest currents encountered in microwave tubes, even under fault conditions (kiloamperes at most) the arc that forms is on the negative electrode surface and the resulting voltage drop between electrodes is from roughly 10 to 35 V, depending on electrode material. Vacuum arcs are discussed later in this section.

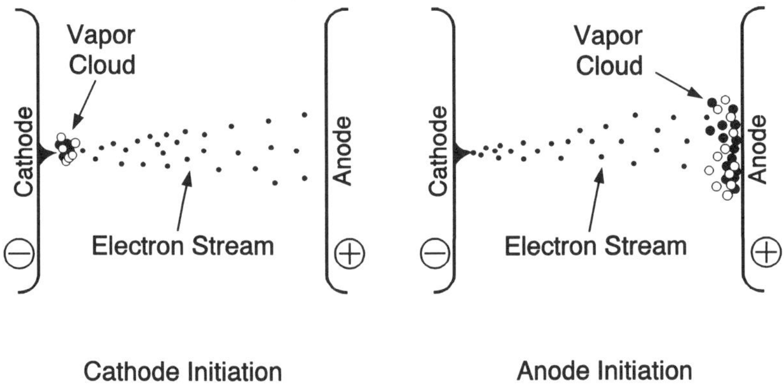

Figure 17-39 Breakdown initiation resulting from field emission from a microprotrusion. Adapted from *High Voltage Technology*, Chap. 4 by R. Hawley, Oxford University Press, London, 1968.

For a large gap (>1 cm), field emission from the negative electrode may produce the release and ionization of vapor from the positive electrode. If the density of the resulting plasma cloud is sufficient, then a vacuum arc identical to the one described in the previous paragraph can form.

As Figure 17-40 shows, the discharge current required to initiate an arc is only a fraction of a milliampere for electrode spacings approaching 1 cm. Figure 17-40 also shows that the value of β plays a role in breakdown as is expected.

It has been clearly established that loose particles in the interelectrode region can result in electrical breakdown of a vacuum gap. Several mechanisms by which the breakdown process can occur have been suggested. Some of these follow.

1. When voltage is applied, a surface charge density is induced in the electrodes and in particles on the surfaces of the electrodes. The particles may be drawn across the gap by the electric field as shown in Figure 17-41. If sufficient vapor is generated when the particles strike the opposing electrodes, breakdown may occur.
2. For large particles, breakdown may occur between a particle in transit and the electrode surface that it approaches. This may lead to breakdown of the gap.

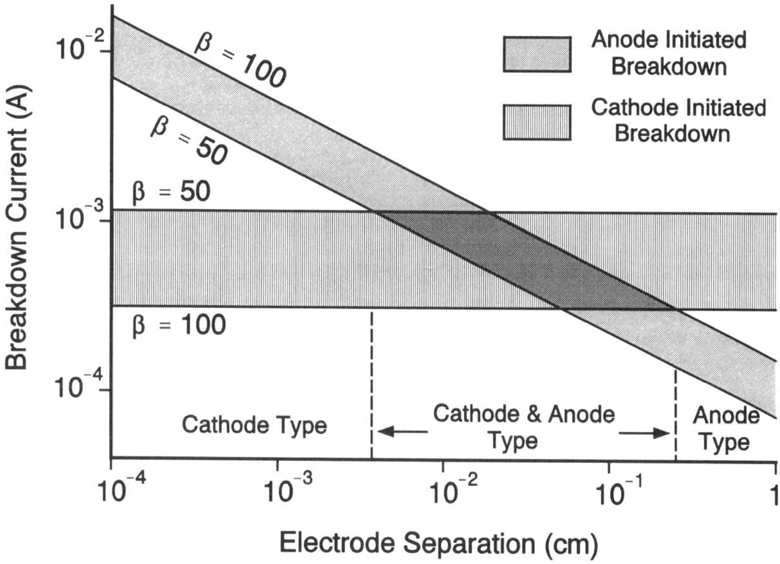

Figure 17-40 Electrode dominance in vacuum arc initiation. From T. Utsumi, Technical Report RADC-TR-66-712, January 1967.

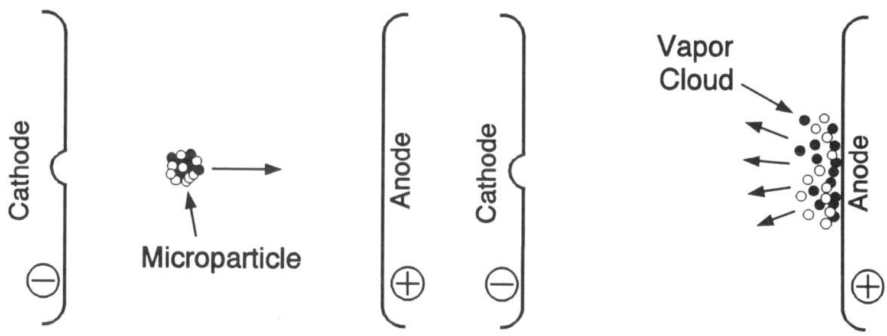

Figure 17-41 Breakdown initiation by a microparticle. Adapted from *High Voltage Technology*, Chapter 4 by R. Hawley, Oxford University Press, London, 1968.

3. When a particle strikes a surface, a crater is produced. The splash rim of material surrounding the crater may have sharp protrusions that field emit and result in breakdown.

A very practical indication of the role of particles in breakdown may be had from a technique sometimes used by the tube industry, apparently to eliminate breakdown caused by particles. An electron gun to be highly stressed is placed in a vacuum chamber, high voltage is applied and it is vibrated until arcing stops. The process may be repeated with varying orientations of the gun. Particles are observed to fall from the gun during the process, and it is presumed that they are the cause of arcing.

In addition to the effect of loose particles, it has been demonstrated that insulating particles like that shown in Figure 17-42 that are partly imbedded in the electrode surfaces are sites for the initiation of breakdown. The bigger the particles are, the lower is the breakdown voltage.

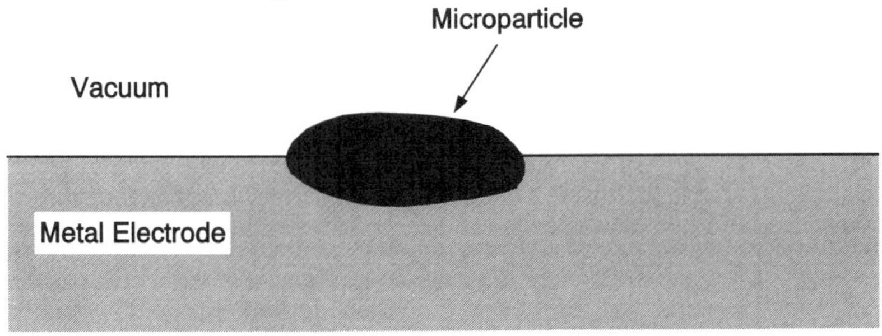

Figure 17-42 Partially embedded dielectric microparticle.

The sources of particles that can lead to breakdown are many and varied. Some examples are

- Debris in the gun from the manufacturing process,
- Careless assembly procedures that cause chipping of insulators or other damage,
- Dust from inadequate cleanliness,
- Cathode evaporants that form loose films on various gun surfaces,
- Use of improper getters,
- Particles from heater potting material,
- Debris that enters the gun through the anode aperture,
- Debris from a previous arc,
- Embedded particles (from a grinding operation, for example) that are exposed by chemical etching.

Regardless of the initiation mechanism, dc breakdown in a vacuum leads to an arc on the negative electrode (at the current levels encountered under fault conditions in a microwave tube). This arc is normally known as a *vacuum arc*, but it is sometimes referred to as a *metal vapor plasma arc*.

In a vacuum arc, the material of the negative electrode is vaporized and ionized by arc spots to form a conducting medium between the electrodes. Vacuum arcs can be established on any reasonably good conductor. The voltage between electrodes during a vacuum arc is dependent on the material of the negative electrode, typically between 10 and 30 V, and varies little with electrode spacing for typical (cms) spacings. As shown in Figure 17-43, the arcing voltage is a slowly varying function of current.

Unlike an arc in gas (such as air at atmospheric pressure), which consists of one large spot anchored in position, many very small arc spots exist in a vacuum arc, as

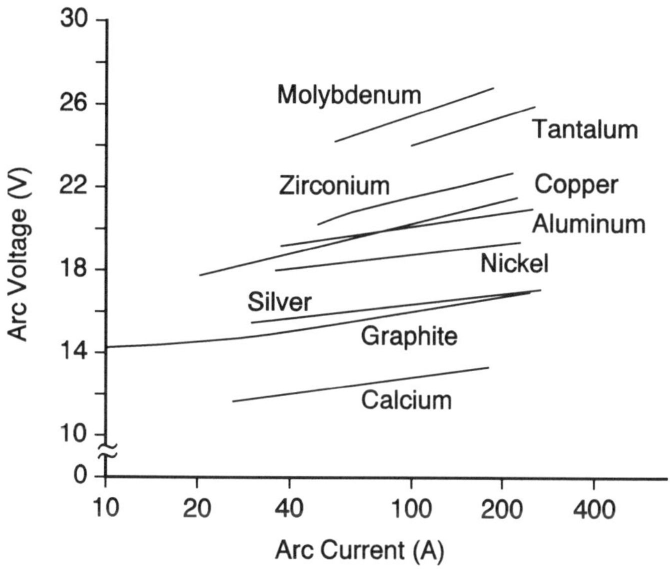

Figure 17-43 The dc volt-ampere characteristics of vacuum arcs. From W.D. Davis and H.C. Miller, *Jour. Appl. Phys.*, 1969.

indicated in Figure 17-44. These spots travel at high speed (meters per second to tens of meters per second) with apparently random motion across the electrode surface. On a macroscopic scale, a surface subjected to prolonged vacuum arcing has the texture of a piece of frosted glass. On a microscopic scale, individual arc tracks somewhat resemble the tracks of insects crawling in random directions across a freshly painted surface.

On a surface consisting of differing metals, a vacuum arc will operate on the low arcing voltage metals. For example, if a calcium film partially covers a copper surface, a vacuum arc will remove the calcium film before attacking the copper. This results from the relatively low arcing voltage of calcium (about 12 V) relative to that of copper (about 20 V). Similarly, the evaporation products from thermionic cathodes will be preferentially removed from nearby grid or control electrode surfaces by a vacuum arc. This is because barium, strontium, calcium, *et cetera* all have similar arcing voltages, which are about one-half (or less) of those of the normal materials of grids and electrodes. As a side note, the arcing may produce additional deposition of the cathode material on critical surfaces such as insulators.

The rate at which material is removed from a surface by a vacuum arc varies somewhat from metal to metal but is generally in the neighborhood of 0.1 mg per Coulomb of charge passed through the arc. Thus, if the time history of the current in a tube during an arc is known, the amount of erosion can be estimated with a reasonable degree of accuracy.

Similarly, if energy input to a vacuum arc in an electron gun is known, the amount

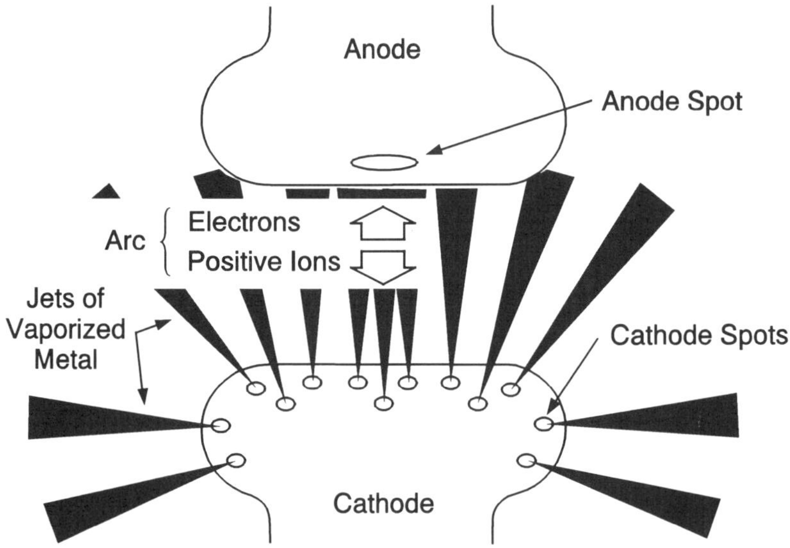

Figure 17-44 Details of a vacuum arc. From *Vacuum Arcs, Theory and Application*, edited by J.M. Lafferty, published by John Wiley & Sons, 1980.

of erosion can be estimated. This is because the arcing voltage is restricted to a reasonably narrow range of values. For example, if a 20 J arc occurs, the charge passed through the arc is estimated to be 1 C using an average arcing voltage of 20 V. For 1 C of charge, 0.1 mg of material is removed. For a density of 10 gm/cm^3 (most electron-gun metals are not significantly different from this value), the volume of electrode material removed is 10×10^{-6} cm^3. For 1 cm^2 of electrode material, the thickness removed is only 10×10^{-6} cm or 0.1 μm. Thus, the amount of material removed by a 1 C or a 20 J arc is extremely small.

Unfortunately, the small amount of material removed by the 20 J arc is not the only factor that must be considered. If the surfaces on which some of the arc material deposits are critical ones, such as insulators, leakage currents or insulator breakdown may result. Furthermore, gas molecules on the arced surfaces or nearby surfaces may be driven off by the arc, producing a gas burst. As an example, the release of a monolayer of gas from 1 cm^2 of area into a volume of 1 liter raises the pressure from 10^{-6} to 10^{-4} torr. This gas must be removed (by adsorption, an appendage pump or a getter) before tube operation can be resumed.

The maximum permissible energy input to a tube during an internal arc in the gun is sometimes specified by the manufacturer. This energy is based largely on the manufacturer's experience with the particular tube type. Maximum permissible energy levels from joules to tens of joules are not uncommon. Knowing the maximum permissible energy level, it is then possible for the tube user to design a protection circuit to interrupt the flow of energy before it becomes excessive.

Before leaving the subject of vacuum arcs, a comment should be made concerning the interruption characteristics of these arcs. As the current through a vacuum arc is reduced, the average arc lifetime becomes shorter and shorter. For example, Figure 17-45 shows that when the current through a tungsten (W) arc approaches 5 A, the average lifetime approaches 100 μs. Thus, as the current is driven toward zero by a protection circuit, the arc will extinguish well before the current reaches zero.

When a vacuum arc extinguishes, it does so very rapidly. In fact, the current is said to "chop," and the chopping time is usually less than 10^{-7} s. If there is significant inductance in the protection circuitry, the voltage generated by an arc that chops can be sufficient to cause insulation failure.

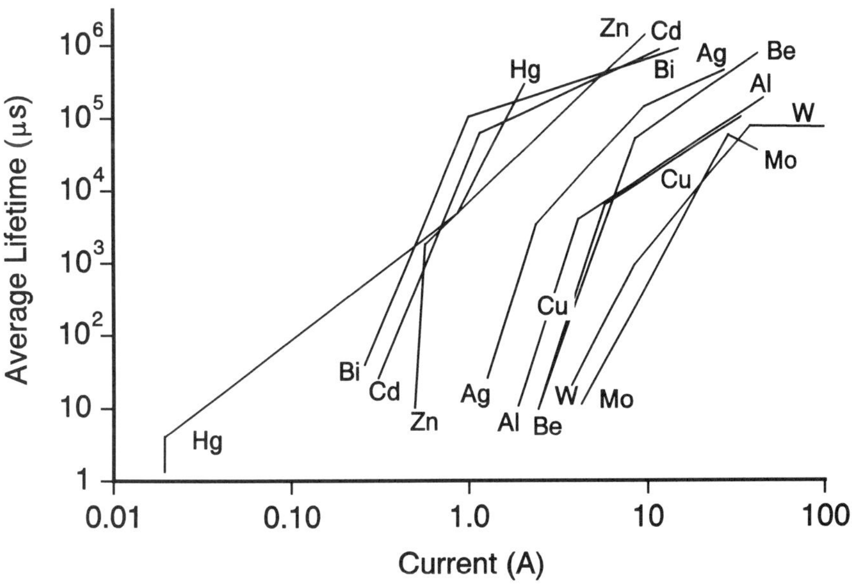

Figure 17-45 Average arc lifetime for pure electrode materials. From J.D. Cobine and G.A. Farrall, *Jour. Appl. Phys.*, December 1960.

Once a vacuum arc has extinguished, the voltage recovery rate across the electrodes between which the arc had occurred can be extremely rapid. The recovery rate depends on the duration and intensity of the arc; however, a value of 10 kV/μs is not uncommon. Thus, it is often necessary to cause only a very brief interruption of current through a vacuum arc before voltage is reapplied to the electrode. In fact, this principle has been used in the development of vacuum arc interrupters [11].

17.5 RF BREAKDOWN IN VACUUM

Kilpatrick's criterion, discussed at the beginning of Section 17.4, is often used as a guideline to the maximum RF voltage that can be applied between electrodes. Very often, however, another kind of discharge develops at voltage levels well below the Kilpatrick level. This discharge, called a *multipactor discharge*, usually occurs when electrons move back and forth across a gap in synchronism with an RF field. If the secondary emission ratio of the gap surfaces is greater than unity, then the number of electrons involved in the process builds up with time.

It is also possible for a multipactor discharge to develop if the electron transit time across a gap corresponds to 3/2, 5/2, *et cetera* cycles of the applied RF voltage. In each case, electrons can receive a net acceleration in crossing the gap and can produce an avalanche build-up of current. The voltage, frequency, f, and gap separation, d, that result in a multipactor discharge are related as shown in Figure 17-46.

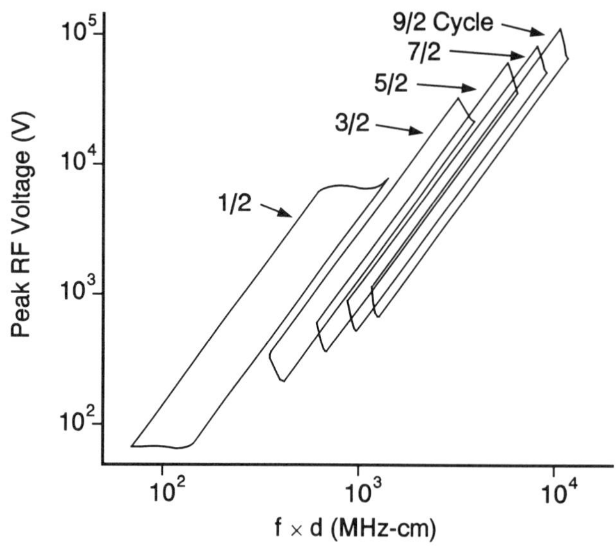

Figure 17-46 Theoretical multipacting zones. From A.J. Hatch and H.B. Williams, *Phys. Rev.*, 1958.

During multipacting, large amounts of energy can be discharged in a small volume with the result that electrodes or RF windows can be damaged or destroyed. Even if excessive damage does not occur, multipacting can load a microwave tube so that the output power is reduced and modulated with a noiselike signal.

The techniques used for suppressing multipacting fall into two general categories:

1. alteration of electron dynamics,
2. reduction of the secondary emission ratio.

The most common ways to alter the electron dynamics are to adjust the magnetic focusing field (within the limits of acceptable tube operation) or to apply RF power rapidly enough so that the voltage rises above the multipacting zone before current has had time to build up.

The secondary emission ratio may be reduced by serrating the cavity drift tube tips on which the multipacting electrons land. Some of the electrons enter the notched regions of the serrations and the secondaries that are produced are trapped within the notches. Titanium-based films have been demonstrated to be successful in suppressing multipacting in the high power klystrons in the Stanford linear accelerator [12].

If multipacting can be avoided, then the next problems to be faced are the vacuum breakdown mechanisms described in Section 17.4. Investigation of the techniques that would permit the use of fields well above the Kilpatrick limit is an active area of research at the present time in the linear accelerator field [13].

17.6 INSULATOR FLASHOVER

One of the weakest points in a vacuum device from an electrical breakdown point of view is the interface between a metal, an insulator and a vacuum. This interface is often called the *triple junction*, and is illustrated in Figure 17-47. Because of electron launching mechanisms at the triple junction and surface effects on the insulator, the breakdown strength of the insulator (actually, its surface) is well below that of a vacuum gap or a solid insulator of equivalent length. Shannon, Philip and Trump have suggested that through proper design, insulators in vacuum without conditioning should achieve 50 kV/cm, and conditioned insulators should be operable at 80 kV/cm [14].

The initiation of surface flashover occurs at the triple junction. In this region, there are at least two situations that must be considered. In one, the insulator is brazed to the metal surface. In the process of metallizing the surface of the insulator and brazing it to metal, it is virtually impossible to avoid the formation of a fillet of braze material on the surface of the insulator. This fillet has a razor-sharp edge with a high field enhancement factor. In the presence of the high field that results when voltage is applied, microprotuberances can grow, causing additional substantial increases in the enhancement factor. As a result, field emission and the generation of microparticles, *et cetera* occur at the triple junction at relatively low applied field levels.

The second situation that must be considered is when the insulator is mechanically held in place but not brazed. In this case, there are voids between the insulator and the metal along the triple junction. The field enhancement factor in these voids may be as high as the relative dielectric constant of the insulator. For aluminum oxide, which is commonly used in these situations, the relative dielectric constant is about 9. This field enhancement factor, coupled with those of microprotuberances, is thought to lead to field emission and microparticle generation.

The second step in insulator flashover is thought to be the positive charging of the insulator in the vicinity of the triple junction. This occurs, as indicated in Figure 17-48,

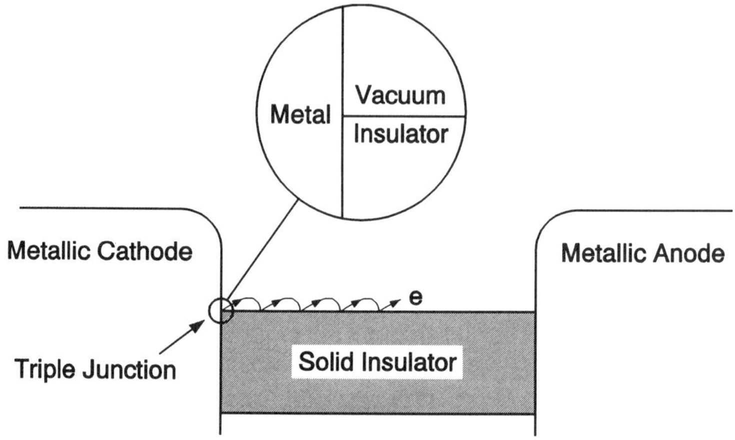

Figure 17-47 Triple junction and hopping mode of electron travel. Adapted from *High Voltage Vacuum Insulation: The Physical Basis*, by R.V. Latham, copyright 1981 by Academic Press.

because field-emitted primary electrons from the triple junction strike the insulator surface causing secondary emission to occur. The secondary emission coefficient for most insulating materials is relatively high (values from 5 to 10 are common, especially if the primary electrons strike the surface at a grazing angle). The secondary emission process is regenerative in that the secondary electrons produced are accelerated by the field at the surface to produce additional secondaries.

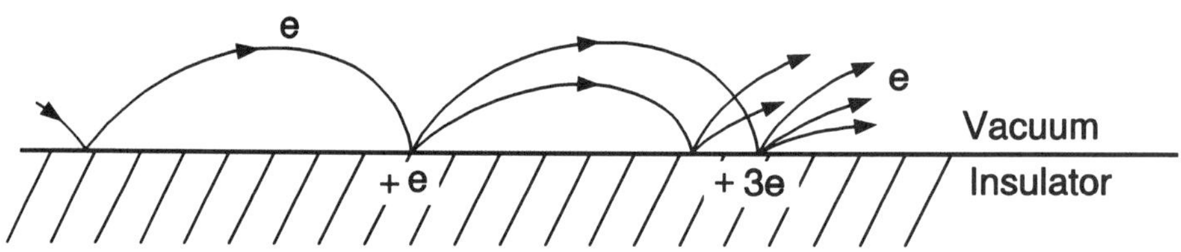

Figure 17-48 Insulator charging by hopping electrons. From *High Voltage Vacuum Insulation: The Physical Basis*, by R.V. Latham, copyright 1981 by Academic Press. Reproduced with permission.

The result of the secondary emission process is that the insulator surface becomes positively charged in the triple junction region, and this increases the field and the rate of electron emission at the triple junction.

The next phase in the flashover process is thought to be the desorption of gas molecules from the insulator surface and subsequent ionization by the hopping electrons. These processes are indicated in Figure 17-49. The positive ions are accelerated toward the cathode and further enhance the field and increase electron emission.

Other processes possibly involving photon emission are thought to occur, and the result is rapid breakdown across the surface of the insulator. The techniques that have been investigated for suppressing insulator flashover center on the prevention of the initial charging of the insulator in the region of the triple junction. Some of the techniques investigated or used are these:

1. The field is reduced at the triple junction by shaping the metal electrodes in this region. On a larger scale, it is common to place the output window for a tube in an enlarged circular section of the waveguide as is shown in Figure 17-50. Not only does this reduce the field at the triple junction, it also facilitates the fabrication of the metal-to-ceramic braze joint.
2. The shape of the insulator, and the angle its surface makes with the metal electrodes, is used to control the trajectories of secondary electrons. The variation of flashover field with insulator-to-metal surface angle is shown in Figure 17-51.
3. A coating of material having a low secondary electron yield (ideally <1) is applied to the surface of the insulator. In addition, this coating is normally slightly conductive so that residual charging effects are minimized.

In concluding this section, an observation by G. Farrall of General Electric Co. is pertinent. He notes that most of the work on insulators in vacuum has been conducted in unbaked systems and in systems sealed with "O" rings. As a result, the phenomena associated with breakdown are bound to be dominated by desorption of gas and a large variety of unknown impurity effects. Finally, Farrall states, "Perhaps the time has come to adopt a more scrupulous and possibly more fundamentally rewarding attitude toward the studies of insulators in vacuum."

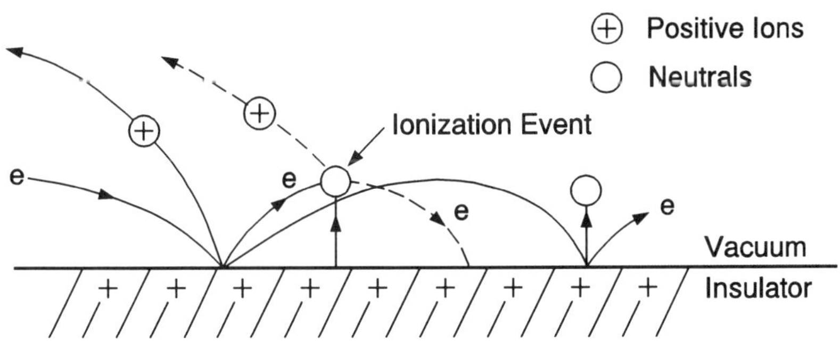

Figure 17-49 Ionization by hopping electrons. From *High Voltage Vacuum Insulation: The Physical Basis*, by R.V. Latham, copyright 1981 by Academic Press. Reproduced with permission.

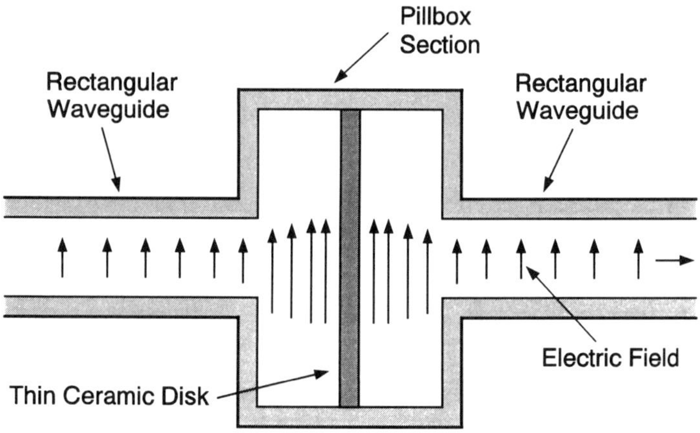

Figure 17-50 Output window in enlarged section of waveguide. From *Microwave Tube Manual* by Varian, Air Force Publication No. T.O.00-25-251, October 1979.

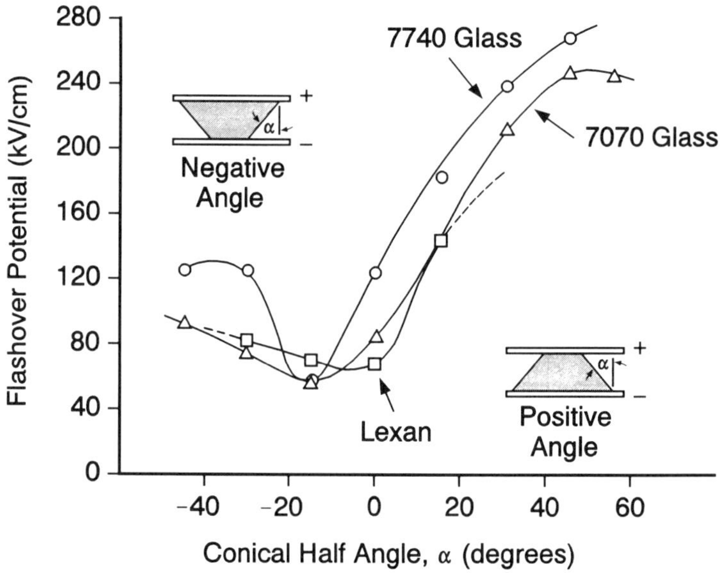

Figure 17-51 Effect of insulator geometry on flashover voltage. From A. Watson and J. Shannon, *Proc. 2nd Int. Symp on Insulation of High Voltages in Vacuum*, Sept. 1966. Courtesy of A. Watson.

REFERENCES

[1] James Dillon Cobine, *Gaseous Conductors, Theory and Engineering Applications*, New York, Dover Publications, 1958.

[2] M.N. Hirsh and H.J. Oskam, *Gaseous Electronics*, Volume 1, Electrical Discharges, New York, Academic Press, 1978.

[3] P.D. Ankrum, *Principles and Applications of Electron Devices*, Scranton, PA, International Textbook Company, 1959.

[4] Richard Woo, "RF voltage breakdown in coaxial transmission lines" *Proc. IEEE*, vol. 57, February 1969, pp. 254–256.

[5] Recommended for further reading: J.M. Lafferty, *Vacuum Arcs, Theory and Applications*, New York, John Wiley and Sons, 1980; R.V. Latham, *High Voltage Vacuum Insulation: The Physical Basis*, New York, Academic Press, 1981.

[6] W.D. Kilpatrick, "Criterion of vacuum sparking designed to include both RF and DC," *Review of Scientific Instruments*, vol. 28, no. 10, October 1957, pp. 824–826.

[7] "Machlett pulse tubes for high voltage, high power, video and RF pulsing," The Machlett Laboratories, Inc., June 1964.

[8] A. Staprans, "Voltage breakdown limitations of electron guns for high power microwave tubes," *Proc. Second International Symposium on Insulation of Voltages in Vacuum*, M.I.T., Cambridge, MA, 1966.

[9] A. Staprans, "Electron gun breakdown," 1985 High Voltage Workshop, Monterey, CA, February 1985.

[10] R.P. Little and W.T. Whitney, "Electron emission preceding electrical breakdown in vacuum," *J. Appl. Phys.*, vol. 34, no. 8, August 1963, pp. 2430–2432.

[11] A.S. Gilmour, Jr., and D.L. Lockwood, "The interruption of vacuum arcs at high dc voltages," *IEEE Trans. Electron Devices*, vol. ED-22, no. 4, April 1975, pp. 173–180.

[12] R.W. Bierce, J. Jasberg, and J.V. Lebacqz, in *The Stanford Two-Mile Accelerator*, edited by R. B. Neal, Benjamin, New York, 1968, Ch. 10.

[13] William Peter, "Vacuum breakdown and surface coating of RF cavities," *J. Appl. Phys.*, vol. 56, no. 5, September 1984.

[14] J.P. Shannon, S.F. Philip and J.G. Trump, "Insulation of high voltages across solid insulators in vacuum," *Proc. First International Symposium on Insulation of High Voltages in Vacuum*, M.I.T., Cambridge, MA, October 1964, p. 281.

Chapter 18

Reliability

The reliability of TWTs (and microwave tubes in general) is of major importance because, if the TWT fails, then, often, the system in which it is being used is no longer operable. TWTs are very complex devices and, in the early years of their development, failure rates were high in some applications. As a result, TWTs and microwave tubes developed a reputation for having high failure rates.

Over the years, design and manufacturing technology for microwave tubes has steadily improved and now the microwave tube industry knows what to do to provide reliable tubes. The major failure modes are well known, and current designs incorporate features to prevent those failures wherever the application allows conservative design and derating. MTBFs (mean time between failures) of tens of thousands of hours are common, and for satellite applications, MTBFs of millions of hours are projected with high confidence.

In preparing this chapter, the author has sought and received extremely constructive comments and material from many sources. In particular, the help of engineers in the Systems Reliability Division of Rome Laboratory, at Teledyne Electronic Technologies (formerly Teledyne MEC), at Hughes Electron Dynamics Division and at the Crane Division of the Naval Surface Warfare Center is acknowledged. The author has made a sincere attempt to provide a balanced view (balanced between users and suppliers) of microwave tube reliability.

In this chapter, we will start with the definition and measurement of failure rate and MTBF. Then the prediction of MTBF based on sample testing and MIL-HDBK-217F will be discussed. MIL-HDBK-217 is the standard reference used to predict the failure rate of components in military systems. Even the most recent revision, MIL-HDBK-217F, predicts MTBFs for TWTs and klystrons that are an average of a factor of well over five below field experience*. The shortcomings of

---

* Bruce Dudley of Rome Laboratory has recently provided the author with a draft of MIL-HDBK-217F, Notice 2, which significantly (by a factor ranging from two to over ten) increases the MTBF predictions for TWTs. Notice 2 predicts MTBFs for TWTs and klystrons that are an average of a factor of less than two below field experience. Notice 2 was prompted by a review of the draft version of this chapter on reliability by Mr. Dudley.

MIL-HDBK-217F are examined and then improved predictive techniques that have been suggested for use are discussed. The chapter ends with a discussion of reliability considerations in the manufacture, use and procurement of TWTs.

18.1 LIFE AND MTBF [1]

Life and MTBF are concepts that are often used to describe the expected behavior of microwave tubes (and other electronic equipment) with time. These two concepts are discrete but are often confused. Life is a period of time, and MTBF describes a failure rate. Also, data concerning tube failures are often improperly interpreted and so values given for life and MTBF are not correct.

If a large number of TWTs were put into service or on life test at the same time, the failure rate (usually expressed as the number of failures per million hours of operating time) as a function of time would appear as shown in Figure 18-1. This curve, often referred to as a *bathtub curve*, is characterized by three regions:

1. An initial high failure rate, often referred to as infant mortality. It should be noted that the vast majority of TWTs are burned in by the manufacturer and this effectively eliminates infant mortality of TWTs in operating systems.
2. A relatively long period during which the occurrence of failures is random and the failure rate is constant.
3. A period in which most of the remaining tubes fail. The expected life corresponds to the point at which the failure rate begins to increase rapidly.

Life and MTBF are defined *during the period in which the failure rate is constant.* Failure during the infant mortality period and during the wear-out period should not be included in the determination of life and MTBF.

During the period of random failures the constant failure rate may be expressed as follows:

$$\frac{\Delta N/N}{\Delta t} = -\lambda \quad \text{or} \quad \frac{\Delta N}{\Delta t} = -\lambda N \qquad (18\text{-}1)$$

where N is the number of TWTs remaining in operation and ΔN is the number of tubes failing in time increment Δt. Thus, $\Delta N/N$ is the percentage of tubes failing in time increment Δt. The minus sign arises because ΔN decreases with time. The constant, λ, is the failure rate. For very small Δt, this expression becomes

$$\frac{dN}{dt} = -\lambda N \qquad (18\text{-}2)$$

The solution to this equation is

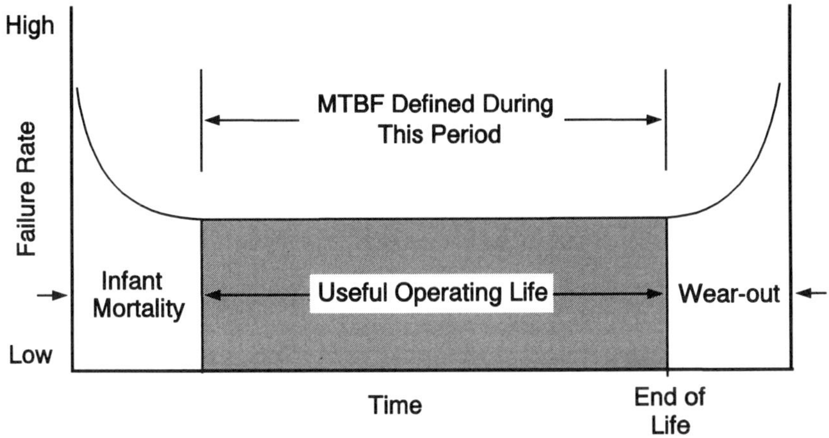

Figure 18-1 Failure rate as a function of time.

$$N = N_o e^{-\lambda t} \tag{18-3}$$

where N_o is the number of tubes placed into operation at time $t = 0$ (at the end of the infant mortality period). Thus, the number of tubes remaining in operation decreases exponentially with time.

Ignoring wearout failures, the average time to failure (that is, the mean time between failures or the MTBF) can be computed as follows. Taking the natural logarithm of both sides of (18-3),

$$\ln N = \ln N_o - \lambda t \tag{18-4}$$

and solving for t,

$$t = \frac{1}{\lambda}\left(\ln N_o - \ln N\right) \tag{18 5}$$

the MTBF is

$$\begin{aligned} \text{MTBF} &= \frac{1}{N_o} \int_0^{N_o} t \, dN \\ &= \frac{1}{N_o} \int_0^{N_o} \frac{1}{\lambda}\left(\ln N_o - \ln N\right) dN \end{aligned} \tag{18-6}$$

This can be integrated to obtain

$$\text{MTBF} = \frac{1}{\lambda N_0}\left(N\ln N_o - N\ln N + N\right)\Big|_0^{N_o} \tag{18-7}$$

or

$$\text{MTBF} = \frac{1}{\lambda} \tag{18-8}$$

Thus, the MTBF is the inverse of the failure rate and is the time constant of the exponential relation that gives the number of tubes still in operation as a function of time. That is, as is shown in Figure 18-2,

$$N = N_o e^{-\frac{t}{\text{MTBF}}} \tag{18-9}$$

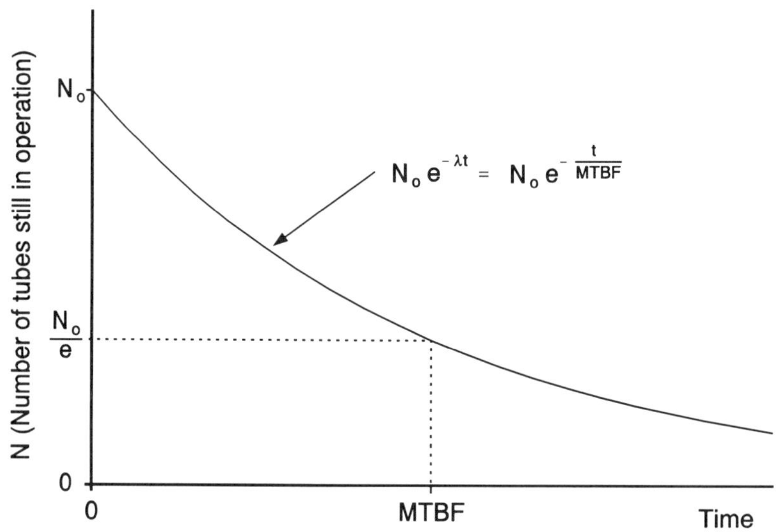

Figure 18-2 Definition of MTBF.

If a large number of tubes were placed in service simultaneously and there were no wear-out failures, $1/e$ or about 37% would still be in operation when the MTBF is reached. Thus, MTBF can and should exceed the average expected life of a TWT by a substantial amount. This is especially true if repair cannot be accomplished as, for example, is the case in space systems. Again, the point should be made that MTBF is defined during the period when the failure rate is constant. Infant mortalities are not included, and wearout failures are not included.

18.2 PREDICTION OF MTBF BASED ON SAMPLE TESTING

MTBF depends on construction technology, the degree of control exercised in assembly and inspection, *et cetera*. As such, MTBF cannot be calculated from design parameters but must be estimated from observation. The only way in which the true MTBF of a tube can be determined is through life testing to end of life, either in the laboratory or in field use in systems where operating time periods and conditions are known. While life testing in the laboratory is sometimes carried out (see [2] for example) costs are high and testing periods can be extremely long (over 17 years in [2]). Concerning the use of field data for determining MTBF, the following statement from [3] is pertinent.

> Unfortunately, most historical or reported MTBF data included both early failures and wearout failures. This occurs because normally only accumulated hours are used in calculating the MTBF, rather than tracking the hours contributed by individual tubes. This calculation method assumes that all failures occurred during the useful operating life period; that none of the failures were due to early or wearout failure mechanisms. This occurs because, at the time of removal from the socket, it is often not known what type of failure, random or wearout, has occurred, *or in many cases whether, in fact, a failure has actually occurred.*

Bazovsky has discussed the use of field data to determine mean time between failures [4]. He considers the failures of a large population of lightbulbs. (This is easy for lightbulbs because they are either on or off.) Only wear-out failures are assumed to occur. As bulbs wear-out and are replaced with new ones, the failure rate of the total population tends to become constant. Then, the bulb life (the mean wear-out life) equals the mean time between failures of bulbs having ages all the way from zero hours to the maximum bulb lifetime. This is quite different than the MTBF of new bulbs just being placed in service. A similar situation exists in using field data for microwave tubes except that, as is clear from the preceding quotation from [3], the reason for failure is often not known, and in many cases, tubes are replaced that have not failed.

The effect on calculations of failure rate and MTBF of including infant mortality and wear-out failures is shown in Figure 18-3. The calculated failure rate can be increased and the MTBF reduced by a significant amount if failure data are used improperly.

An alternative to life testing to end of life or the use of field data is the testing of sample quantities of tubes for a time period that is short compared to the MTBF. Then, based on the failure rate of the samples, MTBF can be predicted with a level of confidence that varies with the number of tubes tested and with the test duration. An example will serve to illustrate the application of this technique. Suppose that each of 100 TWTs has been tested for 1000 hrs and that two failures occurred. If this failure rate of two TWTs per 100,000 TWT-hrs of operation continued for the useful life of the tubes, the MTBF would be 100,000/2 or 50,000 hrs. (Actually, based on 18-9, the

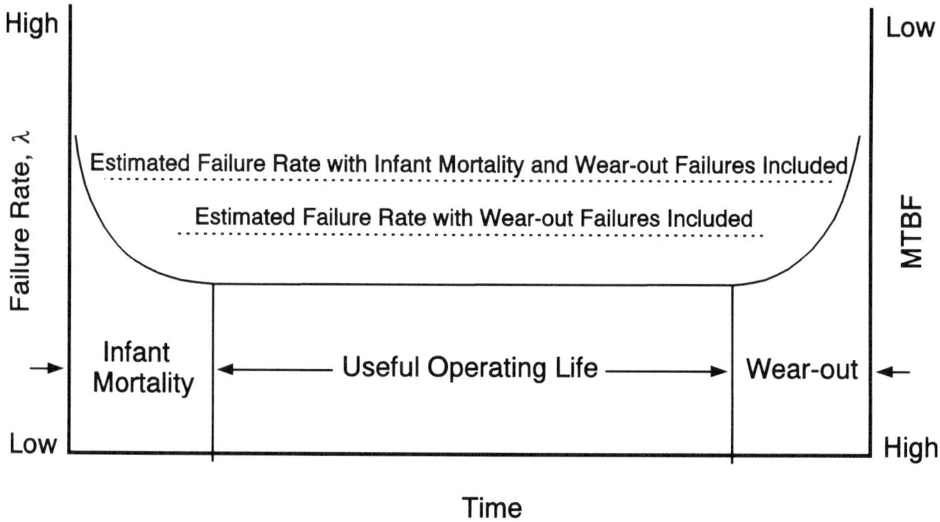

Figure 18-3 Effect on reliability calculations of including infant mortality and wear-out failures. From the Reliability Working Group of the Electronics Industries Association, Microwave Division, January 1991.

MTBF would be 49,498 hrs. This is slightly below the 50,000 hr value because of the exponential nature of *N versus t*.)

Because of the relatively short duration of the test relative to the calculated MTBF (only 2%) and the small number of failures, we can not have a high level of confidence that the MTBF of 50,000 hrs would actually be achieved if all tubes were tested to failure. For example, if 1 second after the 1000 hr test period, another failure occurred, the failure rate would be more like 3/100,000 TWT-hrs and the estimated MTBF would be 33,333 hrs. If the second failure occurred 1 second before the end of the 1000 hr test, the failure rate could be considered to be 1/100,000 TWT hrs and the estimated MTBF would be 100,000 hrs. Thus, it is important to attach a measure of confidence with an MTBF calculation that is based on sample testing. This can be done using the following expression to predict MTBF [5].

$$\text{MTBF} = \frac{2\tau}{\chi^2} \tag{18-10}$$

where τ is the total test time accumulated (100,000 hrs in the preceding example) and χ^2 is a quantity that arises from statistical theory. The value of χ^2 depends on two parameters. One is the confidence level, C, expressed as a fraction. Thus, for example, C might be a confidence level of 0.9 or 90% that the calculated MTBF can be achieved. The other parameter is the number of failures, r, experienced during the test (two failures in the preceding example). Values for χ^2 for various values of C and r are given in Table 18-1.

Table 18-1

Values of χ^2

r	Value of C							
	.005	.01	.02	.05	.10	.20	.30	.50
1	.207	.297	.429	.711	1.064	1.649	2.195	3.357
2	.676	.872	1.134	1.635	2.204	3.070	3.828	5.348
3	1.344	1.646	2.032	2.733	3.490	4.594	5.527	7.344
4	2.156	2.558	3.059	3.940	4.865	6.179	7.267	9.342
5	3.074	3.571	4.178	5.226	6.304	7.807	9.034	11.340
6	4.075	4.660	5.368	6.571	7.790	9.467	10.821	13.339
7	5.142	5.812	6.614	7.962	9.312	11.152	12.624	15.338
8	6.265	7.015	7.906	9.390	10.865	12.857	14.440	17.338
9	7.434	8.260	9.237	10.851	12.443	14.587	16.266	19.337
10	8.643	9.542	10.600	12.338	14.041	16.314	18.101	21.337
11	9.886	10.856	11.992	13.848	15.659	18.062	19.943	23.337
12	11.160	12.198	13.409	15.379	17.292	19.820	21.792	25.336
13	12.461	13.565	14.847	16.928	18.939	21.588	23.647	27.336
14	13.787	14.953	16.306	18.493	20.599	23.364	25.508	29.336

r	Value of C							
	.70	.80	.90	.95	.98	.99	.995	.999
1	4.878	5.989	7.779	9.488	11.668	13.277	14.860	18.465
2	7.231	8.558	10.645	12.592	15.033	16.812	18.548	22.457
3	9.524	11.030	13.362	15.507	18.168	20.000	21.955	26.125
4	11.781	13.442	15.987	18.307	21.161	23.209	25.188	29.588
5	14.011	15.812	18.549	21.026	24.054	26.217	28.300	32.909
6	16.222	18.151	21.064	23.685	26.873	29.141	31.319	36.123
7	18.418	20.465	23.542	26.296	29.633	32.000	34.267	39.252
8	20.601	22.760	25.989	28.869	32.346	34.805	37.156	42.312
9	22.775	25.038	28.412	31.410	35.020	37.566	39.997	45.315
10	24.939	27.301	30.813	33.924	37.659	40.289	42.796	48.268
11	27.096	29.553	33.196	36.415	40.270	42.980	45.558	51.179
12	29.246	31.795	35.563	38.885	42.856	45.642	48.290	54.052
13	31.391	34.027	37.916	41.337	45.419	48.278	50.993	56.893
14	33.530	36.250	40.256	43.773	47.962	50.892	53.672	59.703

At the 90% confidence level, the MTBF calculated for the preceding example (two failures in 100,000 TWT-hrs of testing) is

$$\text{MTBF (90\%)} = \frac{2 \times 100{,}000}{10.645} = 18{,}800 \text{ hrs} \tag{18-11}$$

Thus, with 90% confidence, we can expect an MTBF of at least 18,800 hrs.

For this example of two failures in 100,000 TWT-hrs of testing, predicted MTBF is plotted as a function of confidence level in Figure 18-4. Note that, for small values of confidence level, very large MTBFs are predicted and that predicted MTBF decreases rapidly as higher confidence levels are required. Over a wide range ($\approx$50% to 90%) the decrease in MTBF is nearly linear with increasing confidence level. Finally, note that the confidence level of achieving a failure rate of two failures/100,000 TWT-hrs throughout the life of a TWT, based on the two failures that occurred in the first 1000 hrs of testing is only about 30%.

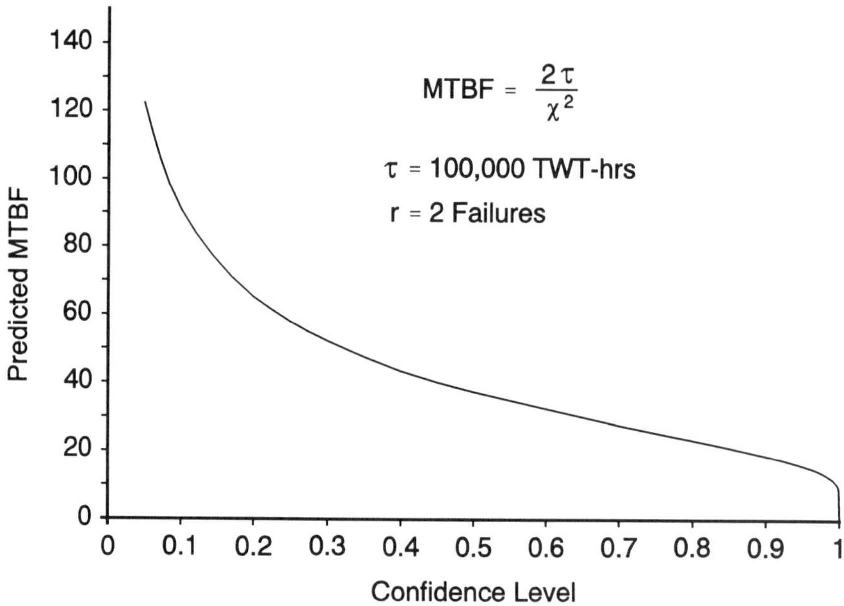

Figure 18-4 Predicted MTBF as a function of confidence level.

18.3 PREDICTION OF MTBF BASED ON MIL-HDBK-217F

MIL-HDBK-217F is the military handbook for the reliability prediction of electronic equipment [6]. MIL-HDBK-217 was originally developed by the Air Force's Rome Air Development Center, now Rome Laboratory, in 1962. Over the years, the handbook has gone through many revisions, the latest version being 217F, which, with Notice 1, was published July 10, 1992. (Notice 2 is pending approval.) The purpose, as stated in the handbook, is

> The purpose of this handbook is to establish and maintain consistent and uniform methods for estimating the reliability (i.e., the reliability of a mature design) of military electronic equipment and systems. It provides a common basis for reliability predictions during acquisition programs for military electronic systems and equipment. It also establishes a common basis for comparing and evaluating reliability predictions of related or competitive designs. The handbook is intended to be used as a tool to increase the reliability of the equipment being designed.

Contractors to the Department of Defense are required to use the handbook to estimate reliability of products. Then those estimates are compared to the reliability specifications that are required for the products. Competing designs may also be evaluated. Producers of some nonmilitary electronic equipment also use the handbook because it provides a convenient and standard way to predict reliability.

There has been substantial controversy concerning MIL-HDBK-217 and its use. The following comments are from [7].

> Adherents defend MIL-HDBK-217 as a necessary standardized medium for assessing reliability and comparing designs. They maintain that it is based on careful analysis of part and system reliability data. Any problems arising with the handbook, they say, are caused by misinterpretation or misuse of it.
>
> Opponents of MIL-HDBK-217 say that it is inaccurate, leads to costly overdesign, actually prevents higher reliability levels from being achieved, and does not address the true causes of failures. Moreover, they say, the handbook's models are out of date and do not reflect emerging technologies.

Microwave tubes were not included in the original version of MIL-HDBK-217. Gradually, as the handbook was revised, various tubes were included along with models for predicting reliability. The model for microwave tubes in revision 217F was originally developed by the Space and Communications Group of Hughes Aircraft Company to correct deficiencies of version 217C [8]. The Hughes model was less pessimistic than the 217C model but still gave extremely pessimistic predictions of microwave tube reliability. For example, the Electronic Industries Association (EIA) compiled data [3] showing that revision 217E predicted reliabilities that were a factor of over eight below actual field data (average for 36 different applications of TWTs, coupled cavity TWTs and klystrons; the model was unable to predict reliability for four tube types and applications). Largely as a result of the EIA report, factors used in the model were changed somewhat in the most recent revision (217F) but the predicted reliability is still a factor of over five below field experience*.

While it is clear that additional effort is needed to further improve the handbook model, Rome Laboratory points out that the upgrading of 217 models is directly proportional to the availability of funds to do so, and those funds are very scarce. This is indeed unfortunate when one considers the importance of MIL-HDBK-217 and the very large impact that it can have (in both technical and monetary terms) on military procurement.

The large difference between the low reliability predicted for microwave tubes by 217F and the much higher reliability demonstrated by field experience is of major concern to the tube industry. The reason for this concern can be understood by considering again the following lines from the stated purpose of MIL-HDBK-217:

- "*common basis* for reliability predictions during acquisition programs,"
- "*common basis* for comparing and evaluating reliability predictions of related or competitive designs."

The problem is that system designs using very different technologies are sometimes evaluated using MIL-HDBK-217. This can lead to a violation of the *common basis* presumption. Consider, for example, the evaluation of two RF systems, one of which uses microwave tubes and the other of which uses solid-state RF amplifiers. Assume that the MIL-HDBK-217 prediction for the reliability of the solid-state amplifiers is close to reality while that for the microwave tubes is a factor of five below reality. Then, the solid-state system may be selected even though it is more expensive and technically inferior to the microwave-tube based system. In a nutshell, we must make certain that we are "comparing apples with apples" when using MIL-HDBK-217. A technology must not be placed at a disadvantage by erroneous reliability predictions.

The relation given in MIL-HDBK-217F for predicting failure rate, λ_p, is

$$\lambda_p = \lambda_b \pi_E \qquad \text{Failures}/10^6 \text{ hrs} \tag{18-12}$$

where λ_b is called the *base failure rate* and π_E is the *environment factor*. The base failure rate is a function of power and frequency and is given by+

$$\lambda_b = 11\,(1.00002)^P (1.1)^F \tag{18-13}$$

---

* The draft version of Notice 2 of 217F contains significant changes to the factors contained in the 217 model. With these changes, the predicted reliability will be a factor of less than two below field experience.

+ In the draft version of Notice 2 of 217 F, the equation for base failure rate is changed to

$$\lambda_b = 11\,(1.00001)^P (1.1)^F$$

where P is the rated power in watts (peak, if pulsed), $0.001 \leq P \leq 40{,}000$ and F is the operating frequency in GHz, $0.3 \leq F \leq 18$. Figure 18-5 shows how base failure rate varies with power and frequency.

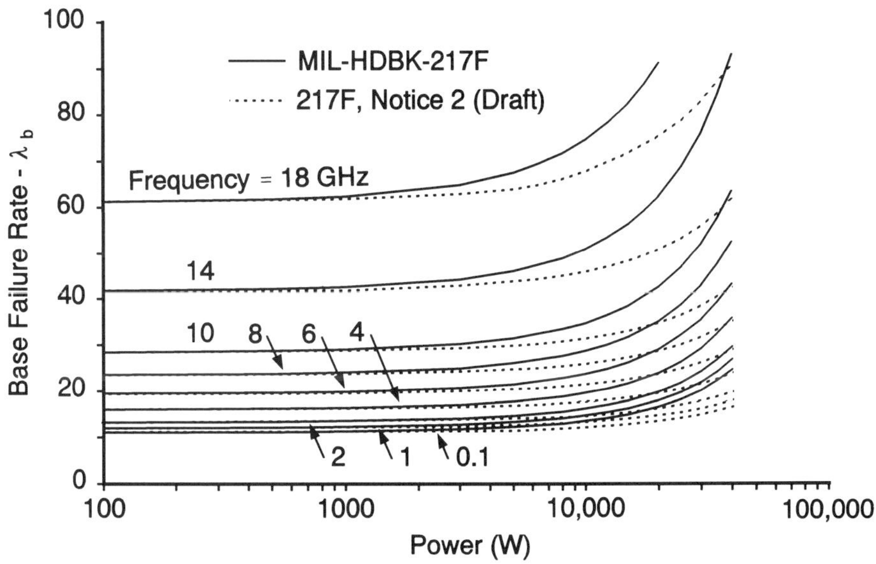

Figure 18-5 Base failure rate as functions of power and frequency.

The environment factor depends on the application of the system in which the TWT is used. Descriptions of the various environments are given in Table 18-2. These environments range from those that impose little stress on the equipment (ground, benign) to cannon launched systems, which produce extreme stress. Values for π_E vary from component to component and are given for TWTs in Table 18-3.

18.4 MIL-HDBK-217F SHORTCOMINGS [9]

As indicated in the previous section, the EIA compiled reliability data for 40 different applications of TWTs, coupled cavity TWTs and klystrons. These data were based on experience in systems. In general, where data for large numbers of tubes were available, a confidence level of 90% was used. For satellite applications, the confidence level depended on the number of TWTs for which data were available and the time in service. Confidence levels of 60% and 90% were used.

Bar graphs showing actual MTBFs from the EIA data (updated results from [9] are used here) and the predictions of MIL-HDBK-217F are given in Figures 18-6 to 18-9. Note that MTBFs range from millions of hours for TWTs in satellite applications to

Table 18-2

Environmental Symbol and Description

(From MIL-HDBK-217F)

Environment	π_E Symbol	Description
Ground, Benign	G_B	Nonmobile, temperature and humidity controlled environments readily accessible to maintenance; includes laboratory instruments and test equipment, medical electronic equipment, business and scientific computer complexes, and missiles and support equipment in ground silos.
Ground, Fixed	G_F	Moderately controlled environments such as installation in permanent racks with adequate cooling air and possible installation in unheated buildings; includes permanent installation of air traffic control radar and communications facilities.
Ground, Mobile	G_M	Equipment installed on wheeled or tracked vehicles and equipment manually transported; includes tactical missile ground support equipment, mobile communication equipment, tactical fire direction systems, handheld communications equipment, laser designators and range finders.
Naval, Sheltered	N_S	Includes sheltered or below deck conditions on surface ships and equipment installed in submarines.
Naval, Unsheltered	N_U	Unprotected surface shipborne equipment exposed to weather conditions and equipment immersed in salt water. Includes sonar equipment and equipment installed on hydrofoil vessels.
Airborne, Inhabited, Cargo	A_{IC}	Typical conditions in cargo compartments which can be occupied by an aircrew. Environment extremes of pressure, temperature, shock and vibration are minimal. Examples include long mission aircraft such as the C130, C5, B52, and C141. This category also applies to inhabited areas in lower performance smaller aircraft such as the T38.
Airborne, Inhabited, Fighter	A_{IF}	Same as A_{IC} but installed on high performance aircraft such as fighters and interceptors. Examples include the F15, F16, F111, F/A 18 and A10 aircraft.

Table 18-2 (cont'd)

Environmental Symbol and Description

(From MIL-HDBK-217F)

Environment	π_E Symbol	Description
Airborne, Uninhabited, Cargo	A_{UC}	Environmentally uncontrolled areas which cannot be inhabited by an aircrew during flight. Environmental extremes of pressure, temperature and shock may be severe. Examples include uninhabited areas of long mission aircraft such as the C130, C5, B52 and C141. This category also applies to uninhabited areas of lower performance smaller aircraft such as the T38.
Airborne, Uninhabited, Fighter	A_{UF}	Same as A_{UC} but installed on high performance aircraft such as fighters and interceptors. Examples include the F15, F16, F111 and A10 aircraft.
Airborne, Rotary Winged	A_{RW}	Equipment installed on helicopters. Applies to both internally and externally mounted equipment such as laser designators, fire control systems, and communications equipment.
Space, Flight	S_F	Earth orbital. Approaches benign ground conditions. Vehicle neither under powered flight nor in atmospheric reentry; includes satellites and shuttles.
Missile, Flight	M_F	Conditions related to powered flight of air breathing missiles, cruise missiles, and missiles in unpowered free flight.
Missile, Launch	M_L	Severe conditions related to missile launch (air, ground and sea), space vehicle boost into orbit, and vehicle re-entry and landing by parachute. Also applies to solid rocket motor propulsion powered flight, and torpedo and missile launch from submarines.
Cannon, Launch	C_L	Extremely severe conditions related to cannon launching of 155 mm. and 5 inch guided projectiles. Conditions apply to the projectile from launch to target impact.

Table 18-3

Environmental Factor—π_E

(From MIL-HDBK-217F and MIL-HDBK-217F, Notice 2 (Draft))

Environment	π_E 217F / Notice 2 (Draft)	Environment	π_E 217F / Notice 2 (Draft)
G_B	1.0 / 0.5	A_{UC}	11 / 6.0
G_F	3.0 / 1.5	A_{UF}	18 / 9.0
G_M	14 / 7.0	A_{RW}	40 / 20
N_S	6.0 / 3.0	S_F	0.10 / 0.05
N_U	21 / 10	M_F	22 / 11
A_{IC}	10 / 5.0	M_L	66 / 33
A_{IF}	14 / 7.0	C_L	1000 / 500

about 1000 hrs for some ECM applications. The highest predicted MTBF (based on MIL-HDBK-217F) is slightly over 500,000 hrs (for satellite TWTs) and the lowest is about 200 hrs for a coupled cavity TWT. For Notice 2, the highest predicted MTBF is slightly over 1,000,000 hrs and the lowest is just over 1000 hrs.

The ratios of actual MTBFs to those predicted by MIL-HDBK-217F are displayed in Figure 18-10. This ratio varies from 65:1 for a TWT in a radar application down to 0.3 for a TWT in an ECM application. From the scatter of the data plotted in Figure 18-10, it is clear that there is little correlation between the predictions of MIL-HDBK-217F and the results obtained from field experience with microwave tubes. (If there were a high degree of correlation, the data would be clustered about unity. Instead, the average ratio is over five. For Notice 2, the ratio is under two.)

The EIA study referred to earlier in this chapter [3] was directed toward correcting deficiencies in MIL-HDBK-217E, and some improvement was, in fact, realized with the publication of revision 217F. With the release of Notice 2, which is presently in draft form, a significant improvement will be made in the factors (λ_b and π_E) used in the 217 model. However, the *model* is still grossly inadequate, and most of the discussion in the EIA report concerning the shortcomings of the *model* are still relevant. Thus, it is appropriate to quote from the EIA report:

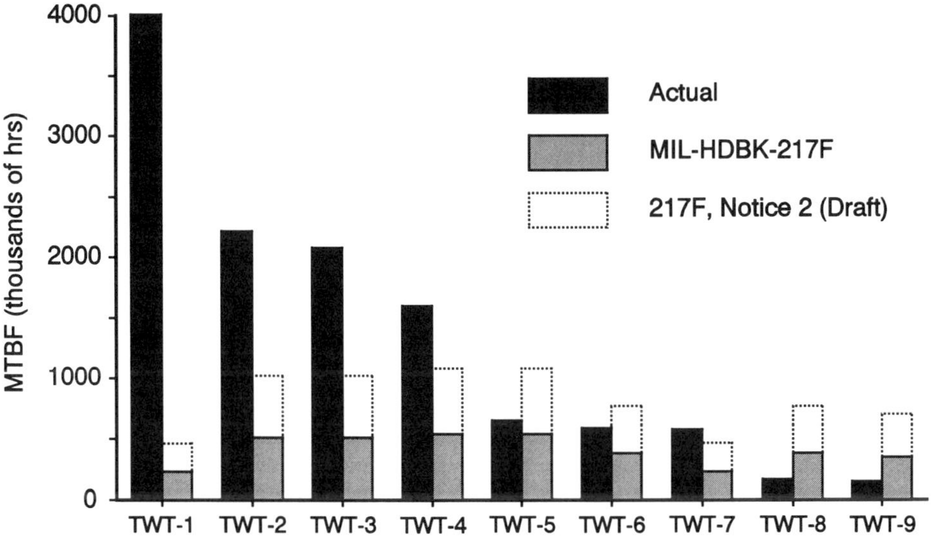

Figure 18-6 Space TWT reliability, actual *versus* MIL-HDBK-217F.

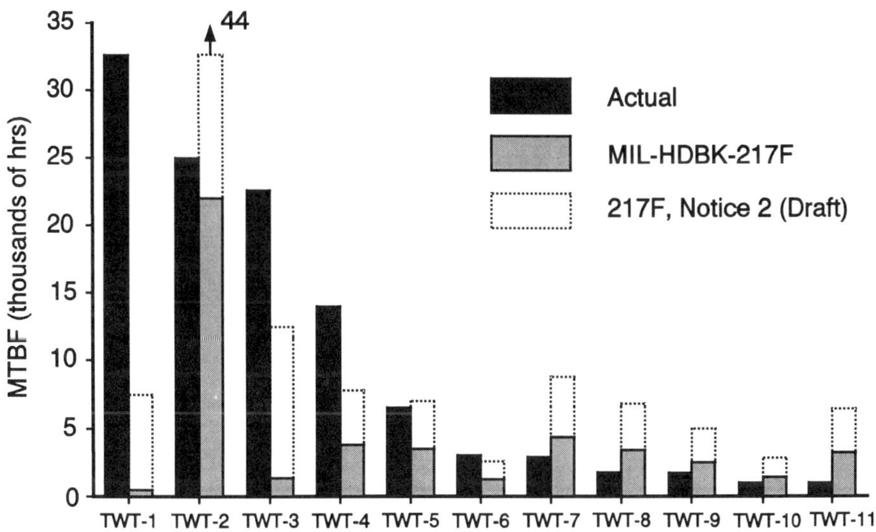

Figure 18-7 Radar and ECM TWT reliability, actual *versus* MIL-HDBK-217F.

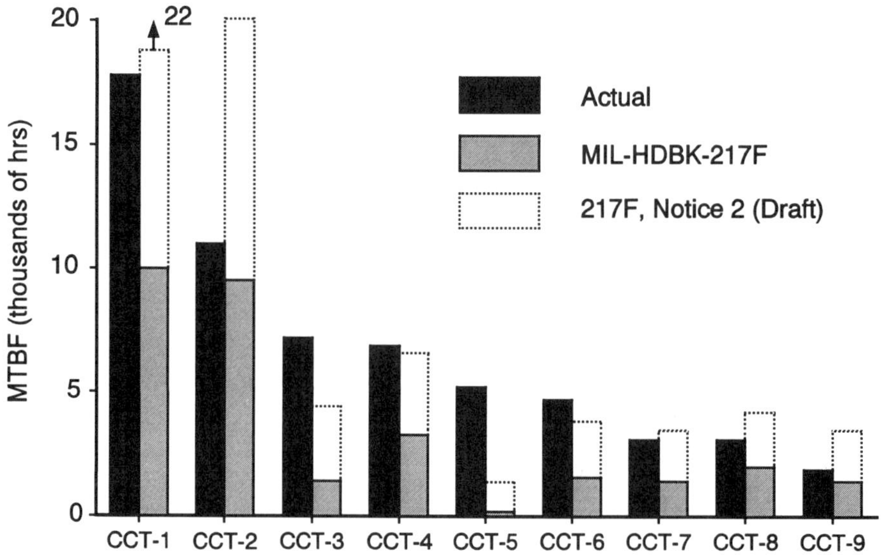

Figure 18-8 Coupled cavity TWT reliability, actual *versus* MIL-HDBK-217F.

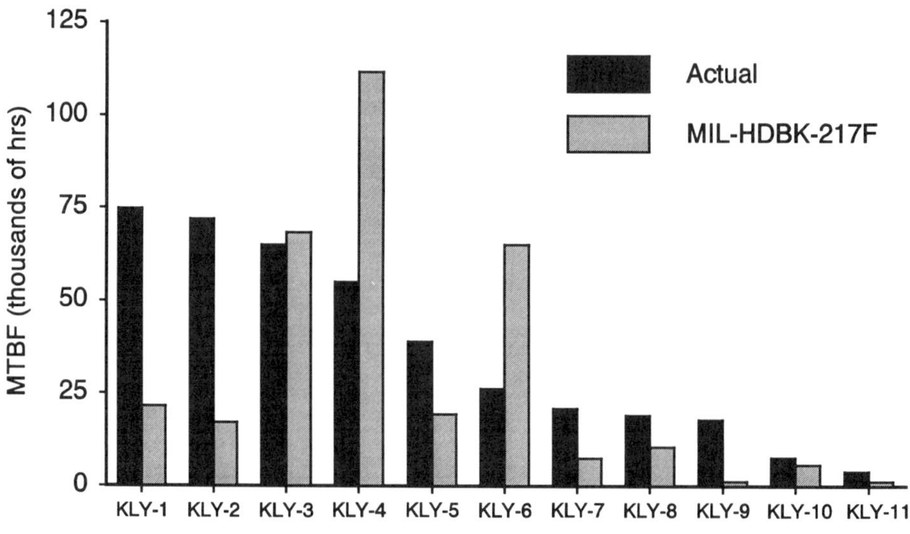

Figure 18-9 Klystron reliability, actual *versus* MIL-HDBK-217F.

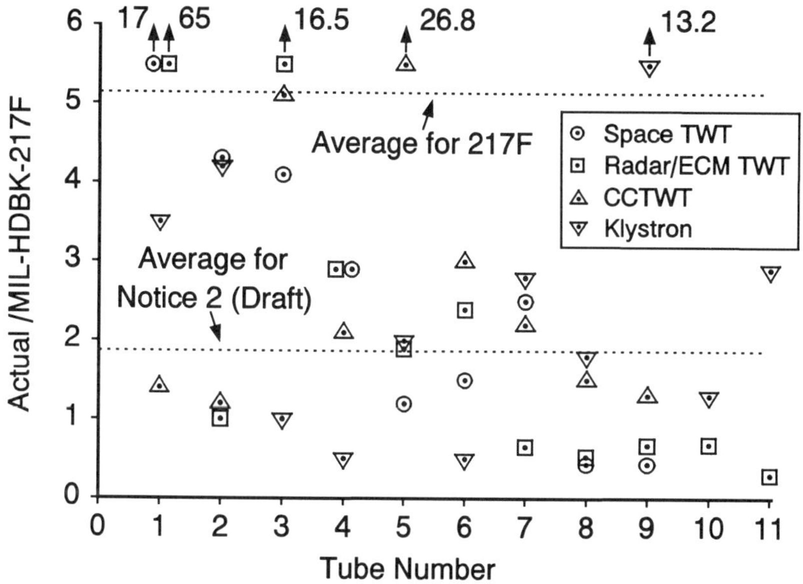

Figure 18-10 Ratio of actual MTBF to MTBF predicted by MIL-HDBK-217F.

> There are many reasons for the consistently pessimistic MTBF values predicted by MIL-HDBK-217E. The basic reason is that the model is far too simplistic. . . . . It does not address the failure rate drivers associated with system requirements and tube design, nor their interaction. In general a tube is designed for a specific system; it is not designed as a generic device that will interface well with all possible systems. This simple model omits RF circuit structure, thermal stress level, electrical stress level, cathode type and cathode current loading. These elements are interactive and play a key role in tube reliability.
>
> The model does not address derating due to thermal or electrical stresses; nor does it address the effects of on/off cycling. A tube operating continuously, whether CW or pulsed, will experience far fewer failures than a tube cycled on/off with short operational duty cycles.

The following statements are from the conclusion of the EIA report:

> The most significant fact concerning the prediction capabilities of the MIL-HDBK-217E model for microwave tubes is that *it does not work*. The model is too simple to adequately describe the complexities of the microwave system/tube interface. As shown from both field demonstrated performance and the physical nature of the microwave tube, the predictive reliability model

described by the handbook is not a reliable approach to describing expected tube performance.

The procedure now in place is misleading to system designers. When the use of MIL-HDBK-217E is mandated, it may result in the incorrect choice of a microwave device. The consequences are unnecessarily high procurement costs and excessive system maintenance. Much better guidelines are needed to match tube and system characteristics for higher system reliability.

While this previous paragraph refers to revision 217E, the same statements apply to revision 217F.

18.5 IMPROVED PREDICTIVE MODELS

From the discussion in the previous section it is clear that an improved model is needed to bring the predictions of MIL-HDBK-217 into line with reality. Government engineers and, in particular, engineers at Rome Laboratory agree that a new model is needed. However, as pointed out earlier in this chapter, the upgrading of 217 models is directly proportional to the availability of funds to do so, and those funds are very scarce.

A model similar to that proposed by Audette [10] or that used by Lalli and Speck [11] would be far superior to the present 217 model. The Audette model, sometimes referred to as the *experience TWT failure rate model* is based on a detailed analysis of 52,000 TWTs used in airborne electronic countermeasures systems manufactured by Sanders Associates. These TWTs accumulated over 2,000,000 device hrs in equipment testing per MIL-STD-781. Failures were classified by failure mode and the element to which the failure mechanism was related. Then, a model was established that took into consideration TWT failures resulting from failures of individual tube elements. Seven tube elements are included in the model as indicated in Figure 18-11. The TWT failure rate, λ_p, is then given by

$$\lambda_p = M_V \pi_L \sum \text{Element Failure Rates} \qquad (18\text{-}14)$$

where M_V is the cathode voltage multiplier and π_L is the learning factor. Note that power is *not* a factor in this predictive model. In the conclusion to his paper, Audette states: "None of the reliability determining failure mechanisms found during Sander's programs were ever determined to be directly related to the peak power output of the TWT."

The failure rate for each of the seven elements consists of a base failure rate and a multiplier. Values for the multiplier depend on whether the tube operates CW or pulsed. In addition, a multiplier is applied to some of the element failure rates to account for the number of grids or anodes used in the tube. The final failure rate model is

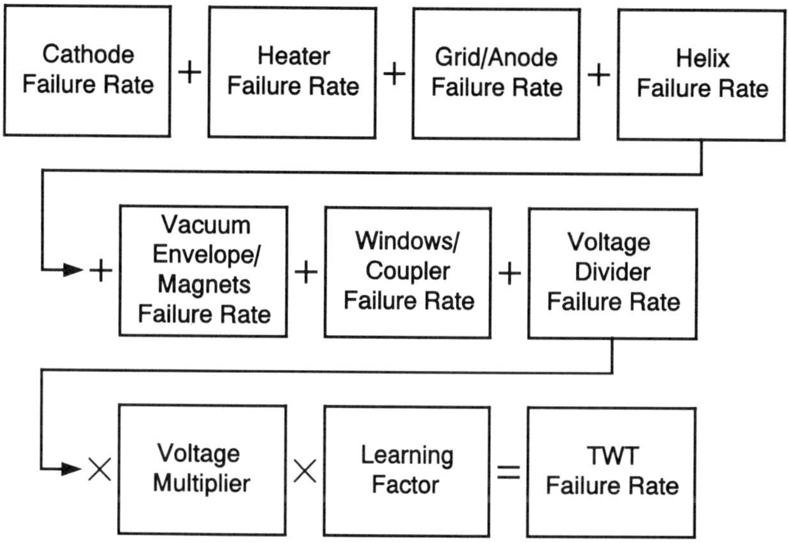

Figure 18-11 Audette failure rate model.

$$\lambda_p = (\lambda_c m_c + \lambda_H m_H + \lambda_G m_G N_G + \lambda_{He} m_{He} + \lambda_{Ve} m_{Ve} + \lambda_W m_W + \lambda_V m_V N_G) M_V \pi_L \quad (18\text{-}15)$$

where

λ_c	=	Cathode failure rate
λ_H	=	Heater failure rate
λ_G	=	Grid/anode failure rate
λ_{He}	=	Helix failure rate
λ_{VE}	=	Vacuum envelope/magnets failure rate
λ_W	=	Windows/couplers failure rate
λ_V	=	Voltage divider network failure rate
N_G	=	Number of grids/anodes or voltage dividers
M_V	=	Cathode voltage multiplier
π_L	=	Learning factor

and

$m_c \ldots m_V$ = Element multipliers depending on tube type

Values for the various base element failure rates and multipliers were derived from data by Audette. As a final result it was claimed that the correlation of the predicted failure rates with actual rates was about 80%. The correlation of MIL-HDBK-217B (available at that time) with actual rates was only 5%. When sharply decreasing failure rates for newly developed TWT sources were taken into consideration, it was claimed that the correlation with the Audette model increased to 98% *versus* 4% for MIL-HDBK-217B.

The philosophy used by Lalli and Speck [11] in predicting reliability was similar to that of Audette. Based on experience and available data, failure rates for various elements of a TWT were estimated. Low, mean and upper values of failures rates were tabulated for each element. These rates depended on design and application stresses and material strengths. Mean failure rates were used and were multiplied by an application factor. In this way, actual part stresses were taken into consideration in an effort to make the total failure rate as accurate as possible. This technique was used to predict MTBFs for satellite TWTs.

18.6 RELIABILITY FACTORS IN THE DESIGN AND MANUFACTURE OF TWTS

The following are the inherent wear-out mechanisms in microwave tubes:

1. Cathode depletion, which is predictable from the Longo model. Modern cathodes are predicted to operate for years to hundreds of years depending on the current density requirement. In most modern microwave tubes, cathode depletion should not be the cause of end of life.
2. Build-up of conductive material, such as barium or metal oxides on internal ceramic surfaces, which can lead to excessive electrical leakage or high voltage breakdown. The severity of this problem is reduced by using relatively low temperature cathodes, such as the M type, which have an impregnant evaporation rate an order of magnitude lower than the B type.
3. Metal fatigue due to thermal expansion and contraction of internal tube elements such as heater leads and supports and the connection of the center conductor of the RF output connector to the helix.
4. Leakage or permeation of gases into the vacuum envelope.

These wear-out mechanisms are well understood by the TWT industry and can usually be controlled through careful design and the proper use of materials. Still, TWT failure often occurs long before it is expected on the basis of a knowledge of the known wearout mechanisms. The question that must be addressed is why this should be the case. The following are some of the more important factors that contribute to premature failure:

1. It must be realized that TWTs are made up of a large number of parts. For example, some coupled cavity TWTs contain more than 1000 parts.
2. The design and manufacture of TWTs requires the application of more technical disciplines than most other electronic devices.
3. The manufacturing process is labor intensive. There is very little automation. This is largely because of the wide variety of types and small number of tubes per type that are required.

4. Most TWTs utilized in military systems are operated near their thermal, voltage and microwave power limits (especially in ECM applications where it seems as though there's never sufficient performance to satisfy requirements).
5. Many TWT failures are the result of a chain reaction originating with a failure in one of the associated system components.
6. Users of TWTs often do not understand the devices and misuse them in installation or application. As a result, TWTs are often "killed" by the users.

For these reasons, a seemingly endless array of things can go wrong in the manufacture and use of TWTs. The author has given short courses on microwave tubes and TWTs to well over 1000 people. The comment is often made by class participants that, considering the complexity of tubes, it is a wonder that they can be made to work as well as they do.

The microwave tube industry has had nearly 50 years of experience in developing the methods and technologies for designing, developing and manufacturing TWTs. In the Executive Summary for the EIA report [3] the following statement is made:

> The major failure modes of high power microwave tubes are well known and current designs incorporate features to prevent such failures wherever the application allows conservative design and derating. The microwave tube industry knows what to do to provide reliable tubes.

When the reliability data are examined, the large MTBFs that have been demonstrated for tubes in some applications provide justification for this statement.

As a specific example of a case in which high reliability is being achieved and in which major improvements have occurred, consider the microwave tubes for the Navy's AEGIS weapon system. The chart in Figure 18-12 lists the original specifications for MTBFs and the assessments of operational MTBFs in 1983 and in 1993 [12]. The tubes listed are crossed-field amplifiers (CFA), switch tubes (SWT) (which are not microwave tubes) and TWTs. In all cases, except for one of the switch tubes, the MTBF exceeds the original specification by a significant amount. Also, except for the 40 W TWTs, reliability increased a great deal during the ten year period from 1983 to 1993. Along with the reliability increases, large cost reductions were achieved. The cost per operating hour for tubes decreased from $4.55 in 1983 to $0.61 in 1992. These improvements resulted, largely, from the procedures used by the Navy for managing the acquisition and use of the tubes. Some of these procedures are discussed in following sections.

In contrast to the Navy's experience with the AEGIS weapon system, there have been applications where MTBFs were low and where improvements were clearly needed. One of the major messages of the presentation "Airborne ECM helix TWT reliability" by L. DesSureault and presented by L. Schwartz at the 1988 Microwave Power Tube Conference was [13]: *There has not been any "real" improvement in TWT reliability in 16 years.*

	Specified	Assessed 1983	Assessed 1993
S/D CFA	5,000	6,000	45,000
D/D CFA	5,000		30,000 (Engineering Estimate)
S/D SWT	3,000	1,800	2,800
D/D SWT	5,000		8,700
10 kW TWT	5,000	5,000	12,000
CWI TWT	2,500	1,300	5,100
40 W TWT	5,000	12,000	12,000

S/D = Single Duty, D/D = Double Duty, CWI = CW Illuminator

Figure 18-12 Reliability improvement of tubes in the AEGIS weapon system. From R. Walker and S. Sims, presented at the American Society of Naval Engineers Conference, October 1993.

The other message was that TWTs are far more reliable than MIL-HDBK-217 predicts. These statements were based on analyses of TWT failures in about 1500 electronic countermeasures systems over a 16 year period. A breakdown of failure modes from this study is shown in Figure 18-13. Note that poor RF performance was not a major reason for failure. Instead approximately two-thirds of the failures (the connected pieces A, B, C and D) resulted from process control problems. HV breakdown included leakage and potting problems. Connections were to electrodes and, internal and external to the vacuum envelope and included welding, crimping and soldering. Grid-cathode shorts resulted from problems such as loose particles and barium deposition on ceramics. The "other" category included tubes with leaks or that were gassy and problems introduced by the systems manufacturer. The statement from Reference 13 concerning these failures was:

> The big surprise was that the failure driver was *not* RF performance but process and material control; better yet, product design and process compatibility. You can have the best design but if it's not producible and processes aren't controlled you end up with these dominant causes of failures.

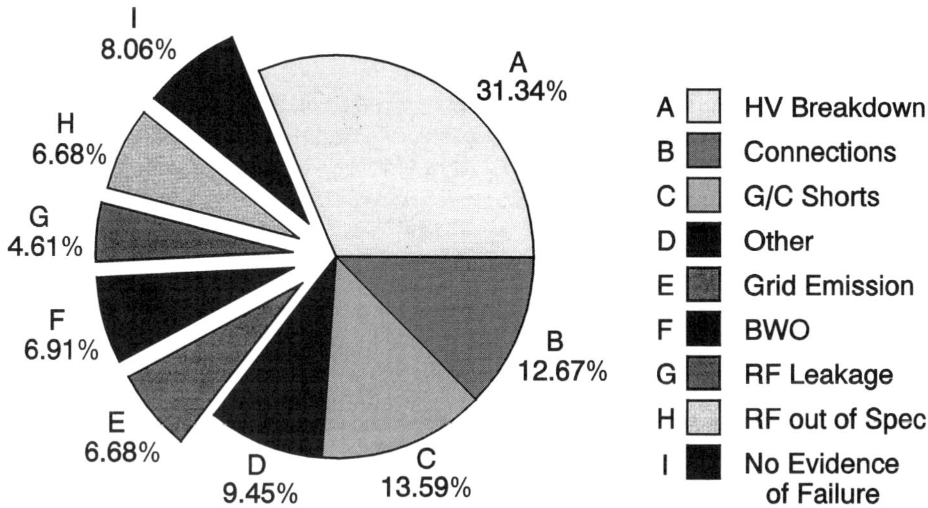

Figure 18-13 Failure modes from DesSureault study; process control dominates.

There have been many explanations given for the disappointing results of the DesSureault study. For example, if the tube specifications were not changed during the course of the study, then tubes continued to be built to old specifications. It was noted by one government official that, due to the current acquisition policy by the Department of Defense, a tube manufacturer will do nothing that will increase the price of a tube regardless of the benefit. The cheapest price sells tubes, not reliability. Whatever the reason for the disappointing results, many people think that the DesSureault paper served as a "wake-up call" to the acquisition agencies and the tube industry to talk less and *do more* to improve reliability. The following are some of the areas in which major improvements or changes have been made.

18.6.1 Improved Cathodes

As described in Chapter 5, there has been enormous improvement in cathodes in recent years. While B-type cathodes and even oxide-coated cathodes are still used in some old tube designs, virtually all new designs use cathodes based on M-type construction. With their lower operating temperatures, these cathodes have lifetimes an order of magnitude greater than the old B-type cathodes. The lower temperatures also reduce the probability of electrical breakdown and reduce thermal stresses on related elements such as the grid and heater assemblies.

18.6.2 Improved Magnets

All modern PPM focused tube designs use samarium cobalt magnets. As pointed out in Appendix D, these magnets have an energy product that is over a factor of four above that for the alnico magnets used in older tube designs. These enhanced magnet characteristics permit improved beam focusing, which, in turn, reduces the thermal stress on the RF circuit. Also, these magnets are extremely stable, and this makes it possible for a tube to survive extreme temperature variations without changes in magnet characteristics.

18.6.3 The Use of Getters

Evaporable getters have been used for decades to help maintain the vacuum level in simple vacuum tubes as well as in cathode ray tubes. As described in Appendix C, nonevaporable getters have found use in microwave tubes in recent years. These getters are particularly useful in helping to ensure that a tube will operate properly after standing idle for an extended period. The improved vacuum level produced by the getter helps to prevent arcing and other gas-related problems when a tube is turned on after having been idle for an extended period.

18.6.4 Improved Test Equipment and Procedures

There have been major improvements in the capabilities of test equipment in recent years. The speed with which data can be obtained has increased significantly, and data analysis capabilities have improved markedly. As a result, errors in tube operation are detected more quickly, and so corrective action can be taken much sooner than was possible a few years ago.

As pointed out in [12], a problem that continues to exist is that equipment and procedural test set ups for some TWT parameters cannot be calibrated on a standard traceable to the National Bureau of Standards. Some measurements, including phase and, in some cases, power, are calibrated using a special set of tubes. These, referred to as *gold tubes*, are taken from standard production and used to define a standard. These tubes are shuttled between vendor and government facilities for test-set calibration. For the most part, this has solved problems such as that of "East Coast watts" and "West Coast watts" in which East Coast systems manufacturers were rejecting tubes shipped to them by West-Coast manufacturers because of claimed power deficiencies.

18.6.5 Environmental Stress Screening

Major improvements in TWT reliability have resulted from the increased use of environmental stress screening (ESS) [14]. ESS subjects tubes to random vibration and temperature cycling *during* production. *Every tube* is screened for defects that have occurred during production. This is in contrast to qualification testing, in which prototypes are checked to uncover defects in design *before* being produced in quantity. By testing during production with ESS, manufacturing defects can be identified and corrected so that, as shown in Figure 18-14, the number of flaws remaining in delivered units is greatly reduced. When the results of ESS are fed back for use in modifying designs, the reduction in flaws is even greater.

Proper ESS does not cause defects, it brings out latent flaws through thermal cycling and gentle vibration. The cost of equipment necessary for ESS may be significant but can be far outweighed by benefits gained. For example, with warranties now being required with most tube procurements, warranty costs can be reduced by identifying and correcting most defects before the tubes are shipped.

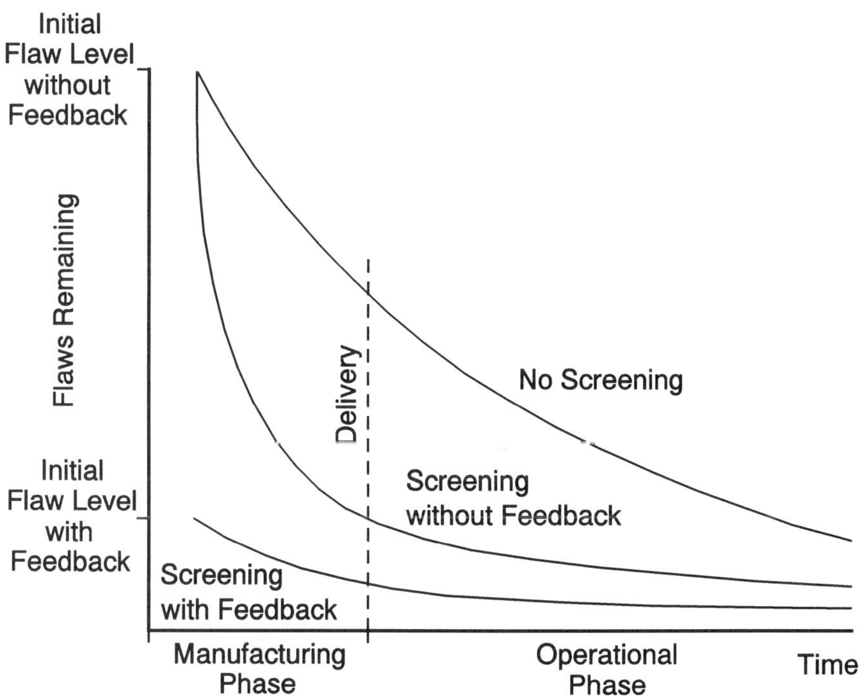

Figure 18-14 Effectiveness of environmental stress screening. From Wayne Tustin, *IEEE Spectrum*, December 1986. © IEEE 1986.

18.6.6 Robust Design and the Use of Taguchi Loss Functions

Historically, good quality in military procurements has meant that the minimum specification standards have been met. As a result, tube manufacturers had little incentive to do better than meet the minimum specifications. Deming [15] has pointed out the lack of logic with meeting the minimum specifications as follows:

> There is obviously something wrong when a measured characteristic barely inside a specification is declared to be conforming; outside it is declared to be nonconforming. The supposition that everything is all right inside the specification and all wrong outside does not correspond to this world.

To correct problems resulting from meeting minimum specifications, many procurement agencies now require warranties and also require manufacturers to "meet specs plus margin." As a result, tube manufactures must do more than just meet specifications. Again quoting from Deming:

> A better description of the world is the Taguchi loss function in which there is minimum loss at the nominal value, and an ever-increasing loss with departure either way from the nominal value.

The Taguchi approach to improved reliability is now being used by some procurement agencies (see [16] for example). Taguchi loss functions equate higher quality with lower variation of the characteristics of a product about a target. The use of these functions permit an objective and numeric rating of quality to be assigned. For procurement agencies, this permits quality to be used as an award criteria. With quality based on the use of Taguchi loss functions to help determine procurement awards, it is essential for the tube manufacturers to understand and use these same quality determining criteria in manufacturing. In [16], Jeffrey Snow of the Crane Division of the Naval Surface Warfare Center gives a straightforward description of Taguchi loss functions.

For simplicity, assume, first, that tube selection depends only on a single tube characteristic, such as RF power output. Assume, also, that there are minimum and maximum power levels beyond which the tube is of no use. (The minimum power specification might result from the minimum power needed for a radar to detect targets at a given range. The maximum power might result from the RF breakdown characteristic of the output RF system.) In Taguchi terms, the minimum power is the lower specification limit (LSL) and the maximum power is the upper specification limit (USL). Between the LSL and the USL there is a target (T) value of the specification that the manufacturer is attempting to meet. The Taguchi loss function for this requirement is shown in Figure 18-15. This particular loss function, which is a quadratic function and symmetrical about the target value, is referred to as a *two-sided function*.

It should be noted that there are two-sided specifications with preference, which means that the target value is not midway between LSL and USL. Thus, either the lower or upper limit is given preference. For example, it might be desirable to have as much power output as possible from a tube without causing breakdown. As a result, T would be set closer to USL than to LSL. There are also one-sided functions for which T is set at either LSL or USL. An example is weight specification. Here, there is usually a maximum weight, and this could be both T and USL.

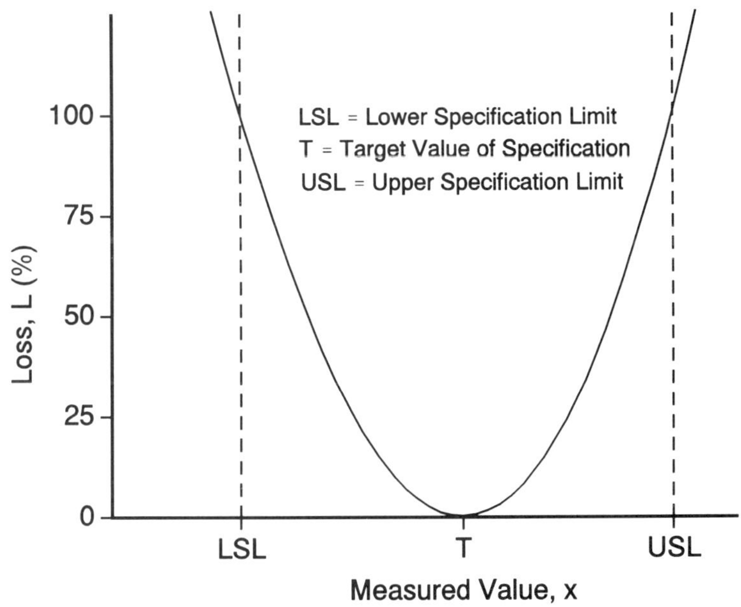

Figure 18-15 Taguchi loss function. From Jeffrey M. Snow, *Naval Engineers Journal*, January 1993.

Returning to Figure 18-15, x is the measured value of interest (power in the example in the preceding paragraphs). Losses are expressed in terms of the tube's purchase cost. At T, there is no loss, and the vendor receives full credit for the tube. At LSL or USL, the tube is considered to be useless, and so the loss is 100%.

To consider an example of the application of the Taguchi loss function to a batch of tubes, we must first write the equations for the quadratic in Figure 18-15. That is,

$$L = k(x - \mathrm{T})^2 \tag{18-16}$$

where

$$T = \frac{\mathrm{USL} + \mathrm{LSL}}{2} \tag{18-17}$$

and k is such that L is 100% when x is USL or LSL. Thus,

$$k = \frac{400\%}{(\text{USL} - \text{LSL})^2} \tag{18-18}$$

Now, assume that the RF power limit specifications for a tube are

$$\text{LSL} = 8 \text{ kW} \quad \text{and} \quad \text{USL} = 12 \text{ kW} \tag{18-19}$$

Then,

$$T = 10 \text{ kW} \tag{18-20}$$

and

$$k = 25\% \tag{18-21}$$

Assume that power measurements on 15 tubes are as shown in column 2 of Figure 18-16. Then, the calculated losses are as shown in column 3. The average loss, AL, of 41% represents the quality rating for the tubes. Clearly, tubes with lower values of AL have higher quality. The usefulness of this rating is that numeric values can be applied to tubes from several production runs, or from several vendors. These ratings can then be used as part of a procurement process.

Tube No.	x (kW)	L (%)
1	8.4	64
2	8.0	100
3	9.0	25
4	9.0	25
5	9.7	2
6	10.2	1
7	9.6	4
8	8.5	56
9	8.8	36
10	8.7	42
11	8.4	64
12	8.6	49
13	8.1	90
14	8.5	56
15	9.9	0
		Average Loss, AL = 41%

Figure 18-16 Example of calculation of average loss.

Of course, the selection of a microwave tube for an application depends on far more than one characteristic. As a result, loss functions must be defined for all relevant characteristics. Then, in calculating an overall average loss, the results for the various characteristics may need to be weighted according to importance to the final application for the tube.

18.6.7 Total Quality Management

Finally, and this may be the most important item of all, there is a basic change (some people say a cultural change) taking place in thinking about how quality can best be improved through better management practices. The overall approach is referred to as *total quality management* (TQM). TQM is occurring in industries throughout the United States including the microwave tube industry. The changes being instituted by TQM have far more fundamental and far greater impact than the slogans "zero defects" and "do it right the first time," which have been used in the past.

TQM was first practiced in Japan. It is interesting that TQM is based on principles set forth by an American, W. Edwards Deming, but these principles were first recognized and put into use by the Japanese. As stated in [17], "the goal of TQM is to create a system of management procedures that focuses on customer satisfaction and transforms the corporate culture so as to guarantee continual improvement." Thus, TQM focuses on three critical elements:

1. Customer expectations,
2. Full participation,
3. Continuous improvement.

Item 2, full participation, is based on the premise that nobody understands a process and its shortcomings better than the person working on it. So to make TQM work, organizations must involve everyone in team efforts to investigate ways to improve processes.

TQM is a management style geared toward customer satisfaction. When TQM is coupled with the other factors for improving reliability that are discussed in this section, the result should be a very high level of customer satisfaction with microwave tubes.

18.7 RELIABILITY FACTORS IN THE APPLICATION OF TWTS

One of the major problems in increasing the reliability of microwave tubes is poor feedback, or the complete lack of feedback from the field concerning tube operating times and conditions of failure. As an example, it is reported in [18] that a company specializing in tube repair received 60 TWTs from the field for rework. The main reasons for failure listed in the accompanying documentation were "not working," "unknown," "no RF" or "no _ good."

The problem of the lack of feedback has been addressed by the Navy in connection with the AN/SPY-1 system used on AEGIS ships. Individual tubes are carefully tracked from date of manufacture to end of life, including repair periods where appropriate. Conditions of failure are documented and fed back to the manufacturer so that corrective action can be taken. This procedure, along with careful management of inventories, production schedules, *et cetera*, has resulted in significant reductions in failure rates for some items.

While some microwave tube failures can be attributed to problems in tube design and manufacture, many result from inadequate system design and poor protection. Then, in addition, there are failures that have resulted from external damage, clearly due to mishandling or abuse. RF connectors and coolant fittings are damaged, coolant lines are clogged, leads are broken, *et cetera*. Furthermore, in many cases, it is estimated that from 25 to 40% of "defective" tubes were not defective after all. Something else in the system failed, but the tube was replaced first, probably because of its reputation as being failure prone. Again, referring to [18], a case is reported where 1464 "unusable" tubes that had been returned from the field were retested, 655 or 45% were OK.

If the truth were known, it could well be that, in some applications, over 50% of all failures are the fault of the users rather than the tube manufacturers. This is not to say that the manufacturers should not continue their efforts to build more reliable tubes, but it is essential to realize that the "bullet-proof" tube is a virtual impossibility and that intense efforts must be made to better inform the users about the tube characteristics.

While Figure 18-17 may be an exaggeration of the user situation, it is symbolic of the lack of understanding on the part of many users of the extreme complexity and fragile nature of some of the elements of many microwave tubes.

The study "Reliability Design for High Power Tubes" addressed many factors relating to reliability in applications of microwave tubes [19]. The final report was written for the designer of microwave systems of sufficient power to require microwave tubes.

18.8 SHELF-LIFE CONSIDERATIONS

Shelf life or *storage life* is the life of a device that is not in operation. Shelf life is important for nearly all applications of microwave tubes because it is inevitable that, after manufacture, the tubes will not be put into immediate operation. Instead, the tubes may be placed in storage for several years before being used. Reference 20 reports a detailed study of shelf-life considerations. Before discussing the results and recommendations of [20] we should ask why a tube should fail if it is not being used.

Figure 18-17 User abuse is often a major factor when the failure rate of microwave tubes is excessive. From *On Target*, January 1983.

At the beginning of this section, four tube wear-out mechanisms are listed. The only one of these that may occur during storage is the last one, which is leakage or permeation of gases into the vacuum envelope. With modern manufacturing techniques, leaks are rarely a problem. If a leak does occur, it is usually detected during tube screening and testing, and so tubes with leaks very rarely leave the factory. Permeation is rarely a problem because the tube industry has long known what materials to use and how to use them in constructing tubes.

Other sources of gas that may occur are contamination and virtual leaks. Contamination may occur if cleanliness and bake-out procedures are not followed properly. As a result of modern procedures for the cleaning of incoming parts and the maintenance of cleanroom discipline in assembly areas, contamination is rarely found to be a source of gas in a tube. Virtual leaks, on the other hand, are harder to control. The most serious virtual leaks are tiny pockets of gas that are trapped in or between materials during manufacture. For example, ceramics contain small voids. A microscopic crack in the surface of the ceramic can open one of these voids and lead to a very small leak (a virtual leak) that continues almost indefinitely.

The effects of gas in a tube have been discussed in many places in this book. To summarize,

1. Gas can cause arcs when a tube is turned on.

2. Gas can poison the cathode. The resulting degraded emission leads to poor tube performance.
3. Gas becomes ionized by the electron beam when the tube is turned on. Ions cause several difficulties with tube operation.

The largest data base used by Buck for the conclusions and recommendations of [20] is that contained in [21]. Over 1000 tubes were taken out of storage and tested. After careful consideration of the various failures reported for these tubes in [21], Buck concludes that the only significant cause of TWT failure in storage is the accumulation of gas from virtual leaks or possibly from contamination. He points out that the effects of gas can be ameliorated by the following three simple solutions:

1. For applications where the tube is disposable (short life applications), run the cathode at a higher than normal operating temperature to minimize the effects of any possible cathode poisoning.
2. Put a chemical (nonevaporable) getter in the tube to immobilize any gas that might evolve in storage.
3. Test the tube before and after a holding period to detect "almost good" tubes and remove them from the population to be stored. A good tube should show no change over a period of weeks. Buck suggests six weeks as an appropriate holding period.

18.9 THE IMPACT OF PROCUREMENT PRACTICES ON RELIABILITY

Several unusual characteristics of the microwave tube industry affect government procurement practices, and those practices, in turn, have a direct and very significant impact on the tube industry and the reliability of the tubes it produces. To understand this interaction, consider the following:

1. The market for microwave tubes is almost exclusively Department of Defense (DoD) applications.
2. While microwave tubes are absolutely essential to the operation of many military systems, the size of the market for them is small. It is less than $500 million, which, as one senior tube manager points out, is smaller than the golf-cart industry.
3. There are only four major U.S. suppliers of tubes. A few smaller companies produce or repair selected tube types.

Thus, the tube industry is a small industry that is, for all practical purposes, a captive of the DoD. Production benefits, quality growth, technical evolution and cost considerations normally imposed by commercial trade are absent.

Historically, the procurement procedure used by DoD has been relatively straightforward. If two suppliers meet the minimum specifications, the low cost supplier wins the contract. While the tube companies attempt to and usually do supply high quality products, there has normally been no *contractual requirement or incentive* to do so. However, as pointed out by Snow [16]:

> Getting the best value for the dollar is within the law and regulations. A contracting officer can award accordingly. Unfortunately, contracting officers often hesitate to award to the higher cost supplier. The statement that a supplier has higher quality and is worth an extra 15% sparks a reasonable question from the DoD procurement world of "How do I know?".

The most successful program for answering the question of "How do I know?" and, in addition, for improving quality and reducing cost is that developed in connection with the Navy's AEGIS weapons system. This program was established in 1979 to ensure that AEGIS tubes "were designed to meet system technical requirements and exhibit a high degree of reliability, while maintaining reasonable acquisition cost" [12]. One measure of the success of this program was shown by the reliability improvement data contained in Figure 18-12. Another measure of the success is the large cost reductions that have been achieved. The cost per operating hour for tubes has decreased from $4.55 in 1983 to $0.61 in 1992.

There are eight elements to the AEGIS program (formally referred to as the AEGIS Weapons System Acquisition Engineering Program for Microwave Tubes). These are

1. Stringent qualification,
2. Complete configuration management,
3. In-plant technical representatives,
4. Warranties,
5. Continuous production,
6. Surveillance testing,
7. "Cradle to grave" tracking and assessment,
8. Use of quality criteria in contracting.

In the author's opinion, two of these items can be singled out as having a particularly large impact on reliability and cost. One is the warranty requirement. As stated in [12], the warranties require tubes to meet the specified performance for a stated period of time. The time is based upon the estimated operational life of a tube, but in no case exceeds 24 mos. The warranty is used as an incentive to the contractor to build good tubes so as to minimize their warranty costs. The most important feature of the warranty is to provide a requirement for in-depth analysis of tubes failing for production related problems. The vendor must not only replace the faulty tube, but must investigate the root cause of failure and implement corrective action.

The other element of particular importance is the use of quality criteria in contracting. This element is discussed in detail in [16] and directed specifically at the answer to the contracting officer's question of "How do I know?" (that a supplier has higher quality and is worth a higher price). Taguchi loss functions are used to quantitatively rate quality. Then both purchase cost and quality are used as contract award criteria. Proposals are evaluated and weighted with a 60/40 technical to cost ratio. Technical scores are derived from surveillance testing results (element six of the list of eight elements) and evaluated using Taguchi loss functions. The loss functions are based on characteristics that are most critical to system performance.

REFERENCES

[1] The author is indebted to P.M. Lally of Teledyne Electronic Technologies (formerly Teledyne MEC) for some of the material used in this section.

[2] P. Barker and W.R. House, "Space tube life testing," *Microwave Journal*, October 1992, pp. 102–106.

[3] "Microwave tube reliability, actual versus predicted, a realistic appraisal," prepared by the Reliability Working Group of the Electronic Industries Association, Microwave Tube Division, January 1991.

[4] Igor Bazovsky, *Reliability Theory and Practice*, Englewood Cliffs, NJ: Prentice-Hall, Inc., 1961.

[5] This relationship is established in "Estimation from life test data" by B. Epstein, *IRE Trans. on Reliability and Quality Control*, April 1960.

[6] *Military Handbook, Reliability Prediction of Electronic Equipment*, MIL-HDBK-217F, Notice 1, 10 July 1992.

[7] George F. Watson, "MIL reliability: a new approach," *IEEE Spectrum*, August 1992, pp. 46–49.

[8] J.H. Engleman, J. Kennedy and S.R. Wood, "Traveling wave tube failure rates," RADC-TR-80-288, Final Technical Report, November 1980.

[9] S.F. Kaisel, "Update; microwave tube reliability prediction," 1992 Microwave Power Tube Conference, Monterey, CA.

[10] D. Audette, "TWT reliability in airborne equipment," *Proc. IEEE 17th Annual Spring Reliability Seminar*, 1979, pp. 63–83.

[11] V.R. Lalli and C.E. Speck, "Traveling-wave tube reliability estimates, lift tests, and space flight experience," *Proc. 1977 Annual Reliability and Maintainability Symposium*, pp. 12–20.

[12] Randall Walker and Suzanne Simms, "Acquisition engineering for AEGIS weapons system microwave tubes," presented at the Naval Engineers Conference, San Diego, CA, October 1993.

[13] L. DesSureault, "Airborne ECM helix TWT reliability," presented by L. Schwartz at the 1988 Microwave Power Tube Conference, Monterey, CA.

[14] Wayne Tustin, "Recipe for reliability: Shake and bake," *IEEE Spectrum*, December 1986, pp. 37–42.

[15] W.E. Deming, *Out of the Crisis*, Cambridge, Massachusetts Institute of Technology, Center for Advanced Engineering Study, 1986.

[16] Jeffrey M. Snow, "Rating quality and selecting suppliers using Taguchi loss functions," *Naval Engineers Journal*, January 1993, pp. 51–57.

[17] B. Smith, "Making war on defects," *IEEE Spectrum*, September 1993, pp. 43–50.

[18] Howard Romaker and Edward Daniszewski, "Air force vacuum electronics reliability and maintainability," *Microwave Journal*, February 1992, pp. 79–82.

[19] John A. Collins, Charles M. Eallonardo, James W. Hansen and Vernon W. Lym, "Reliability design criteria for high power tubes," Final Technical Report RADC-TR-88-304, February 1989.

[20] Eric Buck, "Investigation of long term storage impact on traveling wave tube life," Final Report, Contract Number N00024-92-C-4051.

[21] R.F. Spires, "Final engineering report for the — TWT screening analysis," Contract Number F09603-86-G-3044-0020, August 15, 1991.

Appendix A

Useful Constants and Conversions

Constants

Boltzmann constant	k	1.38×10^{-23} J/K
electronic charge	e	1.6×10^{-19} C
electron rest mass	m	9.1×10^{-31} kg
electron charge-to-mass ratio	η	1.76×10^{11} C/kg
permittivity of vacuum	ϵ_o	8.85×10^{-12} F/m
permeability of vacuum	μ_o	$4\pi \times 10^{-7}$ H/m
Planck constant	h	6.625×10^{-34} J-s
speed of light in vacuum	c	3×10^{8} m/s
hydrogen atom mass	m_h	$1836 \times$ m(electron)

Conversions

1 eV	=	1.6×10^{-19} J
1 Tesla	=	10^4 Gauss
0°C	=	273 K
room temp. (20°C) kT	=	1/40 eV

Appendix B

Glossary of Terms

AM near-carrier noise: AM noise and spurious outputs at frequencies close to the carrier frequency, usually 100 Hz to 500 kHz relative to the carrier. Components not related to the presence of the carrier (signal) are usually caused by ripple on power supply voltages. Interactions in the TWT are usually responsible for components that exist only when a carrier (signal) is present. (*See also* Near-carrier noise.)

Amplitude pulling factor: The amount of gain change experienced when the VSWR of the input source or the load is changed.

Amplitude pushing factor: The amount of gain change experienced as the voltage on a particular tube element is varied, usually expressed in dB/volt.

AM/AM conversion: An abbreviated form of the phrase *amplitude modulation to amplitude modulation conversion*, used to describe a type of distortion in microwave amplifiers. It is the change in the output RF voltage produced by variations in input signal level, usually expressed in dB/dB. An ideal amplifier has an AM/AM conversion of 1 dB/dB. As the amplifier approaches saturation, the amplification process becomes nonlinear and gain compression occurs. At saturation, the AM/AM conversion is 0 dB/dB.

AM/PM conversion: An abbreviated form of the phrase *amplitude modulation to phase modulation conversion*, used to describe a type of distortion in microwave amplifiers. It is the change in phase angle of the output RF voltage produced by variations in input signal level, usually expressed in degrees/dB. An ideal amplifier has an AM/PM conversion of 0 degrees/dB.

AM/PM transfer: An abbreviated form of the phrase *amplitude modulation to phase modulation transfer* used to describe a type of distortion in microwave amplifiers. It is the change in phase angle of the output RF voltage on one signal produced by variations in input signal level on a signal at another frequency, usually expressed in degrees/dB. An ideal amplifier has an AM/PM transfer of 0 degrees/dB.

***b*:** Pierce's velocity parameter, b is a measure of actual beam velocity relative to synchronous beam velocity.

Backstreaming: (1) A condition in which electrons reflected from the collector or secondary electrons from the collector travel "backward" toward the electron gun.

This is an undesirable effect, distorting the primary electron beam and any modulation that may be present. (2) A condition in which gas particles from a vacuum pump travel "backward" into the chamber being pumped.

Backward wave oscillator: Abbreviated BWO, a wideband voltage tunable oscillator related to a traveling wave tube in somewhat the same way that a klystron oscillator is related to a klystron amplifier. It uses a broadband circuit similar to a TWT. Regeneration exists because the backward wave mode on the circuit propagates RF energy in a direction opposite to that of the electron stream. Oscillation occurs when the loop gain is equal to or greater than unity. High power TWTs can experience undesired backward wave oscillations when the backward wave mode has sufficient regeneration.

Beam efficiency (also called RF efficiency): The RF output power divided by the beam power.

Beam power: The product of cathode voltage and cathode current.

Body current: The part of the electron current from the gun that does not arrive at the collector. This current is intercepted by the RF circuit and other electrodes such as the anode in the electron gun.

Body-helix protection: A combination of circuit elements in the TWT power supply, ranging from a simple resistive network to a complex "crowbar" device. The circuitry is designed to prevent damage to the TWT as the result of arcing or unusually high interception current.

Brillouin field: *See* Brillouin focusing.

Brillouin flow: Electron flow resulting from the use of Brillouin focusing.

Brillouin focusing: A magnetic focusing scheme for electron beams in klystrons and TWTs in which the magnetic field that provides focusing is parallel to the direction of electron flow and the electron gun is shielded from magnetic field. The amount of magnetic field to focus a nonthermal beam is called the *Brillouin field* (a nonthermal beam is a beam that contains none of the random components of velocity that result from emission from a hot cathode). In practice, the presence of thermal velocities requires that the actual field be somewhat greater than the Brillouin value. (*See also* Confined flow, PPM.)

Bunching: The result of the interaction of an electron beam with an RF field. Electrons which are accelerated by the field move toward those that have been decelerated, forming a high density region usually referred to as a *bunch*.

BWO: See Backward wave oscillator.

***C*:** Pierce's gain parameter, C is a measure of the effectiveness of the circuit in a TWT in interacting with the electron beam.

Cathode loading: The current density at the emitting surface of the cathode, usually expressed in amperes per square centimeter.

Cavity: A microwave resonant circuit, usually consisting of a hollow metallic enclosure in which RF fields can be established at one or more frequencies.

Centipede structure: A fundamental forward wave coupled cavity structure. Adjacent cavities are coupled by a multiplicity of dual coupling loops.

Charge density: The amount of electric charge contained in a unit of volume, usually expressed in coulombs per cubic meter.

Circuit efficiency: The efficiency with which the slow wave circuit of a microwave tube delivers the RF power generated in the tube to the output connector.

Cloverleaf structure: A fundamental forward wave coupled cavity structure. Each cavity has the shape of a four-leaf clover. Adjacent cavities are magnetically coupled by radial slots and are rotated 45°.

Cold match: The input or output RF impedance match of a TWT obtained when the TWT is not operating (the TWT is electrically "cold").

Collector: The electrode that collects the spent electron beam after it has been used to provide power for microwave amplification or oscillation. The kinetic energy in the spent electron beam is converted to thermal energy (heat) in the collector or is recovered and returned to the power supply by the collector.

Collector efficiency: The efficiency with which the collector recovers power from the spent beam.

Confined flow: A type of beam focusing similar to Brillouin flow except that a small amount of magnetic flux is caused to thread the cathode. More field is required than for Brillouin flow, but the beam is less affected by RF bunching (less scalloping is introduced when RF is applied to the tube and body current is affected less by the application of RF drive). Confined flow is sometimes referred to as *immersed flow*.

Contrawound helix: A helix slow wave structure where two helices, wound in opposite directions, are superimposed into a single structure. This circuit offers substantially higher power than a conventional helix with some sacrifice of bandwidth.

Control grid: An electrode, usually a grid like structure, mounted between the cathode and the anode to control the flow of electrons. An appropriate negative voltage (with respect to the cathode) reduces the electron flow (beam current) to zero or cutoff and an appropriate positive (or, in some cases, less negative) voltage allows current to flow. This electrode is usually some sort of mesh structure.

Coupled cavity tube: A TWT with a slow-wave structure made up of a number of "cavities" electrically coupled by means of coupling holes, slots or loops. This circuit is capable of very high power operation.

Crossed field device: A microwave tube in which the electric field between the cathode and anode is perpendicular to the magnetic field. The resulting average motion of the electrons emitted from the cathode is nearly perpendicular to both the electric and magnetic fields and parallel to the direction of propagation of energy on a nearby slow wave circuit.

Current density: The current per unit area. For electrons in an electron beam, the current density is the beam current divided by the cross-sectional area of the beam. The current density at the cathode surface (cathode loading) is equal to the current divided by the surface area of the cathode and is usually expressed in amperes per square centimeter.

d: Pierce's loss parameter, *d* is a measure of the reduction in gain that occurs in a TWT as the result of circuit loss.

dBc: The level of a spurious output relative to the main signal (carrier) expressed in dB below the carrier power level.

dBm: Power level in dB above a milliwatt; 60 dBm is a power level of 60 dB above a milliwatt and is a kilowatt.

dBW: Power level in dB above a watt; 60 dBW is a power level of 60 dB above a watt and is a megawatt.

Density modulation: The process of controlling the density of electrons emitted from a cathode, usually, by a grid or other control electrode located near the cathode.

Depressed collector: Collector depression is the process of applying a negative potential (with respect to the tube body or "ground") to the collector of a microwave tube to reduce the velocity of the electrons entering the collector. This reduces the kinetic energy in the electrons. Less energy is converted to heat and X rays when the beam impinges on the internal surfaces of the collector. The energy recovered from the electrons is returned to the power supply and the result is that the conversion efficiency of the device is greatly improved. A collector designed to operate in this mode is called a *depressed collector*. Depressed collectors can have one or more depressed stages. Multiple-stage depressed collectors employ velocity sorting techniques to direct the high velocity electrons to the stages having the greatest depression (the greatest amount of retarding field) and the slow electrons to the stages having the least depression.

Dispenser cathode: A cathode that has a body made of porous tungsten. The pores contain the active materials that enhance electron emission. When the cathode is heated to its operating temperature, the material in the pores migrates toward the emitting surface, continuously replenishing the supply of active material at the emitting surface.

Dispersion: A term used to describe the variation in phase velocity of an RF wave with frequency. A non dispersive circuit propagates RF energy at a phase velocity that is independent of frequency.

Drift tube: A section of metal tubing in which an electron beam "drifts" without being accelerated or decelerated by applied electric fields.

Drive: A term to indicate the RF input or RF signal to an electronic device.

Dual mode: Usually refers to devices capable of operating efficiently in both a low power CW mode and a high power pulse mode.

Duty cycle: The fraction of the time during which a device is operating. The term *duty cycle* is usually used with pulsed operation and is the product of pulsewidth and pulse repetition frequency.

DVT (dynamic velocity taper): A variation in the pitch of the helix of a TWT to improve efficiency and linearity. (*See also* Tapered velocity.)

Dynamic range for linear operation: The range of operation of an amplifier in which the power output is nearly proportional to the power input. The lower limit is the input drive level at which the signal input is equal to the noise input plus the noise figure. The upper limit is the input drive level that produces 1 dB or less of gain compression.

Efficiency: The RF output power divided by the sum of all power provided by the power supplies. (*See also* Overall efficiency)

Electronic efficiency: The efficiency with which dc beam power is converted to RF power at the desired frequency within a microwave tube.

Energy product: A measure of the magnetic energy produced in the gap in a permanent magnet. The energy product is the product of B and H at the operating point of the magnet.

Equalizer: A passive device providing selective loss and/or phase shift over an operating band such that the net gain and/or phase shift of the equalized amplifier matches a required profile.

EPC (electronic power conditioner): A power supply-modulator usually associated with space qualified TWTAs. In addition to providing power, EPCs commonly include protection circuits, circuits for receiving commands and circuits for providing telemetry data.

External Q: A measure of the affect on the Q of a resonant circuit by an external circuit attached to that resonant circuit. External Q is the reciprocal of the difference between the reciprocals of the loaded and the unloaded Q.

Ferrule: A drift tube tip in the cavity of a klystron or coupled cavity TWT.

Fine grain: A term used to refer to small variations in gain or phase of an amplifier.

Focus electrode: An element in the electron gun that is used to focus the electrons into a well-defined beam. It is sometimes called the *beam forming electrode* (BFE).

Folded helix circuit: A helix slow wave structure where two helices, wound in the same direction, are superimposed into a single structure. This circuit is sometimes called a *bifilar helix* and offers substantially higher power than a conventional helix with some sacrifice of bandwidth.

Folded waveguide circuit: Another name for a coupled cavity circuit.

Fundamental backward wave circuit: A slow wave circuit used in coupled cavity TWTs. The fundamental mode of the circuit is a backward wave mode. The circuit is operated in a forward space harmonic mode.

Fundamental forward wave circuit: A slow wave circuit used in high power coupled cavity TWTs. The fundamental mode of the circuit is a forward wave mode.

Gain: The ratio of output voltage, current or power to the input voltage, current or power, respectively, in an amplifier stage, receiver or system. Power gain is usually expressed in dB.

Gain compression: The decrease in gain of an amplifier between the small signal gain and the gain at saturation. Typically, gain compression is 6 to 8 dB in TWTs.

Gated emission: Emission from a cathode that is controlled (gated), usually with a grid adjacent to the cathode.

Getter: A device that, after activation, absorbs or adsorbs gases within an electron tube.

Grid: An electrode mounted between the cathode and the anode to control the flow of electrons. When used to control the amount of electron current in a TWT or klystron, the grid is called a *control grid*. The grid electrode may be a cylindrical-shaped ring or a section of wire screen or mesh.

Group delay: The distortion that results when the time delay of a signal being processed through a device is not constant as a function of frequency. Areas of interest are linear delay, parabolic delay and the ripple component. The linear component is the difference in delay between two frequencies within the specified bandwidth. Parabolic delay is determined by fitting the equation of a parabola to a plot of delay versus frequency. Ripple is the maximum peak-to-peak variation relative to a plot of the smoothed curve.

Group velocity: The velocity at which energy advances along an RF circuit. For a coaxial transmission line with no dielectric, the group velocity is the speed of light. For waveguide, the group velocity is less than the speed of light. The group velocity cannot be greater than the speed of light.

Harmonic drive: The inclusion of phase-conditioned harmonic power in the input RF signal to reduce harmonic power, harmonic capture and to improve efficiency at the low end of the operating band of a TWT.

Harmonic interaction: The effect of the harmonic content of the RF input signal on the beam modulation. This is generally undesirable and usually reduces the fundamental power output.

Heater: An electric heating element for raising an indirectly heated cathode to the operating temperature.

Hot match: The RF impedance match at the input or output of a TWT, obtained when the TWT is operating (electrically "hot").

Hub: *See* ferrule.

Immersed flow: *See* Confined flow.

In-line slot structure: A fundamental backward wave coupled cavity structure in which adjacent cavities are electromagnetically coupled by two slots spaced 180° apart in the wall between the cavities. The slots in the walls of adjacent cavity are aligned.

Insertion loss: The reduction in signal strength that results when a passive device is installed (inserted) into an RF transmission line. Insertion loss is caused by power being absorbed in the device and by reflections from discontinuities at the interfaces between the device and the transmission line. Usually expressed in dB.

Intercepting grid: A control grid that is not shielded from electrons emitted by the cathode. The grid intercepts approximately 10 to 15% of the emission current when a positive potential is applied.

Intercept point: On a plot of intermodulation distortion data as a function of RF input drive, the carrier data at small signal drive levels has a slope of 1:1, third order intermodulation products have a slope of 3:1 and fifth order intermodulation products have a slope of 5:1. Extrapolating the slopes for the carriers and the third order products produces an intercept point. The output power at this point is called the *third order intercept point*. Similarly, the fifth order intercept point is the point at which the carrier slope intercepts the fifth order product slope. (*See also* Intermodulation distortion.)

Interdigital line: A slow wave circuit composed of a comb like structure with alternate segments being connected together at one end and remaining segments connected together at the opposite end.

Intermodulation distortion: Distortion resulting from the production of frequencies that are the sum of, and the difference between, frequencies contained in the applied waveform.

Intrinsic efficiency: Efficiency calculated by dividing the RF output power from a microwave tube by the sum of all power provided with the exception of heater power and that provided to the solenoid for focusing the beam.

Ion oscillations: Fluctuations in microwave tube operation resulting from ion effects in the electron beam. These fluctuations are commonly observed as variations in RF power output or as RF signals removed from the carrier frequency by the frequency of ion oscillations in the electron beam.

Iridium: A transition metal sometimes used with dispenser cathodes for enhancing emission.

Klystron: A velocity modulated microwave tube that uses the interaction between an electron beam and the RF energy in cavities to provide signal amplification. In a klystron, the RF interaction takes place at discrete locations along the electron beam. Common types of klystrons are the reflex klystron (an oscillator having only one cavity), two-cavity klystron amplifiers and oscillators and multicavity klystron amplifiers.

Leakage current: Undesirable flow of current between electrodes often resulting from conductive paths through or over the surface of an insulator or from unwanted electron emission from electrode surfaces.

Linear beam tube: A microwave tube (usually a klystron or TWT) in which a long cylindrical electron beam is used.

Linearizer: A device that improves the linearity of the AM/AM conversion curve of an amplifier (causing the slope of power output versus power input to be nearly 1 dB/dB over a wide range of drive levels). Linearizers are usually designed to also reduce the AM/PM conversion (causing the phase shift to remain nearly constant over a wide range of drive levels). The use of a linearizer reduces the intermodulation distortion.

Loaded Q: The Q of a resonant circuit when there is external coupling to the circuit.

Loss buttons: In coupled cavity TWTs, lossy elements used to provide frequency selective loss to inhibit the excitation of higher order modes.

Magnetron: A crossed field microwave oscillator tube containing concentric cylinders; the inner cylinder is the cathode and the outer cylinder is an anode that contains resonant cavities. A strong axial magnetic field causes a cloud of electrons to orbit between the cathode and the anode. The RF voltages across gaps in the resonators modulate the velocities of the electrons. This causes the orbiting electrons to form into "spokes" that rotate around the tube axis. As the spokes of electrons rotate past the resonator gaps, they induce currents that shock excite the cavities. The RF voltages build up to large levels. High power output is obtained at moderately high efficiency.

Metal-ceramic: A term applied to tube assemblies that employ no glass as a part of their vacuum walls. Electrical isolation between elements is achieved by using ceramic insulators. The assembly is integrated by using braze joints between ceramic and metal parts.

Micron: A measure of pressure. One micron is 10^{-3} torr.

Mode interference: *See* Interfering mode.

M-type tube: A crossed field tube.

Multimode: Having the capability of operating efficiently at more than one power level. (*See also* Dual mode.)

Multioctave: Capable of operating satisfactorily over a frequency range of two or more octaves.

Multipactor: A term to denote an electron-RF field interaction in which electrons oscillate back and forth in synchronism with an RF voltage applied between electrodes. Secondary electrons produced when primary electrons strike the electrodes cause the number of oscillating electrons to increase with time. In microwave tubes, multipactor is an undesirable effect and can occur across cavity or drift tube gaps, in output waveguides or on the surface of the ceramic output window. It is detected by observing the output power as the signal level is increased. Multipactor is indicated by the presence of a hysteresis loop as a function of signal level. Increasing the signal level produces a sudden drop in the RF power output followed by low output even when the drive to the amplifier is decreased below the point of initial onset. In high power tubes, there will be heating of the surface or surfaces involved.

Multi stage collector: A collector with several segments, each successive segment being "depressed" more than the preceding segment. This enhances the efficiency of the collector in recovering power from the spent electron beam.

Near-carrier noise: AM noise, PM noise and spurious outputs at frequencies close to the carrier frequency, usually 100 Hz to 500 kHz relative to the carrier. Components not related to the presence of the carrier (signal) are usually caused by ripple on power supply voltages. Interactions in the TWT are usually responsible for components that exist only when a carrier (signal) is present. Special measurement techniques are usually required to measure near-carrier noise because the resolution bandwidth (IF amplifier bandwidth) and local oscillator stability of standard spectrum analyzers are not adequate to provide accurate test data.

Noise figure: The ratio of the signal-to-noise on the output of a device to the signal to noise on the input. It is important because it indicates the amount of noise the amplifier contributes to the signal, and it is an absolute indicator of the sensitivity of the device. Noise figure is usually expressed in dB, and is abbreviated NF or F.

Noise power: The noise generated by a device or amplifier when measured at the output port when the input port is terminated and no RF drive is applied.

Noise power ratio (NPR): A measure of intermodulation when an infinite number of carriers is being amplified. The input to the amplifier is a well-defined bandwidth of random noise with a deep notch in the center. Intermodulation in the amplifier produces intermodulation products that partially fill the notch. The ratio of noise power in the passband to the noise power in the notch, measured at the output of the amplifier, is the NPR.

Nonintercepting grid: A control grid that is mechanically shielded from the cathode by a "shadow grid" in such a way that few electrons are intercepted by the control grid, even when a positive voltage is applied to the control grid (with respect to cathode).

Octave: The interval between two frequencies having a ratio of 2:1.

Osmium: A transition metal sometimes used with dispenser cathodes for enhancing emission.

Outgassing: A term used to describe the release of various gasses from the internal surfaces of a vacuum device during processing and testing.

O-type tube: A linear beam tube.

Overall efficiency: The ratio of RF output power from a microwave tube to the total of all power inputs to the tube.

Overdrive: An input signal level greater than that required for saturation, resulting in decreasing output power.

Oxide coated cathode: A cathode with an emissive surface consisting of oxides of alkaline-earth metals (usually barium, calcium and strontium).

Penultimate cavity: The next to the last cavity of a multi cavity klystron.

Perveance: A numerical constant relating current and voltage in a space charge limited diode. The relation for perveance, usually denoted by P, is

$$P = \frac{V^{3/2}}{I}$$

Perveance depends only on the geometry of the diode. When the voltage is the cathode-to-anode voltage in the electron gun, the perveance is called the *gun perveance*. When the voltage is the cathode to-circuit voltage, the perveance is called the *beam perveance*.

Phase linearity: A term referring to the degree of deviation from a straight line of the phase *versus* frequency characteristic of a device.

Phase pulling factor: The amount by which a change in input or output VSWR changes the phase shift in the TWT.

Phase pushing factor: The amount by which a change in the voltage on an electrode changes the phase shift in the TWT, usually expressed in degrees/volt.

Phase tracking: The closeness or similarity of the phase characteristics of a number of devices. This is an important consideration when power combining the outputs from two or more devices.

Phase velocity: The velocity at which phase fronts for an electromagnetic wave propagate. For a coaxial transmission line with no dielectric, the magnitude of the phase velocity is the speed of light. For waveguide, the magnitude of the phase velocity is greater than the speed of light. Slow wave circuits are designed to provide a phase velocity having a magnitude much less than the speed of light.

PM near-carrier noise: PM (phase modulation) noise and spurious output at frequencies close to the carrier frequency, usually 100 Hz to 500 kHz relative to the carrier. Components not related to the presence of the carrier (signal) are usually caused by ripple on power supply voltages. Interactions in the TWT are usually responsible for components that exist only when a carrier (signal) is present. (*See also* Near-carrier noise.)

Polepiece: A section of ferromagnetic material used to shape the magnetic flux distribution from a solenoid (electromagnet) or permanent magnet.

Power combining: A scheme whereby the outputs from two or more amplifiers are combined to provide a greater output power than available from a single amplifier.

Power curve: A plot of output power versus input drive or input frequency. (For power curve *versus* input drive, *see also* AM/AM conversion.)

PPM (periodic permanent magnet) focusing: A method of focusing a TWT where permanent magnets of opposite polarity are placed side by side along the length of the tube.

Pulse-up ratio: The ratio, usually expressed in dB, between the cw power level and the pulse-power level in a dual mode device.

***QC*:** Pierce's space charge parameter, *QC* is a measure of the strength of space charge forces causing debunching relative to the strength of the circuit field that is bunching the beam.

Reike diagram: A Smith chart based load diagram for microwave oscillators, particularly klystrons and magnetrons.

Residual gas: The gas remaining in an evacuated vacuum tube.

RF efficiency (also called beam efficiency): The RF output power divided by the beam power.

Rhenium: A transition metal sometimes used with dispenser cathodes for enhancing emission.

Ring-bar tube: A TWT with a slow wave structure composed of ringlike segments connected by straight straps or "bars." This device is capable of higher power levels than a conventional helix tube at a significant reduction in bandwidth.

Ring-loop tube: Similar to a ring-bar tube, except that the bars connecting the ringlike segments are replaced by curved straps.

Samarium cobalt: A high energy product magnet material commonly used on microwave tubes.

Saturated gain: The gain of an amplifier operating at saturation.

Saturated power output: The power output of an amplifier operating at saturation.

Saturation: A term used to describe that point on the RF power output *versus* RF power input characteristic where an increase in input power does not produce an increase in output power.

Screen grid: A grid structure placed between the control grid and anode to reduce the capacitive coupling between those two electrodes. In a multimode electron gun, it is the element used to control beam current from the edge of the cathode.

Secondary emission: Emission of electrons resulting from the bombardment of a surface by high energy electrons.

Serrodyne: An operating mode for a TWT in which the input signal is translated to a new frequency at the output. A sawtooth voltage is superimposed on the dc voltage between the cathode and the slow wave circuit, causing the phase shift in the TWT to change in a linear fashion as a function of time. The amplitude of the sawtooth voltage can be adjusted to provide 360° of phase shift during each cycle, causing the output of the TWT to be offset by the frequency of the sawtooth waveform (the "flyback" time of the sawtooth must be very short so the TWT "thinks" that the phase is being offset continuously).

Shadow grid: A grid structure placed between the cathode and control grid and electrically connected to the cathode. This element shields the control grid from interception.

Shot noise: Noise power generation in a vacuum tube resulting from the random nature of the emission of electrons.

Single-stage collector: A TWT or klystron collector that has only one element for collecting electrons. The stage may be operated at ground potential or may be depressed. (*See also* Depressed collector, Multistage collector.)

Slow-wave circuit: Any structure that "slows" the phase velocity of an RF wave below the velocity of light to establish synchronism or near synchronism between the velocities of the wave and an electron beam.

Small signal gain: The gain of an amplifier operating at small signal levels.

Space charge: The electrical charge of electrons in a vacuum environment (as opposed to electrons within the crystal structure of a metal).

Space charge limited operation: Operation of a cathode at an emission density high enough so that the cathode current is limited by the space charge field of the electron cloud near the cathode surface.

Spent electron beam: The electron beam exiting from the interaction region of a linear beam tube.

Spherical diode: A diode constructed to duplicate the spherical geometry of a typical electron gun. Used for the evaluation and analysis of electron emission, analysis of electron gun designs and for realistic cathode life testing.

Stagger tuning: A method of tuning the cavities of a klystron, also called *broadband tuning*. One or more of the cavities is tuned to a frequency(s) above and one or more is tuned to a frequency(s) below the center of the frequency band to provide approximately uniform amplification over a band of frequencies.

Staggered-slot structure: A fundamental backward wave coupled cavity structure in which adjacent cavities are electromagnetically coupled by one opening or "slot" in the wall between the cavities. The slots are rotated 180° from cavity wall to cavity wall.

Synchronous tuning: The tuning of all cavities of a klystron to the same frequency.

Synchronous voltage: The beam voltage in a TWT at which the electron beam has the same velocity as the phase velocity of the RF wave on the circuit.

Tapered termination: A gradual increase in the amount of loss applied to a slow wave structure to control reflections within a TWT.

Tapered velocity: A change in the pitch of a helix, the height of cavities, or some other axial dimensions of a slow wave circuit to change the phase velocity of the RF wave. This is done because the extraction of kinetic energy from the electron beam causes the beam to slow down near the output of the TWT. The taper helps to maintain synchronism between the RF wave and the electron beam, which improves the efficiency of the TWT. (*See also* DVT.)

Temperature limited operation: Operation of a cathode under conditions such that cathode current depends on cathode temperature.

Thermionic emission: The emission of electrons from a surface as the result of heating of the surface.

Tetrode: A vacuum tube having four elements, usually a cathode, control grid, screen grid, and anode. In multi mode TWTs, *tetrode* is a term describing the electron gun (cathode, shadow grid, control grid and screen grid).

Torr: A unit of pressure. Atmospheric pressure is 760 torr. Most linear beam microwave tubes operate at pressures below 10^{-7} torr.

Transfer curve: A curve or family of curves for which output power is plotted as a function of input drive power at one or more beam voltages. Sometimes referred to as a *gain curve*. (*See also* AM/AM transfer.)

Triode: A three element vacuum tube having a cathode, a control grid and an anode.

TWT (traveling wave tube): A microwave tube using a broadband circuit in which an electron beam interacts continuously with a guided electromagnetic field to amplify microwave frequencies.

TWTA (traveling wave tube amplifier): A combination of a power supply, a modulator (for pulsed systems) and a traveling wave tube, often packaged in a common enclosure.

Twystron: A high power microwave tube having an input circuit consisting of klystron cavities and a cloverleaf coupled cavity structure for the output circuit. A Twystron has a nearly flat power output characteristic over a bandwidth as high as 10% or more.

Vacuum envelope: Any structure containing or capable of containing a high vacuum environment. The "body" structure of a microwave tube is usually referred to as the *vacuum envelope*.

Velocity jump: A method of maintaining the proper relationship between beam velocity and circuit phase velocity near the output of a TWT, whereby the beam velocity is caused to increase (as opposed to causing the circuit phase velocity to decrease). The increase in beam velocity is obtained by operating the output portion of the slow wave circuit at a voltage that is positive with respect to the remainder of the circuit.

Velocity modulation: The modulation of the velocities of electrons in a microwave tube as a result of their interaction with RF electric fields.

Velocity resynchronization: Any method for changing the axial velocity of an RF wave or of an electron beam to improve the synchronism between that wave and an electron beam. (*See also* Tapered velocity, Velocity jump.)

Velocity step: *See* Tapered velocity.

Velocity taper: *See* Tapered velocity.

Work function: The minimum energy, usually expressed in electron volts (eV), required for an electron to be thermionically emitted from a cathode.

Appendix C

Vacuum Technology

If a microwave tube is to operate properly, a high or ultra high vacuum must be maintained throughout its life. As a result, special techniques, processes, and materials must be used. The emphasis in this appendix is on vacuum technology. Subjects that will be discussed are

- Units of measurement,
- Ranges of operation,
- Vacuum pumps,
- Vacuum gauges,
- Microwave tube materials,
- Fabrication techniques.
- Vacuum leaks

Several excellent books on vacuum technology are listed with the references at the end of this appendix [1].

C.1 UNITS OF MEASUREMENT

Several years ago, the most commonly used unit of pressure was the millimeter of mercury (mmHg). Atmospheric pressure is 760 mmHg. Now, the accepted units of pressure are the torr and the newton/m^2 (pascal). The relation between these units is

$$1 \text{ torr} \approx 1 \text{ mmHg} \approx 1.33 \times 10^2 \text{ pascal}$$

Another unit of pressure sometimes used at relatively low vacuum (high pressure) levels is the micron:

$$1 \text{ micron} = 10^{-6} \text{ meters of Hg} = 10^{-3} \text{ mmHg} \approx 10^{-3} \text{ torr}$$

Still another unit of pressure used in European literature is the mbar (millibar):

$$1.33 \text{ mbar} = 1 \text{ torr}$$

In microwave tube work, it is rarely and perhaps never necessary to know the pressure to an accuracy better than a factor of two or so. As a result, it is safe to remember that

$$1 \text{ torr} \approx 1 \text{ mmHg} \approx 1 \text{ mbar}$$

C.2 RANGES OF OPERATION

In evacuating a microwave tube, the pressure changes from atmospheric to a value as low as 10^{-8} to 10^{-9} torr. Thus, the pressure is reduced by over eleven orders of magnitude. It is common to divide this very wide pressure variation into four ranges:

760 torr to ~ 1 torr	- Rough vacuum
1 torr to $\sim 10^{-3}$ torr	- Medium vacuum
10^{-3} torr to $\sim 10^{-7}$ torr	- High vacuum
$< 10^{-7}$ torr	- Ultrahigh vacuum

In most electron tubes, the pressure is at the low end of the high vacuum range or the high end of the ultrahigh vacuum range; i.e., from 10^{-7} to 10^{-9} torr. In some cases, for example, some high power klystrons that have been in operation for more than 50,000 hrs, the pressure is even lower.

At atmospheric pressure, the gas density is about 2.7×10^{19} molecules/cm^3. As shown in Figure C-1, density drops linearly with decreasing pressure and, at 10^{-8} torr, is about 3×10^8 molecules/cm^3 . While this density appears to be high, 3×10^8 molecules will form a monolayer on an area of only 10^{-6} cm^2.

The time required to form a monolayer is also shown in Figure C-1. At a pressure of 10^{-8} torr, a monolayer is formed in a time of only about 3 min.

Mean free path length is the mean distance traveled by a molecule between collisions. Note in Figure C-1 that, at 10^{-8} torr, the mean free path length is $\sim 10^4$ m. Thus, the average molecule could pass through a 1-m vacuum chamber 10,000 times without colliding with another molecule.

At 1 torr, on the other hand, the mean free path length is only $\sim 10^{-4}$ m. Thus, molecules collide with each other far more frequently than they collide with the walls of a vacuum system.

At relatively high pressures, where molecules collide with each other far more frequently than with vacuum chamber walls, gas flow is referred to as viscous flow.

At low pressures, where molecules collide with vacuum system walls more

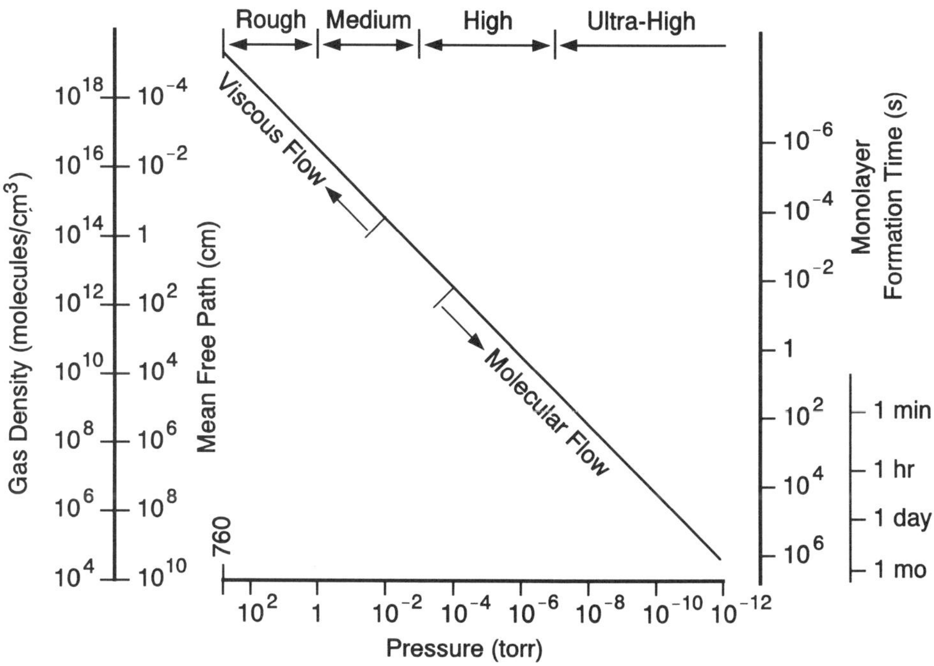

Figure C-1 Relationship of several quantities defining the degree of vacuum.

frequently than with each other, gas flow is referred to as molecular flow. In the viscous flow and molecular flow regions, the analysis of gas flow is relatively straightforward. In the pressure range from $\sim 10^{-2}$ torr to 10^{-4} torr, a transition from viscous to molecular flow occurs, and gas flow behavior is difficult to analyze.

The effects of viscous and molecular flow in vacuum system design are illustrated in Figure C-2. In the viscous flow region, molecules gradually migrate in the direction of the pump because of the change in pressure, ΔP. With molecular flow, molecules collide only with the walls of the chamber and only occasionally are directed toward the pump orifice. As a result, a large pump orifice is required if a high pump speed is to be achieved. For example, if 1-in (2.5 cm) tubulation were used to connect the vacuum system to the pump, the speed would be limited to a few liters per second. If a 6-in (15 cm) diameter orifice were used, the speed could be well over 1000 liters per second.

C.3 VACUUM PUMPS

Vacuum pumps may be divided into two major categories: gas transfer and entrapment. In a gas transfer pump, the gas molecules are forced to move *through* the pump. In an

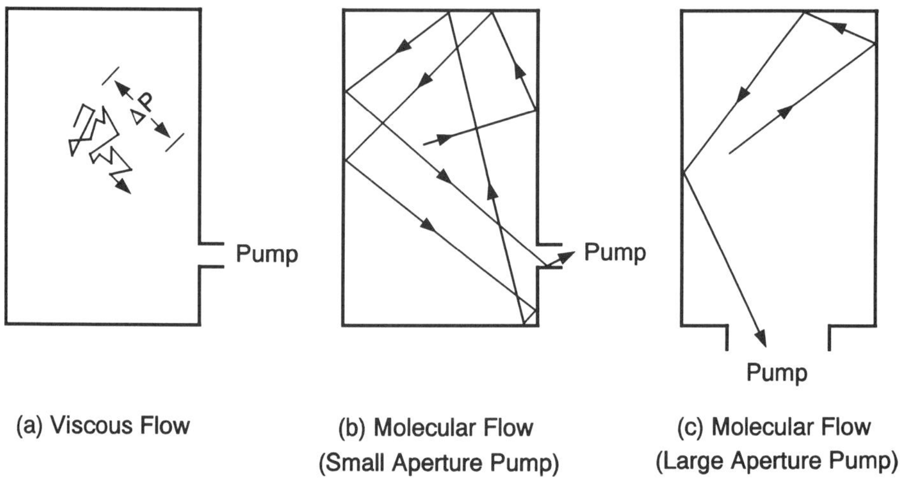

Figure C-2 Comparison of viscous and molecular flow.

entrapment pump, gas molecules combine chemically with or are adsorbed by materials within the pump and are "trapped" by the pump. The pumps most commonly used in the fabrication and operation of microwave tubes are

Gas Transfer

1. Sliding vane rotary pump	Rough and medium vacuum
2. Diffusion pump	High and ultrahigh vacuum
3. Turbomolecular pump	High and ultrahigh vacuum

Entrapment

1. Adsorption pump	Rough and medium vacuum
2. Ion pump	High and ultrahigh vacuum
3. Getter	High and ultrahigh vacuum

In the gas transfer pump category, the diffusion pump and the turbomolecular pump require an exhaust pressure in the medium vacuum range. The sliding vane rotary pump is used in conjunction with the diffusion and turbomolecular pumps to maintain this medium vacuum level. The rotary pump used in this way is frequently referred to as a *backing pump*.

In the entrapment pump category, the ion pump requires a pressure in the low end of the medium vacuum range before it can be started. An adsorption pump is normally used to generate this medium vacuum level. Once an ion pump has been started, it can operate without assistance from other pumps. Getters have been used since the invention of vacuum tubes to maintain the vacuum level after fabrication has been completed. In microwave tubes, nonevaporable getters are often used.

C.3.1 Sliding Vane Rotary Pump

The essential elements of a sliding vane rotary pump are illustrated in Figure C-3. A rotating cylinder containing spring loaded vanes transfers gas from the inlet port to the output port. A high grade vacuum oil is used in the assembly containing the rotating cylinder and vanes. The oil serves as a lubricant and also provides a vacuum seal between moving parts.

A problem with mechanical pumps is the pumping of condensable vapors such as water. If these vapors condense during the compression cycle of pump operation, they may emulsify with the pump oil and cause deterioration of the lubricating properties. Also, the ultimate pressure capability of the pump is reduced because the vapors are released from the pump oil in the inlet side of the pump. This problem may be avoided with the use of a gas ballast device that operates during the compression cycle.

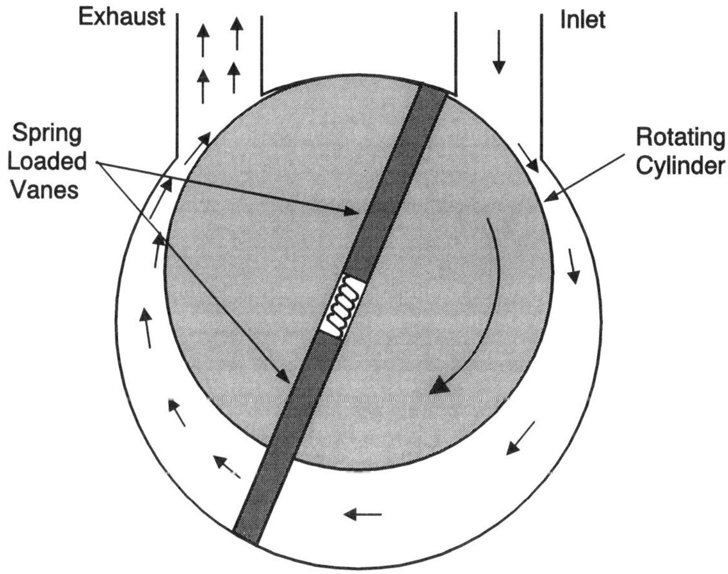

Figure C-3 Essential elements of a sliding vane rotary pump.

With a gas ballast additional air additional air is injected into the pump before compression. With the additional air, a discharge valve is forced open causing gas and vapor to escape before condensation can occur.

Pumping speed curves for sliding vane rotary pumps are shown in Figure C-4. The

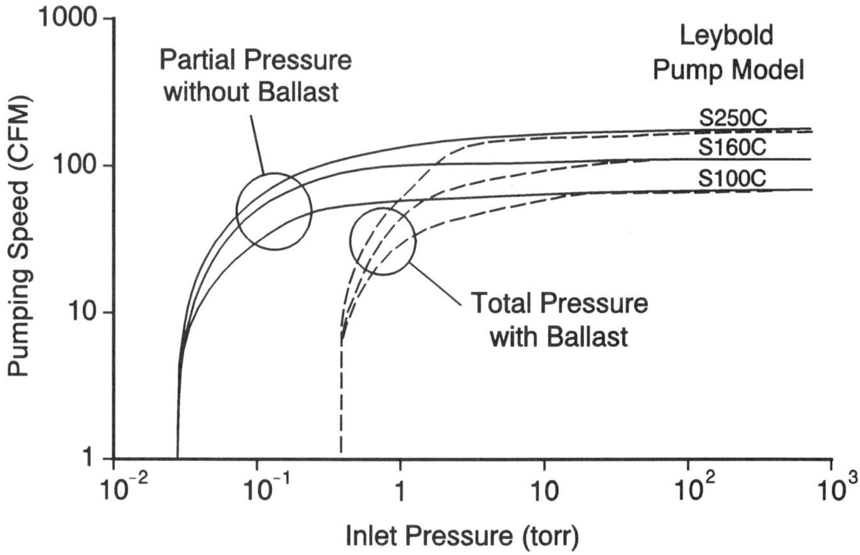

Figure C-4 Pumping speeds of sliding vane rotary pumps. Adapted from Product and Vacuum Technology Reference Book by Leybold Heraeus, Inc.

ultimate pressure is somewhat above 10^{-2} torr. With gas ballast operation, the ultimate pressure is about an order of magnitude higher.

C.3.2 Diffusion Pump

A diffusion pump is normally used in conjunction with a rotary pump to achieve pressures in the high or ultra high vacuum range. An oil-filled diffusion pump is shown in Figure C-5.

A very high quality silicon oil is vaporized by a heating element in the bottom of the pump. The oil vapors are formed into jets by a chimney structure. Collisions of oil molecules with gas molecules cause the gas molecules to diffuse toward the exhaust port. A pressure difference (between inlet and exhaust) of approximately six orders of magnitude can be produced with a diffusion pump, which is normally water cooled. The condensed oil returns to the oil bath in the bottom of the pump.

Typical pumping speed and throughput for a diffusion pump are shown in Figure C-6. Note that speed is constant over a wide range of pressures.

At very low pressures, pressure is limited by diffusion of oil molecules back into the chamber being pumped. This is frequently referred to as *backstreaming*.

A baffle is sometimes used between the pump and the chamber to limit backstreaming. Metallic elements are arranged as indicated in Figure C-7 so that there is no line-of-sight path through the baffle. Baffles are normally cooled, sometimes by

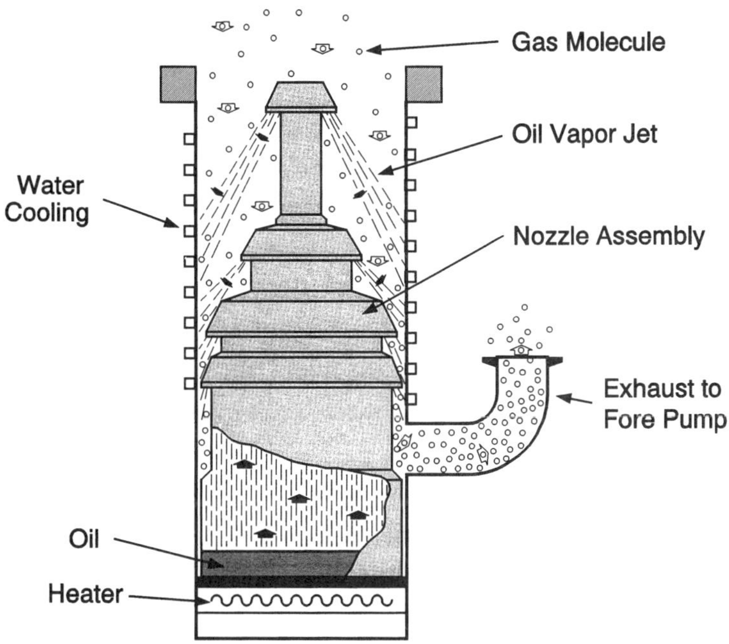

Figure C-5 Oil-filled diffusion pump. Adapted from Product and Vacuum Technology Reference Book by Leybold Heraeus, Inc.

liquid nitrogen, to enhance the condensation of oil vapors. With a baffle it is sometimes possible to reduce the ultimate pressure by over two orders of magnitude.

C.3.3 Turbomolecular Pump

In the turbomolecular pump, a rotor, spinning at a high speed, imparts a tangential velocity to impinging molecules. As indicated in Figure C-8, many rotor discs are used. The gas channels for the discs are connected in series, so that, even though the pressure gradient generated by each disc is small, the pump is capable of maintaining a high pressure ratio. At 16,000 rpm, the ratio of outlet to inlet pressure is 250:1 for hydrogen and 10^6:1 for air. The recommended fore pressure is 10^{-2} torr and the ultimate pressure is 10^{-8} to 10^{-9} torr.

C.3.4 Adsorption Pump

An adsorption pump is capable of evacuating a chamber from atmospheric pressure to the medium vacuum range. A typical pump is shown in Figure C-9. The operating

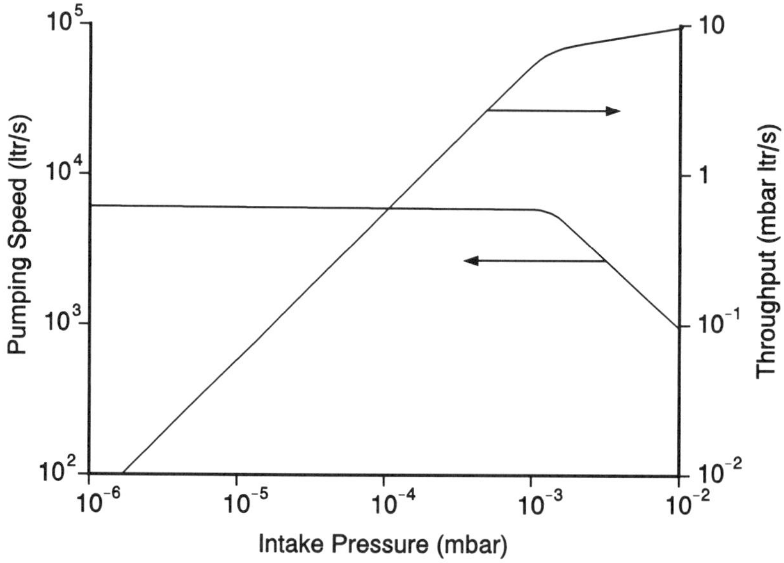

Figure C-6 Typical pumping speed and throughput for a diffusion pump.

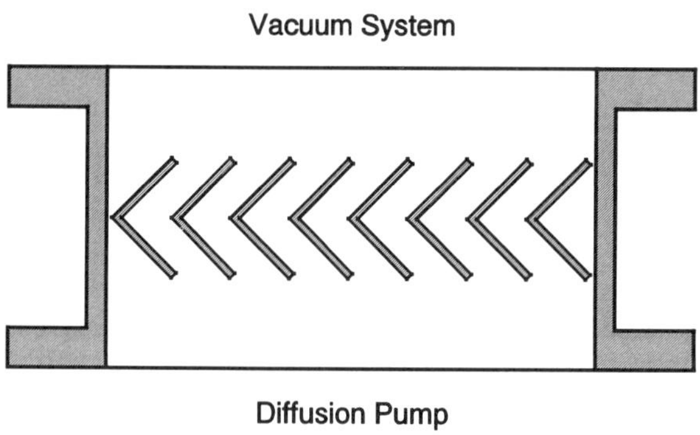

Figure C-7 Chevron-type baffle.

principle is the physical adsorption of gases on the surfaces of molecular sieves. Zeolite, an alkali alumino-silicate, is frequently used. The surface area of the pores is about 10^3 m^2 per gram. The pore diameter of zeolite 13X (about 13 Å) is approximately the size of water vapor, oil vapor and large gas molecules. Large molecules are, therefore, readily adsorbed. Small molecules such as hydrogen, helium and neon are very poorly adsorbed.

The adsorption of gases is dependent on pressure and temperature as shown in Figure C-10. Note that the difference between the quantity of nitrogen adsorbed and the quantity of neon or helium adsorbed is at least four orders of magnitude.

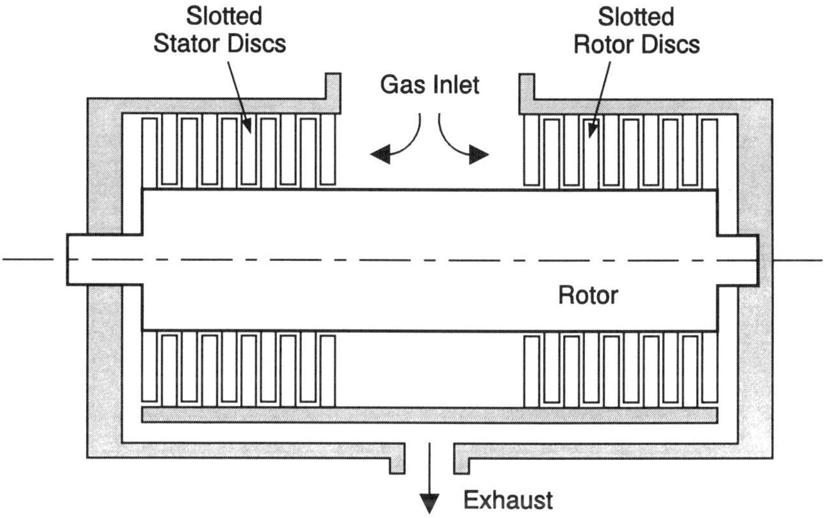

Figure C-8 Turbomolecular pump.

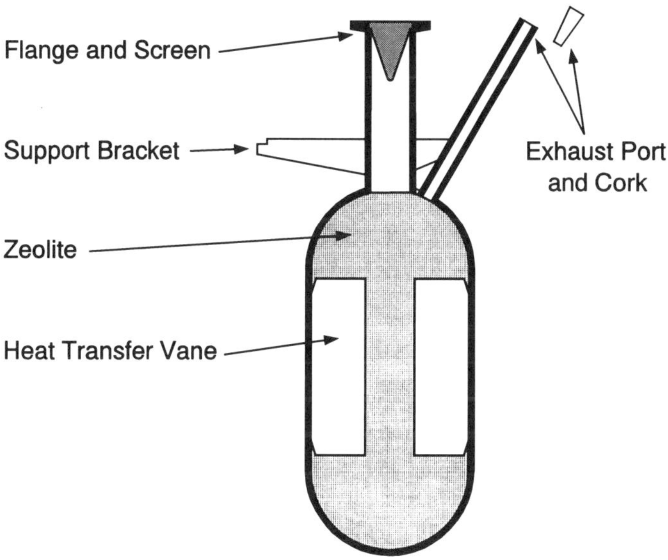

Figure C-9 Typical adsorption pump.

In practice, hydrogen and helium are normally present in the gases to be evacuated from a microwave tube. This is because a hydrogen atmosphere is used during brazing processes and an inert gas atmosphere containing helium is frequently used for inert gas welding. Since hydrogen and helium are normally present, it is necessary to be cautious in using an adsorption pump. Because the pump operates mostly in the viscous

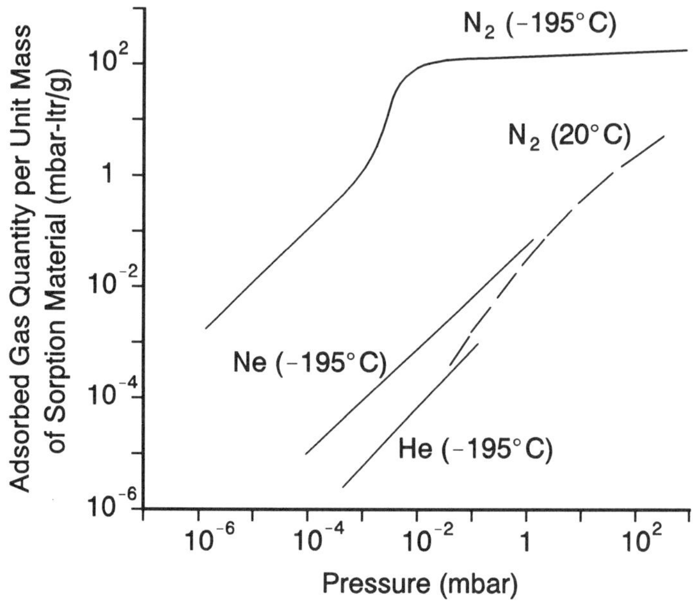

Figure C-10 Adsorption isotherms of zeolite 13X for nitrogen, neon, and helium. Adapted from Product and Vacuum Technology Reference Book by Leybold Heraeus, Inc.

flow range, hydrogen and helium are dragged along with the gases readily pumped by the zeolite. As long as the pumping process takes place fairly quickly, hydrogen and helium can be pulled into the adsorption pump, and then a valve can be closed before the hydrogen and helium diffuse back into the chamber being pumped.

Adsorption pumps are immersed in liquid nitrogen to achieve large pumping capacities. To rejuvenate zeolite after pumping, the pump is warmed to room temperature. Gas is released from the zeolite and escapes from the pump through an exhaust port. If a great deal of water vapor or other condensibles has been pumped, it may be necessary to bake the zeolite for a few hours at 200°C or more.

C.3.5 Ion Pump

Ion pumps are capable of evacuating a chamber from a pressure of about 10^{-2} torr to the ultrahigh vacuum range. The configuration of an ion pump is shown in Figure C-11 and the operation is indicated in Figure C-12. A cellular anode structure is located between two planar cathodes. A magnetic field is applied normal to the cathode planes and parallel to the cell axes. Electrons spiral around the magnetic field lines and ionize gas molecules. Electrons released during ionization participate in further ionization of other gas molecules. During collisions, electrons give up energy and gradually migrate

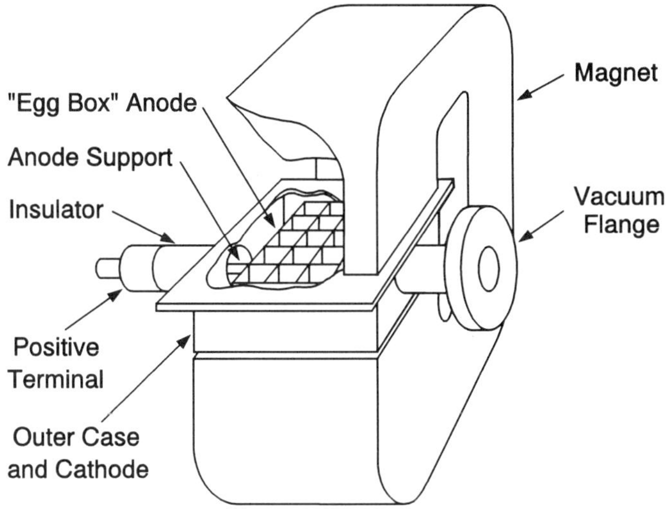

Figure C-11 Typical ion pump.

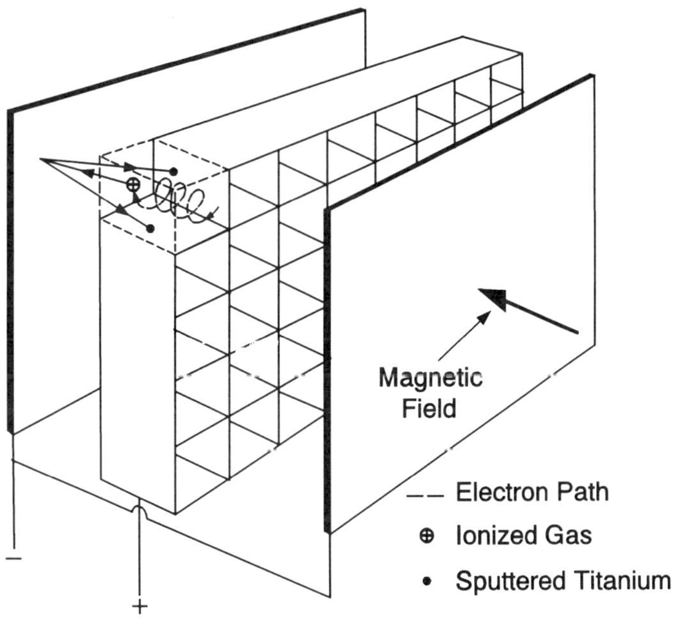

Figure C-12 Operation of diode-type ion pump.

to the anode. The positive ions are accelerated to the negative electrode, which is fabricated from titanium. As a result of ion interaction with the titanium, at least three pumping actions occur as indicated in Figure C-13:

1. Ions sputter titanium cathode material, which deposits and acts as a getter film, adsorbing reactive gases (such as nitrogen, oxygen and hydrogen).
2. The energy of some ions is sufficient for them to be buried in the electrode surfaces, where they are adsorbed by ion implantation. This process is effective for all types of ions including those of rare or inert gases.
3. Enhanced pumping of noble gases is possible with a triode ion pump (Figure C-14) because sputtered titanium from the cathodes buries gas atoms on the target electrode.

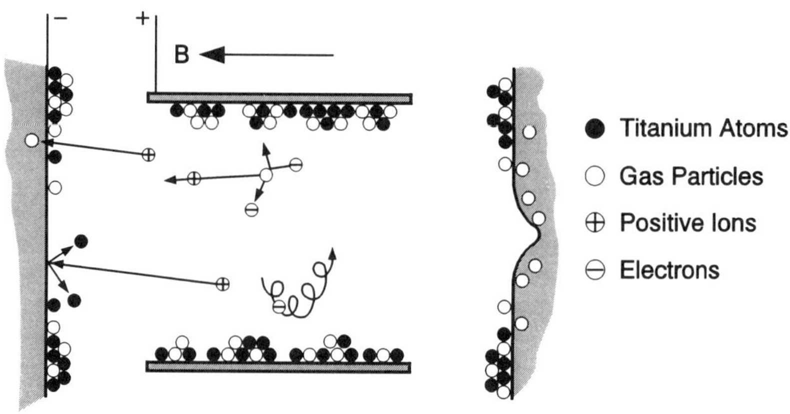

Figure C-13 Pumping processes in a diode-type ion pump. Adapted from Product and Vacuum Technology Reference Book by Leybold Heraeus, Inc.

The pumping speed of an ion pump depends on the type of gas being pumped. Compared with air, the speed for other gases is

Hydrogen	150 to 200%	Oxygen	80%
Methane	100%	Argon	30%
Other light hydrocarbons	80 to 120%	Helium	28%

The voltage of operation is typically about 3.5 kV for small pumps and about 7 kV for large pumps. Ion pumps as small as 0.2 l/s as large as several thousand liters per second are available. Amall pump is often used as an appendage pump, which is an integral part of a microwave tube. As shown in Figure C-15, pump current is directly proportional to pressure and, after calibration, may be used as a measure of pressure.

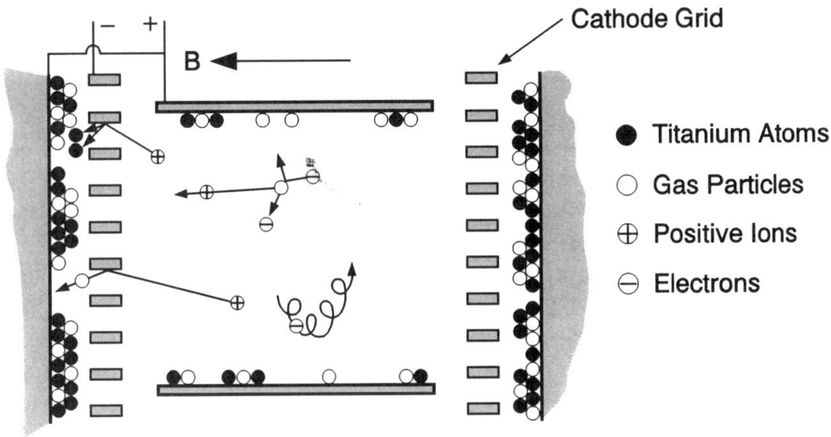

Figure C-14 Pumping processes in a triode-type ion pump. Adapted from Product and Vacuum Technology Reference Book by Leybold Heraeus, Inc.

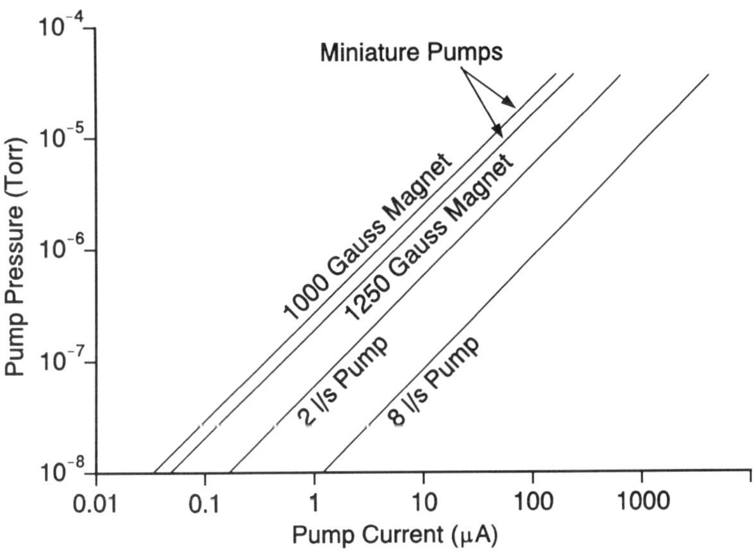

Figure C-15 Ion pump pressure *versus* current.

The magnetic flux density varies somewhat with the size of the pump and is 0.125 tesla (1250 gauss) for an 8 l/s ion pump. The stray magnetic field of the pump is sometimes of concern in electron beam applications and so care must be taken in determining the location of the pump.

C.3.6 Getters

The phenomenon or gettering has long been used in vacuum tubes. In conventional getters, a chemically active metal is evaporated to form a metallic film on the inner surface of the vacuum envelope. Gases that react with the metal are said to be gettered by the evaporated film. Conventional getters have not found application in high power microwave tubes primarily because the evaporated film may cause electrical leakage or breakdown in high voltage structures.

Nonevaporable getters are now commonly used in microwave tubes. The ST172 getters available from SAES GETTERS/USA Inc., 1122 E. Cheyenne Mountain Blvd., Colorado Springs, CO 80906, are particularly attractive for use. These getters utilize a mixture of zirconium and Zr-V-Fe alloy to form a porous sintered structure. This material has excellent strength and compactness and is advertised to be completely free from the release of particulates when subjected to shock and vibration.

While optimum results are obtained after activation at 800 to 900°C, this material performs well after activation at 400 to 500°C. Thus, the ST172 getter can be activated during the normal bakeout process of a microwave tube. Figure C-16 contains illustrations of getters with and without heating elements. The effect of activation temperature on pumping speed and quantity of gas sorbed is shown for carbon monoxide and hydrogen in Figure C-17. At temperatures above room temperature, the sorption capacity for CO is increased significantly as shown in Figure C-18. Other gases of interest in microwave tube applications are CO_2, O_2, N_2, water vapor, argon and helium. The sorption characteristics of ST172 getters for these gases are as follows:

1. CO_2 is about the same as CO.
2. O_2 is somewhat higher than CO.
3. N_2 is approximately one-half of that of CO.
4. Sorption data for H_2O are shown in Figure C-19. Note that the increase in capacity resulting from activation at 800°C, rather than 450°C, is a factor of about five.
5. Inert gases such as argon and helium are not sorbed by ST172 getters.

Some of the microwave tube experience with SAES ST172 getters installed in an electron gun is as follows:

1. Appendage ion pump current, which might sometimes be high, is virtually zero.
2. Arcing in the electron gun is greatly reduced.
3. After extended periods of storage, voltage can be applied immediately to a tube without gassing or arcing;
4. Apparently, the getters are recoverable. After activation and then exposure to air, getters have been reactivated by baking at 450 to 500°C.

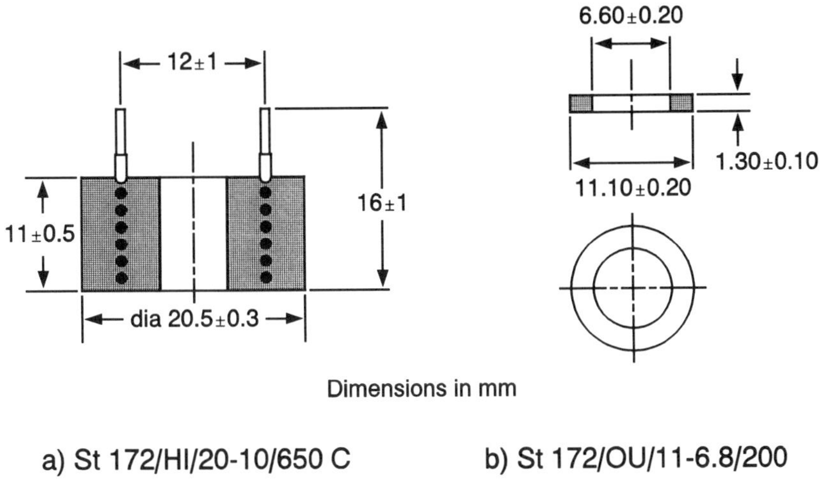

Figure C-16 SAES ST172 getters (a) with and (b) without heating elements (other geometries are also available). Courtesy of SAES GETTERS/USA, Inc.

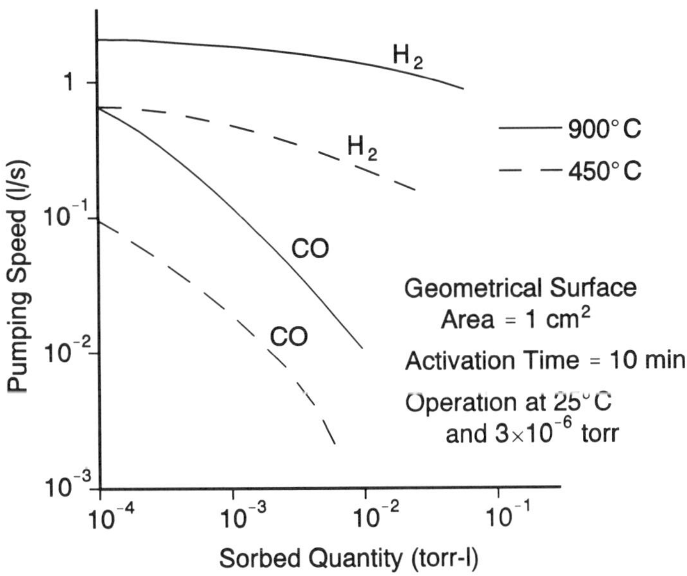

Figure C-17 Effects of activation temperature for SAES ST172 getter material. Courtesy of SAES GETTERS/USA, Inc.

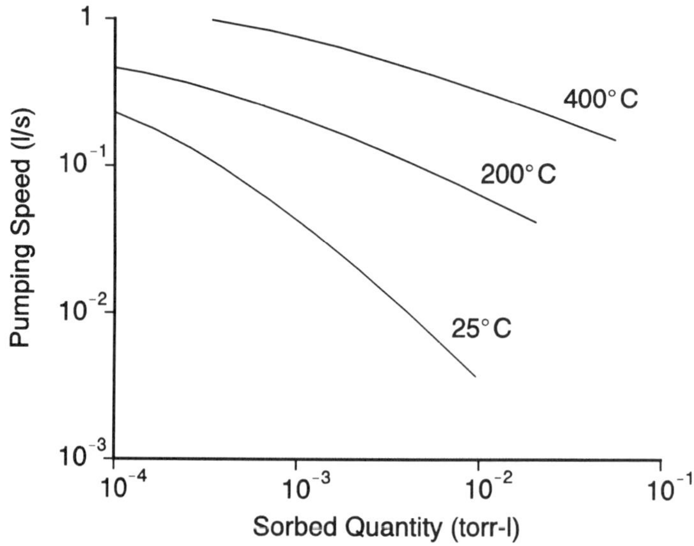

Figure C-18 CO sorption characteristics at different temperatures for SAES ST172 getters. Courtesy of SAES GETTERS/USA, Inc.

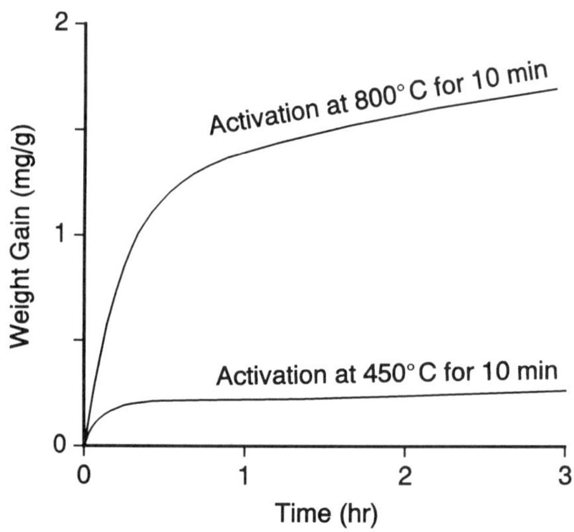

Figure C-19 Room temperature sorption of water vapor for ST172 getters. Courtesy of SAES GETTERS/USA, Inc.

C.4 VACUUM GAUGES

The gauges most commonly used in the microwave tube industry for measuring pressures in the rough to ultrahigh vacuum ranges are

Thermocouple gauge	Rough and medium vacuum
Ionization gauge	High and ultrahigh vacuum
Ion ump current	High and ultrahigh vacuum

The use of ion-pump current as a measure of pressure was discussed in the section on pumps. Thermocouple and ionization gauges will be described in this section.

C.4.1 Thermocouple Gauge

The thermocouple gauge is one of several gauges that uses the thermal conductivity of gas to provide the pressure indication. As indicated in Figure C-20, a filament of wire heated by an electric current is mounted in an envelope attached to the vacuum system. The temperature of the wire depends on heat loss through the gas, which depends on gas pressure.

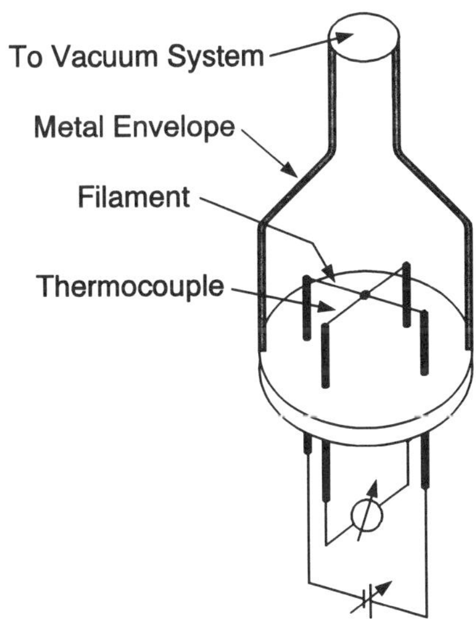

Figure C-20 Thermocouple gauge. Adapted from *Handbook of Thin Film Technology* by L.I. Maissel and R. Glang, copyright 1970 by McGraw-Hill, Inc.

The wire temperature is monitored with a thermocouple, which is calibrated to indicate pressure. Thermocouple gauges operate over the pressure range of approximately 10 to 10^{-3} torr.

C.4.2 Ionization Gauge

Ionization gauges of the type shown in Figure C-21 are commonly used to measure pressures in the high and ultrahigh vacuum ranges.

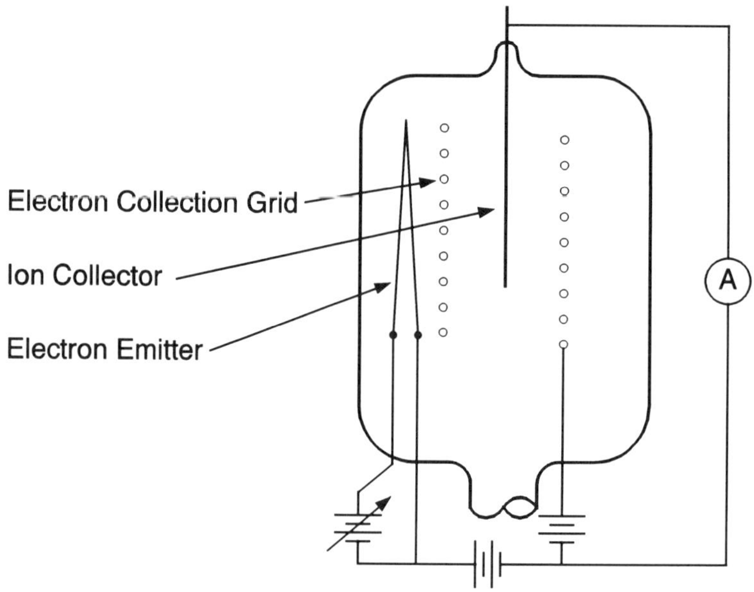

Figure C-21 Ionization gauge.

Electrons, accelerated by a positive grid, produce ions that are collected by an ion collector. The resulting ion current is a measure of gas pressure. These gauges are sometimes referred to as *inverted triode ion gauges*, or *Bayard-Alpert gauges*. In these gauges, a very small diameter wire is used as the ion collector. This is to minimize the xrays produced when electrons strike the positive grid. Some of these xrays impinge on the collector wire, causing it to release electrons. The resulting flow of electron current away from the collector causes an error in pressure indication.

C.5 MICROWAVE TUBE MATERIALS

For proper performance, microwave tubes must operate in the high to ultrahigh vacuum range. At best, an ion pump or getter may be used as an integral part of a tube to help maintain the vacuum level. At worst, no pump is used, and the tube may be required to operate properly after several years in storage.

An extremely important factor in achieving and maintaining an acceptable vacuum level is the selection of the proper materials for the fabrication of a tube. Many of the metallic elements commonly used inside the vacuum envelope of a tube are listed in the chart in Figure C-22.

	Aluminum	Barium	Beryllium	Boron	Calcium	Carbon	Chromium	Copper	Gold	Iridium	Iron	Manganese	Molybdenum	Nickel	Osmium	Palladium	Platinum	Rhenium	Rhodium	Scandium	Silver	Strontium	Thorium	Titanium	Tungsten	Vanadium	Zirconium
Heater	■																	■							■		
Cathode	■	■	■		■					■				■	■		■			■		■	■		■		
Grids						■		■	■				■												■		
Anode								■																			
Cavities								■																			
Helix								■	■				■					■							■		
Support	■		■	■		■																					
Attenuator						■																					
Collector						■		■																			
Envelope							■	■			■			■													
Brazes								■	■			■	■			■	■		■		■						■
Polepieces											■															■	
Window	■		■			■																					
Getter		■																						■		■	■

Figure C-22 Some metals used inside the vacuum envelope of microwave tubes.

The primary factor considered in selecting a metal for use inside the vacuum envelope is vapor pressure. The charts in Figures C-23, C-24 and C-25 give vapor pressure as a function of temperature for the metallic elements.

As an example of the use of these charts, let us consider zinc (see Figure C-23). At low temperatures, the vapor pressure is reasonable (10^{-9} torr at 100°C). At a somewhat elevated temperature (400°C), the vapor pressure is only 10^{-1} torr. This clearly shows why brass (which contains zinc) must not be used in a vacuum system that may be heated. If brass were used and baked at 400°C or higher, zinc vapor would permeate the entire system.

The curve for barium shown in Figure C-25 is of significance in considering the

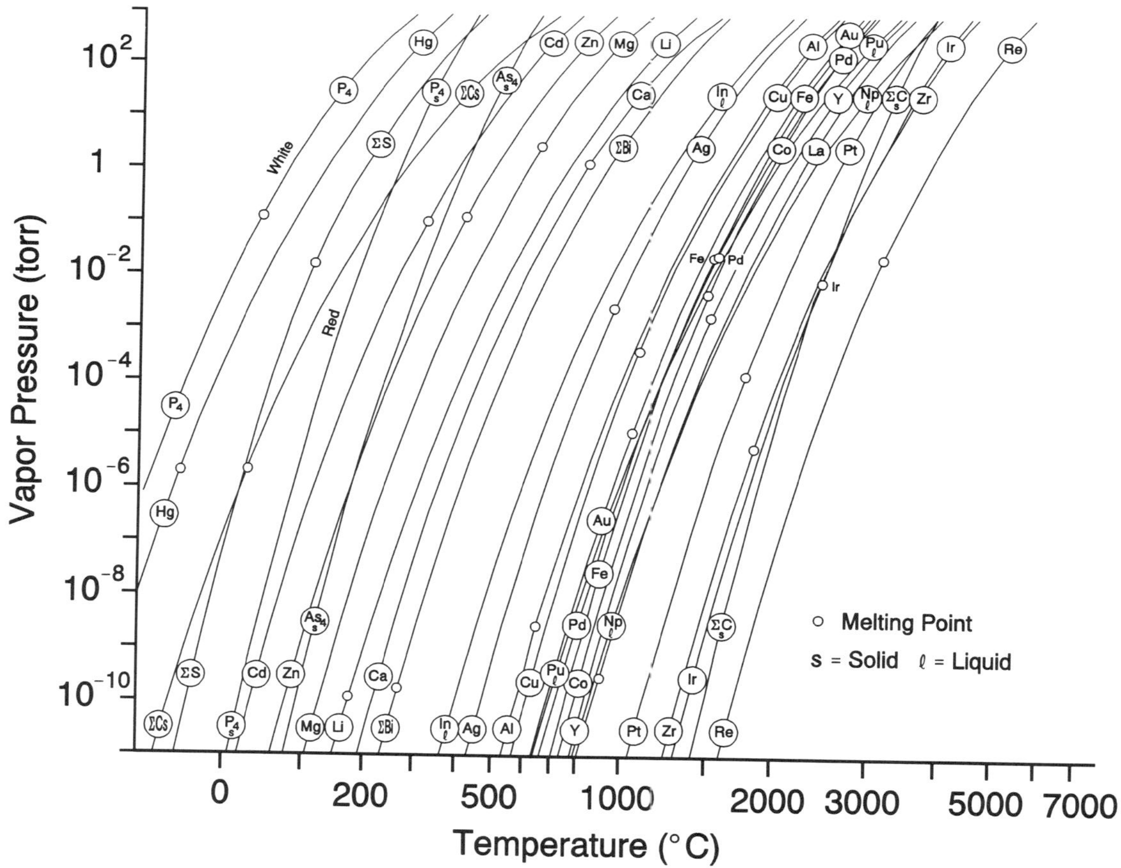

Figure C-23 Vapor pressures of the elements. Adapted from R.E. Honig and D.A. Kramer, *RCA Review*, June 1969, copyright 1969 by Radio Corporation of America.

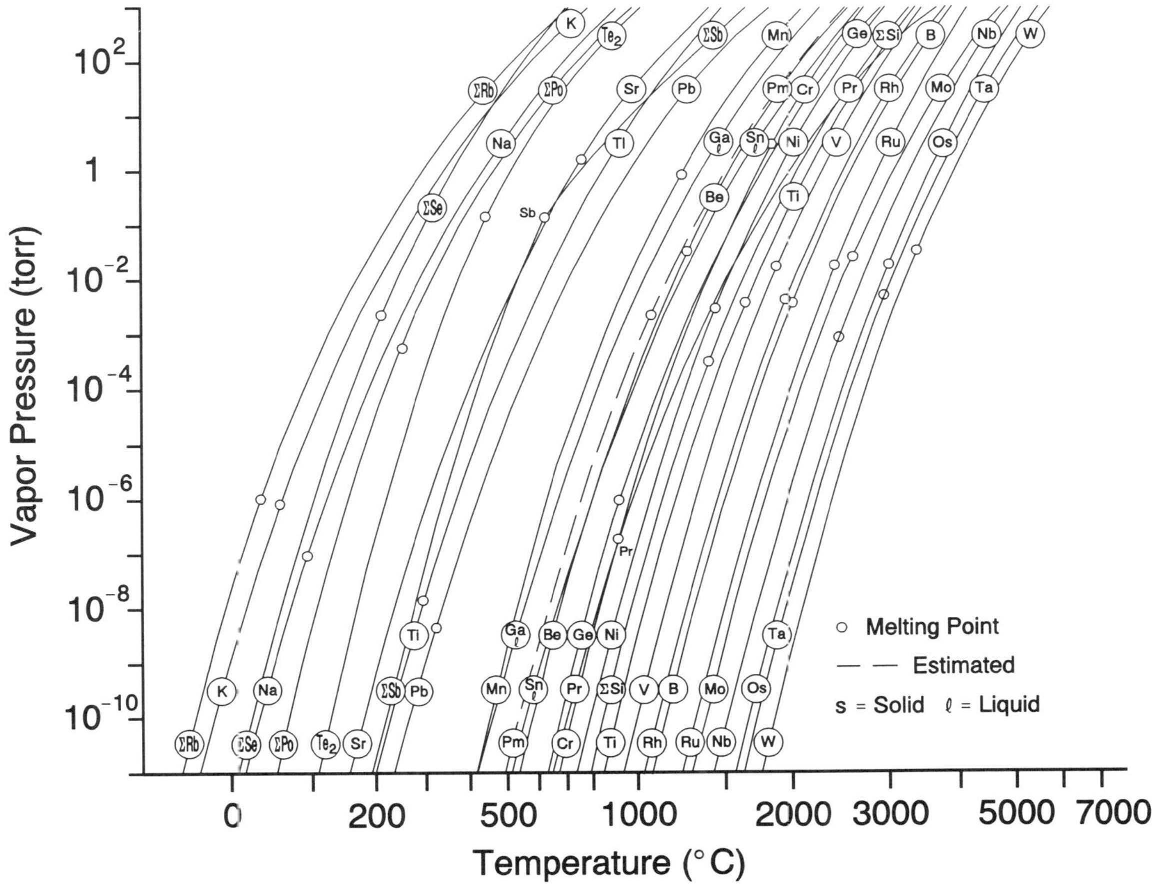

Figure C-24 Vapor pressures of the elements (cont'd).

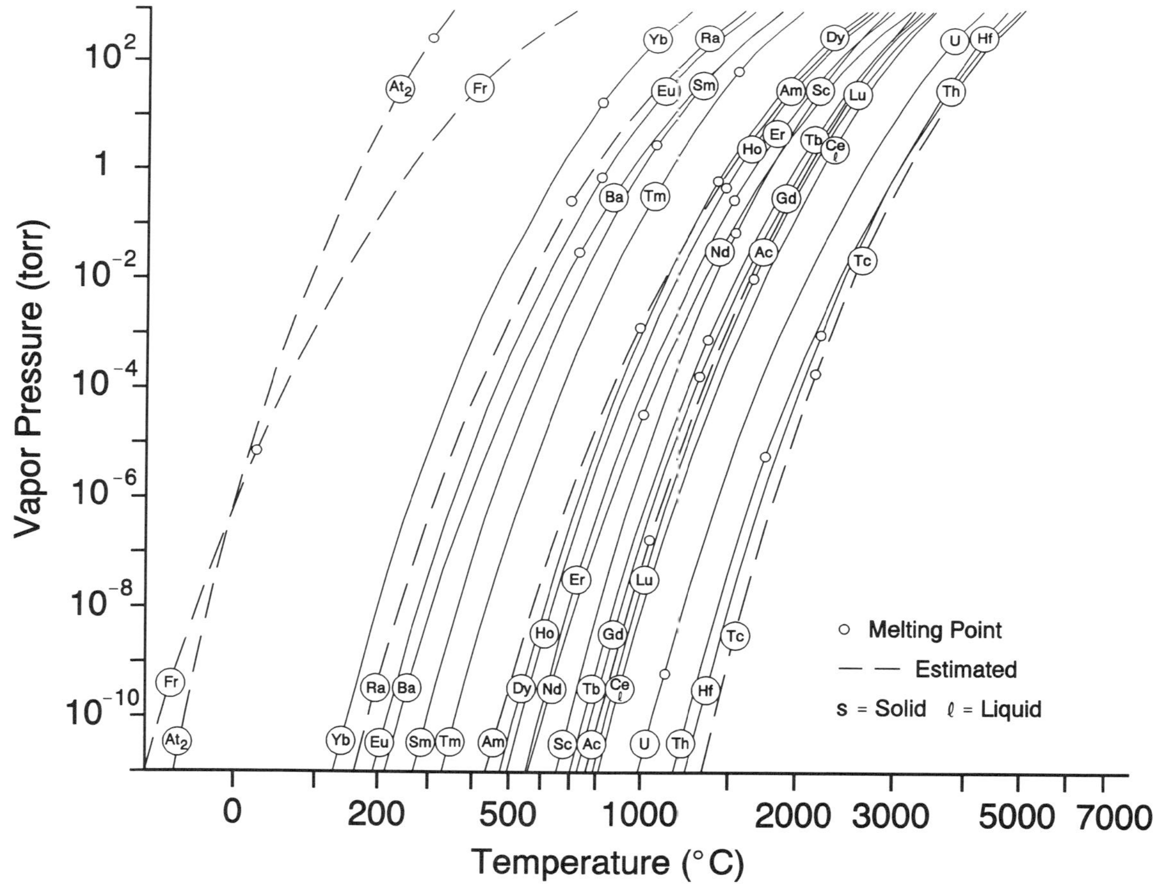

Figure C-25 Vapor pressures of the elements (cont'd).

operation of cathodes. At a cathode operating temperature of 1000°C, the vapor pressure of barium is over 1 torr. Thus, excess barium (over about a monolayer) on the surface of a cathode evaporates very rapidly. This evaporated barium may eventually deposit on insulating surfaces and cause electrical leakage or breakdown. To remove excess barium, a new cathode is usually placed in vacuum and operated at high temperature before it is placed in a microwave tube.

Nonmetallic materials such as the neoprene, rubber and teflon, shown in Figure C-26, are sometimes used in demountable vacuum systems. These materials severely limit the pressure and temperature at which the system can operate.

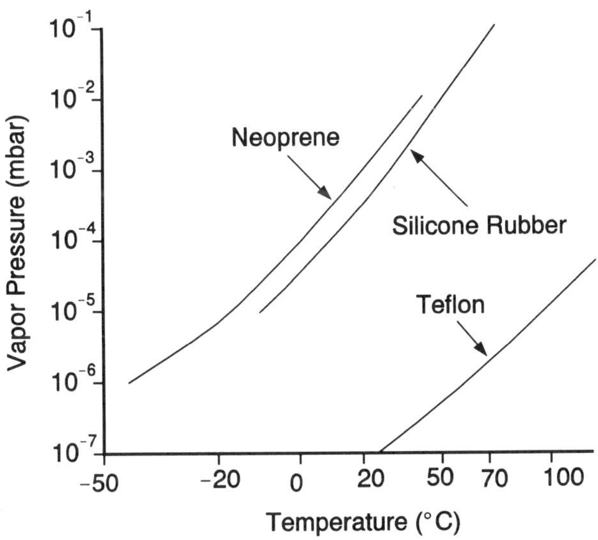

Figure C-26 Vapor pressure of nonmetallic sealing materials. Adapted from Product and Vacuum Technology Reference Book by Leybold Heraeus, Inc.

When a nonmetallic material (such as an insulator, for example) is required in a system that must be baked, one of a very small number of materials is selected. If the nonmetallic material is part of the vacuum envelope, then either aluminum oxide, beryllium oxide or a glass is usually selected. These materials all have extremely low vapor pressures. Inside the vacuum envelope, in addition to aluminum oxide or beryllium oxide, boron nitride or, in very special cases, diamond is used.

Some liquids such as cleaning fluids are sometimes not completely removed before system evacuation. As expected (and shown in Figure C-27), these have a disastrous effect on attainable pressure.

A particularly aggravating problem with water is that, as it vaporizes, it is cooled and may freeze. The vapor pressure may drop to the 10^{-2} to 10^{-3} torr range, where it appears very much like a leak.

Other liquids may have extremely low vapor pressures. Examples are diffusion

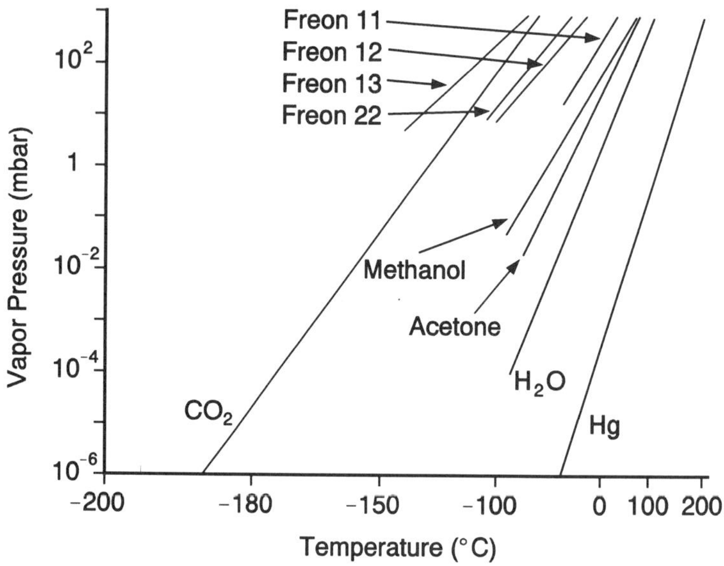

Figure C-27 Saturation vapor pressure of various substances. Adapted from Product and Vacuum Technology Reference Book by Leybold Heraeus, Inc.

pump oils. Still, there is no liquid suitable for use in the high or ultrahigh vacuum environment of a microwave tube. It is also appropriate to point out that no lubricants are suitable for use in a microwave tube. Thus, in the spin tuned magnetron, where a portion of the anode rotates at 4000 rpm, no lubricants are used. Instead, metals are carefully selected, treated and precisely machined so that galling and excessive wear do not occur.

C.6 FABRICATION TECHNIQUES

References [2–4] contain detailed information concerning many of the fabrication techniques and processes used in the microwave tube industry. It is possible here to describe briefly only a few of the important aspects of microwave tube fabrication.

First of all, as was pointed out in the previous section, it is essential that materials to be used are properly selected and cleaned. Then, metals may be joined by welding or brazing. Ceramics may be joined to metals by brazing. In no case is it possible to use fluxes because they may become trapped in joints and then result in virtual leaks, that is, they very slowly leak out of joints and produce contamination.

In the absence of fluxes, inert gas welding is used. A drawing of an inert gas welding torch is given in Figure C-28. Care must be taken in designing weld joints if adequate strength is to be obtained and gas and dirt traps are to be avoided. Examples of the precautions that must be taken in the formation of a welded joint between a tube and a flange are shown in Figure C-29. Whenever possible, the weld should be made on the vacuum side of the joint. Then, weld slag or other contaminants that may be trapped between the mating surfaces of the cylinder and flange will be outside of the vacuum seal.

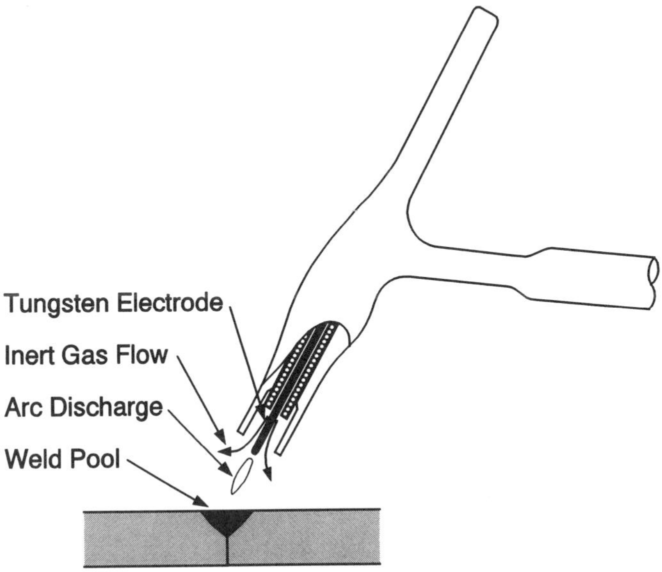

Figure C-28 Schematic of tungsten inert gas (TIG) welding torch. Adapted from *Handbook of Thin Film Technology* by L.I. Maissel and R. Glang, copyright 1970 by McGraw-Hill, Inc.

Similar precautions are necessary for brazed joints. Since fluxes cannot be used, brazes must be performed in a vacuum or reducing atmosphere of hydrogen. Hydrogen is often used rather than a vacuum because the brazing process can be performed much more rapidly. In some cases, such as in the brazing of stainless steel, extremely dry hydrogen must be used to prevent oxidation. Examples of induction heating systems for brazing in hydrogen are shown in Figure C-30.

Ceramic-to-metal seals are made by first metallizing the ceramic and then brazing the metallic part to the metallized surface. Perhaps the most important factor to be considered in the design of ceramic-to-metal seals is the thermal expansion characteristics of the parts. Figure C-31 shows how thermal expansion coefficients vary with temperature. When ceramics and metals having thermal expansion coefficients that differ by a significant amount are used, great care must be exercised in the design of the seal. In some cases the metal element is designed to be flexible in the region of the braze. In a butt ceramic-to-metal seal, as shown in Figure C-32, the metal is

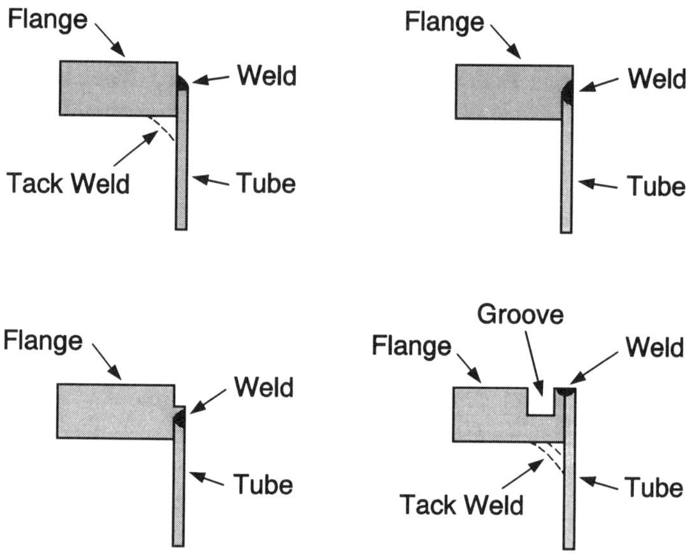

Figure C-29 Common tube-to-flange weld designs. Adapted from *Handbook of Thin Film Technology* by L.I. Maissel and R. Glang, copyright 1970 by McGraw-Hill, Inc.

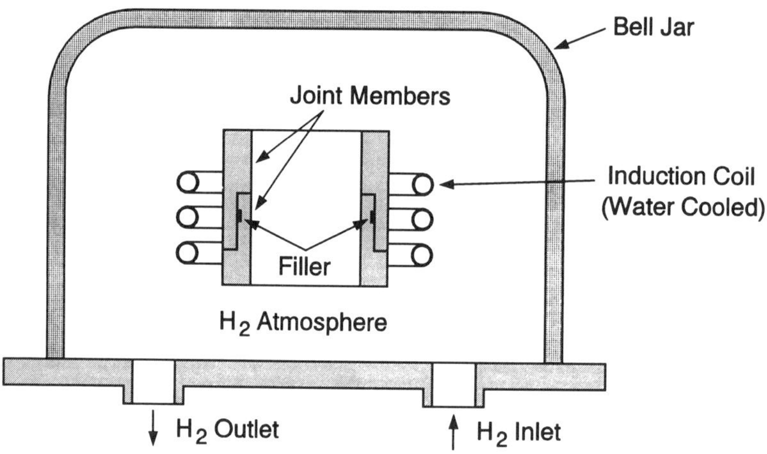

Figure C-30 Brazing by induction heating in hydrogen.

sandwiched between ceramic pieces that are massive enough to force the metal to expand and contract with the ceramic.

Many braze materials are available for use in the construction of microwave tubes. Some of these are listed in Figure C-33 along with various metal couples that can be joined. It should be noted that, in many high power klystrons and TWTs, the use of

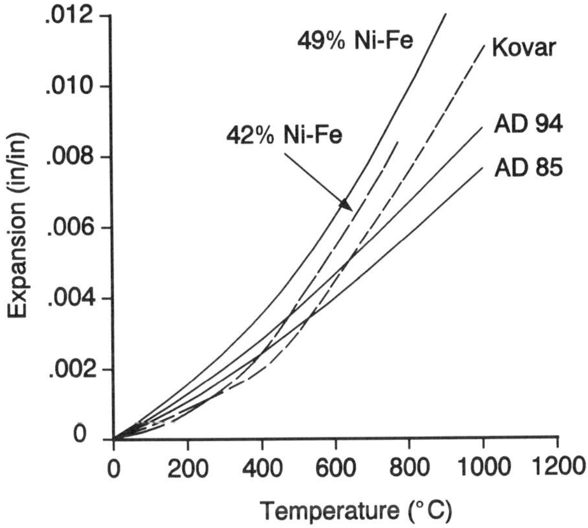

Figure C-31 Thermal expansion characteristics. Adapted from *Handbook of Electron Tube and Vacuum Techniques* by Fred Rosebury, copyright 1965 by Addison-Wesley Publishing Company, Inc.

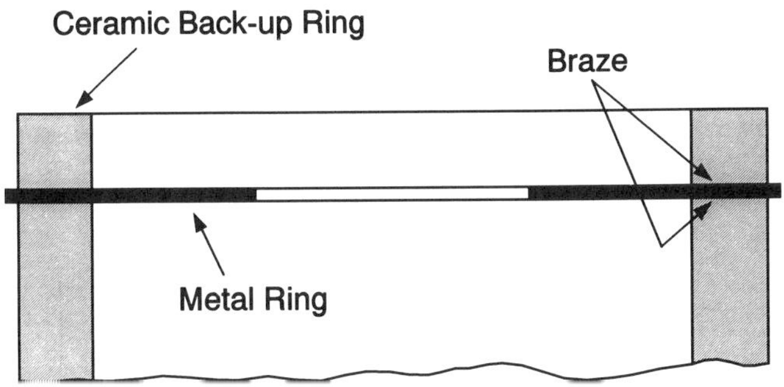

Figure C-32 Butt ceramic-to-metal seal.

alloys containing silver is avoided. This is because, at high temperatures, sufficient vaporization of the silver may occur to degrade tube performance.

After assembly and processing of a metal-to-ceramic vacuum tube has been completed, it is necessary to separate the tube from the pumping station. (An appendage pump may remain as an integral part of the tube.) The technique universally used to separate the tube from the pumping station is illustrated in Figure C-34. Soft copper tubulation is used to make the vacuum connection to the pumping station. This tubulation is molecularly clean on the internal surfaces because it has been baked along with the microwave tube (at 450°C or higher).

Common Brazing Materials

	Brazing-Material Constituents and Composition (wt. %)	Temperature (°C)		
		mp	Liquidus	Solidus
1	Copper (100)	1083		
2	Gold (100)	1063		
3	Pd (10) - Ag (90)		1065	1000
4	Au (35) - Cu (62) - Ni (3)		1030	975
5	B (2.9) - Cr (7) - Si (4.5) - Fe (3) - Ni (82.6)		1000	971
6	Au (40) - Cu (60)		980	950
7	Au (82) - Ni (18)	950		
8	Pd (10) - Ag (58) - Cu (32)		852	825
9	Ag (72) - Cu (28)	780		
10	In (10) - Cu (27) - Ag (63)*		730	685

Metal Couples Which Can be Brazed With the Above Fillers

Metal 2 \ Metal 1	W	Mo	Ni	Monel	Kovar	Stainless Steel	Cu
Copper	4, 7	6, 7, 8, 9	6, 7, 9	4, 9	6, 7, 9	7, 8	8, 9, 10
Stainless Steel	5, 7	3, 5	5, 7	1	2, 7, 9	1, 8	
Kovar	4, 7	7	2, 6, 7	4, 9	2, 9		
Monel	4, 7	4	4, 9	4, 9			
Nickel	4, 7	4, 7	4, 6, 9				
Molybdenum	4, 7, 8	4, 7, 8					
Tungsten	7, 8						

* Used as step braze after No. 9

Figure C-33 Some braze materials used in the construction of microwave tubes. From *Handbook of Thin Film Technology* by L.I. Maissel and R. Glang, copyright 1970 by McGraw-Hill,Inc. Reproduced with permission.

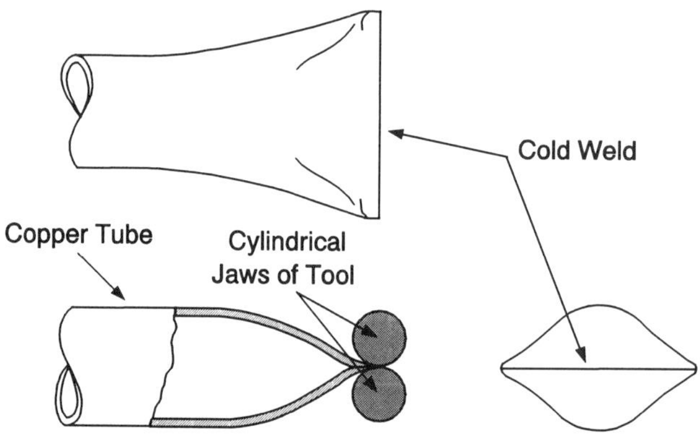

Figure C-34 Method used to squeeze-off (pinch-off) copper tubulation. From *Power Travelling Wave Tubes* by J.F. Gittins, published 1965 by American Elsevier Publishing Company, Inc.

By squeezing the tubulation with a special tool, the internal mating surfaces of the copper can be forced to flow together to form a cold weld. In its crudest form, the tool is similar to a manually operated bolt cutter with the cutting edges replaced by cylindrical steel bars.

Because the manual squeeze-off (pinch-off) operation with the tool that is much like a bolt cutter is highly operator dependent, an automated system is normally employed.

C.7 VACUUM LEAKS [1, 5]

C.7.1 Real Leaks

A real vacuum leak results from a pore in the wall of the vacuum envelope or from permeation of gas through the wall. A pore can result from any one of a variety of mechanical imperfections in the wall. Examples are impurities or inclusions in the wall material or cracks caused by thermal or mechanical stress.

The leak rate for air as a function of pore size (radius and length) is shown in Figure C-35. the slope of 1/3 indicates molecular flow. A hole size on the order of at least 3 molecular diameters (~ 10Å or 10^{-7} cm) is required for pore leakage. Smaller holes are likely to be plugged by large molecules. Thus, pore leakage is unlikely to

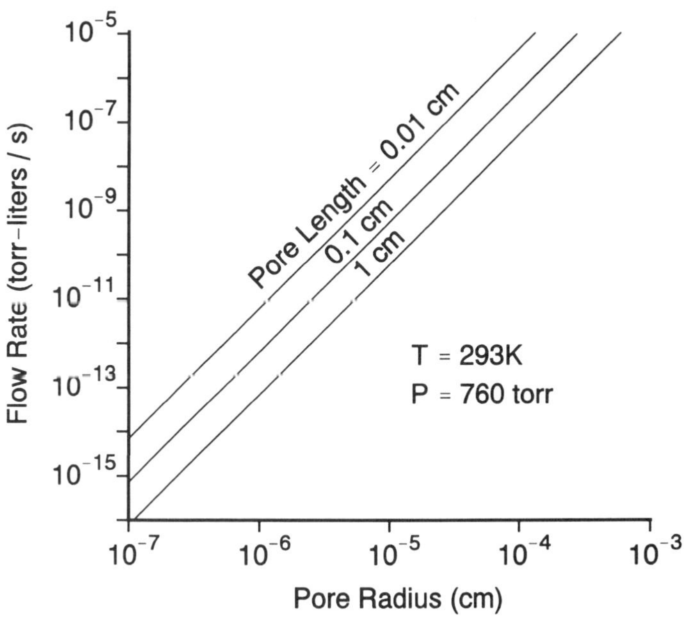

Figure C-35 Flow rate of air as a function of pore size.

occur below a leak rate of about 10^{-15} torr-liters/sec for a 1 mm long pore. (1 torr-liter/sec $\approx$ 1 cm^3/sec at standard temperature and pressure.) At this leak rate, the pressure in a dormant TWT with an internal volume of 0.025 liter and without a getter or pump would increase to about 10^{-6} torr in one year. At this pressure, there is high likelihood that proper tube operation would not be possible. Thus, even the smallest pore leak is unacceptable for most applications.

Permeation results from an electronic interaction between a gas and a solid. This interaction results in the movement of gas molecules from atom to atom through the solid. The rate of gas flow due to permeation depends on the gas and the solid and varies by orders of magnitude with different gas-solid combinations. Permeation for several gas-metal systems are shown in Figure C-36.

The only appreciable permeation rate of interest for vacuum-tube design is hydrogen through iron and iron alloys. Iron is part of the vacuum envelope in many PPM-focused TWTs. The concentration of H_2 in atmospheric air is too low to be of concern. However, hydrogen released from the corrosion of iron by water can be of concern. As a result, surface treatments of iron parts, such as electroplating, which prevent corrosion will minimize the generation and permeation of hydrogen.

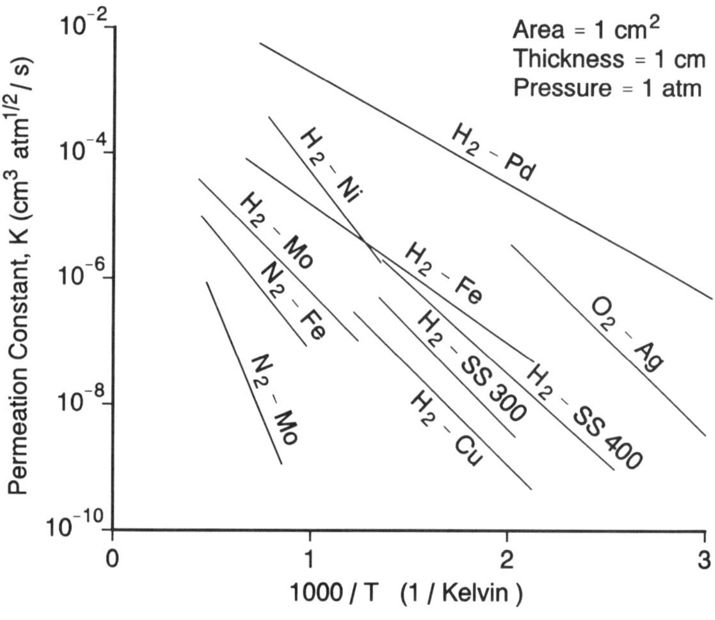

Figure C-36 Permeation rates for several gas-metal systems. Adapted from A. Roth, *Vacuum Technology*, copyright 1990 by Elsevier Science Publishers B.V.

C.7.2 Virtual Leaks

Virtual leaks occur when gases, which are chemically or mechanically trapped inside the vacuum envelope, are released. Some examples of sources of virtual leaks are:

1. A weld joint that is made on the outside (atmospheric side) of the vacuum envelope. Dirt or other contaminants that are trapped between the welded parts on the inside (vacuum side) of the joint release gas which slowly leaks into the vacuum chamber.
2. A screw that is used to fasten a part in an electron gun. the threads of the screw have not been properly vented and so gas that is trapped and not completely driven out during bakeout is released into the vacuum chamber.
3. An inadequately outgassed part inside the vacuum envelope. Gas from the part slowly diffuses into the vacuum chamber.
4. During thermal cycling, a microscopic crack that develops on the inner (vacuum side) of a ceramic insulator. This crack exposes a tiny gas-filled void and the gas slowly leaks into the vacuum chamber.

Virtual leaks are avoided by following good design practice, by eliminating gas traps and by properly processing parts. There is no external test that can detect a virtual leak. The use of a getter or an appendage pump is an inexpensive and straightforward way to prevent gas problems in the event that a virtual leak occurs.

REFERENCES

[1] For additional reading, the following are recommended: "Vacuum Physics and Technology," Vol. 14, *Methods of Experimental Physics*, New York: Academic Press, 1979; S. Dushman and J.M. Lafferty, *Scientific Foundations of Vacuum Techniques*, New York: John Wiley and Sons, 1962; M. Pirani and J. Yarwood, *Principles of Vacuum Engineering*, New York: Reinhold, 1963. R.W. Roberts and T.A. Vanderslice, *Ultra High Vacuum and Its Applications*, Englewood Cliffs, NJ: Prentice-Hall, 1963, A. Roth, *Vacuum Technology*, Amsterdam, The Netherlands: Elsevier Science Publishers B.V., 1990.

[2] F. Rosebury, *Handbook of Electron Tube and Vacuum Techniques*, Reading, MA: Addison-Wesley, 1965.

[3] W.H. Kohl, *Materials and Techniques for Electron Tubes*, New York: Reinhold, 1960.

[4] M. Knoll, *Materials and Processes of Electron Devices*, Berlin: Springer-Verlag, 1959.

[5] Teledyne MEC, "Design considerations for extended dormancy of traveling wave tubes," Technical Memorandum 151, February 1991.

Appendix D
Magnetics

Magnetic fields are used for focusing electron beams in nearly all microwave tubes. These fields may be generated by electromagnets permanent magnets. In an electromagnet, current passing through a wire produces a magnetic field surrounding the wire. In a permanent magnet, electron motion within the material produces the magnetic field. Electromagnets, ferromagnetic materials and permanent magnets are discussed in this Appendix [1].

D.1 MAGNETIC QUANTITIES

In considerations of magnetics and magnetic materials, three quantities are of primary interest:

- Magnetic field intensity, H, is one of the elementary fields in nature.
- Magnetic flux density, B, is the density of magnetic flux lines produced by the magnetic field.
- Magnetic permeability, μ, is the property of a medium that determines how much flux is produced by a certain field level. Flux density is related to field intensity by

$$B = \mu H \tag{D-1}$$

The units commonly used for H, B and μ are given in the following table along with conversion factors. The value for μ in air (free space), μ_o, is

$$\mu_o = 4\pi \times 10^{-7} \quad \text{H/m} \tag{D-2}$$

It is common to use the term *relative permeability*, μ_r, which is the ratio of the permeability of a material to that of free space. That is,

$$\mu_r = \frac{\mu}{\mu_o} \tag{D-3}$$

Quantity	*MKS Unit (a)*	*CGS Unit (b)*	*Conversion*
H	Amperes/meter	Oersteds	$(a) = 79.577 \times (b)$
B	Tesla	Gauss	$(a) = 10^{-4} \times (b)$
μ	Henrys/meter	Gauss/Oersted	$(a) = 1.257 \times 10^{-6} \times (b)$

D.2 ELECTROMAGNETS

A wire carrying a current produces a magnetic field as shown in Figures D-1 through D-3. The direction of the field lines in relation to the direction of current flow can be remembered with the aid of the familiar righthand rule. (The thumb points in the direction of conventional current and the fingers point in the direction of the field lines.)

For microwave tube applications, the current carrying wire is formed into a solenoid, as shown in Figures D-2 and D-3, to produce a reasonably constant axial magnetic field near the center.

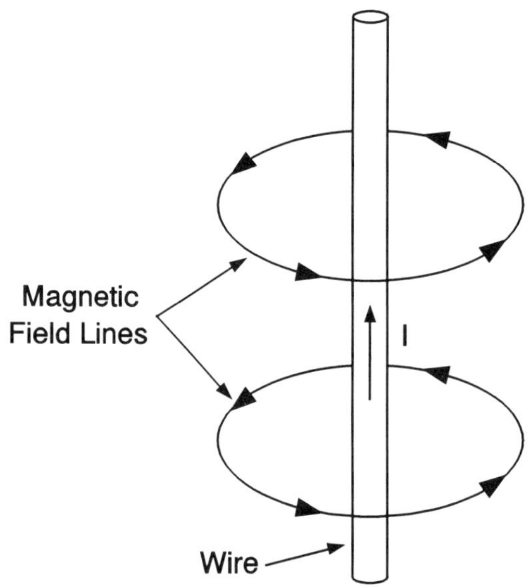

Figure D-1 Magnetic field around a wire carrying a current.

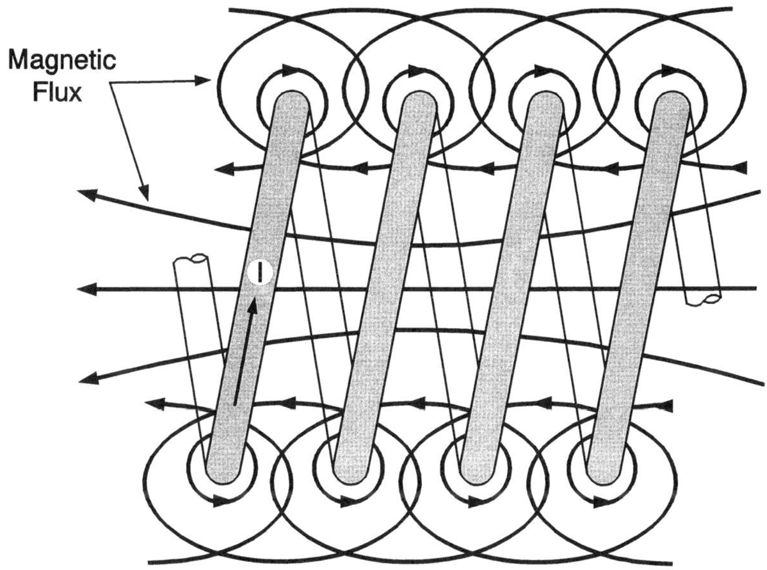

Figure D-2 Portion of solenoid showing magnetic flux linkages.

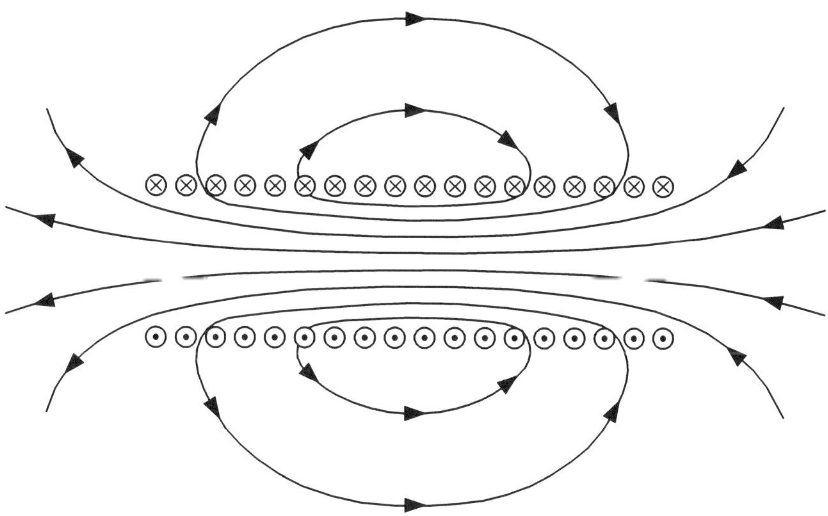

Figure D-3 Magnetic flux pattern for a single-layer solenoid.

Very close to each turn of the solenoid, the flux lines are nearly circular. However, at greater distances, the flux in the region between adjacent turns cancels, so the flux lines around each pair of turns are nearly elliptical. At still greater distances, there is further cancellation of opposing flux lines and so the pattern of Figure D-3 occurs.

In the center of the solenoid, the flux lines are parallel and the flux density is constant. Near the ends of the solenoid, the flux lines diverge and the flux density decreases as indicated by the dashed line in Figure D-4. This large decrease in field near the ends of the solenoid is not acceptable for most microwave tube applications.

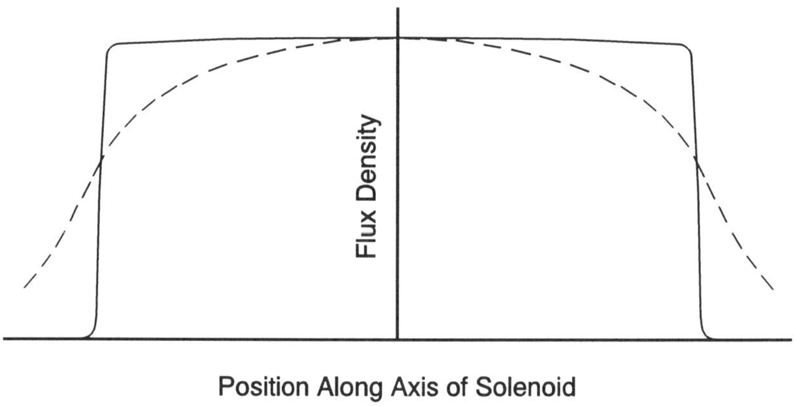

Figure D-4 Solenoid flux density without (dashed curve) and with (solid curve) magnetic pole pieces.

To enhance uniformity of the field on axis where the electron beam is to flow, magnetic pole pieces, as shown in Figure D-5, are used. The pole pieces have high relative permeability ($\mu_r \approx 5000$) and their effect is to strengthen the field near the ends of the solenoid as shown by the solid curve in Figure D-4. The magnetic behavior of the pole pieces is described in the following section.

Figure D-6 shows a typical solenoid for a TWT or klystron. Individual coils of the solenoid are separated to provide space for RF feedthroughs, cooling lines, *et cetera*. Normally, a solenoid is surrounded by a magnetic shield, or magnetic bars, as shown in Figure D-6, to provide a return path for flux lines. These magnetic shields do not enhance the magnetic flux density on the axis of the solenoid. Instead, they confine the magnetic flux lines surrounding a solenoid to a localized region near the solenoid. This helps minimize the possibility that a magnetic object, such as a tool, will accidentally be pulled toward the tube being focused by the solenoid, thus causing damage to, or destruction of, the tube. The magnetic shields or bars also help minimize interactions of the solenoid with nearby electrical circuits.

In many cases, the solenoid used for focusing the beam of a tube is removable from the tube. Sometimes, however, when weight and power utilization must be minimized, the solenoid is made integral with the tube as shown in Figure D-7.

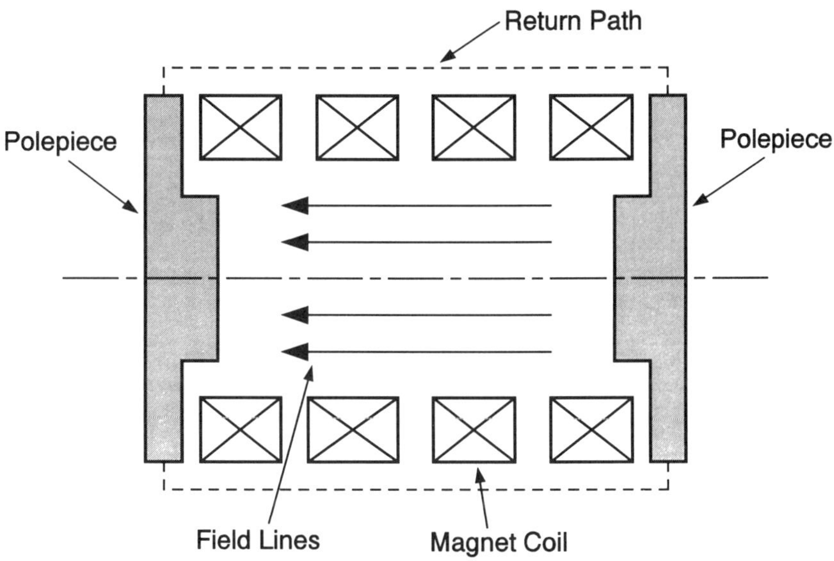

Figure D-5 Focusing solenoid consisting of coils and magnetic polepieces to improve field uniformity.

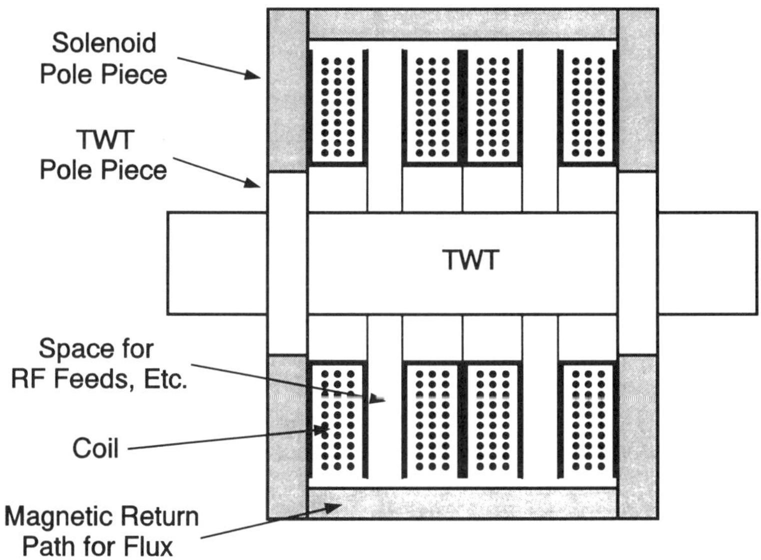

Figure D-6 Typical configuration of a solenoid and a microwave tube.

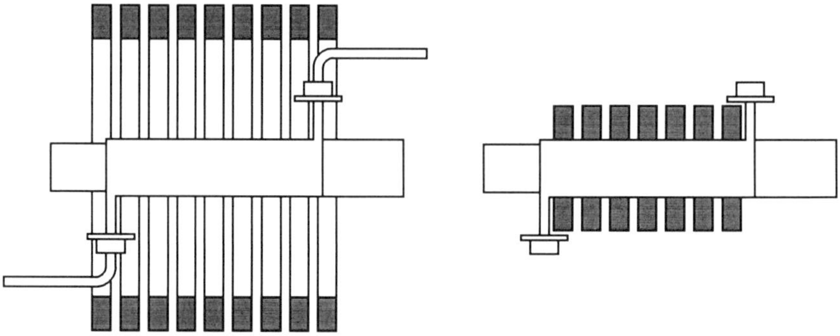

Figure D-7 Size and weight reduction resulting from the use of an integral solenoid.

D.3 FERROMAGNETIC MATERIALS

Magnetic materials can be divided into three categories. These are

- $\mu_r < 1$ (diamagnetic)
- $\mu_r > 1$ (paramagnetic)
- $\mu_r >> 1$ (ferromagnetic)

Diamagnetism is present in all materials. The establishment of magnetic flux is opposed by the material so that the relative permeability is less than unity. This effect is explained by the laws of Faraday and Lenz. When a magnetic field is applied, an electric field is produced. This electric field acts on the electron orbits in a material to produce microscopic reaction currents. The reaction currents produce a magnetic field that opposes the applied field. Diamagnetism is an extremely small effect. In most materials, it is overridden by paramagnetism or ferromagnetism.

Paramagnetic materials have a relative permeability slightly greater than unity. In these materials, the application of an external magnetic field aligns permanent magnetic dipoles of atoms, ions or molecules with the applied field, thus enhancing the field.

In ferromagnetic materials, a specialized condition exists that greatly aids the applied magnetic field. This condition involves the electronic structures of atoms and the distances between atoms. The most stable configuration of the electrons is when their spin magnetic moments are parallel. The result is that complete alignment of the spins occurs within regions called *magnetic domains*. The application of an external field both aligns domains and shifts domain boundaries to produce an enormous field.

Ferromagnetic materials are highly nonlinear. The relation between B and H must be determined experimentally. This is done with a circuit like that in Figure D-8.

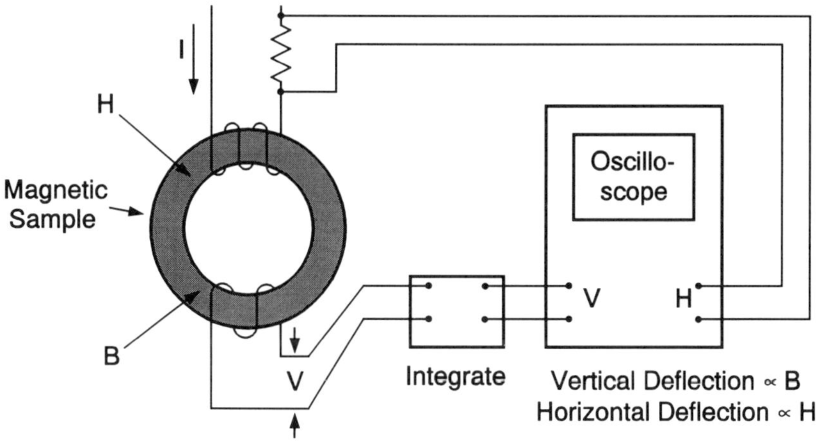

Figure D-8 Circuit for determining the magnetic characteristic of materials.

From Ampere's law, the magnetic field, H, in a magnetic sample is proportional to the current, I. From Faraday's law, the voltage, V, induced in a winding is proportional to the time rate of change of magnetic flux linking the winding. The integral of voltage is, therefore, proportional to flux and magnetic flux density, B.

A signal proportional to B is applied to the vertical input of an oscilloscope. A signal proportional to H is applied to the horizontal input. The result displayed on the oscilloscope is illustrated in Figure D-9.

When current is increased, the righthand portion of the hysteresis curve is followed. When current is decreased, the lefthand portion is followed. At very large current excursions, the shape of the hysteresis curve is relatively independent of the current, and a loop known as the saturation loop is traversed.

Several important characteristics of magnetic materials may be illustrated with the hysteresis loop. Some of these are shown in Figure D-9. When a magnetic field is first applied to a demagnetized sample, the initial magnetization curve is followed. When H is zero, there is a residual flux density, B_r, in the material called the *retentivity*. The flux density, B, is reduced to zero when a field, H_c, called the *coercive force* or *coercivity*, is applied.

At lower current levels, smaller loops are traversed. In fact, as shown in Figure D-10, a material can be demagnetized by gradually reducing the alternating magnetic field to zero. The tips of the hysteresis loops form the "normal magnetization curve" as shown in Figure D-11. This is very similar in shape to the initial magnetization curve. This curve is a reproducible characteristic for a magnetic material.

The relative permeability quoted for a material is based on the slope of the normal magnetization curve as shown in Figure D-12. The maximum value is normally used.

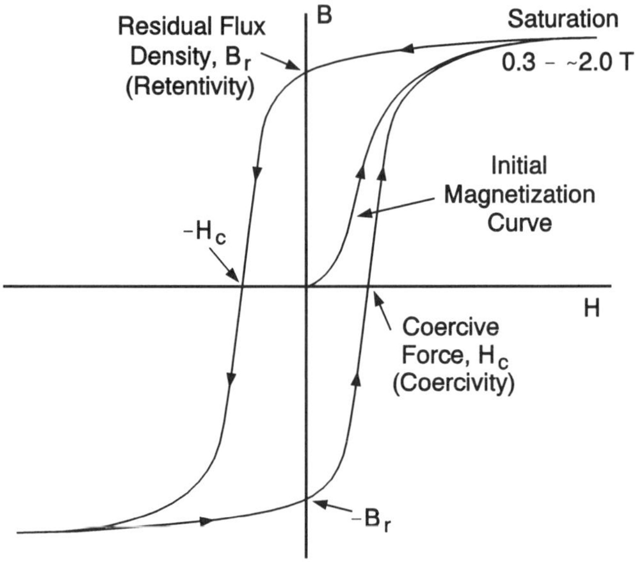

Figure D-9 Hysteresis curve resulting from use of circuit shown in Figure D-8.

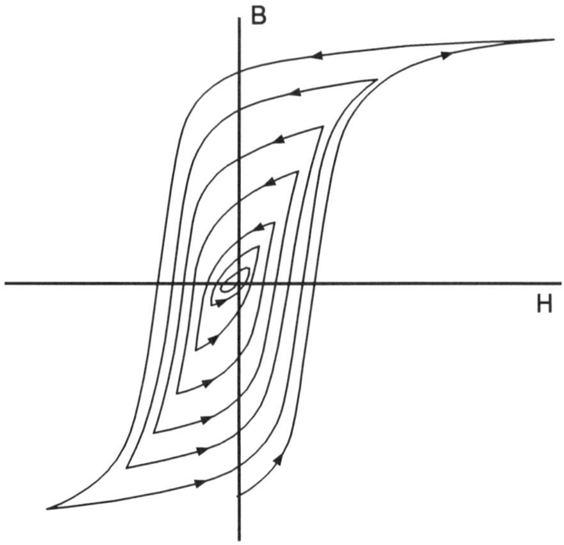

Figure D-10 Demagnetization by reversals of magnetic field.

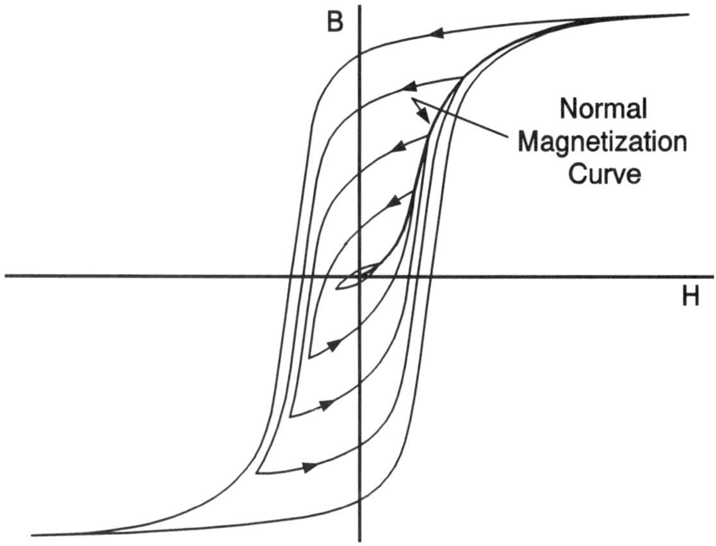

Figure D-11 Normal magnetization curve with relation to hysteresis loop.

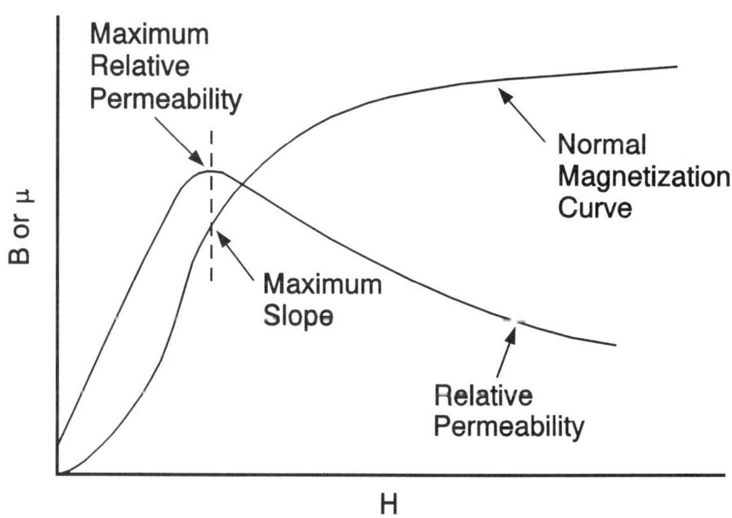

Figure D-12 Relation of relative permeability to normal magnetization curve.

It is significant to note that, at very low values of the magnetic field, the permeability is low. In fact, for most magnetic materials, the permeability at low values of H is so low that they will not shield against stray magnetic fields. This is sometimes of importance in the electron gun of a tube when it is necessary to shield the cathode from stray magnetic flux. At high values of H, the permeability is also low. In fact, when a material is driven well into saturation, the permeability approaches the permeability of free space. This must be kept in mind in the design of magnetic polepieces to be used in regions of high magnetic field. The concentration of flux lines at corners of polepieces, for example, can cause saturation and distortion of the field pattern.

D.4 PERMANENT MAGNETS [2]

In the discussion related to Figures D-8 and D-9, a technique for determining the B–H characteristics of a ring-shaped sample of material was described. When the current through the magnetizing coil was decreased to zero, the sample was left in a magnetized state with a flux density of B_r. Thus, a permanent magnet was formed but all of the flux was contained within the magnetic material, and so there is no way that the flux could be used, for example, to exert a force on electrons.

To make the flux available for use, it is necessary to cut a gap in the ring-shaped sample as shown in Figure D-13(a). Then, ignoring fringe effects, the flux density in the gap that is available for use is the same as in the magnet (because flux lines are continuous).

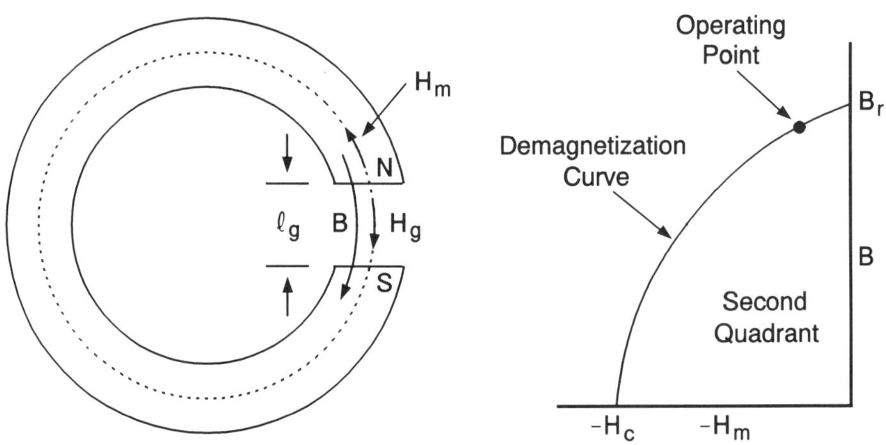

Figure D-13 Effect of gap in ring shaped magnetic sample.

When the gap is cut, north (N) and south (S) magnetic poles are formed in the faces of the magnet. Within the gap, the magnetic field, H_g, that is established is from N to S and is in the same direction as the flux density, B. In the magnet, the field, H_m, is also from N to S but is in the direction *opposite* to B. Thus, H_m tends to demagnetize the magnet. This effect is the same as reversing the current through the coil in Figure D-8. This reverses the field generated by the coil and moves the operating point on the B–H curve to the left of the B axis. Here, the field is opposite to B and so demagnetization occurs.

The operating point on the B–H characteristic for the magnet with a small gap is shown in Figure D-13(b). The curve shown here is the second quadrant of the hysteresis loop and is called the *demagnetization curve*. For a small gap, the operating point is close to the B axis. As the gap is made larger, the operating point moves farther away from the B axis.

When fringing fields are ignored, it is easy to determine the position of the operating point as a function of gap width. Using Ampere's law,

$$\oint \mathbf{H} \cdot d\mathbf{l} = 0 \tag{D-4}$$

where the closed path for integration is the dashed curve in Figure D-13(a). The integral equals zero because no current is contained within the path. Assuming that H_m and H_g are constants, then the integral becomes

$$-H_m \ell_m + H_g \ell_g = 0 \tag{D-5}$$

where ℓ_m is the length of the magnetic sample (along the dashed line from one side of the gap to the other). But, in the gap,

$$B = \mu_o H_g \tag{D-6}$$

so

$$B = \mu_o \frac{\ell_m}{\ell_g} H_m \tag{D-7}$$

This is the equation for a straight line with slope $\mu_o \ell_m / \ell_g$ as shown in Figure D-14. This line, referred to as the *load line* for the magnet, is analogous to the load line for an electrical circuit. The intersection of the load line with the demagnetization curve defines the operating point (B and H_m) for the magnet.

Now, two considerations concerning the design of permanent magnet circuits are immediately evident from Figure D-14:

1. For optimum utilization of the magnet, the gap must be long enough to provide a reasonable gap volume in which interactions with the flux density can take place. On the other hand, the gap must not be so long that the flux density becomes too low.

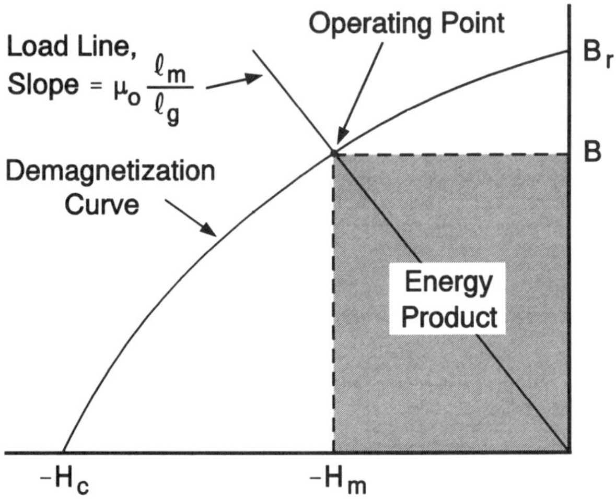

Figure D-14 Use of load line to determine operating point of magnet.

2. A material with a coercivity, $-H_c$, as large as possible as well as with a large retentivity, B_r, should be used to maximize the gap length (and gap volume) for a given flux density. A material with a large coercivity is referred to as a *hard* magnetic material. As shown in Figure D-15, this is in contrast to materials used for transformer cores, which are *soft* magnetic materials. (The area within the hysteresis loop is proportional to part of the power loss in a transformer and so materials are used that have narrow hysteresis loops.)

Magnet materials for TWTs are selected primarily on the basis of the magnetic energy that they can produce in an air (or vacuum) gap. A second important consideration, which will be discussed later, is temperature stability. The energy per unit volume, w, in a magnetic field is

$$w = \int \mathbf{H} \cdot d\mathbf{B} \tag{D-8}$$

Thus, ignoring fringe effects, the total energy, W_g, in the magnetic field in the gap of a magnet is w multiplied by the volume of the gap, or

$$W_g = \ell_g A H_g B \tag{D-9}$$

where A is the cross-sectional area of the gap. Because $H_m \ell_m = H_g \ell_g$ (D-5)

$$W_g = \ell_m A H_m B \tag{D-10}$$

but this is the energy in the field within the magnet. All we are saying here, is that the energy delivered by the magnet is the energy established in the gap. This is analogous to saying that the power delivered by an electrical circuit is the power received by the

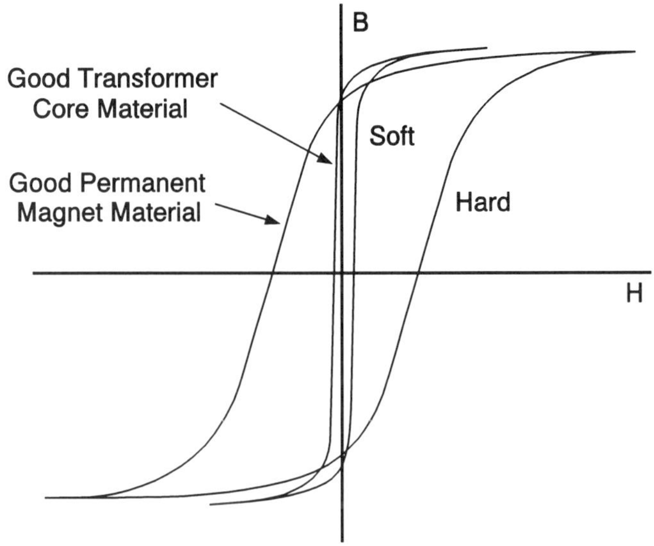

Figure D-15 Hysteresis loops for soft and hard magnetic materials.

load. The energy delivered by the magnet is proportional to the cross-hatched area in Figure D-14. This area is referred to as the "energy product" of the magnet. To maximize the energy product (and maximize the energy established in the gap), the gap length, ℓ_g, must be such that the shaded area is maximized. The variation of energy product with the position of the operating point is shown in Figure D-16.

Rare earth magnets are the magnets of choice for most modern microwave tube applications. The most common material used is samarium cobalt. The advantage of samarium cobalt over some other materials that have been used in the past is shown by the demagnetization curves in Figure D-17. The energy product for samarium cobalt is approximately four times that of alnico V, which is a magnet material that was in common use in microwave tubes 20 years ago. The coercive force is over an order of magnitude greater than that of alnico V, with the result that the samarium cobalt magnets are extremely stable.

Permanent magnets on microwave tubes must be able to operate over a wide temperature range with no significant change in characteristics. Conventional samarium cobalt ($SmCO_5$) has a field strength temperature coefficient of about 0.04%/°C. The temperature coefficient can be reduced by doping, but doping reduces field strength and energy product.

A new class of rare earth magnets (Sm_2CO_{17}) has an energy product as much as twice that of $SmCO_5$. When doped with a material such as gadolinium, a near-zero temperature coefficient can be obtained at field strengths and energy products comparable to undoped $SmCO_5$.

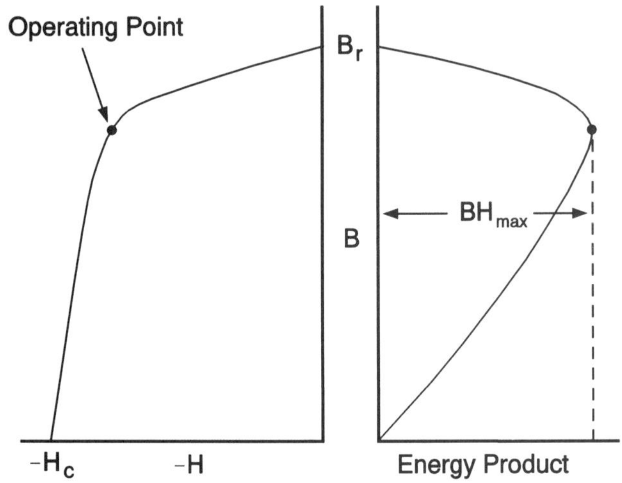

Figure D-16 Demagnetization curve and energy product.

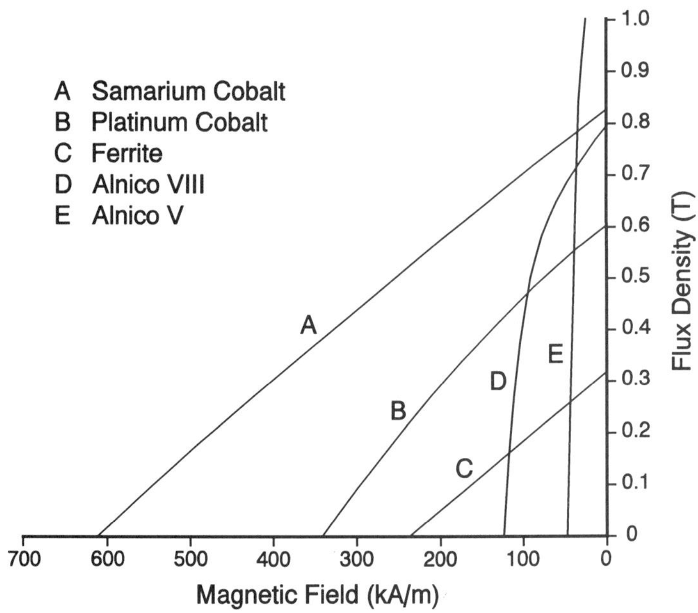

Figure D-17 Comparison of rare earth cobalt magnets with conventional magnets. Adapted from "Rare Earth Cobalt Magnets", Varian Brochure, 1972.

D.5 PERIODIC PERMANENT MAGNETS

When minimum size and weight are essential, periodic permanent magnets are frequently used for focusing electron beams. A PPM structure is shown in Figure D-18. Magnetic lenses are formed that periodically focus the electron beam traveling along the center of the structure. Periodic magnetic focusing of an electron beam is discussed in Chapter 7.

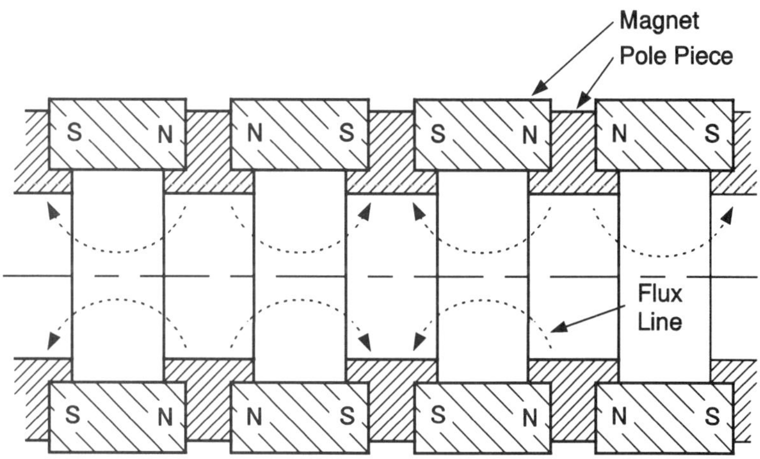

Figure D-18 Periodic permanent magnet structure.

The weight reduction of a periodic permanent magnet system, when compared with a solenoid, can be one to two orders of magnitude. The reason for this weight reduction can be explained by first noting that a solenoid establishes a magnetic field in a volume far larger than that required for focusing the electron beam. Since the weight of the magnet is roughly proportional to the magnetic energy stored in the volume surrounding the magnet, much of the weight of the magnet results from the production of unused magnetic field.

Figure D-19 [3] expands on the reason why the weight of a periodic permanent magnet system can be far smaller than that of a solenoid. Assume, first, that the magnet in (a) has the most efficient design (the magnet weight is minimized) to produce a sufficient magnetic field to focus an electron beam as it travels through the length of the magnet. Next, assume that it is desired to focus the same beam over distance that is three times the length of the magnet in (a). This may be accomplished with the magnet in (b) but, to produce the same magnetic field that was produced by the magnet in (a), it is necessary to increase all dimensions of the magnet by a factor of three. (Remember that the dimensions of a permanent magnet are established to nearly maximize the energy product. A magnet configuration other than that shown in (b) would result in either a reduction in magnetic field or an increase in magnet weight.)

As a result, the magnet in (b) weighs 3^3 or 27 times as much as the magnet in (a), and the energy stored in the field surrounding the magnet in (b) is 27 times that for (a). By comparison, if three magnets like that in (a) were stacked with like poles together as shown in (c), then a field sufficient to focus the beam could be produced over the same length as for magnet (b). The total weight and energy stored for the magnets in (c) would be one-ninth of the weight and energy stored for the magnet in (b).

If the magnet in (b) was replaced by n smaller magnets aligned like those in (c), the weight of the resulting structure would be $1/n^2$ times that of the larger magnet. In practice, the weight saving is not as great as that indicated by this discussion. This is because the field on the axis of the periodic structure does not appear as a square wave with abrupt reversals occurring at magnet interfaces. Instead, the field is more nearly sinusoidal and so the peak value must be increased to produce the same effective field as would be produced by the square wave. As a result, the magnet sizes must be increased to produce the required effective field.

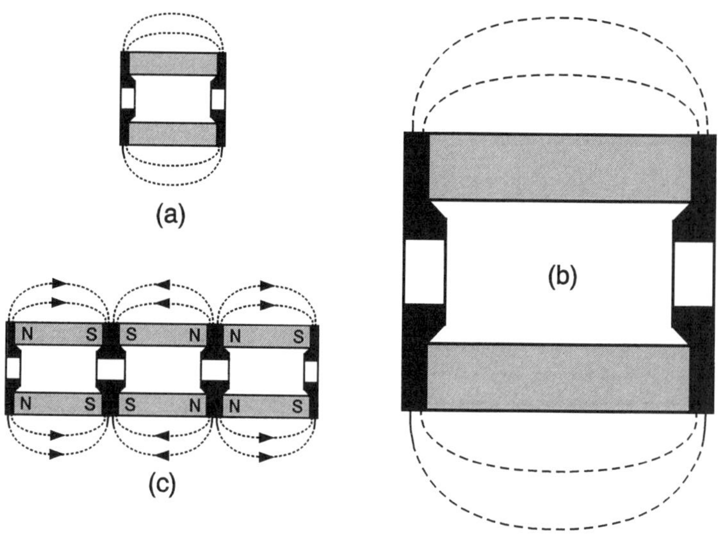

Figure D-19 Size (and weight) reduction resulting from the use of periodic permanent magnets. Adapted from J.T. Mendel *et al.*, *Proc. IRE*, May 1954. © 1954 IRE (now IEEE).

A problem with the ring-shaped magnets illustrated in Figure D-18 and D-19 is that each must be split in order to be placed on the tube. This is one of the factors leading to nonuniformities in the magnetic fields produced by these structures. To obtain acceptable tube performance, it is often necessary to compensate for the nonuniformities by placing magnetic shims on the outside of the magnetic assembly.

A technique for avoiding some of the problems of ring-shaped magnets is illustrated in Figure D-20. Several (often four) pill-shaped magnets are placed between magnetic polepieces and surround the axis of the tube. With this technique, the magnets can be magnetized to the desired level and then placed on the tube without being split. Field uniformity is increased (as compared with the ring-shaped magnets). In addition, magnet cost is reduced because magnets of one size can be used on tubes having a wide range of sizes.

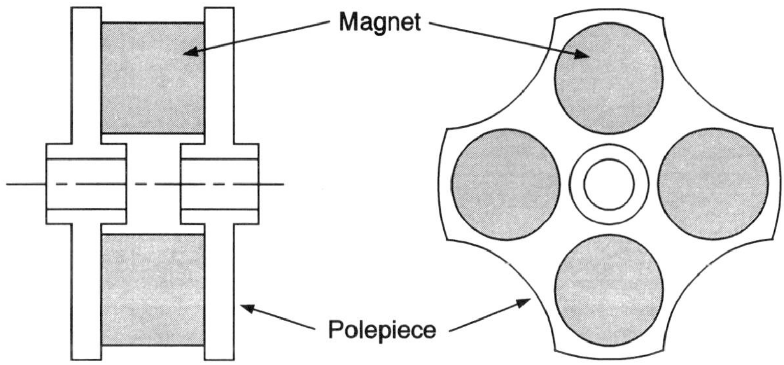

Figure D-20 A PPM configuration using pill-shaped magnets.

REFERENCES

[1] For additional reading, the following are recommended: J.D. Krauss and K.R. Carver, *Electromagnetics*. New York: McGraw-Hill, 1973; W.H. Hayt, Jr., *Engineering Electromagnetics*. New York: McGraw-Hill, 1981; L.M. Magid, *Electromagnetic Fields, Energy, and Waves*. New York: McGraw-Hill, 1972; C.G. Skitek and S.V. Marshall, *Electromagnetic Concepts and Applications*. Englewood Cliffs, NJ: Prentice-Hall, 1982.

[2] For additional reading, the following is recommended: R.J. Parker and R.J. Strudders, *Permanent Magnets and Their Applications*. New York: John Wiley and Sons, 1962.

[3] J.T. Mendel, C.F. Quate and W.H. Yocom, "Electron beam focusing with periodic permanent magnet fields," *Proc. IRE,* May 1954, pp. 800–810.

Author Index

Subject Index

The Artech House Radar Library

David K. Barton, Series Editor

Active Radar Electronic Countermeasures, Edward J. Chrzanowski

Adaptive Signal Processing for Radar, Ramon Nitzberg

Airborne Early Warning Radar, William Morchin

Airborne Early Warning System Concepts, Maurice Long *et al.*

Airborne Pulsed Doppler Radar, Guy V. Morris

AIRCOVER: Airborne Radar Vertical Coverage Calculation Software and User's Manual, William A. Skillman

Analog Automatic Control Loops in Radar and EW, Richard S. Hughes

Aspects of Modern Radar, Eli Brookner *et al.*

Aspects of Radar Signal Processing, Bernard Lewis, Frank Kretschmer, and Wesley Shelton

Bistatic Radar, Nicholas J. Willis

CARPET (Computer-Aided Radar Performance Evaluation Tool): Radar Performance Analysis Software and User's Manual, developed at the TNO Physics and Electronics Laboratory by Albert G. Huizing and Arne Theil

Cryptology: Machines, History, and Methods, Cipher Deavours *et al.*

Cryptology Yesterday, Today, and Tomorrow, Cipher Deavours *et al.*

DETPROB: Probability of Detection Calculation Software and User's Manual, William A. Skillman

Designing Automatic Control Systems, W. L. Rubin

Detectability of Spread-Spectrum Signals, Robin A. Dillard and George M. Dillard

Electronic Countermeasure Systems Design, Richard Wiegand

Electronic Homing Systems, M. V. Maksimov and G. I. Gorgonov

Electronic Intelligence: The Analysis of Radar Signals, Richard G. Wiley

Electronic Intelligence: The Interception of Radar Signals, Richard Wiley

Engineer's Refractive Effects Prediction Systems–EREPS, Herbert Hitney

EREPS: Engineer's Refractive Effects Prediction System Software and User's Manual, developed by NOSC

High Resolution Radar, Donald R. Wehner

High Resolution Radar Cross-Section Imaging, Dean Mensa

Interference Suppression Techniques for Microwave Antennas and Transmitters, Ernest R. Freeman

Introduction to Electronic Defense Systems, Filippo Neri

Introduction to Electronic Warfare, D. Curtis Schleher

Introduction to Sensor Systems, S. A. Hovanessian

IONOPROP: Ionospheric Propagation Assessment Software and Documentation, Herbert Hitney

Kalman-Bucy Filters, Karl Brammer

Laser Radar Systems, Albert V. Jelalian

Lidar in Turbulent Atmosphere, V. A. Banakh and V. L. Mironov

Logarithmic Amplification, Richard Smith Hughes

Machine Cryptography and Modern Cryptanalysis, Cither A. Deavours and Louis Kruh

Mathematical Techniques in Multisensor Data Fusion, David L. Hall

Millimeter-Wave Radar Clutter, Nicholas Currie, Robert Hayes, Robert Trebits

Modern Radar System Analysis, David K. Barton

Modern Radar System Analysis Software and User's Manual, David K. Barton and William F. Barton

Monopulse Radar, A. I. Leonov and K. I. Fomichev

MTI and Pulsed Doppler Radar, D. Curtis Schleher

Multifunction Array Radar Design, Dale R. Billetter

Multisensor Data Fusion, Edward L. Waltz and James Llinas

Multiple-Target Tracking with Radar Applications, Samuel S. Blackman

Multitarget-Multisensor Tracking: Advanced Applications, 2 Volumes, Yaakov Bar-Shalom, editor

Over-The-Horizon Radar, A. A. Kolosov *et al.*

Principles and Applications of Millimeter-Wave Radar, Charles E. Brown and Nicholas C. Currie, editors.

Principles of Traveling Wave Tubes, A.S. Gilmour

Pulse Train Analysis Using Personal Computers, Richard G. Wiley and Michael B. Szymanski

Radar and the Atmopshere, Alfred J. Bogush, Jr.

Radar Anti-Jamming Techniques, M.V. Maksimov *et al.*

Radar Cross Section Analysis and Control, A. K. Bhattacharyya and D. L. Sengupta

Radar Cross Section, Eugene F. Knott *et al.*

Radar Electronic Countermeasures System Design, Richard J. Wiegand

Radar Evaluation Handbook, David K. Barton *et al.*

Radar Evaluation Software, David K. Barton and William F. Barton

Radar Meteorology, Henri Sauvageot

Radar Polarimetry for Geoscience Applicatons, Fawwaz Ulaby

Radar Propagation at Low Altitudes, M. L. Meeks

Radar Reflectivity Measurement: Techniques and Applications, Nicholas C. Currie, editor

Radar System Design and Analysis, S.A. Hovanessian

Radar Technology, Eli Brookner, editor

Radar Vulnerability to Jamming, Robert N. Lothes, Michael B. Szymanski, and Richard G. Wiley

RGCALC: Radar Range Detection Software and User's Manual, John E. Fieldng and Gary D. Reynolds

SACALC: Signal Analysis Software and User's Guide, William T. Hardy

Secondary Surveillance Radar, Michael C. Stevens

Signal Detection and Estimation, Mourad Barkat

SIGCLUT: Surface and Volumetric Clutter-to-Noise, Jammer and Target Signal-to-Noise Radar Calculation Software and User's Manual, Wiliam Skillman

Small-Aperture Radio Direction Finding, Herndon Jenkins

Solid-State Radar Transmitters, Edward D. Ostroff *et al.*

Space-Based Radar Handbook, Leopold J. Cantafio, editor

Spaceborne Weather Radar, Robert M. Meneghini and Toshiaki Kozu

Statistical Signal Characterization, Herbert Hirsch

Statistical Signal Characterization: Algorithms and Analysis Programs Software and User's Guide, Herbert Hirsch

Surface-Based Air Defense System Analysis, Robert Macfadzean

Surface-Based Air Defense System Analysis Software and User's Guide, Robert Macfadzean and James M. Johnson

Statistical Theory of Extended Radar Targets, R. V. Ostrovityanov and F. A. Basalov

VCCALC: Vertical Coverage Plotting Software and User's Manual, John E. Fielding and Gary D. Reynolds

For further information on these and other Artech House titles, contact:

Artech House
685 Canton Street
Norwood, MA 01602
617-769-9750
Fax: 617-762-9230
Telex: 951-659
email: artech@world.std.com

Artech House
Portland House, Stag Place
London SW1E 5XA England
+44 (0) 71-973-8077
Fax: +44 (0) 71-630-0166
Telex: 951-659